T. Kowalczyk, E. Pleszczyńska, F. Ruland (Eds.)

Grade Models and Methods for Data Analysis

Springer

Berlin
Heidelberg
New York
Hong Kong
London
Milano
Paris
Tokyo

Studies in Fuzziness and Soft Computing, Volume 151

Editor-in-chief
Prof. Janusz Kacprzyk
Systems Research Institute
Polish Academy of Sciences
ul. Newelska 6
01-447 Warsaw
Poland
E-mail: kacprzyk@ibspan.waw.pl

Further volumes of this series can be found on our homepage: springeronline.com

Vol 131. J.N. Mordeson, D.S. Malik, N. Kuroki
Fuzzy Semigroups, 2003
ISBN 3-540-03243-6

Vol 132. Y. Xu, D. Ruan, K. Qin, J. Liu
Lattice-Valued Logic, 2003
ISBN 3-540-40175-X

Vol. 133. Z.-Q. Liu, J. Cai, R. Buse
Handwriting Recognition, 2003
ISBN 3-540-40177-6

Vol 134. V.A. Niskanen
Soft Computing Methods in Human Sciences, 2004
ISBN 3-540-00466-1

Vol. 135. J.J. Buckley
Fuzzy Probabilities and Fuzzy Sets for Web Planning, 2004
ISBN 3-540-00473-4

Vol. 136. L. Wang (Ed.)
Soft Computing in Communications, 2004
ISBN 3-540-40575-5

Vol. 137. V. Loia, M. Nikravesh, L.A. Zadeh (Eds.)
Fuzzy Logic and the Internet, 2004
ISBN 3-540-20180-7

Vol. 138. S. Sirmakessis (Ed.)
Text Mining and its Applications, 2004
ISBN 3-540-20238-2

Vol. 139. M. Nikravesh, B. Azvine, I. Yager, L.A. Zadeh (Eds.)
Enhancing the Power of the Internet, 2004
ISBN 3-540-20237-4

Vol. 140. A. Abraham, L.C. Jain, B.J. van der Zwaag (Eds.)
Innovations in Intelligent Systems, 2004
ISBN 3-540-20265-X

Vol. 141. G.C. Onwubolu, B.V. Babu
New Optimzation Techniques in Engineering, 2004
ISBN 3-540-20167-X

Vol. 142. M. Nikravesh, L.A. Zadeh, V. Korotkikh (Eds.)
Fuzzy Partial Differential Equations and Relational Equations, 2004
ISBN 3-540-20322-2

Vol. 143. L. Rutkowski
New Soft Computing Techniques for System Modelling, Pattern Classification and Image Processing, 2004
ISBN 3-540-20584-5

Vol. 144. Z. Sun, G.R. Finnie
Intelligent Techniques in E-Commerce, 2004
ISBN 3-540-20518-7

Vol. 145. J. Gil-Aluja
Fuzzy Sets in the Management of Uncertainty, 2004
ISBN 3-540-20341-9

Vol. 146. J.A. Gámez, S. Moral, A. Salmerón (Eds.)
Advances in Bayesian Networks, 2004
ISBN 3-540-20876-3

Vol. 147. K. Watanabe, M.M.A. Hashem
New Algorithms and their Applications to Evolutionary Robots, 2004
ISBN 3-540-20901-8

Vol. 148. C. Martin-Vide, V. Mitrana, G. Păun (Eds.)
Formal Languages and Applications, 2004
ISBN 3-540-20907-7

Vol. 149. J.J. Buckley
Fuzzy Statistics, 2004
ISBN 3-540-21084-9

Vol. 150. L. Bull (Ed.)
Applications of Learning Classifier Systems, 2004
ISBN 3-540-21109-8

Teresa Kowalczyk
Elżbieta Pleszczyńska
Frederick Ruland (Eds.)

Grade Models and Methods for Data Analysis

With Applications for the Analysis of Data Populations

Springer

Prof. Teresa Kowalczyk
Polish Academy of Sciences
Institute of Computer Science
ul. Ordona 21
01-237 Warszawa
Poland

Dr. Frederick Ruland
The Ohio State University, Retired
Echo Street 822
43758 Malta, OH
USA

Prof. Elżbieta Pleszczyńska
Polish Academy of Sciences
Institute of Computer Science
ul. Ordona 21
01-237 Warszawa
Poland
E-mail: eple@ipipan.waw.pl

ISSN 1434-9922
ISBN 3-540-21120-9 Springer-Verlag Berlin Heidelberg New York

Library of Congress Cataloging-in-Publication-Data

A catalog record for this book is available from the Library of Congress.
Bibliographic information published by Die Deutsche Bibliothek.
Die Deutsche Bibliothek lists this publication in the Deutsche Nationalbibliographie; detailed bibliographic data is available in the Internet at http://dnb.ddb.de

Springer-Verlag is a part of Springer Science+Business Media
springeronline.com

Printed in Germany

Typesetting: data delivered by editors
Cover design: E. Kirchner, Springer-Verlag, Heidelberg
Printed on acid free paper 62/3020/M - 5 4 3 2 1 0

Contents

Acknowledgements:

The authors highly appreciate the valuable comments of Professor Jacek Koronacki and Professor Jan Mielniczuk.

1

Grade Data Analysis — A First Look

Frederick Ruland

> A data set is like a house: It is not enough to know how many rooms and the number of square feet. You want to know how it is put together. You want to see the structure.
>
> For a data set, it is not enough to know the descriptive statistics. You need to know how the data fit together — what the structure looks like.
>
> Grade methods tell you how a data set is put together. They show you the exact structure — not a "theoretical" structure — and grade methods provide you with a picture of this structure.

1.1 "Questions" from clients

If you have experience with data analysis, you may have had the following consulting experiences:

- A client has six clusters of identified subjects. It is suddenly necessary for the client to have only five clusters of the same subjects. The five new clusters bear no subject resemblance to the original six. The client asks why.

- A client is using various methods to collapse categories. (Or perhaps they are making categories from a continuous variable, based on different schemes.) The client indicates that they get completely different results depending upon the scheme used. They ask for your advice.

- A client has "inherited" a number of data sets with similar marginal distributions (means, frequencies, etc.). They want to know what should be considered before combining the data sets.

- A client wants to justify exclusion of an "outlier." Its inclusion or exclusion has serious implications. They want to know what you would suggest.

Would you feel confident in your answers and suggestions for these clients?

We will return to our clients, but first a few words:

1.2 About "Grade Models and Methods for Data Analysis"

This book is written for both the practicing data analyst and those interested in grade theory. Our intention is to provide useful information.

Those readers with a particular interest in correspondence analysis, or cluster analysis, should find this book interesting; however, extensive knowledge of these areas by the reader is not assumed. We will, however, assume that the reader is not familiar with grade theory and methodology.

If you are eating in a restaurant and your friend asks what something tastes like, your mouth-full-of-food answer might be: "chicken."
Your mouth-full-of-food answer to what grade models and methods taste like might be: "curves."

The next question is going to be: "Why curves?"

Although you will learn "why curves" starting in Chapter 2, let us just say here that curves can "store" more information than, for example, a few numerical parameters or a vector of ranks. Of course, you have to know how to use the various grade "tools."

Comment: If the person above is really inquisitive the next question will be: "What's grade?" We will do our best to answer this and other questions throughout the book.

Grade methodology makes no distinction between continuous and discrete data. That is, ***the methodology applies to both continuous and discrete data***, and the underlying ***grade theory is the same for both.*** The theoretical, computational, and "interpretational" principles are the same for continuous and discrete data.

This "same for both" perspective opens the door to a different data analysis approach. However, this approach is not completely new.

Mini History:
At least for continuous populations, the concept of using "grade" to replace that of rank was introduced by Sir Francis Galton (1822-1911).[72]. We will say more about "grade" starting in Chapter 2.

Note: Progress in one area often hinges on progress in another. In the case of grade methodology, the availability of good computers and good graphics played an important part.

1.3 Addressing the practitioner

Let us try to address some possible preliminary questions and concerns of the practitioner:

Grade data analysis — at least on the surface — can be described as a visually oriented method to explain the complexities of data sets. The data sets can be univariate or multivariate — raw data or contingency tables.

Grade models and methods deal with the structure of data. Grade applications relate to trends (latent structures), sub-populations, outliers, and clusters.

If you have applications which use: canonical correlation analysis, cluster analysis, correspondence analysis, discriminant analysis, or factor analysis, then you will find grade applications useful and complementary to these procedures.

The practitioner sections of this book deal with the analysis of data populations. The justification for this focus is given later in the chapter.

What grade methods do:

Grade methods (and grade "tools") are most frequently used to:

- Reveal trends ("functional structures") in the data. These structures involve the columns (e.g., variables), and the rows (e.g., observations) of the data set;
- Detect sub-populations in the data. Different sub-populations are associated with different functional structures. (Often these sub-populations should be studied separately, otherwise, they may mask the real nature of the data;)
- Detect outliers in the data set. (Grade procedures use information from every variable in the data set during the detection of an outlier.) Outlier observations are identified by an "aberrant pattern" on a map of the observations;
- Construct clusters of rows (observations), and also clusters of columns (variables). These clusters are organized around functional trends in the data.

Activities related to the above include: data fusion (merging and creation of "similar" data sets); standardized inter-population comparisons; grade models; and the grade prediction and determination of missing values.

How grade methods "do what they do:"

Grade methods measure the dissimilarity between distributions. (How does this distribution differ from that distribution.)

Note: We are using distribution in a very general sense, and although it is a bit of a stretch, one can almost think of this as: "How does this record differ from that record."

We mentioned curves in Section 1.2. You will see that the shape of these curves will reflect the dissimilarity between distributions.

Comment: In the case of multivariate data, you might think of each record (set of values comprising one observation) as a curve, and the entire data set (population) as a map. While this concept, strictly speaking, may not hold, it may be useful initially.

For those interested in grade theory, we briefly mention below the material covered in the theory chapters. The practitioner can think of these chapters as the "Grade Research and Development" or "Grade Quality Control Department." These chapters deal with "theory," but not "theory about the theoretical." The material has direct application to issues concerning: categorization of continuous data, formation of clusters, determination of trends, detection of outliers, and extraction of sub-populations.

1.4 Addressing the theorist

This book will introduce the theorist to grade models, the infrastructure of parameters, and related grade methods.

Grade models and methods are based on one general concept: concentration, as described in Chapter 3.

(Measurement of "concentration" might be considered a special case of the measurement of variability.)

Chapters 2 through 6 deal with concentration curves. These curves are used to evaluate inequality when comparing two distributions. (A family of curves with fractal properties is discussed in Chapter 6 with respect to problems of optimal categorization.)

Chapters 7 through 11 extend grade models and methods to bivariate distribution tables. The grade concepts which apply to univariate distributions, also apply to bivariate distributions. The concentration curve now becomes the concentration surface. (You will also find an important extension at this point concerning stochastic dependence.) The investigation of dependence in bivariate distributions now becomes the investigation of dependence in grade distributions on the unit square. The grade approach to clustering and prediction is concerned with the regularity of concentration surfaces. The regularity of these concentration surfaces is linked with the determination of the optimal number of clusters.

Chapters 10, 12 and 13 deal with multivariate data and the methods used to describe multivariate distributions. The grade parameters used to describe bivariate distributions serve now to describe multivariate distributions. The chapters exploit the material of previous chapters in the finding of monotone trends between rows (records) and columns (variables). These models and methods solve practical statistical problems (Two case studies appear in Chapter 14.).

Early data analysis techniques, as of necessity, were based on the computational resources available at the time. Grade models and methods - although not born in the computer age - have developed in the computer age.

1.5 Regarding the analysis of data populations

The purpose of using grade methods is to understand the data.

A great body of statistical knowledge and numerous analytical procedures exist for those working with samples. However, suitable methodology is also required for those working with populations.

Examples of data populations:

- records for all patients who received a specific model heart pacemaker;
- maintenance records (or problem reports) for a fleet of planes or vehicles;
- epidemiological tracking information for specific disease cases;
- a database of customers who return a product questionnaire/warranty;
- a database of people awaiting organ transplants;
- all suspects in a large criminal investigation.

(In general, any collection of data where the researcher's interest — for any reason — is in only those data.)

When analyzing data populations, one is usually interested in doing rather specific "things." These "things" usually involve the association of values of the variables (i.e., columns), with values of particular observations (i.e., rows). For example: dividing a population into sub-populations; putting populations together (data fusion); simplifying the "picture" relative to the "importance" of variables. In general, things related to the structure of the data.

One may wonder why "models and methods for populations." There would seem to be no uncertainty when you have the whole population. However, the uncertainty is in knowing what is important (i.e., the proper description, and proper interpretation of the data).

Below you will find procedures designed for working with populations.

1.6 Overview of Grade Data Analysis algorithms

Three algorithms are introduced in this section. We will discuss each algorithm separately, although they are often used together.

Note: A program called GradeStat exists for doing Grade Data Analysis. An overview of this program is presented in Chapter 15. Please contact Olaf Matyja (gradestat@ipipan.waw.pl) for information.

Grade Correspondence Analysis (GCA):

GCA performs correspondence analysis, but with differences. Briefly, results of the GCA can be described not only by parameters of dependence, but also of "regularity" and "representativeness" (see Chapter 9); the latent variables (i.e., trend structures) are based on "overrepresentation values" (see Chapter 7); and the trend structures identified in the data actually exist, as opposed to being theoretical constructs (see Chapter 9). Accordingly, one focus of GCA is on the association of obervations with specific trends.

The results of GCA are usually visualized on, so called, overrepresentation maps (see Chapter 8).

Note: This book, as a matter of course, compares and contrasts grade methods with traditional methods. However, method similarities and differences are not the focus of the book.

Mini GCA Example:

You have information on the topics studied by each student in this year's local graduating class. GCA identifies (based on the structural trends in the data) two sub-populations: science oriented students, and humanities oriented students. GCA also indicates the "importance" priority for each topic relative to the trend orientations.

The Grade Correspondence Analysis material is covered in Chapter 9.

Smooth Grade Correspondence Analysis (SGCA):

The goal of SGCA is to recognize trends in data; decompose the data into the sum of two or more tables representing those trends; and finally, determine the proportional influence of each trend on the observations (or row categories) and variables (or column categories). SGCA is also described in Chapter 9.

Mini SGCA Example concerning influence of trends:

Let us use this year's graduating class from above. Recall: there is a science trend structure and a humanities trend structure. We want to determine the proportional influence of each structure on each student (or row category). Perhaps, student "A" is influenced 70 percent by science and 30 percent by humanities. Student "B" is influenced 20 percent by science and 80 percent by humanities.

Note: "Smooth" (as in Smooth Grade Correspondence Analysis) is related to the data smoothing algorithms used in SGCA.

The Smooth Grade Correspondence Analysis material is covered in Chapter 9.

Grade Correspondence Cluster Analysis (GCCA):

GCCA performs cluster analysis, but with procedural (i.e., grade) differences, and presentation differences. Briefly, the clustering is based on a trend structure, with clusters organized along this structure. The clustering results are displayed on an overrepresentation map which is standardized and comparable across data populations. This map displays the ordered and clustered observations, and also the ordered and clustered columns representing variables. The variables are ordered and clustered based on the trend structure.

Mini GCCA Example:

Note: GCCA is used in combination with GCA.

You have information (e.g., topic grades) on all students at the local medical school. Assume that GCA has already detected two sub-populations: theoretically oriented, and practitioner oriented students.

GCCA orders and clusters the students - and the topics - for each subpopulation. For the theoretical sub-population, it appears that the topics are arranged, for example, according to "organizational aptitude." For the practitioner sub-population, it appears that the topics are arranged according to "deductive aptitude."

The Grade Correspondence Cluster Analysis material is covered in Chapter 10 and 11.

1.7 Returning to the clients from the first page

Regarding the first client with the six clusters:

We will answer by making an analogy: We have a child's crib mobile. It has 6 arms from which a dozen or so different toy circus animals hang in balance. If we now try to balance those animals from only 5 arms, it will likely be necessary to redistribute the animals to maintain the balance. (This is often the case with a traditional clustering approach.)

Grade methods use a different approach to clustering. Let us call it the "circus parade approach." The animals are lined up (linearly ordered) based on the Grade trend and the clusters are formed by the adjacent sections of the parade. If 5 sections (clusters) are desired rather than 6, it is usually just a matter of moving the "cut points." The order of the animals in the parade basically remains the same. (See Chapters 10 and 11.)

Comment: The client will ask which clustering approach is better. We will answer that the circus parade is simpler. (See Chapter 10 to decide for yourself which is better.)

Regarding the second client using various schemes to make categories:

The client may have initially collapsed (or created) categories because of the analytical approach used (i.e., the analytical approach was based on categories), or because of program specifications (i.e., no empty cells, a maximum number of categories, etc.).

Different schemes for creating and collapsing categories result in differing degrees of information loss. The issues might be: whether to make, or not make, categories; the proper number of categories; and the specifics of the categories (e.g., value ranges, nominal composition, etc.).

The client has already discovered:

When you make the categories — you control the results.

Yet, the results are supposed to be derived from data, instead of being influenced by the statistician.

Grade methods use the same approach for continuous and discrete data. Therefore, if you start with continuous data, it is not necessary to create categories. If you start with categorical data, you need to know how to optimize the categories.

The above issues are addressed in a number of chapters, but particularly in Chapter 6.

Regarding the third client who wants to know what should be considered before combining data sets:

We will make another analogy: Say that you have data from two businesses. The average profits for both businesses are the same. However, for one business profits are "on the rise," while for the other, profits are "on the decline."

Data have both an outer and an inner structure. The marginal information (i.e., average profits), is the "outer" structure. The "inner" structure is associated with the rows and columns.

Specific recommendations would depend upon the nature of the client's investigation. However, before data sets are combined, they should have, at least, a similar inner structure.

Related topics can be found in Chapters 4, 9, 10, 12 and 14.

Regarding the fourth client who wants suggestions before excluding a possible outlier:

We will assume that the client is referring to an outlier observation, as opposed to an individual data value.

A basic difficulty in dealing with outliers in classic statistics is that the true multivariate distribution of the data is usually not known. This makes it difficult to decide how unusual the "outlier" is in the distribution.

There exists another approach: The "outlier" is compared with its own predicted values, and also with the values of its "structural neighbors."

The observation in question may be an outlier, or it may belong to a subset of the data with a different structure that should be studied separately.

Related topics can be found in Chapters 2, 9, 12, and 14.

Obvious Disclaimer:
These "Questions from clients" are, of course, staged. However, the questions are plausible and common. We have tried to address the reasons for the client's difficulties, and reference a few grade advantages.

1.8 Conclusion — Chapter 1

The more data you have, the better —
— but only if you understand it.

Grade models and methods provide a different perspective for descriptive and investigative work with populations of multivariate data, and contingency table data.

All grade models and methods are based on approaches that would have been impractical - if not impossible - even 30 years ago.

Grade methods use one consistent approach for both continuous and discrete data, and address both the external and the internal structure of the data.

Grade methods work by comparing distributions. They require essentially no assumptions about the model.

In Chapter 1, we have tried to get you quickly into the role and realm of the grade data analyst. We hope you find the remaining material interesting and useful.

Watch out for those curves!

2

The Grade Approach

Frederick Ruland

The purpose of using grade methods is to understand the data.

2.1 Introduction

Chapter 1 provided a first look at what grade methods do, how they work, why they are useful, and some of their applications. It was stated that grade methods use a different approach to the understanding of data than the traditional (i.e., non-grade) methods. Therefore, to understand grade methods, it is important to understand their approach.

Chapter 2 is divided into two parts. The first part is a ***Quick Start to the understanding of grade concepts***, to get you off and running via examples of the grade approach. The second part provides a foundation for the understanding of concentration curves and other related grade tools.

Grade methods compare and relate distributions: how similar (actually dissimilar) is this distribution to that distribution. In some cases, one of the distributions is a benchmark or reference distribution to which other distributions are compared. In other cases, one distribution might be an "expected" distribution. If we know in what ways two distributions are similar or dissimilar, our understanding of what is "usual" or "unusual" about the data is improved.

Note: Grade methods deal with the ***dissimilarity*** of distributions. However, it may be easier to understand this chapter if, in some cases, the sentence construction is such that the words ***similarity*** and ***similar*** are used.

In this chapter, we will introduce grade methods with the focus on univariate data at the strictly discrete level. There are two reasons for this focus: most grade concepts (univariate and multivariate) can be presented in the univariate context. And, while grade methods also work with univariate and multivariate continuous data (or mixed data), most practical applications are at the discrete level.

The construction and the definition of categories plays an important part in data distributions. Therefore, it is important to understand how the discretizing (i.e., the dividing into categories) of continuous data, and the creating, combining, and also ordering (prioritizing) of categories affect distributions and the eventual interpretation of any data analysis results. An understanding is important, because these defining and construction activities are often under the data analyst's control.

Warning:
A data analyst who has control over the categorizing of the data, has control over the results of the analysis.

2.2 Part 1: Quick start to the understanding of grade concepts

This section gives examples of possible data distribution sources, and introduces the grade approach, its "tools," terminology, and concepts.

The purpose of grade methods is to understand the data; therefore, we will start with the data.

The grade approach compares two distributions.

The grade approach compares two data distributions (i.e., given one distribution, how is a second distribution related to it). The object is to determine the dissimilarity of the two distributions in order to reveal the structural trends in the data.

2.2.1 A simplified case of the grade approach

The "essentials" for the grade approach include:

- Two distributions.
- Same number of categories in each distribution.
- Same meaning of (description for) corresponding categories.
- Same ordering of categories (by meaning) in each distribution.
- One distribution designated as a "relative to" (or reference) distribution.

(2.1)

Note: When we say “two distributions,” this does not restrict the use of grade methods. That is, the “ two” refers to the mechanics of a comparison, and does not limit the number of distributions in a study.

2.2.2 Examples of data distribution sources

A. Two sets of univariate data (e.g., of amount of money spent) distributed by categories:

TABLE 2.1.

Country A: Amount (in millions) spent on five household budget categories.

Category	Food	Housing	Utilities	Transportation	Health care
Amount	60	50	40	30	20

Country B: Amount (in millions) spent on five household budget categories.

Category	Food	Housing	Utilities	Transportation	Health care
Amount	60	60	120	15	45

Let us assume for a moment that you know in what way, and to what degree, the distributions for countries A and B are similar. Then, if two ***other*** countries have the same similarity, you know they have a "budget" relationship like that of countries A and B. (That is, if things have a similar structure, and you understand one, then you have insight into the other.)

Note that you are ***not*** necessarily interested in whether the proportions are the ***same*** between country A and country B. You are interested in how the proportions ***relate*** to each other, between countries. Perhaps, you (as an expert) have a super economic solution for countries: C, D, E, F, etc., ***if*** they have the same similarity relationship to country B, that country A has to country B. (This Chapter will demonstrate how to display the relationship, and later chapters will explain how it is quantified.)

For now, let us continue with our example distributions.

B. Two sets of count (frequency) data – distributed by categories:

TABLE 2.2.

East City School: Frequency of letter grades for geography.

Letter grade	A	B	C	D	F
Frequencies	60	50	40	30	20

West City School: Frequency of letter grades for geography.

Letter grade	A	B	C	D	F
Frequencies	60	60	120	15	45

Note: The same data values are intentionally used in both Tables 2.1 and 2.2.

Table 2.2 displays students' letter grade distributions for a specific classroom topic (e.g., Geography). The two distributions will have some similarity relationship structure.

Your professional interest might be in what other classroom topics exhibit the same letter grade similarity relationship between schools. This could tell you something about the topics, and about the schools. For example, for a large composite study, you might group together all topics having a certain similarity relationship to a targeted grade distribution.

C. Count (frequency) data – totals from a two-way table:

TABLE 2.3. a

Frequencies for car models purchased by repeat car buyers.
(Only total frequencies shown.)

Previous Purchase ▶ Present Purchase ▼	Sub-compact	Compact	Mid-size	Full-size	Super-size	Totals
Sub-compact						200
Compact						250
Mid-size						125
Full-size						25
Super-size						100
Totals	150	200	100	50	200	700

We will call the above: " marginal" information (i.e., coming from the table margins).

Note: Grade analysis of the complete " inner" table of frequencies is covered in Chapter 8, Section 8.2 and 8.6.

The frequencies in Table 2.3a can be presented as in Table 2.3b.

TABLE 2.3. b

Previous model purchased by repeat buyers.

Model	Sub-compact	Compact	Mid-size	Full-size	Super-size
Frequencies	150	200	100	50	200

Present model purchased by repeat buyers.

Model	Sub-compact	Compact	Mid-size	Full-size	Super-size
Frequencies	200	250	125	25	100

Note: The information in Table 2.3b comes from the margins of Table 2.3a. The information in both distributions in Table 2.3b is - in effect - "paired," that is, based on responses from the same individuals. However, the distributions in Table 2.3b could also come from different (i.e. non-paired) sources. For example: models purchased by ***higher*** income families, vs. models purchased by ***lower*** income families.

D. Count (frequency) data – columns or rows from a two-way table:

A question of interest might be: How similar are the distributions (or what is the distributional relationship) of: the present purchase when ***the previous purchase was a "Compact"***, and ***the present purchase when***
the previous purchase was a "Mid-size?"

Note: We could use any two columns or any two rows of the table.

The frequencies in Table 2.4a can be presented as in Table 2.4b.

If the distributions are similar, perhaps, a market researcher for the auto industry will suggest only one marketing promotion for previous compact purchasers and previous mid-size purchasers. If the distributions are not similar, two different promotions, each targeting a specific purchasing group might be suggested.

TABLE 2.4. a
Frequencies for car models purchased by repeat car buyers.
(Only selected column frequencies shown.)

Previous Purchase ▶ Present purchase ▼	Sub- compact	Compact	Mid- size	Full- size	Super- size	Totals
Sub-compact		50	30			
Compact		75	45			
Mid-size		25	10			
Full-size		10	5			
Super-size		40	10			
Totals						

TABLE 2.4. b
Present purchase when previous purchase was a Compact.

Model	Sub-compact	Compact	Mid-size	Full-size	Super-size
Frequencies	50	75	25	10	40

Present purchase when previous purchase was a Mid-size.

Model	Sub-compact	Compact	Mid-size	Full-size	Super-size
Frequencies	30	45	10	5	10

From another point of view: Perhaps one distribution (maybe for the Mid-size) is the "existing company standard," the other distribution being for a new model. The market researchers want to know the functional relationship between the sales of one model, and the sales of the other.

In each of the examples A, B, C, D we deal with two sets of univariate data. ***Our aim will be to compare the distributions appearing in each set.*** This problem is referred to in this chapter and in Chapters 3 to 7.

E. Multivariate data, and the "inner" frequencies of two-way tables:

Both multivariate data and the inner frequencies (i.e., excluding the margins) of two-way tables are analyzed in a similar fashion. This material is covered in Chapters 8 to 14.

2.3 Steps to making a concentration curve

A concentration curve is the visual end result of using grade methods to compare univariate data. In this section, we show how to construct a concentration curve. We then use this curve to highlight some of its uses. In Section 2.6, we give more details relative to the understanding of curves.

Let us illustrate the steps to making a concentration curve using the data from Table 2.1 with the following scenario:

Let us say that you are studying household budgets in country A and country B. You are not interested in the actual amount spent, but only in the distributions of the proportions spent for: food, housing, utilities, transportation, and health care. You would like to know the similarity of the distributions of the two countries. (Or, from another point of view, you would like to have a profile of the relationship of the two distributions.) If — for some theoretical reason — country A and country B ***should*** have the same proportions, then in which areas is A spending too much relative to B.

We will construct a curve showing what the distribution of country A looks like in terms of the distribution of country B. (A expressed in terms of B.)

Grade methods define one distribution in terms of another by means of a concentration curve.

A concentration curve provides information relative to: the concentration (ratio of relative frequencies) of one distribution, in respect to another distribution. Our scenario curve is made up of five straight-line segments. (One segment for each of the five budget categories.)

The following steps are required to construct a concentration curve. (The steps are explained below Table 2.5a and 2.5b.)

1. Normalize the two distributions.
2. Grade transform the values in the two distributions.
3. Designate one of the distributions as the "relative to" or reference distribution. (For now, just assume that it is Country B.)
4. Calculate the concentrations (i.e., slopes) of line segments (one for each category in the distributions).

5. Plot the line segments on a "unit square" coordinate system.

Table 2.5a and Table 2.5b start with the category and amount information from Table 2.1.

TABLE 2.5. a. Amount in Millions Spent for Five Budget Categories by ***Country A***.

Category	Food	Housing	Utilities	Transportation	Healthcare	Total
Amount	60	50	40	30	20	200
Divided by 200	0.30	0.25	0.20	0.15	0.10	
Accumulated	0.30	0.55	0.75	0.90	1.00	

TABLE 2.5. b. Amount in Millions Spent for Five Budget Categories by ***Country B***.

Category	Food	Housing	Utilities	Transportation	Healthcare	Total
Amount	60	60	120	15	45	300
Divided by 300	0.20	0.20	0.40	0.05	0.15	
Accumulated	0.20	0.40	0.80	0.85	1.00	

- First we ***normalize*** the data by dividing each amount by its distribution total (200 for country A, and 300 for country B). Normalizing puts all data values on a 0 to 1 scale.

Note: The word "normalize" often has different meanings in other statistical fields.

- Next is the ***grade transformation***. (This is the transformation for which grade methods are named.) To obtain the grade transformed values, the normalized values are accumulated (here, from left to right) across the categories.

Note: The concept of the grade transformation will be extended in the theory sections of Chapter 3.

Details: Grade transforms are vectors of values of a cumulative distribution function (cdf). A univariate cdf transforms a single random variable into one defined on the interval (0,1).

Comment: Grade is an old term. It was used in reference to frequencies associated with a cumulative distribution.

- We will ***designate*** country B as ***the*** "relative to" or ***reference distribution.***

Note: It is sometimes arbitrary as to which distribution is chosen as the reference distribution. However, it is always necessary for interpretation purposes to indicate which distribution is chosen. In many cases, the reference distribution will be the "standard," or the "control," or the "familiar," or the "expected" distribution.

Note: The choosing of a "reference" distribution may seem confusing. It is frequently the case where either distribution may serve as the reference. For example: Line A is twice as long as line B. (B is used as the reference.) Line B is half as long as line A. (A is used as the reference.) On the other hand, sometimes one distribution is a kind of "standard." Perhaps, it is the "expected," or familiar distribution where "things work," or where things are "in balance," or the distribution about which more is known.

- Next, we calculate the concentrations (slopes) for each category in the distributions. The slopes are the ratios of the distribution proportions.

The proportions for country A are: 0.30, 0.25, 0.20, 0.15, 0.10.

The proportions for country B are: 0.20, 0.20, 0.40, 0.05, 0.15.

Note: By convention, we put the reference distribution in the denominator.

Each of the five corresponding categories has its own individual line segment and slope. The slopes for each of the 5 segments of the concentration curve are:

$$\begin{aligned} &0.30/0.20 = 1.5 \\ &0.25/0.20 = 1.25 \\ &0.20/0.40 = 0.5 \\ &0.15/0.05 = 3.0 \\ &0.10/0.15 = 0.67 \text{ (Rounded)}. \end{aligned}$$

- Plot the line segments having these slopes on a unit square coordinate system. (According to convention, we will put country B, the ***reference distribution***, on the horizontal axis.)

Note: For now, keep the budget categories ordered as they appear in Table 2.5a and 2.5b. (We will return shortly to this "ordering" topic.)

It will help in making the plot if you look at just the "accumulated" coordinates from Table 2.5a and Table 2.5b (shown below in Table 2.5c). The accumulated values are the grade transformed values. (However, remember that the sloping line segments that you will draw are individual "entities," horizontal and vertical "proportions" that you calculated above for each category.)

TABLE 2.5. c. Accumulated proportions to indicate coordinates to facilitate plotting.

Country A proportions (cumulative) (Y axis)	Country B proportions (cumulative) (X axis)
0.30	0.20
0.55	0.40
0.75	0.80
0.90	0.85
1.0	1.0

The reference distribution is associated with the horizontal axis of the unit square.

The concentration curve always starts at coordinates (0,0), and ends at (1,1). The curve is built up from the line segments. Each consecutive category line segment begins where the previous segment ends.

The concentration curve in Figure 2.1 is plotted on a coordinate system called the unit square. Each axis displays the cumulative proportions, and covers the range of 0 to 1.

Note: In addition to the five line segments, each representing one of the five categories, a line referred to as the diagonal (or "identical" distribution line) has been added to the figure.

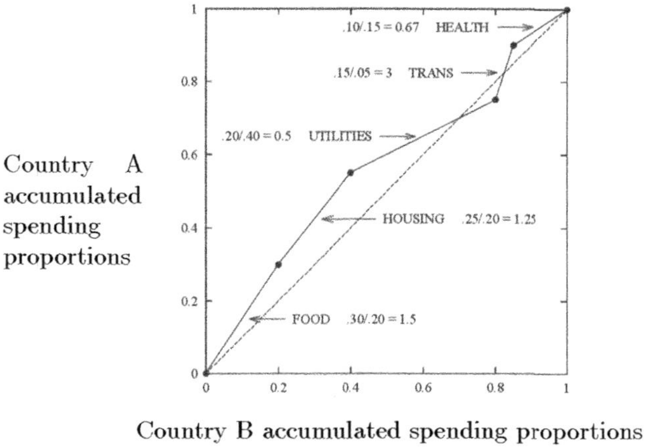

FIGURE 2.1. The concentration curve plotted on the unit square. Distribution of spending for country A relative to country B. (Budget categories are shown in their original sequence from Tables 2.5a and 2.5b.)

Note: The diagonal represents the concentration "curve" that would exist if the proportions for both country A and country B were the same within each category. (For example: 0.3, 0.25, 0.20, 0.15, and 0.10, for both country A and country B.) The diagonal has a slope of 1.0 at each point.

The diagonal has a slope of 1.0 and represents the ratio of equal proportions in each category.

The slope for the first line segment (Food) is 1.5. This means that the ***proportional*** spending for food in country A is 1.5 times the ***proportional*** spending for food in country B.

The value of the slope (i.e., 1.5) is also the "concentration" of distribution A relative to distribution B for the food category. This value is also referred to as the "overrepresentation" value for the food category.

Overrepresentation value

$$= \frac{\textbf{relative proportion of one distribution}}{\textbf{relative proportion of the reference distribution}}$$

for a specific category.

Note: Even when the ratio of the proportions is smaller than 1, the term "over-representation" value is still used.

A category with a line segment having a slope of 1.0 (regardless of whether the line lies exactly on the diagonal or is simply parallel to it) indicates that the relative proportions spent for that category are the same for both countries.

Let us take the housing category as an example. Rather than having a slope of 1.25 as shown, let us assume it has a slope of 1.0. This line would be parallel to the diagonal, but not on the diagonal. The slope of 1.0 would still, however, indicate the same proportion of spending.

From the concentration curve we can see that country A spends proportionally less (i.e., overrepresentation values smaller than 1.0) than country B for utilities (0.5), and health (0.67). Country A spends proportionally more (i.e., overrepresentation values larger than 1.0) than country B for: housing (1.25), food (1.5), and transportation (3.0).

Note: A natural question might be: If we are concerned primarily with the value of the slopes rather than the segment position relative to the diagonal, why make the plot of the concentration curve?

The answer is that we are concerned with both the value of the slope, and the composite position of the complete curve relative to the diagonal. This will be addressed below, but for now notice that it would be possible to compute the area falling between the concentration curve and the diagonal. This area is a kind of index of dissimilarity between the distributions. (But more about this later.)

Note: The concentration curve will never have a negative slope, in that its plotting coordinates come from cumulative distributions. If a zero category exists in one of the distributions, the line segment for that category will be either horizontal (e.g., if a category for country A equaled zero) or vertical (e.g., if a category for country B equaled zero). If both categories equaled zero, there would simply be no category, i.e., only four budget categories in the example.

In our scenario we have five overrepresentation values, one for each of the budget categories: 1.5 (food), 1.25 (housing), 0.5 (utilities), 3.0 (transportation), 0.67 (health).

When working with distributions, depending upon the meaning of their categories, it is either possible to reorder the categories, or it is not. For example, in our scenario the ordering of the budget categories was arbitrary. (Someone simply constructed the table with the categories in that order.) Some examples of distributions having categories that "probably" would not be reordered might include: age categories, time categories, distance categories, measurement categories, scoring or rating categories, and certain positional or location categories.

Let us now reorder the categories in our scenario based on their overrepresentation values (from small to large: 0.5 utilities, 0.67 health, 1.25 housing, 1.5 food, 3.0 transportation), and make another concentration curve.

Note: While it is not absolutely necessary to re-compute Table 2.5c, this is the easiest approach to making the new plot.

TABLE 2.5. d. Accumulated proportions to indicate coordinates to facilitate plotting. *(Based on categories reordered according to overrepresentation values.)*

Country A proportions (cumulative) (Y axis)	Country B proportions (cumulative) (X axis)
0.20	0.40
0.30	0.55
0.55	0.75
0.85	0.95
1.0	1.0

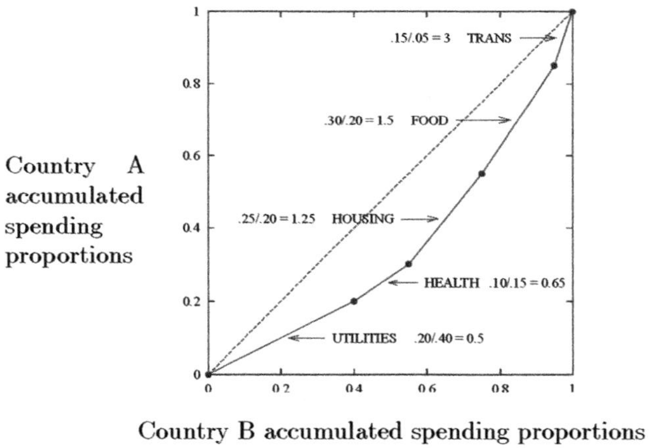

FIGURE 2.2. Concentration curve for the distribution of spending for country A relative to country B. (Budget categories are now reordered by increasing size of overrepresentation values.)

The concentration curve in Figure 2.2 is constructed from the same five line segments as used in Figure 2.1. However, the segments are now ordered from left to right based on the values of their overrepresentation scores (i.e., the slopes are ordered from small to large). When plotted in this ordered fashion a convex curve is produced.

Comment: Reordering, depending upon the specific subject area context, is a kind of "prioritizing" of the categories.

Note: We have also included a diagonal in the figure. Recall that the diagonal represents (e.g., as a reference) the concentration curve of two identical distributions.

The area between the concentration curve and the diagonal is a measure of the dissimilarity between the two distributions (i.e., between country A and country B). This area is used to compute a concentration index. (More concentration index details later.) A greater area is associated with a greater dissimilarity between the two distributions. In other words, the larger the value of the concentration index, the greater the dissimilarity of the distributions.

Details: Grade methodology also addresses the shape of the area, but this is covered in the theory sections.

The concentration index is associated with the area between the concentration curve and the diagonal.

Note: A more comprehensive definition and method of calculating the concentration index is given later. For now, it is enough to associate the area between the concentration curve and the diagonal with the concentration index.

Without going into detail at this point, a few implications based on Figure 2.2 are noteworthy:

- If one simply omits a category (i.e., combines two of the adjacent categories) the curve will move closer to the diagonal. This will decrease the area. As this area represents the dissimilarity of the two distributions, decreasing the area will make the two distributions appear to be more similar.

Comment: Recall that the data analyst often has control over the making and combining of categories.

- If two segments have identical slopes, those segments (i.e., categories) will be adjacent to each other. Segments with the same slope can be combined with no change in the value of the concentration index (i.e., their measure of dissimilarity).

- If the actual length of a line segment is extremely short (not to be confused with a small distance on one axis or the other), the segment will have very little effect on the size of the area (i.e., concentration index value). As we have no extremely short line segments in this example, take this on faith for now.

- Since the plotting area on all unit squares is (by definition) equal to 1, this "standardization" facilitates the comparison of areas (and curves) among different studies and various alternative category configurations.

Note: Differently shaped curves (considering not only shape, but varying degrees of "roughness or smoothness," and degree of asymmetry) can produce the same concentration index. This shape consideration is often ignored by non-grade approaches. A more detailed discussion of the concentration index appears in Chapter 3 Section 3.2.2.

Note: The theoretical chapters go into more detail regarding the concentration index for distributions where it is not possible (or not logical) to reorder the categories. See Chapter 3, 4, and 8.

2.4 Quick Start summary

The purpose of the ***Quick Start*** is to prepare you for the study of multivariate grade concepts and applications via an understanding of grade method concepts with univariate data.

Grade methods use "tools" designed to investigate data populations. In the univariate situation these tools include: the concentration curve, overrepresentation values, the diagonal curve, and the unit square.

The concentration curve plays the most visible role. The curve displays "something" expressed in terms of "something else." The first "something" is a distribution. The second "something" (i.e., "something else") is a second distribution which is often a "reference" or "expected" distribution. The concentration curve displays the dissimilarity of the two distributions, and also allows for the calculation of a measure of this dissimilarity.

For each category in the distributions, an overrepresentation value is obtained. This overrepresentation value is the ratio of the relative frequency of the first distribution to the relative frequency of the second (i.e., reference) distribution. (In some applications, these two distributions can be thought of as the "observed" distribution, and the "expected" distribution.) Grade methods process these overrepresentation values to reveal information about the data.

The diagonal curve profiles two identical distributions. The diagonal is used as a component in the visualization of the concentration index, and also provides a visual baseline reference when interpreting a concentration curve.

The unit square provides the "standardized" axes on which to display the concentration curve. This standardization facilitates the comparison and interpretation of line segments and areas, both within and between various concentration curves.

2.5 Preview of Part 2, and suggestions before your eventual study of the multivariate material

Part 2 revisits the univariate concepts presented in the Quick Start. The approach is still non-mathematical, but from a different point of view. This approach will serve to link continuous and discrete distribution concepts in the context of grade methods.

The examples in Part 2 usually involve the two cumulative distributions: ***space*** in a cylinder, and ***amount*** of a gas which occupies this space. Theoretical partitions are then added, manipulated, and removed to simulate the creation of categories, changes in the size of the categories, and the combining (i.e., merging or aggregation) of categories, respectively. More details are also presented relating to the diagonal, and the concept of concentration.

An applied example is also included for which an interpretation of the concentration curve is presented. Each distribution is, in turn, used as the reference distribution, and the two "different" corresponding concentration curves are contrasted.

Please read these suggestions before eventually starting on the multivariate material:

Comment: Although the multivariate material is not covered until later, this is probably the best place for these suggestions, in that you have just read the Quick Start.

Keep the following in mind as you proceed from the above material (concerning two sets of univariate data) to the material for multivariate data in the general case:

Let us say that we have records consisting of 5 test scores obtained for each of 100 people. Thus we have a data table with 100 rows (records) and 5 columns (variables). This data set can be considered as a two-way table of empirically gathered data (i.e., as an ***empirical distribution***). We may create another distribution — a hypothetical distribution, derived from the margins of this two-way table under the hypothesis that each score is proportional to the respective marginal scores. Using these two distributions, 500 overrepresentation values (i.e., 100×5) can be calculated. This 100×5 table of overrepresentation values forms the so called grade distribution.

The 500 overrepresentation values can be reordered (in two directions) using grade procedures. This reordering procedure is such that the data structure is revealed.

The reordered overrepresentation values are now displayed as a surface on a "grid" (called an overrepresentation map). The size values of the overrepresentations are represented using different shades (white to black). The map displays a "picture" of the data structure (i.e., trends, clusters, subpopulations), while retaining the visual simplicity of only two axes.

The greater portion of this book deals with the theory and solution for reordering the overrepresentation values in the multivariate setting. (The reordering procedure is along the lines of solving a giant Rubik's Cube puzzle, but where each cube face has an infinite range of colors.)

2.6 Part 2: Understanding concentration curves

Note: Your understanding of this section will be better if you have read the Quick Start.

2.6.1 Introduction

The purpose of this section is to explain the "workings" of the concentration curve. An understanding of this curve, with its overrepresentation values (i.e., slopes) is critical to the understanding of overrepresentation plots in the multivariate context.

We are going to look at a number of figures. The objective is to understand where the concentration curve comes from, and what it represents.

A concentration curve provides information relative to the "concentration" of one distribution with respect to a second distribution. The second distribution (i.e., the "with respect to" distribution) usually has some special (i.e., known or established) qualities. The concentration curve profiles the relationship between the distributions.

Comment (bending the theory a little): One distribution could be the members of a big family. The second distribution could be the family car. This second distribution (the space inside the car) is fixed. The first distribution (i.e., the bulk/mass of the family members) will be "concentrated" into the front and back seats. The concentration curve in this case would profile this distributional relationship. (It might represent the degree of passenger comfort after riding for three hours.) This understanding of concentration may be important relative to the understanding of passenger comfort.

In Part 2, we consider a cylinder containing gas (i.e. two distributions). These two distributions deal with the ***amount*** of gas, and the ***space*** inside the cylinder. (As a gas is easily able to get into all the cylinder space, these will make better examples than the family car example.)

We use a cylinder of gas, because it is useful to describe, in theory, what happens when you add, move, and remove partitions. (This is analogous to making, resizing, and combining categories of a distribution.)

Note: Regarding the term "cylinder space:" It turns out that it's a little easier in some figures to refer to the cylinder's "length," or distance between partitions, as opposed to "space." Please accept this occasional technical inaccuracy.

If we consider the cylinder space as a "fixed" reference (i.e., the entire space equals 100 percent) this, in effect, normalizes the distribution. We can do the same for the amount of gas (i.e., total amount equals 100 percent). We can then speak in terms of proportions.

Cylinder activities and their corresponding data analysis activities:

- Inserting cylinder partitions — This is the creating of additional distributional categories for both: gas amount and cylinder space.
- Moving partitions closer together or further apart — This is the changing of the sizes of the categories.
- Removing partitions — This is combining (aggregating) categories for gas amount and cylinder space separately.

2.6.2 *Two identical distributions*

Figure 2.3 shows a cylinder. Inside is a gas, which is distributed uniformly (or equally) in the space. The $(0, 1)$ represents the normalization of the cylinder ***space*** distribution and the enclosed gas ***amount*** distribution. Therefore, the figure shows, in effect, two univariate distributions.

Note: Recall that we divided by the total household budgets (i.e., 200 and 300) to normalize the data in the Quick Start example. For Figure 2.3, because the cylinder length is defined to be from 0 to 1, and because we are using "proportions" of gas, the normalizations are already taken care of.

Because there are no partitions in the cylinder, we have space and gas distributions with only one category. As you move along the length of the cylinder (i.e., the cylinder's space distribution), you are also moving along the length of the gas amount distribution. When you are 50 percent of the way along the one, you are 50 percent along the way of the other. It is a one-to-one relationship. (Two identical or equivalent distributions.)

Details: One can say that the gas has a "uniform density on the interval $(0, 1)$."

FIGURE 2.3. Gas in a cylinder with no partitions.

Note: The amount of gas, and the cylinder space literally share the same distribution.

Figure 2.4 makes it easier to remember that we are dealing with two distributions (i.e., amount of gas, and cylinder space). The figure is constructed from a unit square and a concentration curve which, in this case, is the diagonal.

Let us first check to see if we have all the grade "essentials" from Section 2.2.1, neccessary to proceed from Figure 2.3 to Figure 2.4.

The "essentials" for the grade approach include:

- Two distributions?

  Yes, gas amount and cylinder space
- Same number of categories in each distribution?

  Yes, for now only one category.
- Same meaning of (description for) corresponding categories?

  Yes: "Total".
- Same ordering of categories (by meaning) in each distribution?

  Yes: "Total".
- One distribution designated as a "relative to" (or reference) distribution?

  Yes, cylinder space.

As we have the necessary essentials, we can construct a curve which expresses one distribution in terms of another.

One distribution must serve as the reference distribution. Since the cylinder is easier to visualize, let us make the cylinder's space the reference distribution. Therefore, we are looking at two distributions: the distribution of the gas ***amount*** — relative to — the distribution of the cylinder ***space***.

Note: In the case where there is no obvious reference distribution, either distribution will work. This is analogous to choosing the denominator for a ratio. As long as someone knows which is chosen, they can re-compute the other.

On the horizontal axis of Figure 2.4 we have indicated the position along the length of the cylinder. On the vertical axis, we have the cumulative proportion of gas.

In that we only have one category in each of our two distributions (i.e., the total), 100 percent of the gas is uniformly distributed in 100 percent of the cylinder space.

We draw a line from the coordinates (0,0) to (1,1), which represents the cumulative proportion of gas as we go from left to right in the cylinder. We have 100 percent of the gas, in 100 percent of the cylinder (one cell).

Our diagonal line represents the concentration curve of the gas amount with respect to the cylinder space. It has a slope of 1 at each position in the cylinder. The diagonal we have drawn is also the graph of the cumulative distribution function for, in this case, the gas amount, as well as of cylinder space.

All our figures are being plotted on what is called the "unit square." (This term is used, because both axes go from 0 to 1, and the area of the square is therefore equal to 1.) Using the unit square allows for the comparison of different figures and different areas. The area being equal to 1 is useful for probability applications. For more information see Chapter 3.

When all the proportions of gas exactly equal all the proportions of space (i.e., equivalent distributions), we refer to the diagonal as the equal distribution line. Or, more commonly as the ***diagonal***, or *Diag*.)

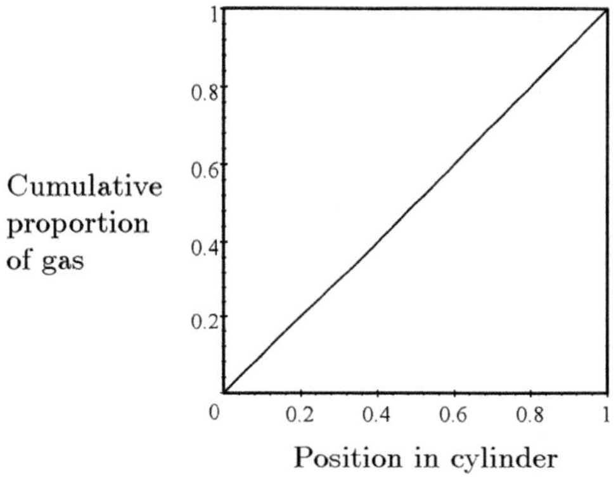

FIGURE 2.4. Cumulative distribution function (cdf) for amount of gas (this cdf is equal to the concentration curve of the gas amount relative to the cylinder space).

The slope for the relationship of gas amount relative to space is: $1.0/1.0 = 1.0$. Therefore, our concentration curve (in this case, the diagonal) has a slope of 1.0. That is: the rise (change in vertical) over run (change in horizontal) equals 1.

Details: We say that the gas amount has "a uniform density on the interval (0,1)." (The slope of the line in Figure 2.4 — a density in mathematical terms — is equal to 1.)

2.6.3 Cylinder with partitions: cells of equal length, gas in equal proportions

Let us put four partitions into our cylinder at equal intervals so as to make five cells of equal size (0.2 of the length). We will also put equal portions (0.2) of gas into the five cells. (This would be equivalent, in practical terms, to making equal length (or size) "categories" for each of two distributions.)

Now we have (using proportions as our measures): 5 cells of 0.2 size, and 5 proportions of gas of amounts equal to 0.2.

Note: Partitioning /discretizing /categorizing/making cut points, have roughly the same meaning.

We still have our "essentials" (see Section 2.2.1): two distributions, the same number of categories (i.e., 5), the same meaning for the categories (in this case just sequential positions along the length of the cylinder), and the cylinder space distribution being the "relative to" distribution.

Each of the five cells occupies 0.2 of the total cylinder space. Each cell contains a 0.2 proportion of the gas. We have a uniform concentration of gas in a partitioned cylinder.

Cell	Gas Amount		Cylinder Space
1	0.20	◀ same ▶	0.20
2	0.20	◀ same ▶	0.20
3	0.20	◀ same ▶	0.20
4	0.20	◀ same ▶	0.20
5	0.20	◀ same ▶	0.20

In Figure 2.4, we had a gas distribution having only one category, and a cylinder space distribution having only one category. We now have five categories in each distribution.

The gas proportions (i.e. in the numerator) are all equal to 0.2. The cylinder space proportions (i.e. in the denominator) are also all equal to 0.2. Therefore the slopes for each of the 5 segments of the concentration curve are: $0.2/0.2 = 1$.

Note: The distribution of gas at this point is no different than that of Figure 2.3.

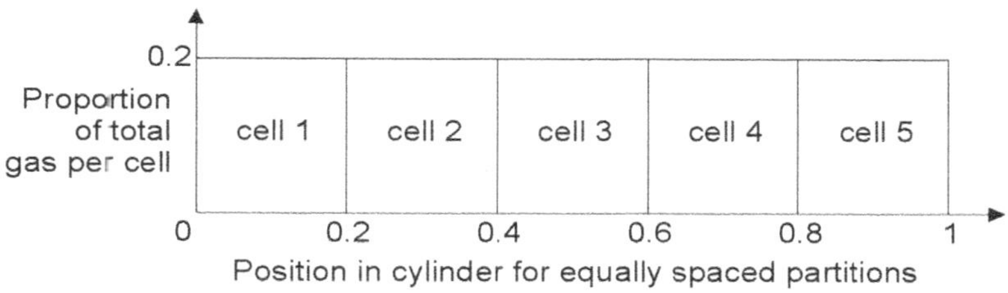

FIGURE 2.5. Gas equally distributed in a cylinder with equally spaced partitions. (With 20% of gas in each cell.)

2.6.4 *Constructing a concentration curve from individual category segments*

In Figure 2.6, we have accumulated the "boxes" from Figure 2.5. As you go from left to right in the cylinder, the 5 space proportions accumulate along the horizontal axis. That is: 0.0, 0.2, 0.4, 0.6, 0.8, 1.0. The amount proportions also accumulate for the gas on the vertical axis. Again we have a diagonal line as in Figure 2.4. Notice that each of the five segments of the concentration curve has an individual slope of 1.0.

When these five line segments are put together, this diagonal line represents the "equal distribution" line. The slopes of 1 in the five "tiny" cells are the same as the slope of 1 in the one big cell in Figure 2.4.

When the proportional amount of gas in each cell is equal to the proportional size of each cell, you get a diagonal line with slope equal to 1.

Note: Rise over run of 0.2 / 0.2 = 1 in each cell.

The diagonal curve is often your "eyeball" reference for other concentration curves.

Details: One says that the density (i.e., slope) of the concentration curve is 1.0.

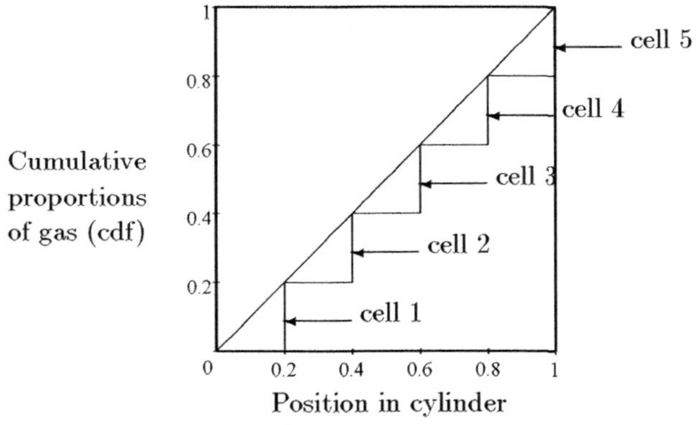

FIGURE 2.6. Gas amount cumulative distribution function (cdf) for cylinder with equally spaced partitions, and with a uniform distribution of gas amount (20% in each cell). Vertical lines represent gas amount proportions in particular cells, horizontal lines represent the lengths of cells.

The diagonal represents the equal distribution line for gas amount and cylinder space.

A "just to make sure" explanation:

We call the diagonal curve the ***equal distribution curve***. However, it is often referred to by the name ***uniform distribution curve.*** The "uniform" refers to the diagonal as being the graph of the cdf of the uniform distribution on interval $(0, 1)$. The "equal" refers to the ***equality between distributions,*** and not to equal (uniform) within distributions. A concentration curve for the following proportions would also be the diagonal:

Cylinder cell	Proportion of gas	Proportion of space
1	0.30	0.30
2	0.25	0.25
3	0.20	0.20
4	0.15	0.15
5	0.10	0.10

2.6.5 When proportions **do not** *correspond between distributions*

Now we will start to manipulate the proportions to demonstrate what happens when the two distributions are not "in sync." Let us keep the cylinder spaces in cells the same size, and vary the gas proportion amounts in each of the spaces.

Figure 2.7 shows a chart representing unequal gas proportions and equal (i.e., 0.2) cylinder cell sizes.

We put the following proportions of gas into the cells from left to right: 0.1, 0.2, 0.4, 0.05, and 0.25.

Let us say that we are concerned with "proportions of gas ***relative to*** cell size."

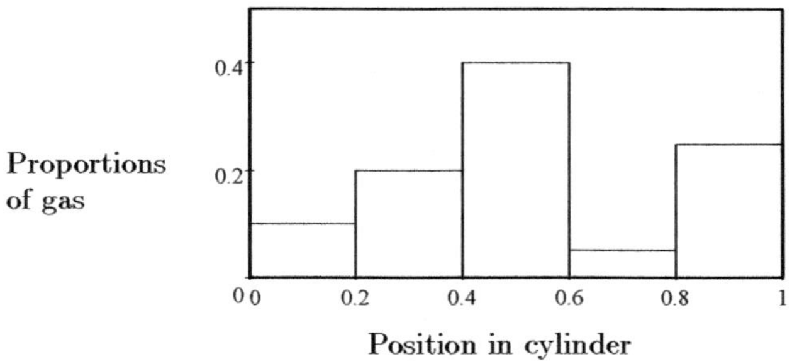

FIGURE 2.7. Unequal gas proportions for equal cell sizes.

Numerator values (proportions of gas): 0.1, 0.2, 0.4, 0.05, 0.25.
Denominator values (cell size): 0.2, 0.2, 0.2, 0.2, 0.2.
The slopes for each of the 5 segments of the concentration curve are:

$$0.1/0.2 = 0.5$$
$$0.2/0.2 = 1.0$$
$$0.4/0.2 = 2.0$$
$$0.05/0.2 = 0.25$$
$$0.25/0.2 = 1.25.$$

Using the values from Figure 2.7, and the information in Section 2.6.6, let us proceed to make Figure 2.8.

2.6.6 Using the concentration curve to introduce the concept of overrepresentation

Figure 2.8 shows the concentration curve for the distributions presented in Figure 2.7.

In the same manner that we took the information from Figure 2.5 and made Figure 2.6, let us now use the information from Figure 2.7 to make Figure 2.8.

The "rectangles" from Figure 2.7 are superimposed on Figure 2.8. This superimposition shows not only the accumulation process, but also that the concentration curve segments are the diagonals of these rectangles.

The gas proportions are: 0.1, 0.2, 0.4, 0.05, 0.25.

The cylinder space proportions are: 0.2, 0.2, 0.2, 0.2, 0.2.

The slopes for each of the 5 segments of the concentration curve are:

$$0.1/0.2 = 0.5$$
$$0.2/0.2 = 1.0$$
$$0.4/0.2 = 2.0$$
$$0.05/0.2 = 0.25$$
$$0.25/0.2 = 1.25$$

Let us plot the concentration curve line segments. It will help in making the plot if we prepare our coordinates by accumulating first. However, remember that these line segments are individual "entities," coresponding to horizontal and vertical "proportions." Each of the five corresponding proportions has its own slope.

Accumulated proportions to compute coordinates for plotting:

Gas proportions (cumulative) (Y)	Cylinder space proportions (cumulative) (X)
0.1	0.2
0.3	0.4
0.7	0.6
0.75	0.8
1.0	1.0

The concentration curve always starts at coordinates $(0, 0)$, and ends at $(1, 1)$.

2.6.7 Overrepresentation

The values of the slopes are the overrepresentation values.

In Figure 2.4 and Figure 2.6, the slopes were always equal to 1, since cells had proportionally the same gas amount as cylinder space.

Recall that our "reference" distribution is cylinder space corresponding to the horizontal axis. Any curve segment with a slope smaller than 1 represents a part of cylinder which has proportionally less gas than space. Any segment with a slope larger than 1 has proportionally more gas than space.

Therefore the first and fourth segments of the curve (counting from the left) correspond to cells which have proportionally less gas than space and have overrepresentation values smaller than 1. The third and fifth correspond to cells which have proportionally more gas than space and have overrepresentation values larger than 1. The second segment has a slope equal to 1, and the second cell has the same proportion of gas, as it has of space.

The 5 segments together form the concentration curve for gas amount relative to cylinder space.

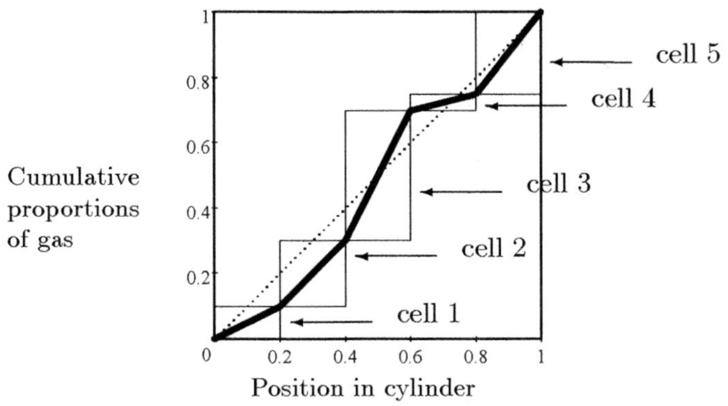

FIGURE 2.8. Cumulative distribution function (cdf) of gas amount with unequal proportions in cells, in a cylinder with equal cell sizes. The cdf (bold line) is also the concentration curve of gas amount relative to cylinder space. The thin vertical lines represent the proportions of gas in particular cells. The thin horizontal lines represent the cell sizes. The dotted line is the diagonal (i.e. "equal distributions line").

Note: The concentration curve is "non-decreasing." That is, it will never have a negative slope in that it represents a "cumulative" distribution. (We cannot say that the curve always goes up, because it is possible to have a zero proportion category for one or the other of the distributions. This results in a line segment which is either vertical (when the horizontal category proportion is zero), or horizontal (when the vertical category proportion is zero).

Details: The slopes are interpreted as "overrepresentations." Even when an overrepresentation value is less than 1.0 (e.g. 0.5), one still uses the prefix ***over*** as opposed to ***under.*** This terminology relates to a theoretical viewpoint. "Under-representation" values could be used, but they would then, likewise, have possible "underrepresentation" values larger than 1.

Special cases or interpretations of overrepresentation:
— **Concentration density**
— **Rise / run (i.e. the slope).**
— **"Observed" / "expected"**
(In that we have a "relative to" concept.)

Let us look at the line segment in the first cell (i.e., position 0.0 - 0.2 on the horizontal axis). The slope is 0.5. That is, in the first cell of the cylinder we have only a 0.5 the gas proportion, relative to the space proportion. (Overrepresentation equal to 0.5.)

The second cell (0.2 - 0.4) has a slope of 1. That is, the amount of gas for this cell is what it would be under a uniform distribution. (Uniform from one distribution to the other.)

The third cell (0.4 - 0.6) has a slope of 2. This indicates that the third cell has 2 times the gas relative to a uniform (equal) distribution. In other words, the proportion of gas is twice the proportion of space.

In the fourth cell (0.6 - 0.8) the slope is 0.25. The proportion of gas is only 0.25 of the proportion of space.

Finally in the fifth cell (0.8 - 1) the slope is 1.25. That is, relative to the cell size, the proportion of gas is 1.25 times more.

* * *

Figure 2.9 is basically a remake of Figure 2.8. In making Figure 2.9, we have copied the concentration curve from Figure 2.8, and eliminated the rectangles. We have then added the diagonal (i.e., ***equal distribution line***) from Figure 2.6, to act as a visual reference.

Notice that the second line segment (0.2 to 0.4) is under the diagonal; however, it has a slope equal to 1, which still represents proportional equality between the two corresponding categories.

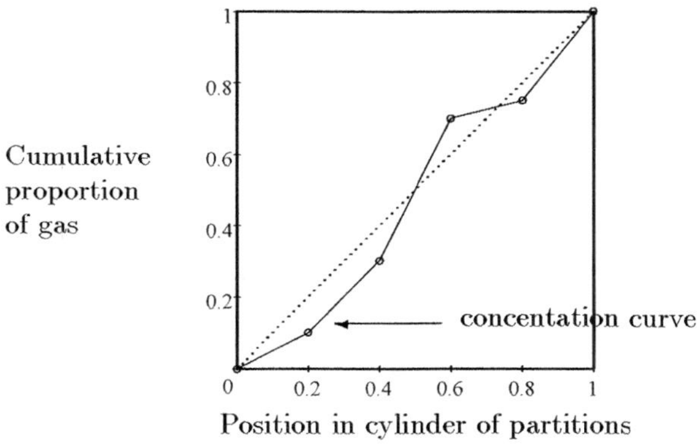

FIGURE 2.9. Concentration curve for gas amount relative to gas space (unequal proportions of gas amount in cells of equal sizes): the curve is the same as in Figure 2.8.

Note: In the present context, it is not whether the slope segments are above or below the diagonal line, but rather which is their relation to the slope of the diagonal - which equals 1.

Details: Notice that the concentration curve crosses the diagonal. In part, this is related to, and is a function of, the ordering of the categories in the distributions. This is addressed in Chapter 4.

2.6.8 When we manipulate both distributions: gas (unequal proportions) and cylinder (unequal cell sizes)

What happens when we manipulate both distributions? To find out, let us change the cylinder cell sizes to: 0.05, 0.3, 0.2, 0.4, and 0.05. (We will use the amounts of gas used in the previous example.)

Figure 2.10 shows both the gas and space distributions.

The gas proportions are: 0.1, 0.2, 0.4, 0.05, 0.25.
The cylinder space proportions are: 0.05, 0.3, 0.2, 0.4, 0.05.

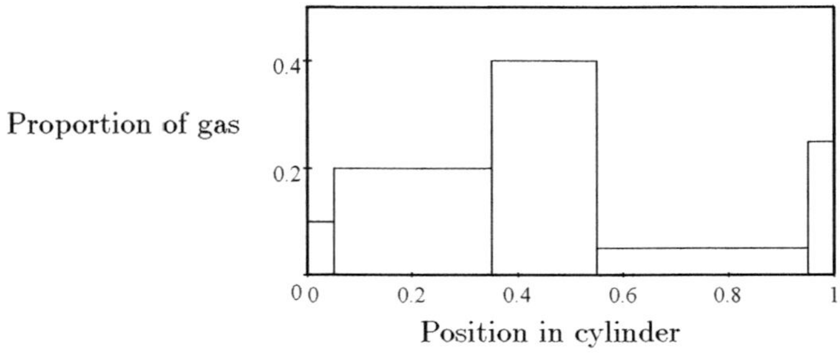

FIGURE 2.10. Unequal gas proportions for unequal partition spacing.

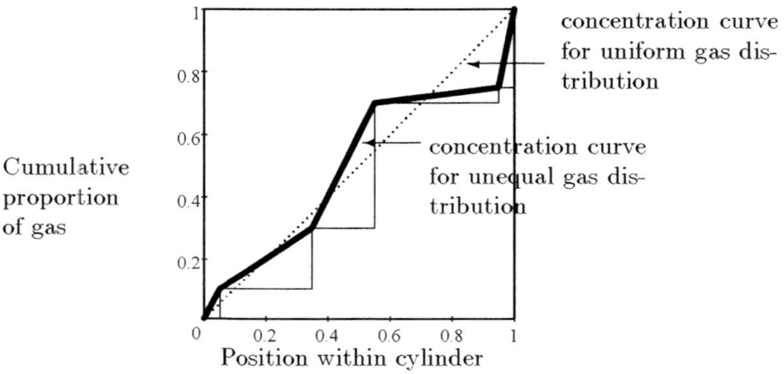

FIGURE 2.11. Concentration curve (bold line) of gas amount relative to cylinder space in the case of unequal cell sizes and unequal proportions of gas amount in cells. (Figure also shows diagonal line. The thin vertical lines represent the proportions of gas in particular cells. The thin horizontal lines represent the cell sizes.)

The slopes for each of the 5 segments of the concentration curve are:

$$0.1/0.05 = 2.0$$
$$0.2/0.3 = 0.666$$
$$0.4/0.2 = 2.0$$
$$0.05/0.4 = 0.125$$
$$0.25/0.05 = 5.0$$

Figure 2.11 is a plot of the concentration curve for the Figure 2.10 information. We have added the diagonal, and also superimposed the "staircase" formed by the tops of the accumulating rectangles of Figure 2.10.

Comment: If you exchange the above numerator values for the denominator values (i.e., you want the gas to be the reference distribution), you will get a similar picture, but it will be rotated or turned 180 degrees along the axis of the diagonal line.

2.6.9 Example application — Winners versus losers in the car sales market

Starting with Section 2.6.2 we have been working with two distributions (i.e. gas amount and cylinder space), where the matching of cathegories was according to ***position*** along a cylinder. Let us now look at distributions where the matching is according to ***meaning*** of the category.

You are a car dealer; the year is 1979. You would like to know the proportional preference for car models ***in 1979, with respect to,*** the proportional preference ***before 1979.*** (Please see Table 2.6.)

TABLE 2.6. Preference for Domestic Car Models (proportions of sales for these models).
From car dealer's viewpoint of sales "winners and losers." (Rounded to two digits.)

When Purchased▼Car Model ▶	Sub-Compact	Compact	Mid-Size	Luxury
In 1979	0.22	0.29	0.24	0.25
Before 1979	0.17	0.20	0.39	0.24

Note: This data are taken from [21] (gathered by Rogers National Research in 1979, reprinted from [43]).

Comment: The ordering of the car models in Table 2.6 is arbitrary (in contrast to the linear "position of the partitions" in our cylinder). If we assume that luxury models are big cars, then the ordering is according to size.

Let us compute our ratios (using purchased "Before 1979" as the denominator, in that "Before 1979" is the "relative to" distribution.):

$$\begin{aligned}\text{Sub-Compact} &: 0.22/0.17 = 1.29 \text{ (Rounded)}\\ \text{Compact} &: 0.29/0.20 = 1.45\\ \text{Mid-Size} &: 0.24/0.39 = 0.62 \text{ (Rounded)}\\ \text{Luxury} &: 0.25/0.24 = 1.04 \text{ (Rounded)}\end{aligned}$$

You want to know which models "In 1979" are "winners" (i.e., overrepresentations >1) and which are "losers" (i.e., overrepresentations <1).

For all models having a slope greater than 1, i.e., sub-compacts, compacts, and luxury, their 1979 proportions increased relative to the pre-1979.

The proportion of sub-compacts sold in 1979 is over-represented relative to the proportion sold before 1979. The 1979 proportion is, e.g., 1.29 times as large as the "Before 1979" proportion.

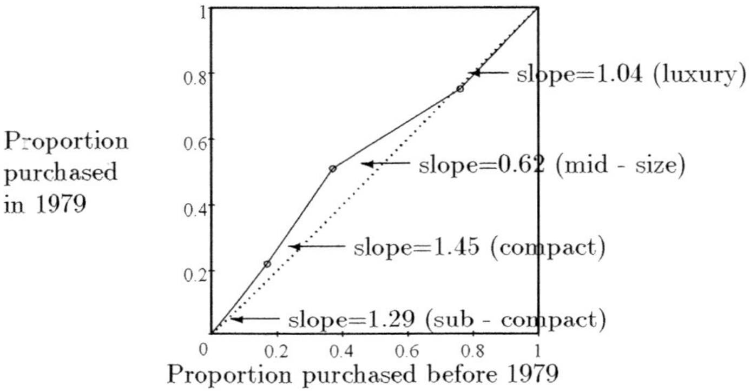

FIGURE 2.12. Concentration curve for preference of car models purchased in 1979 relative to those purchased before 1979

You would use Figure 2.12 (i.e., with the "Before 1979" on the horizontal axis) to determine how the car market was doing in 1979 relative to before 1979. In other words: "Now" relative to "Then."

Regarding Figure 2.12: The slope for sub-compacts is 1.29. The proportion of sub-compacts purchased in 1979 was 1.29 times larger than the proportion purchased before 1979.

The sub-compact proportion in 1979 was 0.22; the proportion before 1979 was only 0.17. The sub-compact proportion has increased in 1979 relative to before 1979.

A slope of 1 (regardless of whether it is exactly on the diagonal line or only parallel to it) would indicate the same proportion of sub-compact buyers in 1979 as before 1979.

Note: We focus on the proportions, and not the actual numbers purchased.

Note: Whether the line segment is above or below the diagonal does not impact on your interpretation of the overrepresentation values. (There is some meaning associated with the line segment being above or below the diagonal, but it will only be apparent after we cover a little more material.)

Be careful that you only mention the proportions when interpreting the results.

2.6.10 Example application — Historic perspective (then vs. now) of car sales market

This example shows what happens to the concentration curve when you change the "with respect to" axis (i.e., when the axes are reversed). We are using the same data used for Figure 2.12.

The year is still 1979. However, you have given up your job as a car dealer. You are now an auto historian. You would like to describe the proportional preference for car models before 1979, with respect to, the proportional preference in 1979. (Please see Table 2.7.)

Your objective is to characterize "before 1979 relative to 1979." You want to make statements like: The proportion of sub-compacts "before 1979" was smaller than the proportion in 1979.

Note: Since we are using the same distributions and just switching the reference, you will have the opportunity to compare the concentration curves of Figures 2.12 and 2.13.

TABLE 2.7. Preference for Domestic Car Models (proportions of sales for these models).
From historian's viewpoint of "how it was ***then***, compared to how it is ***now***." (Rounded to two digits.)

When Purchased▼Car Model ▶	Sub-Compact	Compact	Mid-Size	Luxury
Before 1979	0.17	0.20	0.39	0.24
In 1979	0.22	0.29	0.24	0.25

Note: This is the same as Table 2.6, but with the rows interchanged for consistency in presentation methodology.

Let us compute our ratios (using purchased "In 1979" as the denominators):

$$\begin{aligned}\text{Sub-Compact}: &\ 0.17/0.22 = 0.77 \text{ (Rounded)}\\ \text{Compact}: &\ 0.20/0.29 = 0.69 \text{ (Rounded)}\\ \text{Mid-Size}: &\ 0.39/0.24 = 1.63 \text{ (Rounded)}\\ \text{Luxury}: &\ 0.24/0.25 = 0.96\end{aligned}$$

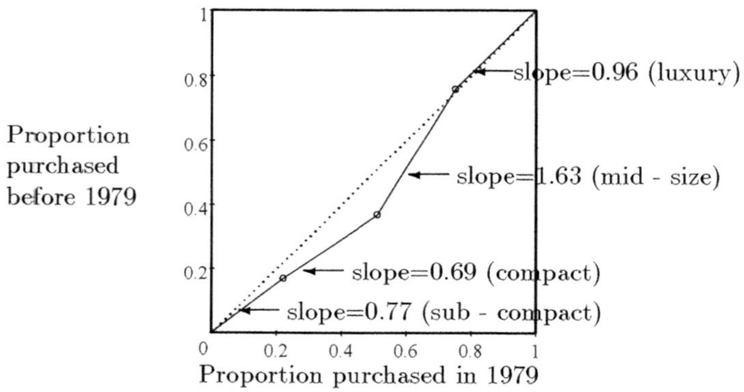

FIGURE 2.13. Concentration curve for preference of car models purchased before 1979 relative to those purchased in 1979.

Figure 2.13 has the horizontal axis as "purchased in 1979" with "purchased before 1979" as the vertical axis. Basically, Figures 2.12 and 2.13 are the same, but one is just rotated 180 degrees along the axis of the diagonal. If a line segment was below the diagonal, it will now be above.

The proportion of sub-compacts purchased before 1979 was smaller than the proportion in 1979. Also, the proportion of compacts was smaller before 1979 than in 1979. The proportion of mid-size, however, was greater (i.e., slope > 1) before 1979 than in 1979. The proportion of luxury cars was nearly the same before 1979 as in 1979 (i.e., almost a slope of 1).

Note regarding the line segment for luxury cars: the proportions are nearly the same ***not*** because the line segment is in close proximity to the diagonal, but rather because it is nearly parallel to it. (It has to do with the ordering of the categories. More about this next.)

Hint relative to the wording of statements and interpretations: First, use the word "proportions." Next, pick a category, for example, "compact cars." Then make sure you have a "relative to" meaning, or "with respect to" meaning in the wording. For example: The proportion of compact cars was smaller before 1979 than in 1979 (i.e., before relative to now).

In the next section, you will see that we are going to start reordering the categories (i.e., the segments).

2.6.11 Reordering (prioritizing) categories — and an introduction to the maximal concentration index

From the standpoint of concentration curve construction, there are basically two kinds of distribution categories: categories you can reorder, and those you cannot reorder. We will focus now on those you can reorder. (The theory sections address both types.)

Comment: With the cylinder, it is logical to assume that the categories are adjacent "discretized" values of a continuous distribution.

Until now, our categories have either been in a fixed ordering (e.g., the adjacent cylinder cells), or in an arbitrary nominal ordering (e.g., the tabulated order of the car models).

However, during the construction of concentration curves, the possibility exists, in some cases, to reorder or prioritize categories based on their over-representation values.

The reordering aids in the visualization and interpretation of the relationships between the two distributions, or, as you will see, facilitates the calculation of something called the maximal concentration index (covered in Chapter 4, Section 4.3.2).

Note: The reordered slopes also provide decision-making information regarding such things as: which categories might be combined (i.e., aggregated).

To see what happens to the concentration curve, we will reorder the categories previously used to make Figure 2.13.

To make Figure 2.14, let us start with the information from Figure 2.13. (This was the historical viewpoint with "proportion purchased in 1979" on the horizontal axis.) We reorder (from smallest to largest) the car model category line segments based on the values of their slopes (i.e., overrepresentation values).

Ordered slopes or overrepresentation values:

$$\begin{aligned} \text{Compact} &: 0.69 \\ \text{Sub-Compact} &: 0.77 \\ \text{Luxury} &: 0.96 \\ \text{Mid-Size} &: 1.63 \end{aligned}$$

The concentration curve for this ordering arrangement is shown in Figure 2.14.

Exactly the same concentration curve construction principles apply regardless of whether the concentration curve is above or below the diagonal, or whether the categories are ordered or not ordered.

Remember in Figure 2.14 to keep the absolute distances the same as in Figure 2.13. (That is, the lengths of the sides of the triangles making up the slopes are the same.) The starting point of each successive sloping line segment is the ending point of the previous line.

Note: The area between the diagonal and the new (reordered) concentration curve is a specific measure of the concentration (i.e., the maximal concentration index value). This index is a numerical measure of the distributional concentration. (The greater the area, the greater the dissimilarity in the two distributions.)

The interpretation of the individual slopes in Figure 2.14 is the same as for Figure 2.13. However, it is worth commenting on the segment for the luxury model category.

The slope for the luxury models is 0.96 or just slightly smaller than 1. This makes the line almost parallel to the diagonal which has a slope of 1. Therefore the proportion of luxury models purchased before 1979 is approximately the same as the proportion purchased in 1979.

Note: In contrast to Figure 2.13, the luxury segment is no longer adjacent to the diagonal. However, the interpretation of the slope is the same.

For mid-size models, there was a greater proportion (slope larger than 1, i.e., 1.63) of mid-size purchased before 1979 than was purchased in 1979.

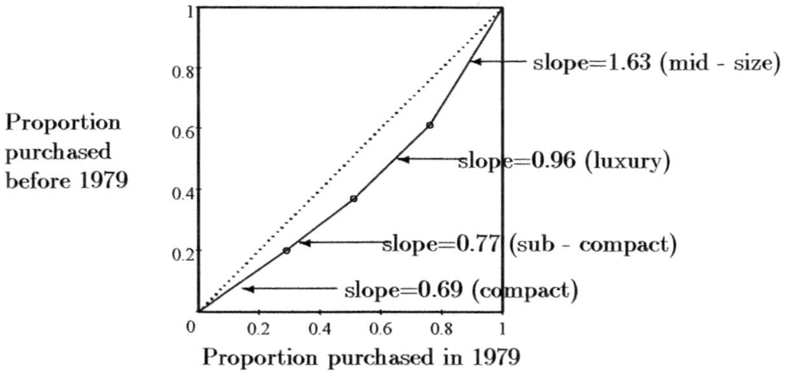

FIGURE 2.14. Concentration curve with car models arranged in order of increasing slopes.

Recall that the area between the diagonal line and the concentration curve in Figure 2.14 is a measure of concentration. More concerning the measurement of concentration will be found in the theory sections (Chapters 3 and 4).

2.6.12 Part 2 summary

The most important objectives of Part 2 are to show where the overrepresentation values come from, and demonstrate the effects of the reordering of their values on the shape of the concentration curve.

Through an understanding of the calculation of the overrepresentation values, and the "assembling" of concentration curves, one comes to realize that these values and curves are not abstract, but represent features of the actual data.

It will be shown in later chapters how areas bounded by these concentration curves will be used to quantify distributional differences. You will see how changes in the categories (i.e., ordering, re-sizing, omission, creation, etc.) affect the distributional differences, the research results, and finally the research conclusions.

2.7 Chapter Summary

Chapter 2 is written for the practitioner. It covers the essentials of the grade data analysis approach. This approach is explained in terms of univariate data to "set the stage" for the analysis of multivariate data.

From the practitioner's point of view, the main usefulness of grade methods is in the analysis of populations of multivariate data. However, to begin learning directly at the multivariate level is to risk misunderstanding a basic point. (The point being: that grade methods are not abstract.)

* * *

The grade methodology is based on the grade transform, which produces the grade distribution. This methodology preserves the features of the data needed for analysis, while providing a suitable normalization of the data. The density of the grade distribution is called overrepresentation. By plotting overrepresentation on the unit square we achieve a convenient graphical representation of the data. This graphical representation is called an overrepresentation map. In the following chapters, you will see how these maps reveal and clarify the link between continuous and discrete data.

So, of critical importance in grade methodology is the shift (via normalization and the grade transform) to the use of overrepresentation values in place of the original data values. These overrepresentation values incorporate: a standardization, a reference distribution, and a data continuity, all of which facilitate the investigation of structures in the data.

* * *

It is suggested that the practitioner reads Chapters 6, 7, 9, 10, 11, 12, 14 and 15. The material in Sections 8.1, 8.2, 8.6, 13.1, 13.2 may also be useful.

3

Univariate Lilliputian Model I

Teresa Kowalczyk, Wiesław Szczesny

3.1 Introduction

This chapter presents the ideas introduced in Chapter 2 in a more formal and systematic way. We show how two random variables X and Y can be compared and how they can be represented by one variable valued in the interval $[0, 1]$ and therefore called "Lilliputian". The set of all Lilliputian variables is called the *Univariate Lilliputian Model* and denoted $\mathbb{ULM}$.

The graph of the cumulative distribution function of $\xi \in \mathbb{ULM}$ coincides with the so called *concentration curve* $C(Y : X)$ which is the plot of the cdf's (cumulative distribution functions) of X and Y. The concentration curve measures how differentiated are X and Y. Pairs of univariate distributions with the same plot $C(Y : X)$ are treated as *equivalently differentiated.* Curve $C(Y : X)$ is the result of the so called *grade transformation* of Y on X applied to the distribution of Y; when this transformation is applied to the distribution of X, it produces the uniform distribution on the interval $[0, 1]$.

The main parameter of a Lilliputian variable ξ is its expectation. The expectation $E(\xi)$ gives rise to another Lilliputian variable $\xi^{(1)}$ associated with ξ, which has the cdf $E(\xi; \xi \leq u)/E(\xi)$ where $E(\xi; \xi \leq u)$ is the expectation of X *on* the interval $[0, u]$ (i.e. $E(\xi; \xi \leq 1) = E(\xi)$). Some essential parameters of ξ refer to $\xi^{(1)}$ and to the comparison of ξ and $\xi^{(1)}$. Specific parameters of Lilliputian variables correspond to space (i.e. *area* of regions) contained in the *unit square*, and are presented geometrically.

The parameters of ξ are then extended onto the set of all equivalent pairs of distributions of X and Y. They are called *grade parameters* of any such pair.

The so called *normal concentration pattern* is introduced as a reference model (which is also a grade parameter since, obviously, grade parameters can be numerical function - valued, etc). The normal concentration pattern enables easy visual comparison of the concentration curve for any pair $(X; Y)$ with its normal counterpart.

Data concerning hypothetical male and female earnings are introduced and examined in the second half of the chapter.

Remark 3.1 *We carefully distinguish the notation* $(X;Y)$ *from* (X,Y). *In the case of* (X,Y) *we deal with a random vector with two components, i.e., with a pair of random variables, each defined on the common probability space, and we deal with the joint distribution of such pair. This case will be considered in Chapters 7–13. The notation* $(X;Y)$ *refers to two random variables* X *and* Y, *which can have nothing to do one with another. Formally it means that each variable is defined on its own measurable space and these two spaces can be quite different.*

Effort has been taken to avoid - as far as possible - difficult probabilistic subtleties. Some references to the theoretical background are given in the Appendix (Sections 3.5.1 and 3.5.2).

The Univariate Lilliputian Model is also considered in Chapter 4, which deals with inequality measures for ratio variables (such as earnings: see explanation in the Appendix to Chapter 4, Section 4.5.1 concerning measurement scales). Moreover, Chapter 4 introduces order oriented concentration curves (with stress laid on maximal concentration curves), and dual concentration curves. Chapters 5 and 6 also refer to the $\mathbb{ULM}$.

Bibliographical remarks for Chapters 3 and 4 are given jointly at the end of Chapter 4.

3.2 Lilliputian variables and their basic parameters

3.2.1 The cdf of a Lilliputian variable

We say that a random variable is *Lilliputian* if it takes values from the interval [0,1] and is continuous or discrete, or discrete-continuous as explained later. The graph of its cdf (*cumulative distribution function*) denoted F_ξ lies on a non-decreasing curve in the unit square which joins points (0,0) and (1,1).

Definition 3.1 *(the Univariate Lilliputian Model). The set of all Lilliputian variables is called the Univariate Lilliputian Model and denoted* ($\mathbb{ULM}$).

Details: F_ξ has left and right limits $F_\xi(u+)$, $F_\xi(u-)$ for each point $u \in (0,1)$, and limits $F_\xi(0+)$, $F_\xi(1-)$. We use the convention everywhere in this book that the cdf of a Lilliputian random variable is right-continuous, i.e. $F_\xi(u) = F_\xi(u+)$. The same applies for cdf's of random variables which are not Lilliputian.

Definition 3.2 *(continuous Lilliputian variable). A Lilliputian variable will be called continuous if its cdf is continuous and its derivative*

$$f_\xi(u) = dF_\xi(u)/du,$$

called the density of ξ, exists almost everywhere relative to Lebesgue measure.

Illustration (three continuous Lilliputian variables):

In Fig. 3.1 we present graphs of cdf's F_1, F_2, F_3 of three continuous Lilliputian random variables ξ_1, ξ_2 and ξ_3, where

$$F_1(u) = u^4, \quad f_1(u) = 4u^3,$$

$$F_2(u) = \begin{cases} \frac{u}{3} & \text{if } 0 \le u < 0.3, \\ 3.5u - 0.95 & \text{if } 0.3 \le u < 0.5, \\ 0.4u + 0.6 & \text{if } 0.5 \le u \le 1, \end{cases} \tag{3.1}$$

$$f_2(u) = \begin{cases} \frac{1}{3} & \text{if } 0 < u < 0.3, \\ 3.5 & \text{for } 0.3 < u < 0.5, \\ 0.4 & \text{for } 0.5 < u < 1, \end{cases}$$

$$F_3(u) = \begin{cases} \frac{u}{4} & \text{if } 0 \le u < 0.4, \\ 0.1 & \text{if } 0.4 \le u < 0.8, \\ \frac{9u}{2} - \frac{7}{2} & \text{if } 0.8 \le u \le 1, \end{cases} \tag{3.2}$$

$$f_3(u) = \begin{cases} \frac{1}{4} & \text{if } 0 < u < 0.4, \\ 0.0 & \text{if } 0.4 < u < 0.8, \\ \frac{9}{2} & \text{if } 0.8 < u < 1. \end{cases}$$

Note that the graphs of F_1 and F_2 in Fig. 3.1 are continuous with no horizontal segments, and that the graph of F_3 contains a horizontal segment.

Definition 3.3 *(a discrete Lilliputian variable). A Lilliputian variable will be called discrete if there is a sequence of points $\{u_i\}$, $u_i \in [0,1]$ for $i = 1, 2, \dots$, and a sequence of probabilities $p_\xi(u_i)$ such that $\sum_i p_\xi(u_i) = 1$. The cdf F_ξ satisfies*

$$p_\xi(u_i) = dF_\xi(u_i) = F_\xi(u_i+) - F_\xi(u_i-) > 0. \tag{3.3}$$

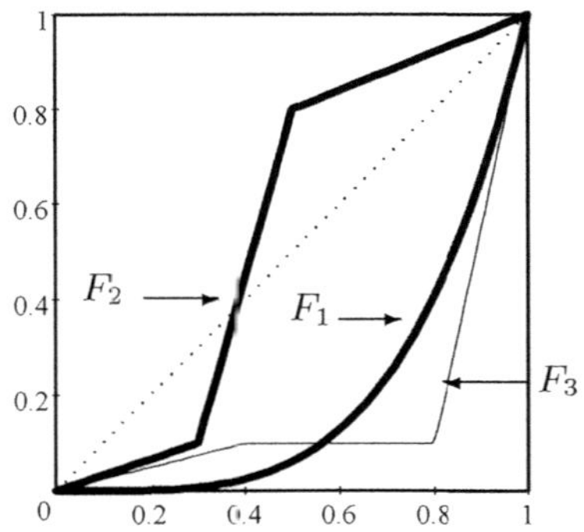

FIGURE 3.1. Examples of graphs of cdf's in the case of continuous Lilliputian variables ξ_1, ξ_2, and ξ_3.

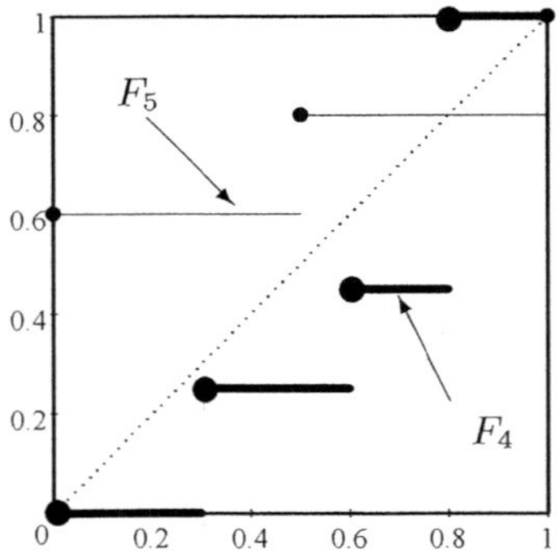

FIGURE 3.2. Examples of graphs of cdf's in the case of discrete Lilliputian variables ξ_4 and ξ_5

Remark 3.2 *$p_\xi(u_i)$ in (3.3) is the density relative to the count measure.*

Illustration (two discrete Lilliputian variables):

In Fig. 3.2 we present the graphs of two cdf's F_4 and F_5 of discrete random variables ξ_4 and ξ_5. In case of ξ_4,

$$F_4(u) = \begin{cases} 0 & \text{if } 0 \le u < 0.3, \\ 0.25 & \text{if } 0.3 \le u < 0.6, \\ 0.45 & \text{if } 0.6 \le u < 0.8, \\ 1 & \text{if } 0.8 \le u < 1, \end{cases} \tag{3.4}$$

so that

$$p_4(0.3) = 0.25,\ p_4(0.6) = 0.45 - 0.25 = 0.20,\ p_4(0.8) = 1 - 0.45 = 0.55,$$

$$F_4(0.3-) = 0,\ F_4(0.3) = 0.25 = F_4(0.6-) = 0.25,$$

$$F_4(0.6) = 0.45 = F_4(0.8-),\ F_4(0.8) = 1 = F_4(1).$$

In case of ξ_5,

$$F_5(u) = \begin{cases} 0.6 & \text{if } 0 \le u < 0.5, \\ 0.8 & \text{if } 0.5 \le u < 1, \\ 1 & \text{if } u = 1, \end{cases} \tag{3.5}$$

so that

$$p_5(0) = 0.6, \quad p_5(0.5) = 0.2, \quad p_5(1) = 0.2,$$

$$F_5(0) = 0.6 = F_5(0.5-), \quad F_5(0.5) = 0.8 = F_5(1-), \quad F_5(1) = 1.$$

Note that the graphs in Fig. 3.2 are formed by horizontal segments only.

An important role is played by Lilliputian variables which are *partly discrete and partly continuous* (shortly: discrete-continuous). The graphs of their cdf's must be supplemented with vertical segments to form a continuous curve joining (0,0) and (1,1) (i.e. the graph of a nondecreasing relation). A simple example is a Lilliputian variable which is continuous on interval (0,1] and has a jump in point 0. We will use the notation:

$$dF_\xi(u) = p_\xi(u_i) \ \text{ if } \ u = u_i, \quad dF_\xi(u) = f_\xi(u)du$$

where f_ξ is the derivative $dF_\xi(u)/du$ for almost all points $u \in (0,1)$ except the points in the sequence $\{u_i\}$. Hence the cdf of a discrete-continuous Lilliputian variable is given by

$$F_\xi(u) = \sum_{u_i \le u} p_\xi(u_i) + \int_0^u f_\xi(t)dt;$$

the sum

$$\alpha_\xi = \sum_i p_\xi(u_i), \tag{3.6}$$

is the sum of lengths of vertical segments of cdf F_ξ, denoted $p_\xi(u_i)$, $i = 1, 2, ...$ and

$$\int_0^1 f_\xi(u)du = 1 - \alpha_\xi.$$

Figures 3.1-3.2 illustrate that the graphs of Lilliputian variables lie on a continuous non-decreasing curve, which can contain vertical and horizontal segments. This curve will be called the *Lilliputian curve.* It has no vertical segments if $\alpha_\xi = 0$, while it contains solely horizontal and vertical segments if $\alpha_\xi = 1$.

The cdf of a Lilliputian variable is *uniquely determined* by the Lilliputian curve, and vice versa. Therefore, the $\mathbb{ULM}$ can be identified with a set of all possible continuous nondecreasing curves in the unit square which join points (0,0) and (1,1). The $\mathbb{ULM}$ can also be identified with the set of all cdf's of Lilliputian variables.

In regard to the last sentence in the above box, we will used the symbol $\mathbb{ULM}$ exchangeably in expressions and notation such as:

random variable ξ belongs to $\mathbb{ULM}$ (symbolically $\xi \in \mathbb{ULM}$),

curve C belongs to $\mathbb{ULM}$ (symbolically $C \in \mathbb{ULM}$),

cdf F_ξ belongs to $\mathbb{ULM}$ (symbolically $F_\xi \in \mathbb{ULM}$).

The Lilliputian curve of ξ_1 in Fig. 3.1 is convex. Generally, a Lilliputian curve is convex if and only if the density f_ξ exists almost everywhere and is nondecreasing (see Fig. 3.1). Lilliputian variables with convex Lilliputian curves play an important role.

Definition 3.4 *(Convex Univariate Lilliputian Model). The set of Lilliputian variables with its cdf lying on a convex Lilliputian curve will be called a Convex Univariate Lilliputian Model and denoted* $\mathbb{ULM}_{CON}$.

3.2.2 The expectation of a Lilliputian variable and the index ar

The expectations (mean values) of Lilliputian variables can be easily presented graphically as certain areas contained in the unit square. This is illustrated in Fig. 3.3 according to which the expectation $E(\xi)$ is equal to the area above the Lilliputian curve attached to ξ, or the area of the unit square minus the area of the region below the Lilliputian curve.

Formally,

$$E(\xi) = \int_0^1 u dF_\xi(u) = 1 - \int_0^1 F_\xi(u) du = \int_0^1 (1 - F_\xi(1-u)) du.$$

In Fig. 3.3, $E(\xi) = A + B + C = \int_0^1 u dF_\xi(u) = 1 - D$, $D = \int_0^1 F_\xi(u) du$.

Another formula for $E(\xi)$ refers to the function *inverse* to F_ξ, denoted F_ξ^{-1} and called *the quantile function* of ξ. This function is obtained immediately from the *inverse Lilliputian curve*, presented in Figs. 3.3 and 3.4 (left).

Details: Let $F_\xi : [0,1] \to [0,1]$ be a nondecreasing function with $F_\xi(0) \geq 0$ and $F_\xi(1) = 1$. Then we put: for $u \in (0,1)$ $F_\xi^{-1}(u) = \inf\{t : F_\xi(t) \geq u\}$, $F_\xi^{-1}(0) = \sup\{t : F_\xi(t) = 0\}$ if $F_\xi(0) = 0$, and $F_\xi^{-1}(0) = 0$ if $F_\xi(0) > 0$, $F_\xi^{-1}(1) = \inf\{t : F_\xi(t) = 1\}$.

> The inverse Lilliputian curve is a curve *symmetric* to the considered Lilliputian curve w.r.t. *Diag*, where *Diag* is the *main diagonal of the unit square* joining (0,0) and (1,1). Clearly, horizontal (respectively, vertical) segments of the Lilliputian curve become vertical (respectively, horizontal) segments of the inverse curve.

In Fig. 3.3 we have: $B + C + D = \int_0^1 F_\xi^{-1}(v)dv$, $A = \int_0^1 v dF_\xi^{-1}(v) = D$, $B = C$.

Generally, we have:

$$E(\xi) = \int_0^1 F_\xi^{-1}(v)dv = 1 - \int_0^1 v dF_\xi^{-1}(v).$$

For continuous ξ,

$$E(\xi) = 1 - \int_0^1 \frac{v}{f_\xi(F_\xi^{-1}(v))} dv.$$

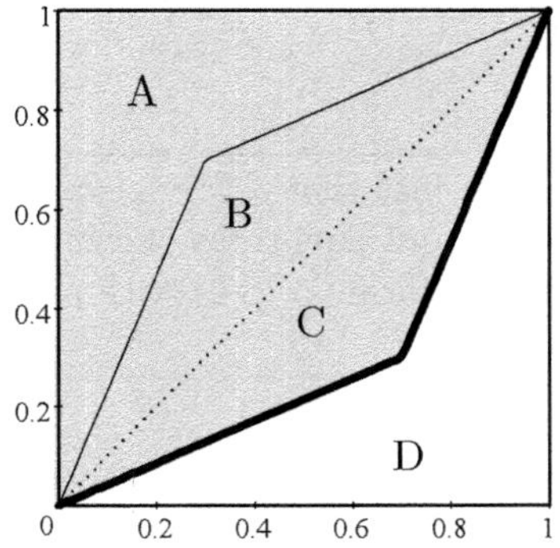

FIGURE 3.3. $E(\xi)$ as area: $A + B + C = B + C + D$

In the case of discrete ξ with points $\{u_i\}$, $i = 1, \ldots, k$, $0 = u_0 \le u_1 < \cdots < u_k \le u_{k+1} = 1$, and with $F_\xi(u_{i-1}) = v_i$ for $i = 1, \ldots, k+1$, we have:

$$dF_\xi(u_i) = F_\xi(u_i) - F_\xi(u_{i-1}) = v_{i+1} - v_i = p_\xi(u_i)$$
$$dF_\xi^{-1}(v_i) = F_\xi^{-1}(v_{i+1}) - F_\xi^{-1}(v_i) = u_i - u_{i-1},$$

and hence

$$E(\xi) = \sum_{i=1}^{k} u_i(F_\xi(u_i) - F_\xi(u_{i-1})) = \sum_{i=1}^{k} u_i(v_{i+1} - v_i)$$
$$= 1 - \sum_{i=1}^{k} F_\xi(u_i)(u_{i+1} - u_i) = \sum_{i=1}^{k}(u_{i+1} - u_i)(1 - F_\xi(u_i))$$
$$= 1 - \sum_{i=1}^{k} v_{i+1}(F_\xi^{-1}(v_{i+1}) - F_\xi^{-1}(v_i)) = 1 - \sum_{i=1}^{k} v_{i+1}(u_i - u_{i-1}).$$

The expectation $E(\xi)$ is the basic parameter of ξ. The parameter varies from 0 (when ξ is concentrated at point $u = 0$) to 1 (when ξ is concentrated at point $u = 1$). Further, it is smaller than 1/2 if (*but not only if*) the Lilliputian curve lies above the *Diag*, greater than 1/2 if (*but not only if*) this curve lies below the *Diag*, and equal to 1/2 if (*but not only if*) the Lilliputian curve lies on the *Diag* (i.e. if we deal with the Lilliputian variable U *uniformly* distributed on [0,1]).

It seems convenient to consider a linear transform of $E(\xi)$ such that the transformed $E(\xi)$ will take values from [-1,1]. This leads to a parameter denoted *ar*, which has a clear interpretation as a departure of ξ from U :

$$ar(\xi) \stackrel{df}{=} \frac{E(\xi) - E(U)}{1 - E(U)} = 2E(\xi) - 1 = 2\int_0^1 (u - F_\xi(u))du. \tag{3.7}$$

> The name "*ar*" refers to the fact that $ar(\xi)$ has a representation as an *area* contained in the unit square as shown in Fig. 3.4 for two cdfs: F_6 defined below by formula (3.9) with $p = 0.7$, $q = 0.3$, and F_2 given by formula (3.1).

The right-hand side of (3.7) can be expressed as the double difference of two areas: that between the *Diag* and the Lilliputian curve situated *below* the *Diag*, and that area between the *Diag* and the Lilliputian curve situated *above* the *Diag* (see Fig. 3.4 (right)):

$$ar(\xi) = 2\int_{u > F_\xi(u)} (u - F_\xi(u))du - 2\int_{u < F_\xi(u)} (F_\xi(u) - u)du.$$

We also have:

$$ar(\xi) = \int_0^1 (1 - F_\xi(1-u) - F_\xi(u))du. \tag{3.8}$$

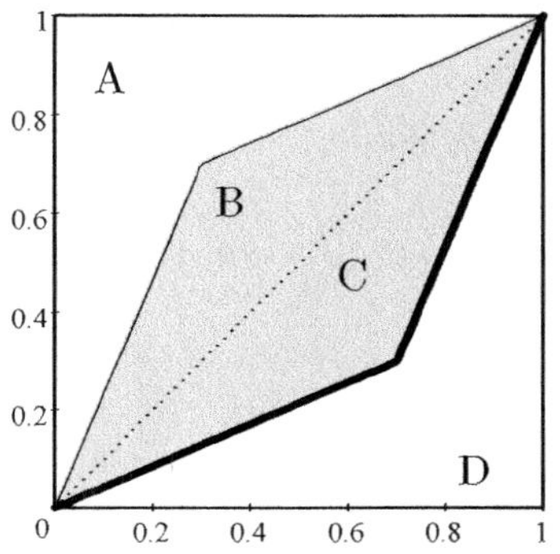

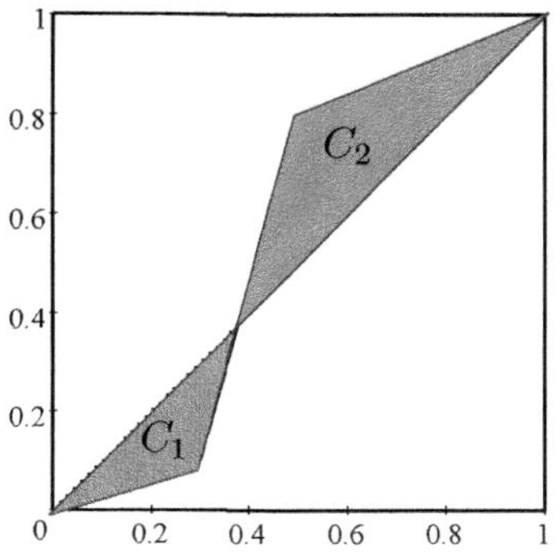

FIGURE 3.4. $ar(\xi)$ as area: $ar(\xi_6) = B + C = 2B = 2C$ (left figure); $ar(\xi_2) = 2C_1 - 2C_2$ (right figure).

To find another important formula for $ar(\xi)$, let us consider first a Lilliputian variable ξ with the cdf given by

$$F_\xi(u) = \begin{cases} \frac{q}{p}u & \text{if } 0 \le u < p, \\ \\ q + \frac{1-q}{1-p}(u-p) & \text{if } p \le u \le 1, \end{cases} \tag{3.9}$$

where $p > q$. This cdf is presented in Fig. 3.4 (left) for $p = 0.7$ and $q = 0.3$. The area of the region below F_ξ is immediately calculated as the sum of the area of a *triangle* equal to $(\frac{1}{2}pq)$ and the area of a *trapezoid* equal to $(\frac{1}{2}(1+q)(1-p))$. Their sum is $\frac{1}{2}\left(1 - (p(1-q) - q(1-p))\right)$. Hence

$$ar(\xi) = p(1-q) - q(1-p) = \begin{vmatrix} p & 1-p \\ q & 1-q \end{vmatrix} = \begin{vmatrix} 1-q & q \\ 1-p & p \end{vmatrix} = p - q. \tag{3.10}$$

Formula (3.10) holds also for $p < q$ which can be deduced from the mirror reflection of the graph of F_ξ defined by (3.9) and shown in Fig. 3.4 (left).

Remark 3.3 *Formula (3.10) is an important special case of the general property: the determinant of a matrix is the (oriented) volume of a parallelogram with sides given by the columns of the matrix.*

An extension of (3.10) for any continuous Lilliputian variable ξ is

$$ar(\xi) = \int_0^1 \int_0^v (f_\xi(v) - f_\xi(u))dudv \tag{3.11}$$

which may be rewritten as

$$\int_0^1 \int_0^v \left[f_\xi(v) f_U(u) - f_\xi(u) f_U(v)\right] dudv, \tag{3.12}$$

showing how diversified the density f_ξ is with respect to a constant density $f_U \equiv 1$. We also have

$$ar(\xi) = \frac{1}{2} E\left[sgn(V-U)(f_\xi(V) f_U(U) - f_\xi(U) f_U(V))\right] \tag{3.13}$$

where U and V are independent uniformly distributed Lilliputian variables, and

$$sgn(v-u) = \begin{cases} -1 \text{ if } v < u, \\ \ \ 0 \text{ if } v = u, \\ \ \ 1 \text{ if } v > u. \end{cases}$$

For any discrete ξ with $F_\xi(u_i) - F_\xi(u_{i-1}) = p_\xi(u_i)$, $0 = u_0 \leq u_1 < \cdots < u_k \leq u_{k+1} = 1$ we can write a counterpart of formula (3.12) using the following notation: for $i = 1, \ldots, k+1$,

$$\begin{array}{lll} p_{2i-1} = u_i - u_{i-1}, \; p_{2i} = 0, & & p_{2k+1} = 1 - u_k, \\ q_{2i-1} = 0, & q_{2i} = p_\xi(u_i), \; q_{2k+1} = 0. & \end{array}$$

The counterpart is given by

$$\begin{aligned} ar(\xi) &= \sum_{j=2}^{2k+1} \sum_{s=1}^{j-1} (p_s q_j - p_j q_s) = \sum_{j=2}^{2k+1} \sum_{s=1}^{j-1} \begin{vmatrix} p_s & p_j \\ q_s & q_j \end{vmatrix} \\ &= \sum_{j=2}^{2k+1} \sum_{s=1}^{j-1} \begin{vmatrix} q_j & q_s \\ p_j & p_s \end{vmatrix}. \end{aligned} \tag{3.14}$$

Remark 3.4 *The increments of u_i's and of probabilities $p_\xi(u_i) = F_\xi(u_i) - F_\xi(u_{i-1})$ are "exchangeable" in the following sense:*

$$ar(\xi) = 2 \sum_{i=1}^{k} u_i (F_\xi(u_i) - F_\xi(u_{i-1})) - 1 = 1 - 2 \sum_{i=1}^{k+1} (u_i - u_{i-1}) F_\xi(u_{i-1}). \tag{3.15}$$

This exchangeability of the Lilliputian variables with their cdfs is the first signal of duality which will be traced till the last chapters of this book (in particular, Chapter 12).

For any discrete-continuous Lilliputian variable ξ,

$$ar(\xi) = \frac{1}{2}\int_0^1 \int_0^1 sgn(v-u)\left[dF_U(u)dF_\xi(v) - dF_U(v)dF_\xi(u)\right]. \qquad (3.16)$$

We also have

$$\begin{aligned} ar(\xi) &= \int_0^1 \int_0^v \left[dF_U(u)dF_\xi(v) - dF_U(v)dF_\xi(u)\right] \\ &= \int_0^1 vdF_\xi(v) - \int_0^1 F_\xi(v)dv \\ &= \int_0^1 F_\xi^{-1}(v)dv - \int_0^1 vdF_\xi^{-1}(v). \end{aligned}$$

Let us stress that the parameter ar serves as a numerical measure of *ordered differentiation of* ξ *w.r.t.* U. The respective order we have in mind is the stochastic order $\preceq_{st}$ restricted to variables from $\mathbb{ULM}$, namely:

Definition 3.5 *(stochastic order). For any two Lilliputian variables ξ, ξ' we say that ξ is stochastically smaller than ξ' ($\xi \preceq_{st} \xi'$) if*

$$F_\xi(u) \geq F_{\xi'}(u) \;\; \text{if} \;\; 0 \leq u \leq 1; \qquad (3.17)$$

and we say that ξ is strictly stochastically smaller than ξ' ($\xi \prec_{st} \xi'$) if

$$F_\xi(u) \geq F_{\xi'}(u) \text{ and there exists } u \text{ such that } F_\xi(u) > F_{\xi'}(u). \qquad (3.18)$$

For instance, in Figure 3.1, $\xi_2 \preceq_{st} \xi_1$ and $\xi_2 \preceq_{st} \xi_3$ but ξ_1 and ξ_3 are not stochastically ordered.

Proposition 3.1 *(i) $\xi \preceq_{st} \xi' \Rightarrow ar(\xi) \leq ar(\xi')$.*
(ii) $U \prec_{st} \xi \Rightarrow ar(\xi) > 0$, $\xi \prec_{st} U \Rightarrow ar(\xi) < 0$.

3.2.3 The first moment Lilliputian variable, its variance, and the Gini Index

Recall that the expectation of a Lilliputian variable is positive if and only if ξ is not identically equal to zero. In this section we exclude such variable from consideration. Then, we may attach to any $\xi \in \mathbb{ULM}$ its "first moment variable" $\xi^{(1)}$, which has the cdf given by

$$F_{\xi^{(1)}}(u) \stackrel{df}{=} \frac{E(\xi;\xi \leq u)}{E(\xi)}, \qquad (3.19)$$

where $E(\xi;\xi \leq u) = \int_0^u tdF_\xi(t)$.

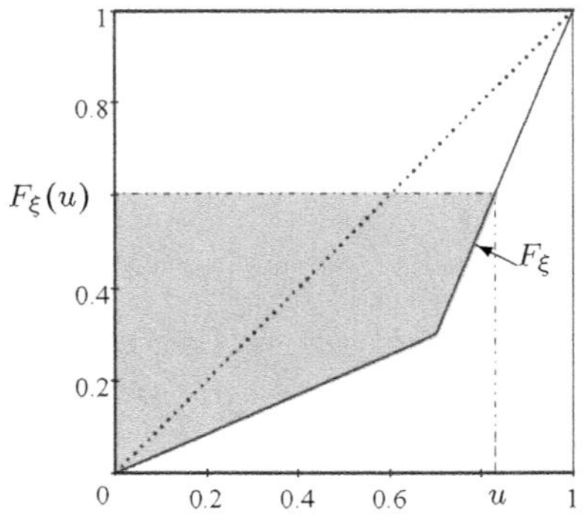

FIGURE 3.5. Graphical presentation of $E(\xi; \xi \leq u)$.

According to common notation, we introduce the symbol $F_\xi^{(1)}$:

$$F_\xi^{(1)}(u) \stackrel{df}{=} F_{\xi^{(1)}}(u). \tag{3.20}$$

Therefore any $\xi^{(1)}$ is a Lilliputian variable.

The values of the numerator of $F_\xi^{(1)}$ in formula (3.19) are represented by the respective areas contained in the unit square. For instance, in Fig. 3.5 the area of the region above the Lilliputian curve but below the horizontal line going through the point $F_\xi(u)$ is equal to

$$E(\xi; \xi \leq u) = \int_0^u t dF_\xi(t) = uF_\xi(u) - \int_0^u F_\xi(t)dt = E(\xi)F_\xi^{(1)}(u).$$

Proposition 3.2 *Let $\xi \in \mathbb{ULM}$. Then*

(i) if $\xi > 0$ then the supports of ξ and $\xi^{(1)}$ are the same,

(ii) $\xi \preceq_{st} \xi^{(1)}$, i.e. $F_\xi^{(1)}(u) \leq F_\xi(u)$, for any $u \in [0, 1]$,

(iii) $E(\xi; \xi \leq u) = E(F_\xi^{-1}(U); U \leq F_\xi(u)) = E[((F_\xi(u) - F_\xi(U)); U \leq u]$,
$E(\xi; \xi < u) = E(F_\xi^{-1}(U); U \leq F_\xi(u-)) = E[((F_\xi(u-) - F_\xi(U)); U \leq u]$,

(iv) Lilliputian curve $C(v) = \int_0^v F_\xi^{-1}(t)dt$ is a continuous convex curve on $[0, 1)$, equal to

$$C(v) = E(F_\xi^{-1}(U); U \leq v) =$$
$$= E(\xi; \xi < F_\xi^{-1}(v)) + F_\xi^{-1}(v)(v - F_\xi(F_\xi^{-1}(v)-));$$

if $v \in (0, 1)$ is such that $v = F_\xi(u)$, then

$$C(v) = E(\xi; \xi \leq u) = E(\xi)F_\xi^{(1)}(u) = E(\xi)F_\xi^{(1)}(F_\xi^{-1}(v)).$$

(v) $E(\xi^{(1)}) = \int_0^1 u^2 dF_\xi(u)/E(\xi) = E(\xi^2)/E(\xi)$.

(vi)

$$ar(\xi^{(1)}) = \frac{1}{2}\int_0^1\int_0^1 sgn(v-u)(dF_\xi^{(1)}(v) - dF_\xi^{(1)}(u)) \qquad (3.21)$$

(vii) the right hand side of (3.21) can be rewritten as

$$\frac{1}{2E(\xi)}\int_0^1\int_0^1 sgn(v-u)(vdF_\xi(v) - udF_\xi(u)).$$

Other formulas concerning $\xi^{(1)}$:

$E(F_\xi^{(1)}(\xi)) = E(F_\xi(\xi^{(1)})) - \frac{1}{E(\xi)}\int_0^1 (1-F_\xi(u))F_\xi(u)du$,

$E(F_\xi^{(1)}(\xi)) = 1 - E(F_\xi(\xi^{(1)}-)) = 1 - \frac{1}{E(\xi)}\int_0^1 uF_\xi(u-)dF_\xi(u)$,

$ar(\xi^{(1)} : \xi) = \frac{1}{E(\xi)}\int_0^1 (1-F_\xi(u))F_\xi(u)du$.

The expectation $E(\xi^{(1)})$ is closely related to the most popular measure of variability of values of ξ around $E(\xi)$, called *variance* and denoted $var(\xi)$:

$$var(\xi) = E(\xi - E(\xi))^2 = E(\xi^2) - (E(\xi))^2. \qquad (3.22)$$

Thus,

$$var(\xi)/E(\xi) = E(\xi^{(1)}) - E(\xi) = \frac{ar(\xi^{(1)}) - ar(\xi)}{2}. \qquad (3.23)$$

Let ξ and ξ' be two independent variables with the same cdf F_ξ. The formula for $var(\xi)$ which is congruent with formula (3.11) for $ar(\xi)$ is given by:

$$var(\xi) = \frac{E(\xi-\xi')^2}{2} = \frac{1}{2}\int_0^1\int_0^1 (u-v)^2 dF_\xi(u)dF_\xi(v). \qquad (3.24)$$

It follows that

$$var(\xi) = \int_0^1\int_0^v (v-u)^2 dF_\xi(u)dF_\xi(v) = \int_0^1 (u - E(\xi))^2 dF_\xi(u).$$

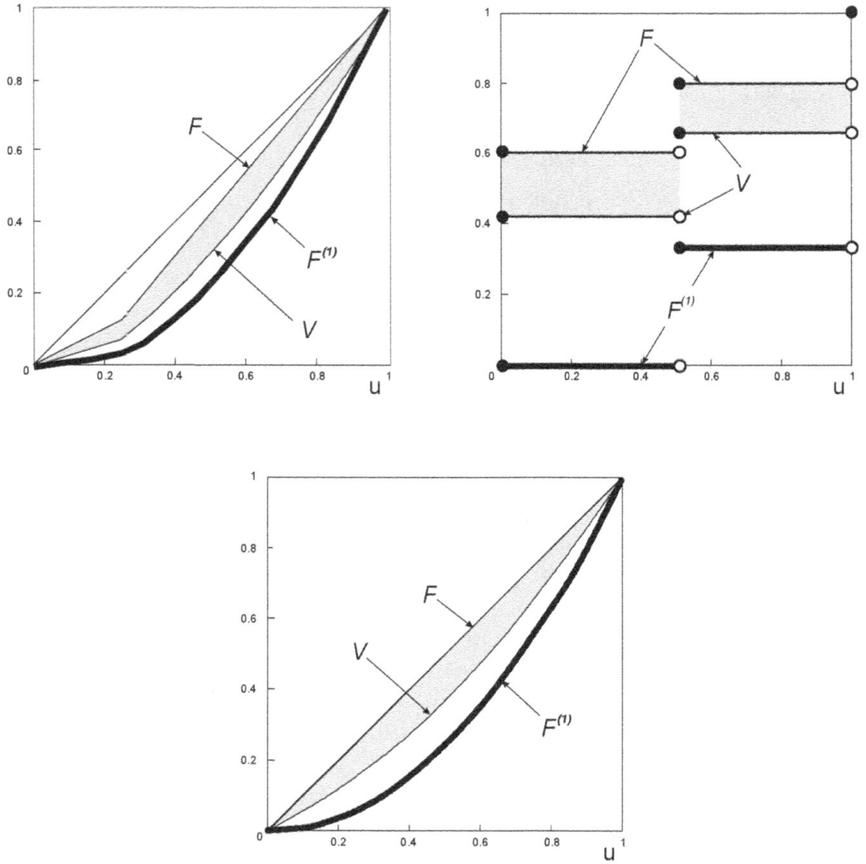

FIGURE 3.6. Visualisation of $var(\xi)$ as a shaded area for ξ continuous (upper left), for ξ discrete (upper right), and for the uniformly distributed Lilliputian variable U (bottom), for which $var(U) = 1/12$.

The last expression presents $var(\xi)$ as mean squared distance from $E(\xi)$.

The graphical interpretation of the $var(\xi)$ as an area of the region contained in the unit square is shown in Fig. 3.6. The area taken into account lies between cdf's of ξ and cdf denoted $V_\xi(u)$, defined as $V_\xi(u) = (1 - E(\xi))F_\xi(u) + E(\xi)F_\xi^{(1)}(u)$. Then

$$
\begin{aligned}
Var(\xi) &= \int_0^1 (F_\xi(u) - V_\xi(u))du = E(\xi) \int_0^1 (F_\xi(u) - F_\xi^{(1)}(u))du \\
&= \int_0^1 E(E(\xi) - \xi; \xi \leq u)du = \int_0^1 \int_0^u (E(\xi) - t)dF_\xi(t)du.
\end{aligned}
$$

The square root of variance is called the *standard deviation* and the ratio of the standard deviation and the expectation $E(\xi)$ is called the *coefficient of variation*.

Another popular measure of the variability of ξ is called *the Gini Index*, and is denoted $Gini(\xi)$. The Gini Index averages the absolute differences for all pairs of values (while $var(\xi)$ averages squared differences and also measures variability around $E(\xi)$) (compare (3.24)). Indeed, let ξ and ξ' be two independent variables with the same cdf F_ξ, and let

$$\delta(u,t) = |u-t| \tag{3.25}$$

be a "distance" assigned to each pair (u,t). Then

$$\begin{aligned}
E_{\xi,\xi'}\delta(\xi,\xi') &= \frac{1}{2}E|\xi-\xi'| = \frac{1}{2}\int_0^1\int_0^1 |u-t|dF_\xi(t)dF_\xi(u) \\
&= \int_0^1\int_0^u (u-t)dF_\xi(t)dF_\xi(u) \\
&= E(\xi)(E(F_\xi(\xi^{(1)})) - E(F_\xi^{(1)}(\xi))) \\
&= \int_0^1 F_\xi(u)(1-F_\xi(u))du = E(F_\xi(U)) - E(F_\xi^2(U)) \le E(\xi).
\end{aligned}$$

This expression is analogous to that for $Var(\xi)$ (formula (3.24)) when $(u-v)^2$ is replaced by $|u-v|$. The Gini Index $Gini(\xi)$ normalizes this expression to values from [0,1], i.e. divides the expression by $E(\xi)$. So we have

$$Gini(\xi) = \frac{E|\xi-\xi'|}{2E(\xi)}, \tag{3.26}$$

which is analogous to the coefficient of variation.

Gini Index is zero when $\xi \equiv E(\xi) > 0$ with probability one, i.e., when variable ξ is concentrated at a positive point. If $\xi = U$ then

$$E_{U,U'}\delta(U,U') = \frac{1}{2}\int_0^1\int_0^1 |u-t|dudt = \frac{1}{6}, \quad Gini(U) = 1/3 = 4var(U).$$

For ξ continuous, $Gini(\xi)$ can be presented in a way *congruent with formulas* (3.13) and (3.21) for $ar(\xi)$ and $ar(\xi^{(1)})$:

$$\begin{aligned}
Gini(\xi) &= \frac{1}{2}E\left(sgn(V-U)(f_\xi^{(1)}(V)f_\xi(U) - f_\xi^{(1)}(U)f_\xi(V))\right) \\
&= \frac{1}{2E(\xi)}E\left(sgn(V-U)(V-U)f_\xi(U)f_\xi(V)\right) \qquad (3.27)\\
&= \frac{1}{2E(\xi)}E\left(|V-U|f_\xi(U)f_\xi(V)\right).
\end{aligned}$$

The geometrical interpretation of $Gini(\xi)$ is shown in Fig. 3.7 as the area between cdf F_ξ and F_ξ^2 divided by the area above cdf F_ξ.

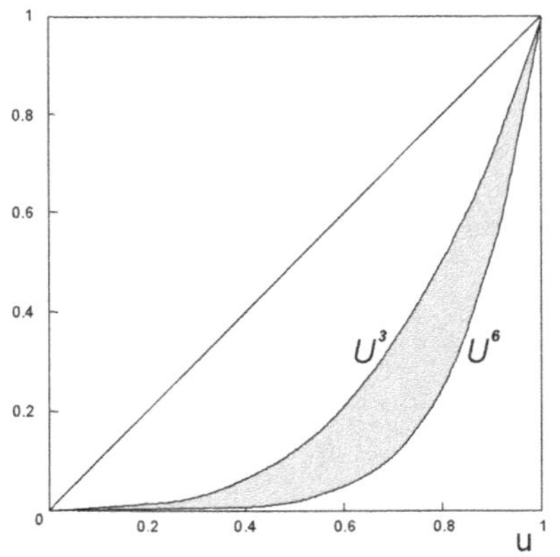

FIGURE 3.7. Graphical presentation of $Gini(\xi)$ for $\xi = U^{1/3}$ as the area between curves $F_\xi(u) = u^3$ and $F_\xi^2(u) = u^6$, divided by the area above $F_\xi(u)$

We introduce now the *smoothed function of* ξ, defined as

$$\tilde{F}_\xi(\xi) = \frac{F_\xi(\xi+) + F_\xi(\xi-)}{2} \tag{3.28}$$

where $F_\xi(\xi+)$ is a random variable which takes on the values of the right limit of $F_\xi(u)$ at point $\xi = u$, while $F_\xi(\xi-)$ is a random variable which takes on the values of the left limit of $F_\xi(u)$ at point $\xi = u$. Then we have:

$$\begin{aligned} Gini(\xi) &= E(F_\xi(\xi^{(1)})) - E(F_\xi^{(1)}(\xi)) \qquad (3.29)\\ &= \frac{1}{E(\xi)}\int_0^1 tF_\xi(t)dF_\xi(t) - \frac{1}{E(\xi)}\int\int_{t\le u} tdF_\xi(t)dF_\xi(u)\\ &= \frac{1}{E(\xi)}\int_0^1 tF_\xi(t)dF_\xi(t) - \frac{1}{E(\xi)}\int_0^1 t(1-F_\xi(t-))dF_\xi(t)\\ &= \frac{1}{E(\xi)}\int_0^1 t(F_\xi(t)+F_\xi(t-))dF_\xi(t) - 1\\ &= \frac{2cov(\xi,\tilde{F}_\xi(\xi))}{E(\xi)} = \frac{cov(\xi,F_\xi(\xi)) + cov(\xi,F_\xi(\xi-))}{E(\xi)}\\ &= 2(E(\tilde{F}_\xi(\xi^{(1)})) - E(\tilde{F}_\xi^{(1)}(\xi^{(1)})) = ar(\tilde{F}_\xi(\xi^{(1)}))\\ &= 2(E(\tilde{F}_\xi(\xi)) - E(\tilde{F}_\xi^{(1)}(\xi))) = -ar(\tilde{F}_\xi^{(1)}(\xi)). \end{aligned}$$

Other interpretations of the Gini Index associated with measuring discontinuity and inequality will be given in Sec. 3.2.4 and Sec. 4.2.2.

3.2.4 Discontinuity measures

Parameter α_ξ (formula (3.6)) is an important measure of the discontinuity of any discrete-continuous Lilliputian variable ξ : it evaluates the rate of the discrete part. But this is a measure unsuitable to compare discrete variables (all of them have $\alpha_\xi = 1$). However there are other measures of discontinuity, obtained from various transformations of ξ which map ξ into another element of $\mathbb{ULM}$.

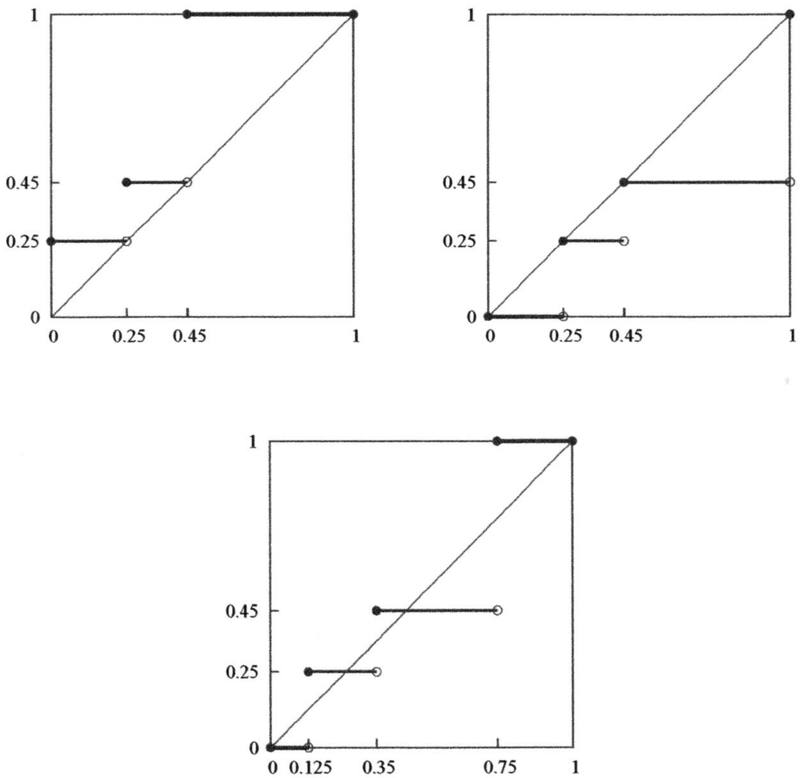

FIGURE 3.8. Cdf's of variables: $F_\xi(\xi-)$ (upper left), $F_\xi(\xi)$ (upper right) and $\tilde{F}_\xi(\xi)$ (down)

The first measure of discontinuity is obtained from transformations of ξ into $F_\xi(\xi)$ and into $F_\xi(\xi-)$. The Lilliputian variables $F_\xi(\xi)$ and $F_\xi(\xi-)$ belong to $\mathbb{ULM}$ since they are valued in [0,1]. The cdfs of $F_\xi(\xi-)$ and $F_\xi(\xi)$ for discrete Liliputian variable ξ_4 given by formula (3.4)) are presented in Fig. 3.8 - upper left and upper right.

We have:

$$F_{F_\xi(\xi)}(u) = \begin{cases} 0 & \text{if } 0 \leq u < 0.25, \\ 0.25 & \text{if } 0.25 \leq u < 0.45, \\ 0.45 & \text{if } 0.45 \leq u < 1, \\ 1 & \text{if } u = 1. \end{cases}$$

$$F_{F_\xi(\xi-)}(u) = \begin{cases} 0.25 & \text{if } 0 \leq u < 0.25, \\ 0.45 & \text{if } 0.25 \leq u < 0.45, \\ 1 & \text{if } 0.45 \leq u \leq 1. \end{cases}$$

Hence

$$\begin{aligned} ar(F_\xi(\xi)) &= 2\int_0^1 (u - F_{F_\xi(\xi)}(u))du \\ &= \int_0^1 (F_\xi(u) - F_\xi(u-))dF_\xi(u) = \sum_i p_i^2, \end{aligned} \tag{3.30}$$

$$\begin{aligned} ar(F_\xi(\xi-)) &= 2\int_0^1 (u - F_{F_\xi(\xi-)}(u))du \\ &= -\int_0^1 (F_\xi(u) - F_\xi(u-))dF_\xi(u) = -\sum_i p_i^2; \end{aligned} \tag{3.31}$$

and the differences under both integrals are zero when ξ is continuous.

Remark 3.5 *The cdf's of $F_\xi(\xi)$ and of $F_\xi(\xi-)$ lie on the Diag when F_ξ is a continuous distribution.*

> The index $ar(F_\xi(\xi))$ is a *numerical measure of the discontinuity of* ξ; its supplement to one (which measures *closeness to continuity*) is known in the statistical literature as the *Gini-Simpson Index,* denoted here by $GS(\xi)$.

Now, instead of F_ξ we will make use of the smoothed function $\tilde{F}_\xi$ (see formula (3.28)).

Illustration (two discrete Lilliputian variables):
The cdf of $\tilde{F}_\xi(\xi)$ for ξ given by formula (3.4) is equal to

$$F_{\tilde{F}_\xi(\xi)}(u) = \begin{cases} 0 & \text{if } 0 \le u < 0.125, \\ 0.25 & \text{if } 0.125 \le u < 0.35, \\ 0.45 & \text{if } 0.35 \le u < 0.75, \\ 1 & \text{if } 0.75 \le u \le 1. \end{cases} \tag{3.32}$$

The graph of this cdf is presented in Fig. 3.8 (bottom).

For the smoothed function used as a transformation of ξ, the index ar is not applicable as a measure of discontinuity since it is identitically equal to zero:

$$\begin{aligned} ar(\tilde{F}_\xi(\xi)) &= 2E(\tilde{F}_\xi(\xi)) - 1 \\ &= \frac{1}{2}\left[2E(F_\xi(\xi)) - 1 + 2E(F_\xi(\xi-)) - 1\right] \\ &= \frac{1}{2}ar(F_\xi(\xi)) + \frac{1}{2}ar(F_\xi(\xi-)) = 0. \end{aligned}$$

Therefore, a counterpart of the Gini-Simpson measure is obtained when ar is replaced by the Gini Index. The index $Gini(\xi)$ is compared with $Gini(U)$ as follows. Due to (3.29),

$$Gini(\xi) = \frac{2cov(\xi, \tilde{F}_\xi(\xi))}{E(\xi)}.$$

Since $\tilde{F}_U(U) = U$, $Gini(U) = 1/3$, and $\eta \stackrel{df}{=} \tilde{F}_\xi(\xi)$ is equal to $\tilde{F}_\eta(\eta)$ (i.e. variables η and $\tilde{F}_\eta(\eta)$ have the same distribution), then

$$Gini(\eta) = 4var(\tilde{F}_\xi(\xi)) = 1/3 - 1/3\sum_i p_i^3.$$

It follows that the second discontinuity measure sought is equal to:

$$Gini(\tilde{F}_\xi(\xi))/Gini(U) = 3Gini(\tilde{F}_\xi(\xi)) = 1 - \sum_i p_i^3. \tag{3.33}$$

Both discontinuity measures will be used in this book to construct upper bounds for some measures of dependence (see Chapter 8, formulas (8.16), (8.17), (8.27), (8.28) so that they can serve to normalize these measures.

3.3 The main equivalence relation which creates the Univariate Lilliputian Model

3.3.1 Preliminary definitions and examples

Let X and Y be random variables valued in $\mathbb{R}$ with cumulative distribution functions F_X and F_Y. We assume that both variables are continuous, discrete, or discrete-continuous as defined in Sec. 3.2.1 for Lilliputian variables, the only difference now being the interval [0,1] is replaced by the real line.

Definition 3.6 *(concentration curve). The curve*

$$C = \{(F_X(x), F_Y(x)); x \in \mathcal{R}\} \tag{3.34}$$

completed if necessary by points $(0,0)$, $(1,1)$ *and by linear interpolation, is called the concentration curve of* F_Y *to* F_X *(or of* Y *to* X*) and denoted by*

$$C(F_Y : F_X) \quad or \quad C(Y : X).$$

Curve C defines a cdf $C(u)$, $u \in [0,1]$ such that the points $(u, C(u+))$ lie on curve C. The random variable with this cdf, denoted ξ, belongs to $\mathbb{ULM}$. So the concentration curve $C(Y : X)$ is equal to the Lilliputian curve of ξ.

On the other hand, for any Lilliputian variable ξ with cdf F_ξ there exist sets of pairs $(X; Y)$ with concentration curves which are all equal to the graph of F_ξ or become equal to it when the vertical segments are removed.

Remark 3.6 *The notation* $(X; Y)$ *refers to two separate random variables* X *and* Y *with cdf's* F_X *and* F_Y, *and has nothing to do with a vector* (X, Y) *defined by its joint cumulative distribution function* $F_{X,Y}$.

Clearly, the graph of F_ξ lies on the concentration curve $C(\xi : U)$ where U is uniform on [0,1]. Variable ξ is discrete-continuous; in the notation introduced in Section 3.2 the derivative of F_ξ in point u is $f_\xi(u)$, while $dF_\xi(u) = p_\xi(u)$ for $u \in \{u_i\}$.

Remark 3.7 *The curve defined by formula (3.34) does not depend on the choice of left or right continuity of the two cdf's.*

It is important to specify the properties of X and Y : certain properties will create a Lilliputian variable which is continuous while other properties will create a Lilliputian variable which is discrete.

Details: In the definition of curve C it sufficies to take into account only those points x which belong to the sum of *supports* of X and Y; then, the transformed points $(F_X(x), F_Y(x))$ are linearly connected. Recall that the *support* of a real-valued random variable $X : \Omega \to \mathcal{R}$ which is discrete-continuous is the set of points $\{X(\omega); \omega \in \Omega\}$ such that either the probability of the discrete component is positive or there exists a positive derivative of F_X w.r.t. the Lebesgue measure. If both variables are continuous and have positive densities on the same subset of the real line, or both are discrete and have positive probabilities on the same sequence of points, then their supports are equal.

Assume first that X and Y are discrete with the same supports. Then, it follows from definition (3.34) that the concentration curve is continuous, i.e. has no vertical segments. Therefore, the cdf of the actual Lilliputian variable ξ can be written as $C(u) = \int_0^u f_\xi(t)dt$, where $C = F_\xi$.

A pair of *discrete* variables X and Y may be represented by a *continuous* Lilliputian variable ξ.

To illustrate this, we provide example pairs of cdf's of discrete X and Y which are represented by the continuous Lilliputian variable ξ. Let $(X;Y)$ be the pair of discrete random variables with the support

$$S = \{x_1, \dots, x_k\}, \quad x_i \geq 0$$

and probability vectors

$$p = (p_1, \dots, p_k), \quad q = (q_1, \dots, q_k).$$

Their cdf's are:

$$F_X(x) = \begin{cases} 0 & \text{if } x < x_0, \\ S^X_{i-1} & \text{if } x_{i-1} \leq x < x_i, \quad i = 1, ..., k, \\ 1 & \text{if } x \geq x_k, \end{cases} \tag{3.35}$$

where $x_0 = 0$, $S^X_i = p_1 + \cdots + p_i$, $S^X_0 = 0$,

$$F_Y(x) = \begin{cases} 0 & \text{if } x < x_0, \\ S^Y_{i-1} & \text{if } x_{i-1} \leq x < x_i, \quad i = 1, ..., k, \\ 1 & \text{if } x \geq x_k, \end{cases} \tag{3.36}$$

where $S^Y_i = q_1 + \cdots + q_i$, $i = 1, \dots, k$, $S^Y_0 = 0$.

Then the curve $C(Y : X)$ is a piecewise linear continuous function:

$$C(Y : X)(u) = \frac{q_i}{p_i}(u - S_{i-1}^X) + S_{i-1}^Y \tag{3.37}$$

for $u \in (S_{i-1}^X, S_i^X)$, $i = 1, \dots, k$.

If $(X; Y)$ were replaced by $(X'; Y')$ which are *continuous* on $[x_0, x_k]$ with densities

$$f_{X'}(x) = \frac{p_i}{x_i - x_{i-1}}, \quad f_{Y'}(x) = \frac{q_i}{x_i - x_{i-1}}, \quad \text{for } x \in (x_{i-1}, x_i),$$

the concentration curve will remain the same:

$$C(Y' : X') = C(Y : X).$$

Generally, if variables X and Y are discrete-continuous with the same supports of the discrete components and the same supports of the continuous components, then the variable ξ representing $(X; Y)$ is continuous with positive density.
If the supports of X and Y are *disjoint*, the resulting Lilliputian variable ξ referring to pair $(X; Y)$ is discrete.

Illustration (pair of exponential variables):

Let cdf's $F^{(\lambda)}$ $(\lambda > 0)$ and $F^{(\lambda,a)}$ $(a > 0)$ be, respectively, exponential distributions with parameter λ and shifted exponential with parameters (λ, a), i.e.

$$F^{(\lambda)}(z) = \begin{cases} 0 & \text{if } z \le 0, \\ 1 - \exp(-\lambda z) & \text{if } 0 \le z < \infty, \end{cases} \tag{3.38}$$

$$F^{(\lambda,a)}(z) = \begin{cases} 0 & \text{if } z \le a, \\ 1 - \exp(-\lambda(z-a)) & \text{if } a \le z < \infty. \end{cases} \tag{3.39}$$

Then the concentration curve $C(F^{(\lambda)} : F^{(\lambda,a)})$ fulfils

$$C(F^{(\lambda)} : F^{(\lambda,a)})(u) = \begin{cases} F^{(\lambda)}(a) = 1 - \exp(-\lambda a) & \text{if } u = 0, \\ 1 - (1-u)\exp(-\lambda a) & \text{if } 0 \le u \le 1. \end{cases} \tag{3.40}$$

Remark 3.8 *The jump at point $u = 0$ of Lilliputian cdf lying on curve $C(Y : X)$ (as happens for the curve (3.40)) manifests the fact that there exists a set $A_0 = \{z \in \mathcal{R}\}$ such that $F_Y(z) > 0$ for $z \in A_0$ (i.e. a set A_0 has positive probability mass w.r.t. variable Y) and $F_X(z) = 0$ for all $z \in A_0$ (i.e. a set A_0 has probability mass equal to zero w.r.t. variable X). Similarly, the jump at point $u = 1$ (as happens for the curve $C(1/Y : 1/X)$) manifests the fact that there exists a set $A_1 = \{z : z \in \mathcal{R}\}$ such that $F_X(z) = 1$ for all $z \in A_1$ (set A_1 has probability mass equal to zero w.r.t. variable X) and $F_Y(z) < 1$ for some $z \in A_1$ (i.e. a set A_1 has positive probability mass w.r.t. variable Y).*

Illustration (Earnings Data):

We will consider an example based on hypothetical male and female earnings, presented in Table 3.1. There we have three pairs of variables, denoted

- $(W; M)$ when one deals with 43 microcategories of earnings
- $(W_8; M_8)$ when one deals with 8 midcategories (aggregated microcategories of earnings)
- $(W_4; M_4)$ when one deals with 4 macrocategories (aggregated categories)

The effects of aggregation will be considered during the course of analyzing the earnings data.

The concentration curves for the three pairs of male-to-female earnings are shown in Fig. 3.9 (left) while the curves for female-to-male earnings are shown in Fig. 3.9 (right).

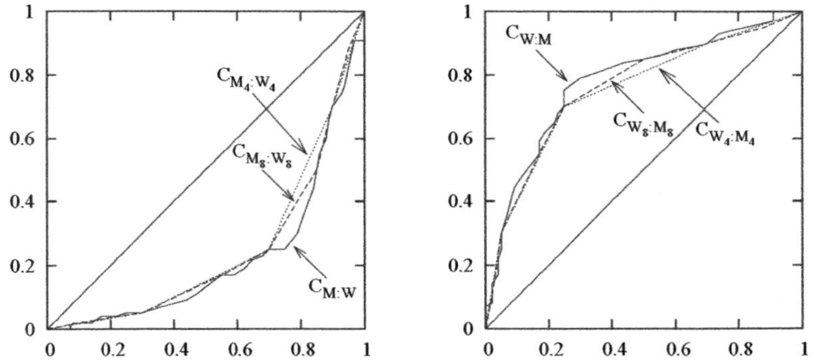

FIGURE 3.9. Curves $C(M : W)$, $C(M_8 : W_8)$, $C(M_4 : W_4)$ for male-to-female earnings (left) and $C(W : M)$, $C(W_8 : M_8)$, $C(W_4 : M_4)$ for female-to-male earnings (right).

If all the male earnings are higher than all the female earnings, i.e. all males are paid higher than females, then the male-to-female concentration curve would lie on two sides of the unit square: the horizontal segment linking points (0,0) and (1,0), and the vertical segment linking (1,0) and (1,1). In a reverse situation, the curve would consist of the remaining two sides of the unit square: that linking (0,0) and (0,1) and that linking (0,1) and (1,1).

No vertical or horizontal segments appear in curves $C(M_4 : W_4)$ and $C(M_8 : W_8)$ so that the related Lilliputian variables representing $(W_4; M_4)$ and $(W_8; M_8)$ are continuous. In $C(M : W)$, there are a few vertical and a few horizontal segments. They refer to categories in which the supports of M and W are disjoint, i.e. either

TABLE 3.1. Average male and female earnings: $(M; W)$ referring to microcategories, $(M_8; W_8)$ referring to midcategories, $(M_4; W_4)$ referring to macrocategories

earnings in microcategories	number of females	number of males	midcategories	macrocategories
$z_{1(43)} = 550$	2	0		
650	2	0		
750	3	0	$z_{1(8)} = 758$	
850	1	2		
950	2	0		$z_{1(4)} = 1084$
$z_{6(43)} = 1050$	4	0		
1150	3	2		
1250	5	0	$z_{2(8)} = 1254$	
1350	3	1		
1450	5	0		
$z_{11(43)} = 1550$	6	2		
1650	8	2		
1750	3	2	$z_{3(8)} = 1736$	
1850	4	3		
1950	4	3		$z_{2(4)} = 1932$
$z_{16(43)} = 2050$	4	0		
2150	3	2		
2250	3	3	$z_{4(8)} = 2246$	
2350	3	1		
2450	2	2		
$z_{21(43)} = 2550$	5	0		
2650	4	5		
2750	2	6	$z_{5(8)} = 2765$	
2850	3	8		
2950	1	6		$z_{3(4)} = 2942$
$z_{26(43)} = 3050$	1	5		
3150	2	5		
3250	0	4	$z_{6(8)} = 3226$	
3350	1	2		
3450	1	4		
$z_{31(43)} = 3550$	3	4		
3650	1	3		
3750	1	5	$z_{7(8)} = 3711$	
3850	0	3		
3950	0	3		$z_{4(4)} = 3963$
$z_{36(43)} = 4050$	2	3		
4150	3	0		
4250	**0**	**0**		
4350	0	4	$z_{8(8)} = 4303$	
4450	0	1		
4550	0	2		
4650	**0**	**0**		
$z_{43(43)} = 4750$	0	2		

the number of males is zero (and there appear horizontal segments) or the number of females is zero (and there appear vertical segments).

Remark 3.9 *The two microcategories with zero males and zero females, namely 4250 and 4650, are not taken into account in the construction of the curve. Note that with more disaggregated earnings it could happen that only such categories having frequency one or zero would be observed. Then, the concentration curve would contain solely horizontal and vertical segments.*

3.3.2 Equivalent pairs of random variables

It follows from the examples in the previous section that any two univariate distributions P and Q, or the cdf's F and G related to them, can be transformed into *one* Lilliputian cdf $C(Q : P)$ defined on the interval [0,1], as well as into the related Lilliputian curve C in the unit square. This statement can be also formulated in terms of the three involved random variables X, Y and ξ where X and Y are distributed according to P and Q, and ξ has the cdf $C = C(Y : X)$.

Consequently, we introduce the following *equivalence relation* in the set of all pairs of univariate distributions:

Definition 3.7 *(C-equivalence). Pair $(P; Q)$ is said to be C-equivalent to pair $(P'; Q')$ iff*

$$C(Q : P) = C(Q' : P').$$

This will be written as $(P; Q) \equiv_C (P'; Q')$.

The notation $\equiv_C$ will be also used with $(P; Q)$ replaced by the respective pair of random variables $(X; Y)$; we will write

$$(X; Y) \equiv_C (X'; Y') \quad \text{if} \quad C(Y : X) = C(Y' : X').$$

Any Lilliputian variable ξ *represents* the whole set of pairs of real valued random variables which are C-equivalent to $(U; \xi)$. This means that we have the set of pairs $(X; Y)$ such that $C(Y : X) = C_\xi$, where

$$C_\xi \overset{\text{df}}{=} C(\xi : U). \tag{3.41}$$

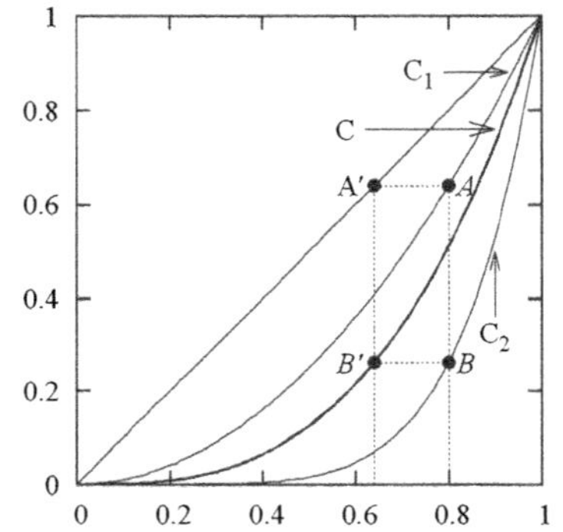

FIGURE 3.10. Graphical illustration of construction of the concentration curve $C = C(C_2 : C_1)$ for two continuous Lilliputian variables with cdfs C_1 and C_2

Fig. 3.10 provides a graphical illustration of how $C(\eta : \xi)$ can be constructed for a pair of *Lilliputian variables* $(\xi; \eta)$ with cdf's F_ξ and F_η.

Let C_1 and C_2 be the respective Lilliputian curves. For convenience, let ξ and η be continuous and strictly stochastically ordered: $\xi \prec_{st} \eta$, so that $C_2 = F_\eta$ lies below $C_1 = F_\xi$; Moreover, let them belong to $\mathbb{ULM}_{CON}$.

The procedure is the following: at any point t on the bottom side of the unit square we draw a vertical line which crosses C_2 and C_1, and we obtain the points $A = (t, C_1(t))$ and $B = (t, C_2(t))$. Next, we draw the horizontal line from point A which crosses the line *Diag* at the point $A' = (u, u)$. The interval $[A, B]$ is then transferred to the line which joins A' and the bottom side of the square at point u; hence point B is transformed into B', and the lengths of the intervals AB and $A'B'$ are equal. So, geometrically, we transpose the segments which join the curves C_1 and C_2 (along a vertical line) to be "hung" on the diagonal *Diag*. We then claim that curve $C(\eta : \xi)$ goes through B'.

The justification of our claim is immediate. Clearly, $B' = (u, C_2(t))$ where t, by construction, is equal to $C_1^{-1}(u)$. Hence $C(u) = C_2 \circ C_1^{-1}(u)$ and

$$C = \{(C_1(t), C_2(t)); t \in (0, 1)\}.$$

It is seen from Fig. 3.10 that for continuous $\xi, \eta \in \mathbb{ULM}$ the assumption $C_1 \leq_{st} C_2$ is not essential, and the same construction provides the respective curve $C(\eta : \xi)$ in the general case. It is only necessary to note that for t satisfying $F_\xi(t) < F_\eta(t)$ the segments BA and $B'A'$ have the same length. So we obtain point B' which lies on the curve $C(\eta : \xi)$ below the *Diag*.

The graphical construction of curve $C = C(C_2 : C_1) = C(\eta : \xi)$ for $C_1, C_2 \in \mathbb{ULM}$ shows that:

(*i*) curve C_1 is replaced by line *Diag* due to a suitable transformation of the cdf of ξ, (3.42)

(*ii*) curve C_2 is replaced by curve C, if *the same* transformation which was used to transform C_1 into *Diag* is applied to the distribution of η. (3.43)

3.3.3 Grade transformations of univariate distributions

If ξ and η (Lilliputian variables with cdf's C_1 and C_2) are both continuous, then the function $C_1(\xi_1)$ is a Lilliputian variable *uniformly distributed on* [0,1]. Therefore function C_1 is the sought transformation of ξ_1 since

$$Pr(C_1(\xi) \le u) = Pr(\xi \le C_1^{-1}(u)) = u = Diag(u), \tag{3.44}$$

while the cdf of $C_1(\eta)$ is given by:

$$Pr(C_1(\eta) \le u) = Pr(\eta \le C_1^{-1}(u)) = C_2 \circ C_1^{-1}(u) = C(\eta : \xi)(u). \tag{3.45}$$

But if ξ is discrete, or partly discrete and partly continuous (i.e. $\alpha_\xi > 0$), then it is not possible to obtain the uniform variable U as function of ξ. Therefore we have to use another probabilistic tool, known as the *grade monotone probability transition function.* This tool is presented in the Appendix 3.5.1 in formula (3.57). *It is not a function;* although it looks complicated, it is in fact quite simple.

This grade monotone probability transition function is applied to the distribution of ξ (not applied to the random variable itself). It is only slightly more complicated to apply *the same* grade monotone probability transition function to the distribution of η.

Up to now, we talked about pairs of *Lilliputian* variables ξ and η because it was convenient for graphical presentation. However, these considerations can be immediately extended to any pair of random variables X and Y. The only difference is that step (i) mentioned in (3.42) transforms the cdf F_X into F_U (i.e. transforms the graph of F_X into the *Diag*), and in step (ii) in (3.43) *the same grade monotone probability transition function* is applied to F_Y. The formulas (3.57) and (3.61) in Section 3.5.1 in the Appendix refer to this general case.

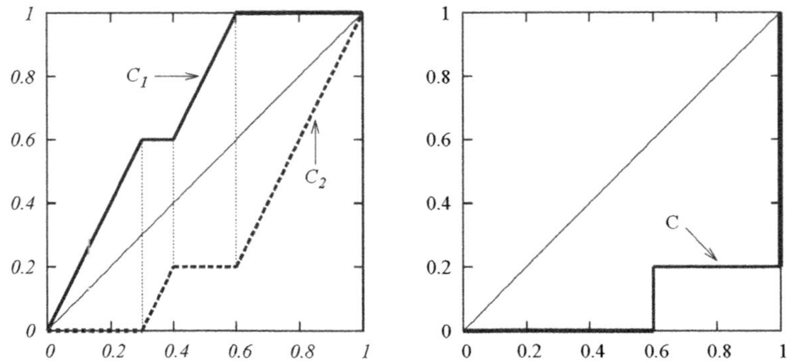

FIGURE 3.11. Graphical illustration of grade transformation: $C_1 = F_{F_Z^*(X)}$, $C_2 = F_{F_Z^*(Y)}$ (left), $C = C(C_2 : C_1)$ (right)

According to the notation introduced in the Appendix (Section 3.5.1), the grade monotone probability transition function is denoted F_Z^* for arbitrary chosen random variable Z. The variable $F_Z^*(X)$ denotes a variable with the cdf obtained from F_X due to F_Z^*. The variable $F_Z^*(Z)$ is denoted Z^*.

Illustration (pair of exponential variables):

In the case of two continuous variables X with cdf $F^{(\lambda,a)}$ (formula (3.39)) and Y with cdf $F^{(\lambda)}$ (formula (3.38)) we obtain $(F^{(\lambda,a)})^*(Y) = F^{(\lambda,a)}(Y)$ and this variable has a cdf lying on the curve $C(Y : X)$ defined by formula (3.40).

It is important to note that we can proceed in a slightly different way, using *three* steps instead of *two*. The first step is now to apply *any chosen grade monotone probability transition function to both X and Y*. So X and Y will be now replaced by a pair $(\xi; \eta)$ of two Lilliputian variables, and according to property (3.62) in the Appendix (Section 3.5.1), this pair still has the same concentration curve (i.e. pair $(\xi; \eta)$ is $C-$equivalent to pair $(X; Y)$). The next two steps are then applied to $(\xi; \eta)$ as indicated above. This idea is shown in Fig. 3.11 for X and Y, where X is a random variable taking on values a_1, a_2 with probabilities 0.6, 0.4, respectively, and Y is a random variable taking on values b_1, b_2 with probabilities 0.2, 0.8, and $a_1 < b_1 < a_2 < b_2$. Let $F_Z = (F_X + F_Y)/2$. First we obtain continuous cdfs of variables $F_Z^*(X)$ and $F_Z^*(Y)$ equal to $C_1 = C(X : Z)$, $C_2 = C(Y : Z)$, respectively (they are shown in Fig. 3.11 left. Then we take plot $C = C(C_2 : C_1)$ which is shown in Fig. 3.11 right.

The passage from any real-valued random variable X to a Lilliputian variable ξ by means of a grade monotone probability transition function will be called a *grade transformation* of F_X onto F_ξ. Therefore, the set of grade monotone probability transition functions as introduced in Appendix (Section 3.5.1) will be called the set of *grade transformations* of univariate distributions. The passage from any real-valued random variable X to the *uniform* Lilliputian variable U (i.e., to X^*) will be called a *uniform grade transformation* of X,

We will try to explain why the operations on the cdf's of univariate distributions are called *grade*. This refers to the fact that for any *continuous* variable X, variable $F_X(X)$ is a random variable which is uniformly distributed on the interval [0,1]. But, in early statistical literature (see Mini History in Chapter 1 Section 1.2), the expression $F_X(x)$ was called the *grade of* x (since it showed the "level" or "position" of x in the considered population of values of variable X).

The mapping of F_X into the *Diag* described in the Appendix (Section 3.5.1) is a "fuzzy" analogue of the transformation F_X performed on a continuous variable X.

Remark 3.10 *For any X with cdf F_X, let F_X^* denote the grade monotone probability transition which, applied to F_X, provides the cdf of the uniform Lilliputian variable U. So we will use the notation $F_X^*(X)$ later where $F_X^*(X)$ is a uniform variable U, and the notation $F_X^*(Y)$, where $F_X^*(Y)$ is a Lilliputian variable which results from applying F_X^* to F_Y.*

Illustration (pair of exponential variables):

In the case of X with cdf $F^{(\lambda,a)}$ and Y with cdf $F^{(\lambda)}$ we obtain $F^{(\lambda,a)*}(Y) = F^{(\lambda,a)}(Y)$, and this variable has a cdf lying on the curve $C(Y:X)$.

Remark 3.11 *Any set of $C-$equivalent pairs $(X;Y)$ is fully described by all the pairs consisting of F_X and of $C(Y:X)$, where $C(Y:X)$ is the same in all pairs while F_X changes (and F_Y is produced accordingly, i.e. it results from $C(Y:X)$ and F_X). Usually we are interested in $C(Y:X)$ only, treating F_X as an additional parameter (possibly even as a* ***nuisance parameter****) in a problem under consideration. Therefore problems and methods concerning pairs of univariate distributions, which refer to $C(Y:X)$ only, will be* *called **grade**. These problems and methods remain unchanged when any grade monotone probability transition function is applied to **both** elements of the pair. This invariance also applies to some parameters describing two univariate distributions: this will be considered in Section 3.4.*

3.4 Grade parameters

3.4.1 The parameter ar

Parameters of Lilliputian variables can be used to introduce parameters for each pair of random variables $(X;Y)$ which is $C-$equivalent to the pair $(U;\xi)$. In particular, we extend the definition of parameter ar into the set of pairs of real-valued variables $(X;Y)$, putting

$$ar(Y:X) \stackrel{df}{=} ar(\xi) = ar(\xi:U).$$

Details: The expectation of ξ (but <u>not</u> of X) is also a *grade* parameter.

Illustration (Earnings Data):
The concentration curves for the male-to-female earnings $C(M:W)$, $C(M_8:W_8)$ and $C(M_4:W_4)$ are presented in Fig. 3.9. The respective values of ar, equal to

$$ar(M:W) = 0.5574, \quad ar(M_8:W_8) = 0.5205, \quad ar(M_4:W_4) = 0.505,$$

slightly diminish with progress in aggregation (from micro- to macrocategories).

Various formulas for $ar(Y:X)$ in terms of F_X and F_Y are:

$$\begin{aligned} ar(Y:X) &= 2\int_0^1 (\tilde{F}_X(x_u) - \tilde{F}_Y(x_u))du \qquad (3.46)\\ &= 1 - 2\int_0^1 \tilde{F}_Y(x_u)du = 1 - 2E(\tilde{F}_Y(X))\\ &= 2E(\tilde{F}_X(Y)) - 1 \end{aligned}$$

and

$$ar(Y:X) = \frac{1}{2}\int_{-\infty}^{\infty}\int_{-\infty}^{\infty} sgn(y-x)(dF_X(x)dF_Y(y) - dF_X(y)dF_Y(x)) \qquad (3.47)$$

where $\tilde{F}$ is defined by (3.28) and $dF_X(x) = f_X(x)dx$ when there exists a density $f_X(x)$ w.r.t. Lebesque measure and $dF_X(x) = p(x) = F_X(x) - F_X(x-)$ when x is an atom of F_X.

There is a certain difficulty with $ar(\eta : \xi)$ when both ξ and η are Lilliputian variables. We then have two Lilliputian curves, C_1 and C_2, and we are prompted by mathematical tradition to consider

$$\int_0^1 (C_1(u) - C_2(u))du \tag{3.48}$$

as the distance of C_2 from C_1. Geometrically, when C_2 lies below C_1, this expression is equal to the area between the two curves.

Let $C = C(\eta : \xi)$ and let ζ be the respective Lilliputian representative of $(\xi; \eta)$. We have

$$ar(\eta : \xi) = 2\int_0^1 (u - C(u))du.$$

Is there a relation between these two areas: the area between C_1 and C_2 and the area between C and *Diag*? When are they equal? Equality of

$$\frac{1}{2}ar(C_2 : C_1), \quad \text{and} \quad \int_0^1 (C_1(u) - C_2(u))du, \tag{3.49}$$

i.e., equality of $ar(C_2 : C_1)$ and the difference $ar(C_2) - ar(C_1)$, would provide a desirable marriage between mathematical tradition and geometrical expressions for concentration, as well as a desirable inner congruence of index *ar*. This, however, is rarely satisfied, although departures from equality are usually rather small.

Illustration (three continuous Lilliputian variables):
In the example considered in Fig. 3.1, $ar(F_2 : F_1)/2 = -0.4022$, and $\int_0^1 (F_1(u) - F_2(u))du = -0.355$.

Proposition 3.3 *(i) If* $C_1, C_2 \in \mathbb{ULM}$ *then*

$$\begin{aligned} ar(C_2 \ : \ C_1) &= \int_0^1 (C_1(C_2^{-1}(u)) - C_2(C_1^{-1}(u)))du \\ &= 2\int_0^1 (\tilde{C}_1 \circ C_1^{-1}(u) - \tilde{C}_2 \circ C_1^{-1}(u))du \\ &= 1 - 2\int_0^1 \tilde{C}_2 \circ C_1^{-1}(u)du. \end{aligned}$$

(ii) if $C_1 = C(V : Z)$ *and* $C_2 = C(W : Z)$ *where* V, W, Z *are any variables such that* $F_Z = \alpha F_V + (1 - \alpha)F_W$ *for some* $\alpha \in [0, 1]$ *then*

$$ar(C_2 : C_1) = ar(C_2) - ar(C_1).$$

In this case curves C_1 *and* C_2 *satisfy* $\alpha C_1 + (1 - \alpha)C_2 = Diag$.

(iii) if both curves C_1 *and* C_2 *are symmetric so that* $u - C_i(u) = 1 - u - C_i(1 - u)$, $i = 1, 2$, *then* $ar(C_2 : C_1) = ar(C_2) - ar(C_1)$.

Remark 3.12 *It is proved in (3.63) in the Appendix that for any variable Z with a cdf which is a convex combination of cdf's of X and Y, then it holds that*

$$ar(Y : X) = ar(Y : Z) - ar(X : Z).$$

Proposition 3.3 (ii) provides a sufficient condition for $\xi, \eta \in \mathbb{ULM}$ under which $ar(\eta : \xi) = ar(\eta : U) - ar(\xi : U) = ar(\eta) - ar(\xi)$.

The stochastic order in $\mathbb{ULM}$ introduces a respective order of pairs of univariate distributions: if ξ and ξ' are Lilliputian variables and $\xi \preceq_{st} \xi'$ then, for any $(X;Y)$ represented by ξ and any $(X';Y')$ represented by ξ', pair $(X;Y)$ is said to be stochastically smaller than pair $(X';Y')$. This will be denoted $(X;Y) \preceq_{st} (X';Y')$.

We introduce the *convex order* $\preceq_{con}$ as follows:

$$(X;Y) \preceq_{con} (X';Y') \text{ if } (X;Y) \preceq_{st} (X';Y')$$

and the curve $C(C(Y' : X') : C(Y : X))$ is convex.

Proposition 3.4 *For any two pairs of random variables*
(i) $(X;Y) \preceq_{st} (X';Y') \Rightarrow ar(C(Y' : X') : C(Y : X)) \geq 0.$
(ii)

$$(X;Y) \preceq_{con} (X';Y') \Rightarrow ar(C(Y' : X') : C(Y : X)) = Gini(f_\xi(U))$$

where ξ has cdf lying on curve $C_\xi = C(C(Y' : X') : C(Y : X))$.

3.4.2 Normal concentration pattern

The traditional model used by statisticians to consider how two univariate distributions are differentiated is that of two homoscedastic normal distributions $N(m_1, \sigma)$ and $N(m_2, \sigma)$, where m_1 and m_2 are the means while σ is the common standard deviation. It is well known that the *shapes* of both densities are identical, so that the distributions differ only in where they are situated on the real line according to the values of m_1 and m_2. The exact positions of m_1 and m_2 should not influence measures of differentiation of $N(m_1, \sigma)$ and $N(m_2, \sigma)$, and this is a postulate put usually on various measures of differentiation in use.

This postulate is satisfied by the concentration curve:

$$C(N(m_2,\sigma):N(m_1,\sigma))(u)=\Phi(\Phi^{-1}(u)-\mu),\quad u\in(0,1)$$

where Φ is the cdf corresponding to $N(0,1)$ and

$$\mu=(m_2-m_1)/\sigma.$$

Therefore,

$$C(N(m_2,\sigma):N(m_1,\sigma))=C(N(\mu,1):N(0,1)). \tag{3.50}$$

Curve $C(N(\mu,1):N(0,1))$ will be called the *normal concentration curve with parameter* μ and will be denoted CN_μ.

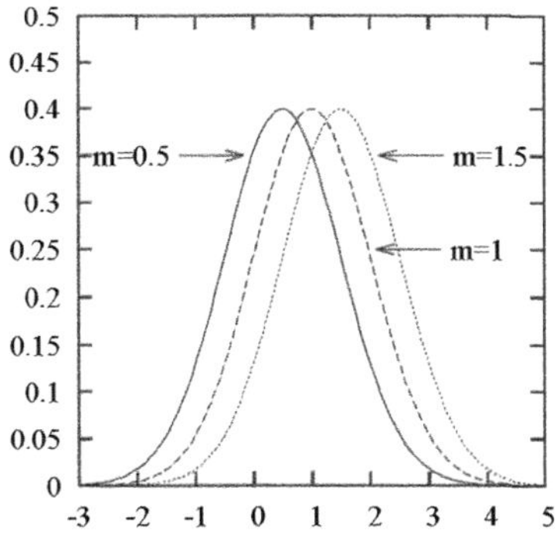

FIGURE 3.12. Densities of $N(m,1)$ for $m=0.5,1,1.5$

If $m_2=m$, $m_1=0$, and $\sigma=1$, then $\mu=m$. The densities of distribution $N(m,1)$ are shown in Fig. 3.12 for a few values of m, while Fig. 3.13 displays the normal concentration curves CN_m for these m's.

All normal concentration curves belong to $\mathbb{ULM}_{CON}$ and are stochastically ordered w.r.t. increasing μ; the curves are placed below the *Diag* if $\mu>0$ and above the *Diag* if $\mu<0$. Curves CN_μ are symmetric w.r.t. $Diag^\perp$, that is to the diagonal of the unit square which joins (0,1) and (1,0) and thus is perpendicular to *Diag*. When μ is replaced by $-\mu$, the respective concentration curve $CN_{-\mu}$ is equal to CN_μ^{-1}.

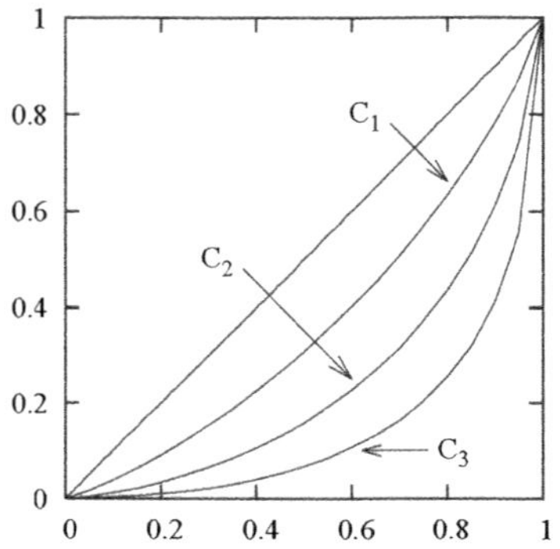

FIGURE 3.13. Normal concentration curves $CN_m = C(N(m,1) : N(0,1))$ for $m = 0.5$ (C_1), $m = 1$ (C_2), and $m = 1.5$ (C_3)

The concentration index ar of the normal concentration curve with parameter μ can be expressed explicitly as:

$$\begin{aligned} ar(CN_\mu) &= ar(N(\mu,1) : N(0,1)) = 1 - 2\int_{-\infty}^{\infty} \Phi(t-\mu)d\Phi(t) \\ &= 1 - 2\Phi\left(\frac{-\mu}{\sqrt{2}}\right) = 2\Phi\left(\frac{\mu}{\sqrt{2}}\right) - 1. \end{aligned} \quad (3.51)$$

This is an increasing function of μ.

To any pair $(X;Y)$ we can attach the respective normal concentration curve CN_μ with the value of μ such that

$$ar(CN_\mu) = ar(Y : X). \quad (3.52)$$

The normal concentration curve attached to the pair $(X;Y)$ will be denoted $NOR(Y : X)$. That curve is a (function-valued) *grade* parameter of pair $(X;Y)$.

A comparison of $NOR(Y : X)$ with $C(Y : X)$ provides a quick visual perception of the main differences.

Illustration (Earnings Data):

Consider the pair of earnings $(W_4; M_4)$. Since $ar(M_4 : W_4) = 0.505$, it follows from formula (3.51) that

$$\mu = -\sqrt{2}\Phi^{-1}(0.2475) = \sqrt{2}\Phi^{-1}(0.7525) = 0.963$$

is the value of μ such that the normal concentration curve $NOR(M_4 : W_4)$ is equal to CN_μ with $\mu = 0.963$. Both curves: $C(M_4 : W_4)$ and its normal counterpart $NOR(M_4 : W_4)$ are presented in Fig. 3.14.

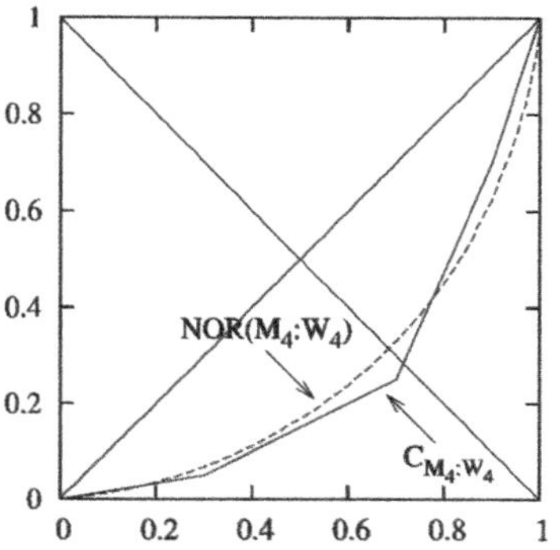

FIGURE 3.14. Concentration curve $C(M_4 : W_4)$ and the respective normal concentration curve $NOR(M_4 : W_4)$

A visual comparison of $C(M_4 : W_4)$ and $NOR(M_4 : W_4)$ shows that these curves are very similar, but $C(M_4 : W_4)$ is not quite symmetric with respect to $Diag^{\perp}$ (the line joining (0,1) and (1,0)).

The specific character of this asymmetry is the strong domination of males concerning higher earnings (the density of the right part of the curve $C(M_4 : W_4)$ is increasing more than the same part of the curve $NOR(M_4 : W_4)$).

3.4.3 Likelihood ratio and local concentration

Definition 3.8 *(likelihood ratio of Y to X). Let P_X and Q_Y be measures induced by random variables X and Y, respectively. The likelihood ratio $h_{Y:X}$ is defined on the set $\frac{dP_X}{d(P_X+Q_Y)}(z) > 0$ by*

$$h_{Y:X}(z) = \frac{dQ_Y}{d(P_X + Q_Y)} / \frac{dP_X}{d(P_X + Q_Y)}(z). \tag{3.53}$$

From this definition it follows that $h_{Y:X}(X)$ is a random variable which is finite with P_X-probability one. This variable is a function-valued *grade* parameter.

Let $\overline{x}_u = sup\{x : F_X(x) \leq u\}$, $u \in (0, 1)$ be the upper quantile of F_X. Then (by the Lebesgue decomposition of probability measures) we obtain

$$C(Y : X)(t+) = \int_0^t h_{Y:X}(\overline{x}_u)du + Q_Y\{z : F_X(z) \leq t \cap \frac{dP}{d(P+Q)}(z) = 0\}$$

which implies the equality

$$C(Y : X)(t+) = F_Y(\overline{x}_t-) + (t - F_X(\overline{x}_t-))h_{Y:X}(\overline{x}_t).$$

Let ξ be a Lilliputian variable with cdf lying on a concentration curve $C(Y:X)$. Then, $h_{\xi:U}(U)$ is equivalent to $h_{Y:X}(X)$:

$$h_{\xi:U}(u) = f_{\xi}(u) = h_{Y:X}(x) \tag{3.54}$$

where:

$u \in (F_X(x-), F_X(x))$ if x is a discontinuity point of F_X,

$u = F_X(x)$ if x is a continuity point of F_X and F_Y and there exist the densities $f_Y(x)$, $f_X(x)$ (with $f_X(x) > 0$).

Definition 3.9 *(local two points concentration). Let X and Y be both discrete or both continuous and let Y be absolutely continuous with respect to X (i.e., equality $P_X(A) = 0$ for $A \in \mathcal{B}(\mathcal{R})$ implies $P_Y(A) = 0$). Let ν be the Lebesgue measure in the case of continuous X and Y, or the count measure in the case of discrete measures. Then, for any pair of points $x, y \in \mathbb{R}$, the expression*

$$sign(y-x)\left(\frac{dF_X}{d\nu(x)}(x)\frac{dF_Y}{d\nu(y)}(y) - \frac{dF_X}{d\nu(y)}(y)\frac{dF_Y}{d\nu(x)}(x)\right), \tag{3.55}$$

will be called a local two points concentration of Y w.r.t. X.

The expression (3.55) is a grade parameter which is equal to *local two points concentration* of ξ w.r.t. U, namely to

$$sign(v-u)(f_{\xi}(v) - f_{\xi}(u)) \tag{3.56}$$

where f_{ξ} is density of ξ which exists for almost all $u \in [0,1]$ due to of absolute continuity of Y w.r.t. X.

Definition 3.10 *(increasing likelihood ratio). Pair of variables $(X;Y)$ is said to have an increasing likelihood ratio of Y to X if local two points concentrations of Y w.r.t. X for all points x, y satisfying $x < y$ are positive.*

This property holds for all pairs which are C-equivalent to $(X;Y)$ with an increasing likelihood ratio. In other words, the indicator of this property is a *grade* parameter.

All pairs $(X;Y)$ with an increasing likelihood ratio of Y to X are represented by a Lilliputian variable with a *convex* cdf (which implies that a continuous ξ has an *non-decreasing density*).

3.5 Appendix

3.5.1 Monotone grade probability transition function

Let F_Z be the cdf of a random variable Z. If Z is continuous, then it is known that variable $F_Z(Z)$ is uniformly distributed on [0,1]. After such a transformation, the values of Z from the interval $[-\infty, z]$ are mapped to the interval $[F_Z(-\infty), F_Z(z)] = [0, u]$ so that variable $F_Z(Z)$ takes on values from the interval $[0, u]$ with probability $F_Z(z) = u$, i.e. has a uniform distribution. This transformation can be rewritten by means of the *probability transition function* $G_Z([0, u]; z)$;

$$G_Z([0,u];z) = \begin{cases} 1 & \text{if } F_Z(z) \leq u, \\ 0 & \text{if } F_Z(z) > u. \end{cases}$$

The variable obtained when G_Z is applied to F_Z has a uniform distribution, since

$$\int_0^1 G_Z([0,u];z)dF_Z(z) = \int_{F_Z(z)\leq u} dF_Z(z) = u.$$

If cdf F_Z is *not continuous*, then it is not true that for each $u \in (0, 1)$ there exists a point z_u for which $F_Z(z_u) = u$. But for any $u \in (0, 1)$ there always exists point z_u for which $F_Z(z_u-) \leq F_Z(z_u)$, hence we take just a part of the probability mass of point z_u which ensures that the probability of $[0, u]$ will be equal to u. This means that we "randomize" this mass using the following monotone grade probability transition function F_Z^* :

$$F_Z^*([0,u];z) = \begin{cases} 1 & \text{if } F_Z(z) \leq u, \\ \frac{u - F_Z(z-)}{F_Z(z) - F_Z(z-)} & \text{if } F_Z(z-) < u \leq F_Z(z), \\ 0 & \text{if } F_Z(z-) > u. \end{cases} \tag{3.57}$$

The cdf at point u of the variable $F_Z^*(Z)$ obtained as a result of applying F_Z^* to Z is given by

$$\int_{-\infty}^{\infty} F_Z([0,u];z)dF_Z(z) = u \tag{3.58}$$

i.e. $F_Z^*(Z)$ is uniform on [0,1].

Clearly, G_Z is a special case of F_Z^*, suitable when Z is continuous.

Let us apply F_Z^* to F_X. We then obtain a random variable denoted $F_Z^*(X)$ such that the value of the cdf of this variable in point $u \in [0,1]$ is equal to

$$\int_{-\infty}^{\infty} F_Z^*([0,u];x)dF_X(x). \tag{3.59}$$

Hence, the random variable $F_Z^*(Z)$ is uniform on [0,1]. We will use the notation

$$Z^* = F_Z^*(Z). \tag{3.60}$$

If variables X and Z have the same support and are continuous, then the Lilliputian variable $F_Z^*(X)$ has cdf on [0,1] equal to $F_X(F_Z^{-1}(u))$ for $u \in [0,1]$ and density equal to $\frac{f_X}{f_Z}(F_Z^{-1}(u))$ (in those points u for which the density exists).

If variables X and Z have the same discrete support, then variable $F_Z^*(X)$ is continuous with a piecewise linear cdf (see example cdf's (3.35) and (3.36)) and the related formula (3.37)).

Particularly important is the case when the cdf F_Z is equal to a convex combination of the cdf's of random variables X and Y. Then variable $F_Z^*(X)$ (and also $F_Z^*(Y)$) have continuous cdf's on (0,1) (possibly with horizontal segments).

For any cdf's F_Z, F_X, the cdf of Lilliputian variable $F_Z^*(X)$ lie on the curve $C(X:Z)$.

The important properties of the grade transformation are the following:

$$C(Y:X) = C(F_X^*(Y):X^*) = C(Y^*:F_Y^*(X)), \tag{3.61}$$

for any cdf's F_X, F_Y, and F_Z

$$C(Y:X) = C(C(Y:Z):C(X:Z)) = C(F_Z^*(Y):F_Z^*(X)). \tag{3.62}$$

3.5.2 Properties of concentration measures

Proposition 3.5 *Properties of index ar.*

(i) Let F_X and F_Y be arbitrary convex combination of cdf's $\{F_i\}$ and $\{G_j\}$, respectively, i.e. $F_X = \sum_i \alpha_i F_i$, $F_Y = \sum_j \beta_j G_j$, where $\alpha_i \geq 0$, $\beta_j \geq 0$, $\sum_i \alpha_i = 1$, $\sum_j \beta_j = 1$. Then

$$ar(F_Y : F_X) = \sum_i \sum_j \alpha_i \beta_j ar(G_j : F_i).$$

(ii) For any random variables X, Y, Z

$$ar(Y : X) = ar(C(Y : Z) : C(X : Z)).$$

(iii) Let cdf F_Z be a convex combination of cdf's F_X and F_Y, i.e. for some $\alpha \in [0,1]$, $F_Z = \alpha F_X + (1-\alpha) F_Y$. Then

$$ar(F_Y : F_X) = ar(F_Y : F_Z) - ar(F_X : F_Z). \tag{3.63}$$

Proof of (iii). Using property (i) we obtain

$$ar(F_Y : F_Z) = (1-\alpha) ar(F_Y : F_X), \quad ar(F_X : F_Z) = -\alpha\, ar(F_Y : F_X)$$

which implies equality (3.63).

4

Univariate Lilliputian Model II

Teresa Kowalczyk, Elżbieta Pleszczyńska and Wiesław Szczesny

4.1 Introduction

Chapter 3 dealt with measures of concentration which refer to a certain initial ordering of the data. In the example concerning male-to-female earnings, data were ordered according to increasing earnings. For earnings aggregated to four macrocategories, the concentration curve $C(M_4 : W_4)$ relating male earnings M_4 to female earnings W_4 happened to be convex. This indicated a clear monotone trend: the higher the earnings macrocategory, the stronger the domination of male earnings over female earnings. Departures from convexity were noted for less aggregated data earnings.

Monotone trend occurs in the case of a convex concentration curve. However, it is always possible to *reorder* data so that the curve *becomes convex after reordering*. The modified curve, called the *maximal concentration curve* and denoted $C_{\max}$, measures *maximal* concentration of one variable w.r.t. the other, and indicates the corresponding trend.

It could be a surprise to most readers that measures of maximal concentration are closely related to *measures of inequality of a ratio variable,* namely to the *inequality curve* (called usually the *Lorenz curve*) and the *inequality index* (called usually the *Gini Index*). This link will be described in detail.

Therefore, the chapter starts with a short study of ratio variables and the concentration curves related to each of them (Section 4.2.1). Earnings (denoted: Z for males and females when considered together, M for males only, W for females only) serve as examples of *ratio* data sets.

For any data set, say the one containing values of Z, the distribution of earnings is compared with the partition of the total amount of money spent on employee earnings. Therefore we deal with two cdf's: F_Z and $F_Z^{(1)}$, where $F_Z^{(1)}(z)$ is the part of total earnings spent on people with earnings not exceeding z. The concentration curve for this pair of cdf's serves to measure, for any z, *the inequality* between the proportion of total earnings up to z (i.e., $F_Z^{(1)}(z)$) and the proportion of people receiving those earnings (i.e., $F_Z(z)$).

Sections 4.2.2 and 4.2.3 supplement Section 4.2.1 using a more formal mathematical description.

Earnings represent variables which are known as ratio variables appearing in practical applications (length, weigth, heigth etc). There also exist ratio variables which appear in statistical infrastructure and serve to describe important probabilistic concepts. A concept called *overrepresentation variable* is introduced in Section 4.2.4.

Section 4.2.4 builds a bridge between measures of inequality of Z and measures of concentration applied to a pair of random variables. This bridge is illustrated in the case of the male and female earnings M and W (the concentration curve $C(M : W)$ was presented in Fig. 3.9 in Chapter 3). We are interested in the overrepresentation of males earnings over female earnings.

The first step in defining this concept is to introduce the ratio of the proportion of men to the proportion of women with the same category of earnings. This function is called the likelihood ratio. It has to be calculated separately in three cases: when earnings are expressed in macrocategories, midcategories, and microcategories. Let us start with macrocategories. Then, the likelihood ratio, say $h_{M_4:W_4}$, will be applied to the female earnings W_4; the resulting random variable $h_{M_4:W_4}(W_4)$ will be called the overrepresentation of M_4 over W_4. The same can be done for midcategories, and then we obtain $h_{M_8:W_8}(W_8)$. The overrepresentation variables obtained in this way are measured on the absolute scale. They are described by their Lorenz Curves and Gini Indices which measure *inequality* of overrepresentation.

Now we proceed to microcategories; then the likelihood ratio $h_{M:W}$ takes on value $+\infty$ for such microcategories for which there is a positive proportion of men with such earnings but a zero proportion of women. Therefore we have to treat these microcategories in a special way which leads to a variable with the distribution which is a slightly modified distribution of $h_{M:W}(W)$. Generally, the overrepresentation variable $OV_{Y:X}$ has the distribution obtained from the distribution of $h_{Y:X}(X)$ after this modification whenever applicable (formula (4.17)).

It is important that the overrepresentation of one variable over another is invariant on a grade transformation applied simultaneously to both variables.

The reciprocal of overrepresentation is called *underrepresentation*. Section 4.2.5 presents diagram visualizing links existing between $OV_{Y:X}$, $OV_{X:Y}$, $1/OV_{Y:X}$, $1/OV_{X:Y}$.

Section 4.3 deals with order oriented concentration curves. This is a key concept for the remainder of the book. It concerns distributions of X and Y transformed by the same order function (formula (4.20).

The opposite concentration curve (for the opposite order) is introduced in Section 4.3.1, and the maximal concentration curve $C_{\max}(Y : X; \varphi)$ (with the order corresponding to the likelihood ratio is introduced in Section 4.3.2. The latter curve is convex, and lies below the curve $C(Y : X; \varphi)$ for any other φ (Proposition 4.3).

Overrepresentation variables have Lorenz curves *equal to* $C_{\max}(Y : X)$ (Theorem 4.3). This means that *maximal concentration* of Y on X can be expressed as *inequality of overrepresentation* of Y on X. It is an important example of integrity of grade infrastructure: the concepts of overrepresentation and of inequality are perfectly matched with the concept of concentration.

The last topic of this chapter covered in Section 4.4 refers to a basic *duality* between two opposite points of view, which are both relevant in a current statistical investigation. A typical example of these two opposite outlooks is *wealth* in contrast to *poverty.* This is presented in Section 4.4.1 in which dual concentration curve and dual Lorenz curve are defined. Then, consequently, it is possible to define the dual random variable (Section 4.4.2). A variable X^D is dual if its Lorenz curve is dual to the Lorenz curve of X.

Usually investigators collect data with the intention of measuring some aspect of one or the other of these two viewpoints. When the viewpoint is in terms of wealth, the focus might be on earnings as a convenient observable characteristic of wealth. When the viewpoint is poverty, the distance from some established level (perhaps from the established income poverty level) provides information with a direct focus on poverty.

A diagram which presents various dualities connected with the concept of overrepresentation (introduced in Sec. 4.2.5) is developed in Sec. 4.4.3. This diagram in its final form presents graphically interrelations among main concepts and parameters appearing in the Univariate Lilliputian Model. As mentioned in Section 4.4.4, it could be particularly useful for data analysts interested in interpopulation comparisons.

A short introduction to measurement scales is given in the Appendix, Section 4.5.1.

4.2 Lorenz Curve and Gini Index

4.2.1 Ratio variables and related concentration curves

This section provides an informal introduction to the inequality measures: Lorenz Curve and Gini Index, based on a dataset concerning information on earnings, education and gender. A formal approach to the inequality measures is given in the next section; rudiments of measurement are discussed in the Appendix in Section 4.5.1. We intend to make clear when the inequality measures are meaningful in practice and when they are not (in spite of being widely used).

In this section we introduce the notation needed in the next section and explain its origin. This notation helps to indicate the connection between measurements on the ratio scale and the set of so called ratio random variables.

Illustration (Earnings and Education Data):

Table 3.1 in the previous chapter refers to a set Ω_0 consisting of people who are described by *two* features: *gender* (S) and *earnings* (Z). These earnings are on three levels of aggregation: there are 43 microcategories (with values $z_{i(43)}$, $i = 1, \ldots, 43$), 8 midcategories (with values $z_{i(8)}$, $i = 1, \ldots, 8$), and 4 macrocategories (with values $z_{i(4)}$, $i = 1, 2, 3, 4$). To distinguish among three levels of aggregation, we label Z_8 the mean earnings which would be received by each person in the corresponding midcategory if money were equally partitioned); and analogously we label Z_4 the mean earnings corresponding to macrocategories $z_{i(4)}$, $i = 1, ..., 4$.

We will introduce now the next feature, namely *level of education* denoted T. It is assumed that there are six levels of education which form categories of T, labeled 1 (lowest) through 6 (highest). We also attach label 1 for females and label 2 for males. Then variables S, T, Z, Z_8, Z_4 are functions which attach positive numbers to each member of Ω_0 :

$$S : \Omega_0 \to \{1, 2\}, \quad T : \Omega_0 \to \{1, 2, ..., 6\}, \quad Z, Z_8, Z_4 : \Omega_0 \to \mathbb{R}^+,$$

where $\mathbb{R}^+$ is the set of non-negative numbers. So any person $\omega \in \Omega_0$ is described by three numbers: $S(\omega)$, $T(\omega)$, $Z(\omega)$ (e.g. sequence (1, 4, 950) describes a *woman* with 4th *level of education* who *earns* 950).

Functions S, T, Z exemplify three important types of measurements: *nominal, ordinal,* and *ratio.* These terms are in common use in statistics and data analysis; a formal approach to measurement scales is given in the Appendix (Sec. 4.5.1). The important concept introduced there is that of *admissible function* which is of the form ($\varphi(x) = ax$, $x \in \mathbb{R}^+$, $a > 0$) for the ratio scale, an increasing function $\varphi : \mathbb{R} \to \mathbb{R}$ for the ordinal scale, a one-to-one function $\varphi : \mathbb{R} \to \mathbb{R}$ for the nominal scale. The measurement scales and admissible functions are essential concepts of *measurement theory,* which should be an integral part of data analysis and statistics.

In statistics, functions defined on Ω_0 (such as S, T, Z) are characterized by their probability distribution. In data analysis, sets of objects are usually finite and these "probabilities" are expressed by the number of elements belonging to the considered subset, divided by the number of elements belonging to the whole set. For non-finite sets of objects, the probability measures are introduced in the way provided in each introductory textbook on probability and statistics.

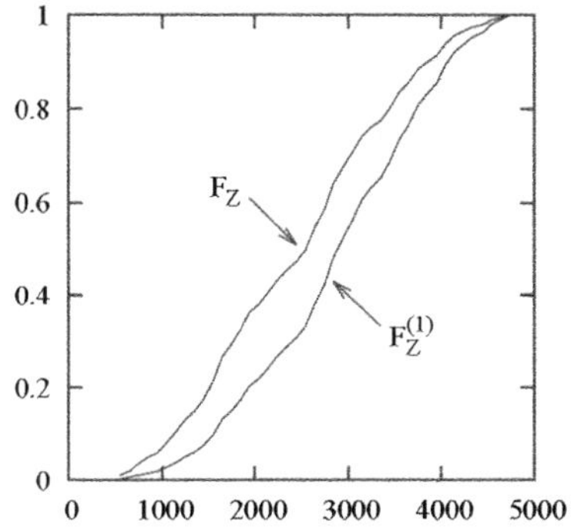

FIGURE 4.1. The cdf's of variable Z (earnings) and its first moment $Z^{(1)}$

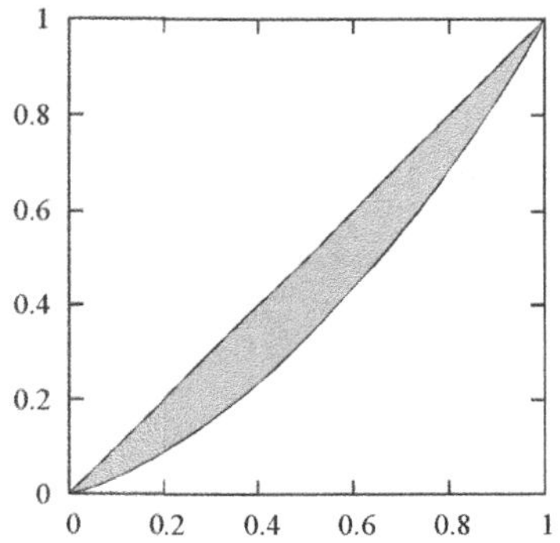

FIGURE 4.2. Lorenz Curve L_Z of earnings Z. The shaded area is one-half of the Gini Index of Z.

Thus a *random variable* which is ratio "in the strict sense" should be a function $Z : \Omega_0 \to \Omega \subset \mathbb{R}^+$ which fulfills the requirements of a measurement on a ratio scale (in particular, for any $a > 0$ the function aZ is also a measurement on a ratio scale, prescribed to the same pair of relational structures) and which, moreover, is considered jointly with its cdf F_Z.

The "rationess" of a random variable (say earnings Z) *should* influence the methods by which a ratio dataset is analyzed and how the underlying distribution is described, with stress laid on invariance due to the set of admissible functions. But such an approach is very rarely observed in practice.

Illustration (Earnings Data):

For earnings Z, Ω_0 is finite and consists of 200 people; since the distribution of Z is discrete, F_Z is presented in Fig. 4.1 with jumps at each of 43 microcategories. This implies that earnings Z maps Ω_0 onto the set Ω consisting of 43 microcategories ($\Omega \subset \mathbb{R}^+ \subset \mathbb{R}$).

The "probability" p_i attached to i^{th} microcategory z_i shows the chance of drawing a person with microcategory z_i *when drawing a person from* Ω_0 *at random.* Probabilities p_{38} and p_{42} are equal to zero (no person is attached to these microcategories).

This random drawing can be modelled as follows: each person will be represented by a segment of length $1/N$, where $N = 200$ is the total number of people in Ω_0, and these 200 segments will be located in the interval (0,1), firstly ordered according to non-decreasing earnings, and then aggregated for people with the same earnings microcategory. The result is visualized in Figure 4.3 (upper part).

Then, a random number generator is used to draw at random (uniformly) a value from [0,1]. If the result is, say, $u = 0.12$ which lies in the interval $(0.105, 0.13)$ then it is derived from Table 3.1 that the microcategory $z_8 = 1250$ has been drawn according to F_Z.

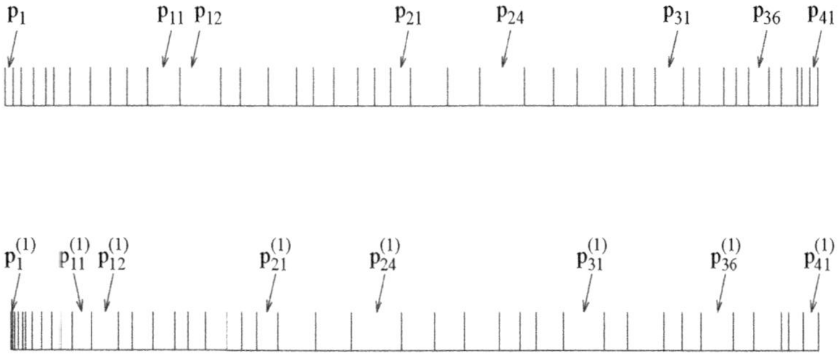

FIGURE 4.3. Graphical presentation of random drawing of earnings Z according to F_Z (upper) and $F_Z^{(1)}$ (lower), using a random number from interval (0,1)

Let us now introduce another random variable which also maps Ω_0 into the set of microcategories in the same way (i.e. the same microcategory is assigned to the same person as before), but which has a distribution other than F_Z. This variable will be called *the first moment variable for* Z and will be denoted $Z^{(1)}$. Its cdf, denoted $F_Z^{(1)}$, is defined by

$$F_{Z^{(1)}}(z) = \sum_{\{j:\sum z_j \le z\}} p_j^{(1)}, \tag{4.1}$$

where

$$p_j^{(1)} \stackrel{df}{=} \frac{N_j z_j}{\sum_i N_i z_i} = \frac{p_j z_j}{\sum_i p_i z_i} \tag{4.2}$$

(N_j is the number of people with earnings z_j, $p_j = N_j/N$). The numerator of $p_j^{(1)}$ represents the total earnings paid to these people from Ω_0 who receive z_j, the denominator is the grand total of all earnings; hence $p_j^{(1)}$ is the proportion of the former to the latter.

A random drawing according to this distribution can be modeled similarly as in case of the distribution of Z, only now each person is represented by the segment of length equal to his/her earnings divided by $\sum_i (N_i z_i)$, and hence category z_j is represented by $p_j^{(1)}$ (see Figure 4.3). In other words, the distribution in Ω_0 is now not uniform but it is related to the joint earnings of persons who belong to a certain subset. This distribution is then transferred by the function Z to form a distribution of $Z^{(1)}$ on the real line. Analogous procedures lead to variables $Z_8^{(1)}$ (with probabilities specified in Table 4.1) and $Z_4^{(1)}$.

Variable $Z^{(1)}$ is called *the first moment variable for* Z because the denominator of $p_j^{(1)}$ is the expectation (called also *first moment*) of Z while the numerator is the part of this expectation related to microcategory z_j, so we deal with the partition of total expectation. Consequently, $F_{Z^{(1)}}(z)$ can be expressed as

$$F_{Z^{(1)}}(z) = \frac{E(Z; Z \leq z)}{E(Z)} = \frac{\int_0^z t dF_Z(t)}{E(Z)}, \qquad 0 \leq z < +\infty. \tag{4.3}$$

Since the distribution of $Z^{(1)}$ is induced solely by the distribution of Z, the notation $F_Z^{(1)}$ is commonly used instead of $F_{Z^{(1)}}$:

$$F_Z^{(1)} \stackrel{df}{=} F_{Z^{(1)}}, \tag{4.4}$$

and random variable $Z^{(1)}$ with this cdf is called the *first moment variable for* Z.

The Lilliputian curve refered to a ratio random variable Z is equal to the concentration curve of $F_Z^{(1)}$ w.r.t. F_Z. This curve is called the *Lorenz Curve* of Z and denoted L_Z :

$$L_Z = C(Z^{(1)} : Z).$$

The Lorenz Curve is also called the *Inequality Curve* of Z, since it represents the inequality suffered by members of Ω_0 (e.g., the inequality in respect to their participation in earnings symbolized by values of Z).

Equal participation occurs when Z is concentrated in *one* positive point belonging to $\mathbb{R}^+$ (e.g., everybody earns the same). Since then $F_Z^{(1)} = F_Z$,

$$\text{equal participation} \Rightarrow L_Z = Diag.$$

The concentration index $ar(L_Z)$, equal to $ar(Z^{(1)} : Z)$, will be called the inequality index of Z. This index is equal to the Gini Index of Z :

$$ar(L_Z) = Gini(Z).$$

Illustration (Earnings Data):

The Lorenz Curves for earnings W, W_4, M and M_4 are presented in Fig. 4.4. The respective probability vectors are given in Table 4.1.

We see from Fig. 4.4 that $L_M \approx L_{M_4}$, $L_W \approx L_{W_4}$, so that the effect of aggregation is almost neglible, while inequalities for women are larger than inequalities for men.

The Gini Indices for male and female earnings (on different levels of aggregation) are given in Table 4.2 .

Remark 4.1 *Different aggregation of a given ratio variable leads to different Lorenz Curves (although the differences may be small). Another situation which results in usually just slightly different curves arises in statistics when different finite random samples are drawn from the same (possibly infinite in size) population Ω_0. If Z maps Ω_0 into $\Omega \subset \mathbb{R}^+$, and if the distribution F_Z of Z and its first moment distribution $F_Z^{(1)}$ are described by L_Z, then the Lorenz curves corresponding to particular random samples (not necessarily disjoint) tend to L_Z when their sizes increase to infinity.*

TABLE 4.1. Probability vectors for earnings M_8 (males), W_8 (females) Z_8 (males and females) and for the respective first moment variables $M_8^{(1)}$, $W_8^{(1)}$, $Z_8^{(1)}$,

Cate-gory	Mean earnings			Probability					
	M_8	W_8	Z_8	M_8	W_8	Z_8	$M_8^{(1)}$	$W_8^{(1)}$	$Z_8^{(1)}$
1	850	740	758	0.02	0.10	0.06	0.006	0.04	0.018
2	1217	1260	1254	0.03	0.20	0.115	0.012	0.12	0.057
3	1775	1718	1736	0.12	0.25	0.185	0.071	0.21	0.128
4	2288	2223	2246	0.08	0.15	0.115	0.061	0.16	0.103
5	2810	2690	2765	0.25	0.15	0.200	0.234	0.20	0.220
6	3225	3230	3226	0.20	0.05	0.125	0.215	0.08	0.160
7	3739	3610	3711	0.18	0.05	0.115	0.225	0.09	0.169
8	4383	4110	4303	0.12	0.05	0.085	0.176	0.10	0.145
means	2996	2040	2518						

TABLE 4.2. Gini Indices (concentration indices for the respective Lorenz Curves) for all considered earnings variables

variable	M	W	Z	M_8	W_8	Z_8	M_4	W_4	Z_4
Gini Index	.166	.248	.231	.161	.242	.226	.148	.225	.214

The reader can observe that the cdf $F_Z^{(1)}$ might be calculated whenever

$$Z : \Omega_0 \to \Omega \subset \mathbb{R}^+ \quad \text{and} \quad 0 < E(Z) < +\infty. \tag{4.5}$$

One may also rightly comment that requirements (4.5) are not sufficient to treat Z as a *measurement on the ratio scale.*

Illustration (Education Data):

The level of education T satisfies (4.5) and so there is no *technical* objection against calculating the Lorenz Curve L_T : the distribution of T is given in Table 4.3, it is accompanied by the distribution of $T^{(1)}$, and the related Lorenz Curve is presented in Fig. 4.5. But we know that T is just an *ordinal* variable, so that *any arbitrarily chosen numbers* could be assigned to the six education levels provided that they were properly *ordered.* If these numbers were chosen to be, say, 1,2,5,6,7,10 instead of 1,2,3,4,5,6, the Lorenz Curve of the corresponding random variable denoted T' would be quite different from L_T, as shown in Fig. 4.5. The difference between L_T and $L_{T'}$ follows from the fact that the function

$$(\varphi(i) = t_i', \quad i = 1, \dots, 6) = (1, 2, 5, 6, 7, 10)$$

is not admissible for the measurement on the ratio scale, and so T' is not a ratio measurement scale. Since T is ordinal, the Lorenz Curve L_T *is meaningless* although formally calculable.

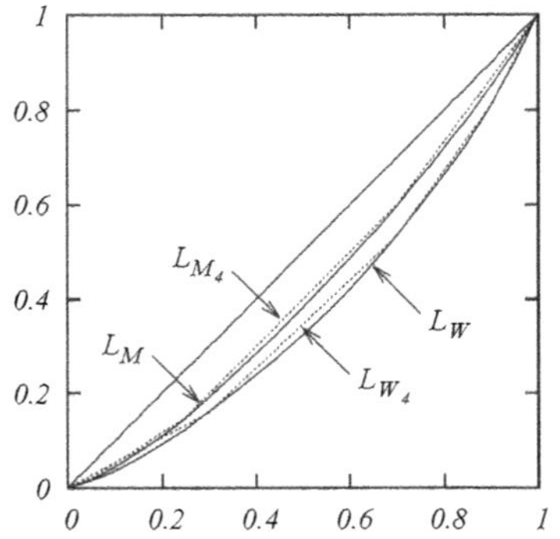

FIGURE 4.4. Lorenz curves for earnings W, W_4, M, M_4.

Illustration (Earnings Data):

Please note that Z_8 and Z_4 are ratio *in the strict sense,* but they will *not* be ratio if values $Z_{i(8)}$ and $Z_{i(4)}$ were replaced by i. This is so because the ratio of, say, $z_{1(8)}$ and $z_{8(8)}$ is 758/4303 which is not equal to the ratio 1/8, and the latter ratio has no interpretation at the empirical level.

More often than not, the statistical infrastructures used by statisticians and/or data analysts are not rigorously associated with the requirements concerning measurement scales. This is why there are only tiny links between measurement theory and statistics (as well as data analysis).

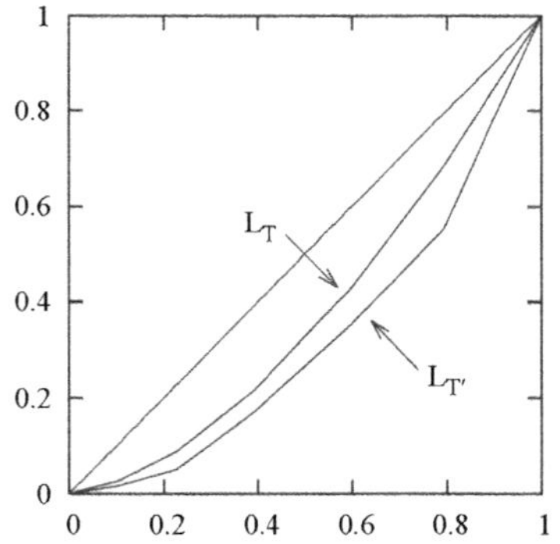

FIGURE 4.5. Lorenz Curves L_T and $L_{T'}$

TABLE 4.3. Cdf's of T and $T^{(1)}$

value of T and $T^{(1)}$	$\Pr(\{T=i\})$	$\Pr(\{T^{(1)}=i\})$
1	0.105	0.027
2	0.12	0.062
3	0.165	0.127
4	0.205	0.211
5	0.200	0.257
6	0.205	0.316
sum	1	1

The set of random variables satisfying (4.5) will be denoted $\overline{\mathbb{RATIO}}$. The line over the word $\mathbb{RATIO}$ is meant to suggest that set $\overline{\mathbb{RATIO}}$ is an *extension* of the set of variables which are ratio *in the strict sense;* any variable X which is ratio *in the strict sense* belongs to $\overline{\mathbb{RATIO}}$ but not vice versa.

Note that set $\overline{\mathbb{RATIO}}$ contains all Lilliputian variables with the exception of a variable concentrated at point zero. Recall that the Lilliputian first moment variable $\xi^{(1)}$, its cdf $F_\xi^{(1)}$ and density $f_\xi^{(1)}$ have been introduced in Sec. 3.2.3. These concepts are now extended onto the set $\overline{\mathbb{RATIO}}$.

4.2.2 First moment distribution and Lorenz curve

This section contains a more formal approach to the considerations of the previous section. It may be omitted by readers who found the information from Section 4.2.1 sufficient to understand what a Lorenz Curve and Gini Index are, and the same concerns Section 4.2.3 while the material of Section 4.2.4 could be reduced to the illustrative examples: this will be enough to understand the important overrepresentation and underrepresentation diagrams in Section 4.2.5.

Definition 4.1 *(the first moment distribution). For any $Z \in \overline{\mathbb{RATIO}}$, the distribution with the cdf*

$$\left(\frac{E(Z;Z\le z)}{E(Z)}, \quad z\in\mathbb{R}^+\right),$$

where

$$\frac{E(Z;Z\le z)}{E(Z)} = \frac{1}{E(Z)}\int_0^z x\,dF_Z(x), \tag{4.6}$$

will be called the first moment distribution of Z; a random variable with this cdf will be called the first moment random variable w.r.t. Z and will be denoted $Z^{(1)}$. The cdf of $Z^{(1)}$ will be denoted $F_Z^{(1)}$.

Variable $Z^{(1)}$ belongs to $\overline{\mathbb{RATIO}}$ if the second moment of Z exists, since

$$E(Z^{(1)}) = E(Z^2)/E(Z).$$

If Z is discrete with values $z_1, ..., z_k$ and probabilities $p_1, ..., p_k$ then $Z^{(1)}$ is discrete with values z_i and probabilities $z_i p_i / E(Z)$, $i = 1, ..., k$. If Z is continuous with density $f(z)$ then $Z^{(1)}$ is continuous with density

$$f_{Z^{(1)}}(z) = \frac{z f(z)}{EZ}. \tag{4.7}$$

Generally, $Z^{(1)}$ has the same support as Z if $Z > 0$ with probability 1.

Definition 4.2 *(The Lorenz Curve). Let $Z \in \overline{\mathbb{RATIO}}$. The concentration curve $C(Z^{(1)} : Z)$ is called the Lorenz Curve of Z and denoted L_Z.*

L_Z is convex since its density is nondecreasing: its value at point u is equal to $z_u/E(Z)$, where z_u is the uth quantile of Z i.e., $z_u = \inf\{z : F_Z(z) \geq u\}$. This is seen from (4.6) and (4.7).

Remark 4.2 *The pair $(Z; Z^{(1)})$ is C-equivalent to $(U; \zeta)$ where ζ is a Lilliputian variable with cdf equal to L_Z. Please note that L_Z (the Lorenz Curve of Z) IS NOT EQUAL to L_ζ (the Lorenz Curve of ζ) i.e., $C(Z^{(1)} : Z) \neq C(\zeta^{(1)} : \zeta)$.*

Example illustrating Remark 4.2. Let Z be a random variable taking values $1/4$ and $3/4$ with probabilities 0.2 and 0.8, respectively. Then $E(Z) = 13/20$, and $Z^{(1)}$ is valued $1/4$ and $3/4$ with probabilities $1/13$ and $12/13$, respectively. The Lorenz Curve $L_Z = C(Z^{(1)} : Z)$ is calculated as follows:

$$L_Z(u) = \begin{cases} \frac{5}{13}u & \text{if } 0 \leq u \leq 0.2, \\ \frac{15}{13}u - \frac{2}{13} & \text{if } 0.2 \leq u \leq 1. \end{cases} \tag{4.8}$$

The random variable ζ with cdf equal to L_Z has $E(\zeta) = 73/130$, and $\zeta^{(1)}$ has cdf equal to

$$F_\zeta^{(1)}(u) = \begin{cases} \frac{25}{73}u^2 & \text{if } 0 \leq u \leq 0.2, \\ \frac{75}{73}u^2 - \frac{2}{73} & \text{if } 0.2 \leq u \leq 1, \end{cases}$$

and consequently the Lorenz Curve $L_\zeta = C(\zeta^{(1)} : \zeta)$ is of the form

$$L_\zeta(u) = \begin{cases} \frac{169}{73}u^2 & \text{if } 0 \leq u \leq \frac{1}{13}, \\ \frac{169}{219}u^2 + \frac{52}{219}u - \frac{2}{219} & \text{if } \frac{1}{13} \leq u \leq 1. \end{cases} \tag{4.9}$$

The Lorenz Curves L_Z and L_ζ given by (4.8) and (4.9) are presented in Figure 4.6.

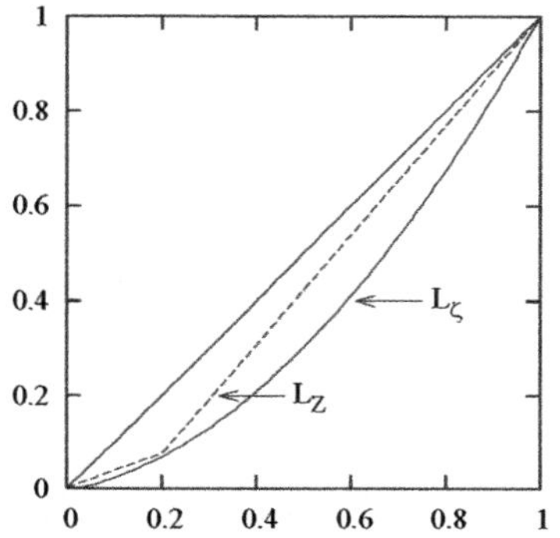

FIGURE 4.6. Lorenz Curves L_Z and L_ζ

The Lorenz Curve is invariant w.r.t. transformations admissible for a ratio scale due to the following fact:

Proposition 4.1 *Let $Z, Z' \in \overline{\mathbb{RATIO}}$. Then*
(i) $L_Z = L_{Z'}$ iff $F_Z(0) = F_{Z'}(0)$ and there exists $a > 0$ such that $F_{(Z'|Z'>0)} = F_{(aZ|Z>0)}$.
(ii) $ar(L_Z) = Gini(Z)$.

The proof of Proposition 4.1 is in the Appendix, Section 4.5.2.

Definition 4.3 *(two points local inequality). Let $Z \in \overline{\mathbb{RATIO}}$ and let $z < z'$, $z, z' \in \mathbb{R}^+$. Local inequality for $(Z; z, z')$ is defined as*

$$f_Z(z)f_Z^{(1)}(z') - f_Z(z')f_Z^{(1)}(z) = \frac{(z'-z)f_Z(z)f_Z(z')}{E(Z)}, \tag{4.10}$$

which for discrete variable Z changes to

$$\frac{(z'-z)p(z)p(z')}{E(Z)}. \tag{4.11}$$

It follows that local inequality is a special case of local concentration (definition 3.9), we remind that $Z^{(1)}$ is absolutely continuous w.r.t. Z).

Local inequality is non-negative (positive) which means that the pair $(Z; Z^{(1)})$ has an increasing likelihood ratio (see definition 3.10).

According to formula (3.29), the Gini Index for $Z \in \overline{\mathbb{RATIO}}$ can be expressed by means of local inequalities:

$$Gini(Z) = \int_0^{\infty} \int_0^{z'} \left(dF_Z(z) dF_Z^{(1)}(z') - dF_Z(z') dF_Z^{(1)}(z) \right). \tag{4.12}$$

So for a continuous Z

$$Gini(Z) = \int_{-\infty}^{\infty} \int_{-\infty}^{z'} \left(f_Z(z) f_Z^{(1)}(z') - f_Z(z') f_Z^{(1)}(z) \right) dz dz',$$

and a similar formula exists for discrete Z.

The Gini Index averages the local inequalities, while the shape of the Lorenz Curve shows where the local inequalities are larger and where they are smaller. The latter is illustrated by the shapes of Lorenz Curves obtained before through analyzing the earnings data (see Fig. 4.4).

4.2.3 Lorenz Curves with horizontal and/or vertical segments

Illustration (Extended Earnings Data):

Suppose that the set of 200 employees with earnings presented in Table 3.1 is supplemented by 5 owners of the enterprise. The enlarged set will be denoted $\Omega_0{}'$. For each member of this set, at the end of a year, it is calculated how much money he/she received during the year and how much time he/she spent in the office. The *money received/time spent ratio*, denoted Z, is calculated for each person. For those owners who never came to the office, $+\infty$ is put as the value of Z.

Obviously, the persons with infinite value of Z (at most five owners, possibly less) can be separated from the others. Then one deals with a variable Z *truncated to finite values*, which will be denoted $Z\mid_{\mathbb{R}+}$. This variable is ratio *in the strict sense* and belongs to set $\overline{\mathbb{RATIO}}$.

The set of random variables which are valued in $\{\mathbb{R}^+ \cup \{+\infty\}\}$ and belong to set $\overline{\mathbb{RATIO}}$ when truncated to $\mathbb{R}^+$, will be denoted $\overline{\mathbb{RATIO}_{\infty}}$.

The expectation of any $Z \in \overline{\mathbb{RATIO}}_\infty$ might be infinite and therefore the first moment distribution of Z cannot be introduced according to definition 4.1. As an obvious extension of this definition for $Z \in \overline{\mathbb{RATIO}}_\infty$, we put:

$$F_Z^{(1)}(z) = \begin{cases} (1-\alpha)F_{Z|_{\mathbb{R}+}}^{(1)}(z) & \text{if } z < +\infty, \\ 1 & \text{if } z = +\infty, \end{cases} \tag{4.13}$$

where

$$\alpha = \Pr(Z = +\infty) = 1 - \Pr(Z < +\infty). \tag{4.14}$$

For $Z \in \overline{\mathbb{RATIO}}_\infty$, the concentration curve $C(F_Z^{(1)} : F_{Z|_{\mathbb{R}+}})$ will be called the *Lorenz Curve of Z* and denoted L_Z. Clearly, $\alpha > 0$ implies that the Lorenz Curve satisfies $L(1-) = 1 - \alpha$ and therefore there appears a vertical segment joining point $(1, 1-\alpha)$ with point $(1,1)$ at the right end of Lorenz curve L_Z.

A horizontal segment $[0, u_0]$ for $u_0 > 0$ appears in the Lorenz Curve (at its left end) if $F_Z(0) = u_0 > 0$, since then $F_Z^{(1)}(0) = 0$.

Now, we turn to the reciprocal $1/Z$ of a variable Z from $\overline{\mathbb{RATIO}}_\infty$ (in the Extended Earnings Data variable $1/Z$ it is *the time spent/money received ratio*). We have

$$F_{\frac{1}{Z}}^{(1)}(z) = \frac{E(\frac{1}{Z}; \frac{1}{Z} \le z)}{E(\frac{1}{Z})}.$$

If $\alpha = \Pr(Z = +\infty) > 0$ then $F_{\frac{1}{Z}}(0) = \Pr((1/Z) = 0) > 0$, $F_{\frac{1}{Z}}^{(1)}(0) = 0$ and the Lorenz Curve $L_{1/Z}$ has a horizontal segment at the left end, which refers to the vertical segment of Lorenz Curve L_Z at the right end.

Fig. 4.7 shows the shape of L_Z for a variable $Z \in \overline{\mathbb{RATIO}}_\infty$ such that $\beta = \Pr(Z = 0) > 0$ and $\alpha = \Pr(Z = +\infty) > 0$.

It follows that if $Z \in \overline{\mathbb{RATIO}}$ and $\Pr(Z = 0) > 0$ then $1/Z$ results in the value $+\infty$ with positive probability, so that $1/Z \in \overline{\mathbb{RATIO}}_\infty$.

Important examples of variables from $\overline{\mathbb{RATIO}}_\infty$ (called *overrepresentation* and *underrepresentation*) will be given in the next sections.

The Lorenz Curve is a *specific concentration curve* attached to any variable from $\overline{\mathbb{RATIO}}_\infty$.
Following the approach according to which equal concentration curves induce equivalent pairs of variables (definition in Sec. 3.3.2), we will treat *ratio variables with equal Lorenz Curves as equivalent.*

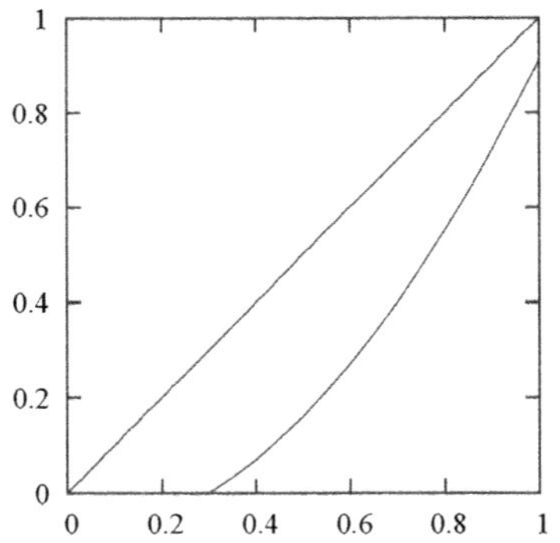

FIGURE 4.7. Lorenz Curves L_Z for a variable $Z \in \overline{\mathbb{RATIO}}_\infty$ when $\beta = \Pr(Z = 0) > 0$ and $\alpha = \Pr(Z = +\infty) > 0$

Definition 4.4 *(Lorenz equivalence). Let Z and Z' belong to $\overline{\mathbb{RATIO}}_\infty$. We say that Z and Z' are equivalent if $L_Z = L_{Z'}$.*

Proposition 4.2 *For $Z, Z' \in \overline{\mathbb{RATIO}}_\infty$, $L_Z = L_{Z'}$ iff $F_Z(0) = F_{Z'}(0)$, $\Pr(Z = +\infty) = \Pr(Z' = +\infty)$ and there exists $a > 0$ such that the cdf of $(Z'|Z' > 0)$ is the same as that of $(aZ|Z > 0)$.*

4.2.4 The variable called overrepresentation and its Lorenz curve

This section builds a bridge between measures of inequality of Z from $\overline{\mathbb{RATIO}}_\infty$ and measures of concentration applied to a pair of random variables. We shall illustrate this bridge in the case of the male and female earnings M and W.

Illustration (Earnings Data):
The concentration curve $C(M : W)$ was presented in Fig. 3.9 in Chapter 3. Recall that $W = (Z|S = 1)$, $M = (Z|S = 2)$ (and the same convention will be used later for Z_8 and Z_4). The Lilliputian variable ξ with cdf defined by curve $C(M : W)$ is partly continuous and partly discrete. Its density f_ξ satisfies

$$\int_0^1 f_\xi(u)du = 1 - \alpha_\xi = 0.81,$$

while the total length of *vertical* segments (see formula (3.6)) of $C(M : W)$ is

$$\alpha_\xi = \sum_i p_\xi(u_i) = 0.04 + 0.03 + 0.03 + 0.09 = 0.19.$$

The total length of the five *horizontal* segments of $C(M:W)$ (at which $f_\xi(u) = 0$), is equal to

$$\beta_\xi = \Pr(\{f_\xi(U) = 0\}) = 0.35. \tag{4.15}$$

Probabilities α_ξ, β_ξ and density f_ξ will be used now to define a random variable called *overrepresentation* of ξ relative to U, denoted $OV_{\xi:U}$.

Let ξ be a Lilliputian variable with the cdf lying on a concentration curve $C(Y:X)$. The likelihood ratio of ξ to U was defined in Chapter 3 by formula (3.54). So $h_{\xi:U}$ is a function from [0,1] onto $\mathbb{R}^+$. Its value at u shows how ξ *overrepresents* U at this point (since $f_U(u) \equiv 1$). The likelihood ratio of ξ to U is a (function-valued) parameter in $\mathbb{ULM}$.

Remark 4.3 *Our aim in this section and the next one is to compare ξ and U using the likelihood ratio $h_{\xi:U}$. This will be obtained by comparing the transforms $h_{\xi:U}(U)$ and $h_{\xi:U}(\xi)$. These transformations - and their extensions to any random variables X and Y - underline the concepts of overrepresentation and underrepresentation of one variable relative to another.*

Now, let us take into account $h_{\xi:U}(U)$. Note that this random variable takes on *finite* values with probability 1. In case of $\alpha_\xi = 0$ this variable is excellent *to represent the overrepresentation of ξ relative to U* . If however $\alpha_\xi > 0$, we have to take this fact into account in our definition of overrepresentation. Therefore, the cdf of the variable $OV_{\xi:U}$, called *the overrepresentation of ξ relative to U*, is given by:

Definition 4.5 *(overrepresentation of ξ relative to U). A random variable from $\overline{\mathbb{RATIO}}_\infty$ is called the overrepresentation of a Lilliputian variable ξ relative to U, and denoted $OV_{\xi:U}$, if*

$$F_{OV_{\xi:U}}(z) = \begin{cases} (1-\alpha_\xi)F_{h_{\xi:U}(U)}(z) & \text{if } z < +\infty, \\ 1 & \text{if } z = +\infty, \end{cases} \tag{4.16}$$

where $\alpha_\xi = \sum_i p_\xi(u_i) = 1 - \int_0^1 f_\xi(u)du = 1 - E(f_\xi(U))$.

It is clear that we can write $F_{f_\xi(U)}$ instead of $F_{h_{\xi:U}(U)}$. Thus $OV_{\xi:U}$ is equal to variable $f_\xi(U)$ with probability $1-\alpha_\xi$ and takes the value $+\infty$ with probability α_ξ (i.e. $OV_{\xi:U}|_{\mathbb{R}} = f_\xi(U)$). Hence

$$F_{OV_{\xi:U}}(z) = (1-\alpha_\xi)F_{f_\xi(U)}(z) + \alpha_\xi I_{\{z=+\infty\}}(z)$$

where $I_{\{z=+\infty\}}(z) = 0$ if $z < +\infty$, and $I_{\{z=+\infty\}}(z) = 1$ if $z = +\infty$.

Although definition 4.5 looks complicated, it turns out to be quite simple when $\alpha_\xi = 0$:

> If $\alpha_\xi = 0$ then
>
> $$OV_{\xi:U} = h_{\xi:U}(U) = f_\xi(U)$$
>
> (since $\alpha_\xi = 0$ implies that ξ is continuous).

Variable $f_\xi(U)$ is a grade parameter (see Section 3.4.3), namely it is attached to each pair $(X;Y)$ which is C-equivalent to $(U;\xi)$. Generally, it was said in Section 3.4.3 that variable $h_{\xi:U}(U)$ is equivalent to $h_{Y:X}(X)$ for any pair $(X;Y)$ which is C-equivalent to $(U;\xi)$. (The definition of $h_{Y:X}$ was given in Section 3.4.3 by formula (3.8)).

Details: We have the equality $h_{Y:X}(x) = h_{\xi:U}(u) < +\infty$ if $F_X(x) - F_X(x-) > 0$ and $u \in (F_X(x-), F_X(x))$ or if x is a point of continuity of F_X and F_Y for which there exist densities $f_X(x) > 0$, $f_Y(x) \geq 0$ and $u = F_X(x)$. This legitimizes the definition of the overrepresentation of Y on X which is analogous to Definition 4.5 (i.e., the overrepresentation of $OV_{\xi:U}$).

Definition 4.6 *(the overrepresentation variable). A random variable from* $\mathbb{RATIO}_\infty$ *is called the overrepresentation of* Y *relative to* X *and denoted* $OV_{Y:X}$ *if*

$$F_{OV_{Y:X}}(z) = \begin{cases} (1-\alpha_{Y:X})F_{h_{Y:X}(X)}(z) & \text{if } z < +\infty, \\ 1 & \text{if } z = +\infty, \end{cases} \tag{4.17}$$

where $\alpha_{Y:X} = 1 - E(h_{Y:X}(X)) = P_Y(h_{X:Y}(Y) = 0)$.

We can rewrite 4.17 as

$$F_{OV_{Y:X}}(z) = (1-\alpha_{Y:X})F_{h_{Y:X}(X)}(z) + \alpha_{Y:X} I_{\{z=+\infty\}}(z)$$

> Formulas (4.16) and (4.17) for $F_{OV_{\xi:U}}$ and $F_{OV_{Y:X}}$ are identical. So, $OV_{Y:X}$ and $OV_{\xi:U}$ treated as variables on a real line are identical. Please keep in mind that
>
> $$U = F_X^*(X), \quad \xi = F_X^*(Y),$$
>
> where F_X^* is the uniform grade transformation for X (see Section 3.3.3). In general, overrepresentation of one variable over another is *invariant w.r.t. a grade transformation* when such transformation is applied simultaneously to variables X and Y.

Definition 4.6 refers to the pair $(X;Y)$. When this definition is used for pair $(Y;X)$, we deal with *the overrepresentation of X relative to Y*, denoted $OV_{X:Y}$. The links between $OV_{Y:X}$, $OV_{X:Y}$, their reciprocals, and first moment variables will be considered in the next section.

The following Remark 4.4 can be omitted by a reader oriented mainly to applications.

Remark 4.4 *Let $(U;\xi) \equiv_C (X;Y)$. Then*

$$\begin{aligned} &OV_{Y:X} = OV_{\xi:U} = OV_{F_X^*(Y):U}, \\ &OV_{\xi:U}|_{\mathbb{R}} = f_\xi(U), \\ &OV_{X:Y} = OV_{U:\xi} = OV_{F_\xi^*(U):U} = OV_{F_Y^*(X):U}, \end{aligned}$$

and

$$f_{F_\xi^*(U)}(u) = \frac{1}{f_\xi(C(U:\xi))(u)}.$$

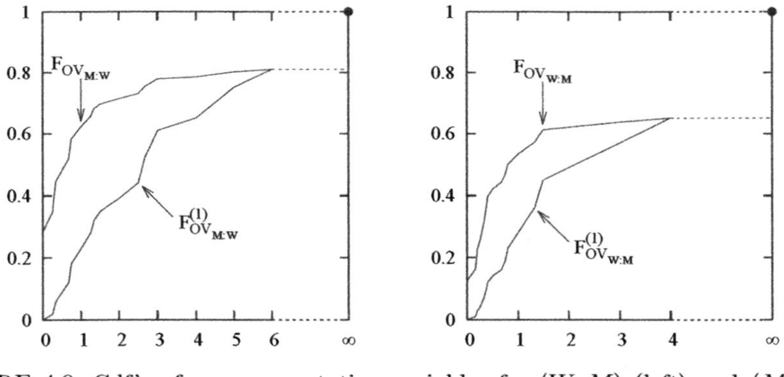

FIGURE 4.8. Cdf's of overepresentation variables for $(W;M)$ (left) and $(M;W)$ (right)

Illustration (Earnings Data):
We will illustrate the contents of this section by presenting the overrepresentation of male-to-female earnings and of female-to-male earnings (i.e. $OV_{M:W}$ and $OV_{W:M}$) using Lilliputian variable ξ corresponding to $(W;M)$.

The cdf's of $OV_{M:W}$ and $OV_{W:M}$ are presented in Fig. 4.8. The cdf of $OV_{M:W}$ has a jump at point zero equal to $(1-\alpha_\xi)\beta_\xi = 0.81 \times 0.35 = 0.2835$, which corresponds to $(1-\alpha_\xi)$ multiplied by the sum of the probabilities of earnings for categories in which *only women have earnings*; the jump at point $x = +\infty$ is equal to $\alpha_\xi = 0.19$.

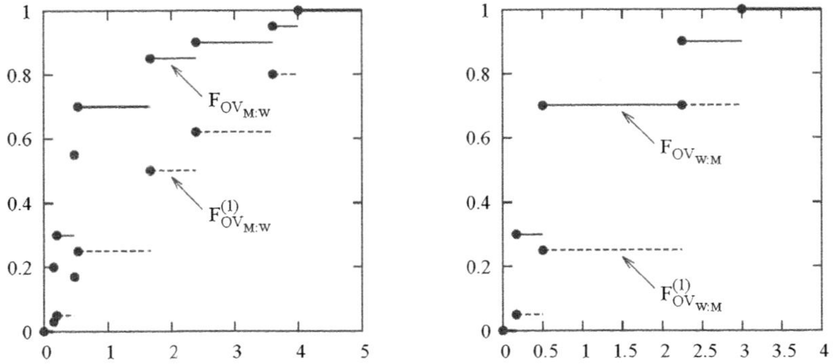

FIGURE 4.9. Cdf's of overrepresentation variables for $(W_8; M_8)$ (left) and $(M_8; W_8)$ (right)

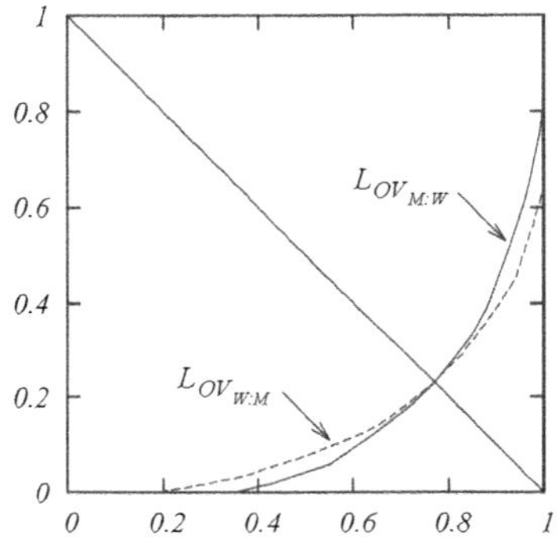

FIGURE 4.10. Lorenz Curve $L_{OV_{M:W}}$ and $L_{OV_{W:M}}$

To compute the Lorenz Curve for $OV_{M:W}$, we calculate first the distribution of $OV^{(1)}_{M:W}$ according to the formula (4.3) (see Fig. 4.8 and also 4.9 corresponding to (W_8, M_8)). The concentration curve $L_{OV_{M:W}} = C(OV^{(1)}_{M:W} : OV_{M:W}|_{\mathbb{R}})$ is shown in Fig. 4.10. It is a convex curve lying below $C(M : W)$, which starts with a horizontal segment of length 0.35, and ends with a vertical segment of length 0.19. A comparison of $L_{OV_{M:W}}$ and $L_{OV_{W:M}}$ is given in Fig. 4.10. It is seen that $L_{OV_{M:W}}$ is symmetric to $L_{OV_{W:M}}$ w.r.t. the diagonal $Diag^{\perp}$.

The differences between the overrepresentation variables, due to different aggregation, were measured by the Gini Indices as shown in Table 4.4. Their values do not depend on the order (male-to-female or female-to-male earnings) but is very sensitive on the aggregation level until aggregation is strong enough.

TABLE 4.4. Gini Indices for overrepresentation variables of male earnings relative to females earnings (or vice versa)

variable	Gini Index
$OV_{M:W}$	0.726
$OV_{W:M}$	as above
$OV_{M_8:W_8}$	0.539
$OV_{W_8:M_8}$	as above
$OV_{M_4:W_4}$	0.505
$OV_{W_4:M_4}$	as above

4.2.5 *Diagram of over- and underrepresentation*

Definition 4.7 *(underrepresentation). The reciprocal of the overrepresentation of Y relative to X will be called the underrepresentation of Y relative to X (denoted* $1/OV_{Y:X}$*).*

The underrepresentation of Y on X belongs to $\overline{\mathbb{RATIO}}_{\infty}$ and is equal to the value $+\infty$ with probability β_{ξ} (i.e. it is $+\infty$ when $OV_{Y:X} = 0$).

Let us consider the over and underrepresentation of Y relative to X *and* of X relative to Y. We obtain in this case four random variables interrelated as shown in the diagram presented in Fig. 4.11 (for pair $(X;Y)$ and for pair $(Y;X)$).

> In the diagram, the columns will be interchanged when
> 1° the word "underrepresentation" is replaced by "overrepresentation" (and vice versa),
> 2° the symbol OV is replaced by $1/OV$ (and vice versa),
> 3° the pair $(X;Y)$ is replaced by $(Y;X)$ (and vice versa).
> To interchange rows, it is only necessary to replace $(X;Y)$ by $(Y;X)$ (and vice versa).

Remark 4.5 *Note in the diagram that a consideration of the over- and underrepresentation of male relative to female earnings is equivalent to the respective consideration of these concepts concerning female-to-male earnings. This is not only a desirable but also a necessary property; otherwise these concepts would be meaningless.*

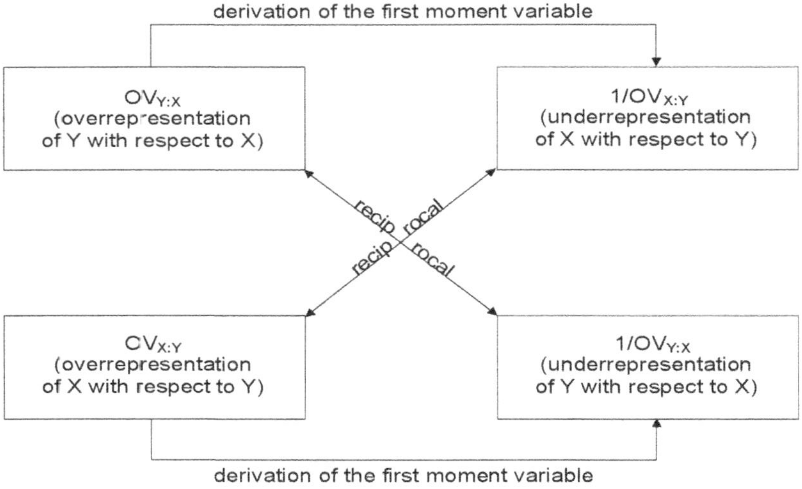

FIGURE 4.11. Diagram of overrepresentation and underrepresentation for the pairs of random variables $(X;Y)$ and $(Y;X)$

The condition $\alpha_\xi = 0$ for a Lilliputian variable ξ representing a considered pair $(X;Y)$ implies that the diagram in Fig. 4.11 is reduced to diagram in Fig. 4.12.

We recall that

$$\begin{bmatrix} h_{\xi:U}(U) & h_{\xi:U}(\xi) \\ h_{U:\xi}(\xi) & h_{U:\xi}(U) \end{bmatrix} = \begin{bmatrix} f_\xi(U) & f_\xi(\xi) \\ \frac{1}{f_\xi(\xi)} & \frac{1}{f_\xi(U)} \end{bmatrix}$$

Remark 4.6 *The diagram in Fig. 4.11 is particularly simple when it refers to the pair $(U;\xi)$, where ξ is a Lilliputian variable such that $\alpha_\xi = 0$ (see Fig. 4.12). Then $h_{\xi:U}(\xi)$ (equal to $f_\xi(\xi)$) is the first moment variable for $h_{\xi:U}(U)$ (equal to $f_\xi(U)$).*

Whenever $\alpha_\xi = 0$, diagram in Fig. 4.11 presents graphically the following statement: *underrepresentation* of one ratio variable relative to a second one, is the first moment variable for *overrepresentation* of the second variable relative to the first one. The general case needs a truncation (qualification regarding infinite values) shown in the following theorem. The theorem clarifies the phrase "derivation of the first moment variable" which appears in the diagram.

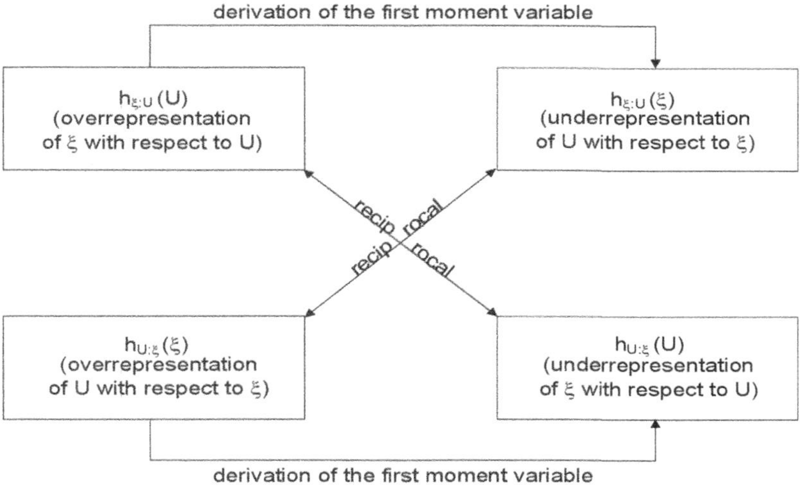

FIGURE 4.12. Diagrams of overrepresentation and underrepresentation for the pair $(U;\xi)$ in the case $\alpha_\xi = 0$

Theorem 4.1 *For any pair $(X;Y)$, its over- and underrepresentation satisfy:*

$$\frac{1}{OV_{Y:X}|_{\mathbb{R}}} = (OV_{X:Y})^{(1)}; \quad \frac{1}{OV_{X:Y}|_{\mathbb{R}}} = (OV_{Y:X})^{(1)}.$$

The four variables in the diagram are all identical if and only if $OV_{Y:X}$ is *concentrated at point 1* since there is then neither *over* nor *under* representation of one variable relative to a second one: their cdf's are identical. This happens iff the distributions of X and Y are identical.

If $\alpha_{Y:X} = 0$, overrepresentation concentrates *around* 1 since 1 is the *expectation of its distribution* (see Corollary 4.2 below). This condition concerning the expectation cannot be preserved under nominal, ordinal, interval or ratio scale: it is preserved only if identity is the unique admissible function. Therefore we treat overrepresentation as a measurement on the absolute scale (cf Section 4.5.1 in the Appendix).

Formally, both: overrepresentation $OV_{Y:X}$ and underrepresentation $1/OV_{Y:X}$ belong to $\mathbb{RATIO}_\infty$. It is therefore possible to describe them using inequality measures: Lorenz Curve and Gini Index.

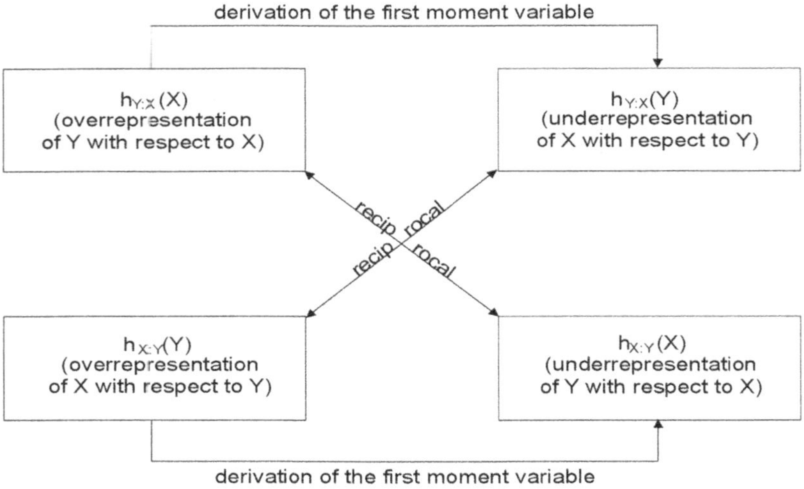

FIGURE 4.13. Diagram of overrepresentation and underrepresentation for the pairs $(X;Y)$ and $(Y;X)$ in the case when $\alpha_{Y:X}=0$

Illustration (Earnings Data)
$OV_{M_4:W_4}$ and $1/OV_{M_4:W_4}$ belong to $\overline{\mathbb{RATIO}}$, $OV_{M_8:W_8}$ and $1/OV_{M_8:W_8}$ belong to $\overline{\mathbb{RATIO}}$, while $OV_{M:W}$ and $1/OV_{M:W}$ belong to $\overline{\mathbb{RATIO}}_{\infty}$).

Corollary 4.2 *(i)* $E_X(OV_{Y:X}|_{\mathbb{R}}) = 1-\alpha_{Y:X}$;
(ii)

$$\begin{aligned} L_{OV_{Y:X}} &= C(OV^{(1)}_{Y:X} : OV_{Y:X}|_{\mathbb{R}}) \\ &= C\left(\frac{1}{OV_{X:Y}|_{\mathbb{R}}} : OV_{Y:X}|_{\mathbb{R}}\right), \end{aligned}$$

$$L_{OV_{Y:X}} = C\left(OV^{(1)}_{Y:X} : \frac{1}{OV^{(1)}_{X:Y}}\right);$$

(iii) $Gini(OV_{Y:X}) = Gini(OV_{X:Y})$.

Remark 4.7 *If* $E_X(OV_{Y:X}) = 1$, *(i.e.* $\alpha_{Y:X}=0$*) then the overrepresentation of* Y *relative to* X *remains constant when* X *is replaced by* $OV_{Y:X}$ *while* Y *is replaced by* $OV^{(1)}_{Y:X}$ *(c.f. formula (4.6)). This means that function* $f : \mathbb{RATIO} \to \mathbb{RATIO}$ *such that* $f(Z) = OV_{Z^{(1)}:Z}$ *is the identity* $f(Z)=Z$*, which is concordant with the fact that* $OV_{Y:X}$ *is a measurement on the absolute scale.*

The link between $OV_{Y:X}$ and $OV_{X:Y}$ will be clarified in Section 4.4 concerning dual variables.

4.2.6 Lorenz Curve and Gini Index for density transform of categorical variables

Overrepresentation $OV_{Y:X}$ transforms any pair of random variables, defined on the same set of objects and measured on arbitrary scales of the same type: e.g. both nominal, both ordinal, etc, into a random variable which can be treated as measured on absolute scale. A similar action can be undertaken in the case of categorical variables which can be transformed by their probabilities into variables measured on absolute scale, and then can be described by parameters suitable for this scale.

Let X be a categorical variable, assuming values $x_1, \dots, x_m$ with probabilities $p_1, \dots, p_m$. Let p_X be the function defined on the set $\{x_1, \dots, x_m\}$, valued in [0,1], such that

$$p_X(x_i) = p_i, \quad i = 1, \dots, m,$$

and let

$$Z = p_X(X).$$

Then Z is a discrete Lilliputian variable measured on the absolute scale, with cdf F_Z given by

$$F_Z(z) = \sum_{\{i: p_X(x_i) \leq z\}} p_X(x_i), \quad z \in [0,1]$$

and with the expectation $E(Z) = \sum_{i=1}^{m} p_X^2(x_i)$.

The first moment variable $Z^{(1)}$ takes values (the same as variable Z) in the interval [0,1] with probabilities $p_Z^{(1)}(z) = z dF_Z(z)/E(Z)$. So the cdf $F_Z^{(1)}$ is equal to

$$F_Z^{(1)}(z) = \frac{E(p_X^2(X); p_X(X) \leq z)}{E(Z)}$$

and so the Lorenz Curve $L_Z = C(F_Z^{(1)} : F_Z)$ is

$$L_Z(u) = F_Z^{(1)}(z_{i-1}) + \frac{z_i(u - F_Z(z_{i-1}))}{E(Z)}, \quad u \in [F_Z(z_{i-1}), F_Z(z_i)] \quad (4.18)$$

for $i = 1, \dots, m(X)$, where $m(X)$ is the number of different values of Z ($m(X) \leq m$). Moreover, putting $q_i = \Pr(Z = z_i)$, $i = 1, ..., m(X)$, according to formula (4.12) we obtain

$$
\begin{aligned}
Gini(Z) &= \frac{1}{2}\sum_{i=1}^{m(X)}\sum_{j=1}^{m(X)} |z_j - z_i| q_i q_j / E(Z) \qquad (4.19)\\
&= \frac{1}{2E(Z)}\sum_{i=1}^{m}\sum_{j=1}^{m} |p_X(x_i) - p_X(x_j)| p_X(x_i) p_X(x_j)\\
&= \sum_{i=1}^{m} p_{j_i}^2 (S_{j_i} + S_{j_{i-1}} - 1)/E(Z)\\
&= \sum_{i=1}^{m} p_{j_i}^2 (S_{j_{i-1}} - Z_{j_{i+1}})/E(Z)\\
&= 2cov(Z, \tilde{F}_Z(Z))/E(Z)
\end{aligned}
$$

where $(j_1, \ldots, j_m)$ is a permutation of numbers $(1, \ldots, m)$ such that $p_{j_1} \leq \cdots \leq p_{j_m}$, and $S_{j_i} = p_{j_1} + \cdots + p_{j_i}$, $Z_{j_i} = p_{j_i} + \cdots + p_{j_m}$.

Variables $p_X(X)$ and their parameters will be used afterwards in Chapters 8,9, and 12.

Remark 4.8 *Note that the transformation p_X represents an important class of transformations of random variables which refer to their generalized densities. For a continuous random variable X with density f_X, the transformed variable $f_X(X)$ is measured on the absolute scale. For a discrete variable we have the transformed variable $p_X(X)$ measured on the absolute scale. Generally, for any discrete-continuous variable, it is natural to denote such transformation $dens_X$. This class of transformations should be perhaps given attention similar as is given to the class of grade transformations (which also lead to variables measured on the absolute scale).*

In the present book we use only either p_X or f_X; the variables $p_X(X)$ and $f_X(X)$ are called density transforms of X; their parameters (such as the Gini Index) are referred to as the Gini Index of the density transform of X, etc.

4.3 Order oriented concentration curves

4.3.1 Basic definitions

Until now, we considered concentration curves for any two cdf's of random variables X, Y. They were denoted $C(Y : X)$, or $C(F_Y : F_X)$, or $C(Q : P)$, where Q and P were the probability measures on the real line and corresponded to Y and X.

Now, we will consider two probability measures P and Q defined on sets Ω_1 and Ω_2 (more precisely, on probability spaces $(\Omega_1, \mathcal{B}(\Omega_1))$ and $(\Omega_2, \mathcal{B}(\Omega_2))$ where $P(\Omega_0 \setminus \Omega_1) = Q(\Omega_0 \setminus \Omega_2) = 0$. A linear order $\preceq$ in the set $\Omega = \Omega_1 \cup \Omega_2$ will be specified.

Remark 4.9 *Ω_1 and Ω_2 are not necessarily disjoint (even they can be equal).*

We are going to define a concentration curve of Q w.r.t. P according to order $\preceq$.

This curve and the corresponding concentration index will be denoted $C(Q : P; \preceq)$ and $ar(Q : P; \varphi)$.

Illustration (Earnings Data):

Sets Ω_1 (set of females) and Ω_2 (set of males) are disjoint. As both sets are finite, *all* subsets are admitted as $\mathcal{B}(\Omega_1)$ and $\mathcal{B}(\Omega_2)$. Each person is "equally probable" or "equally weighted", i.e., $P(\{\omega_i^{(1)}\}) = 1/N_1$, $P(\{\omega_j^{(2)}\}) = 0$, $Q(\omega_j^{(2)}) = 1/N_2$, $Q(\{\omega_i^{(1)}\}) = 0$, $i = 1, \ldots, N_1$, $j = 1, \ldots, N_2$ where N_1 and N_2 are the sizes of Ω_1 and Ω_2.

The order introduced in the dataset by *earnings* Z (defined on the sum of Ω_1 and Ω_2, i.e. when males and females are pulled together) is given by:

$$\omega \preceq \omega' \quad \text{if} \quad Z(\omega) \leq Z(\omega').$$

In this case Z is called *the order function* corresponding to $\preceq$. This function induces two random variables: in (Ω_1, P) and in (Ω_2, Q), denoted W and M. Therefore, in this case,

$$C(Q : P; \preceq) = C(M : W).$$

One might also obtain the concentration curve $C(M : W)$ starting directly from $(\Omega_0, \mathcal{B}(\Omega), P_{\Omega_0})$, where $\Omega_0 = \Omega_1 \cup \Omega_2$, $P_{\Omega_0}(\omega) = 1/N$ and $N = N_1 + N_2$. Then random variable Z is defined on Ω_0, $M = Z|_{\Omega_1}$, $W = Z|_{\Omega_2}$.

Instead of earnings Z, it is possible to use *the rank function* $\textit{RANK}$ which assigns to each person in Ω_0 its position (rank) according to Z.

Due to ties, the values of $\textit{RANK}$ are not necessarily $1, \ldots, N$. A *tie* appears when more than one person is included in certain microcategory of earnings. In this case all persons receive the same *mean value of consecutive ranks* as the value of $\textit{RANK}$.

Therefore, for any $\omega, \omega' \in \Omega_0$,

$$\textit{RANK}(\omega) \leq \textit{RANK}(\omega') \quad \text{iff} \quad Z(\omega) \leq Z(\omega')$$

and therefore $C(M : W) = C(\textit{RANK}|_{\Omega_2} : \textit{RANK}|_{\Omega_1})$.

Remark 4.10 *Please note that the symbol $\textit{RANK}|_{\Omega_1}$ concerns ranks provided jointly for men and women and not ranks provided for women.*

Any function $\varphi : \Omega_0 \to \mathbb{R}$ which satisfies

$$\varphi(\omega) \leq \varphi(\omega') \text{ iff } \omega \preceq \omega',$$

will be called *the order function* w.r.t. $\preceq$. We will use the notation $C(Q : P; \preceq)$ or the notation $C(Q : P; \varphi)$, where

$$C(Q : P; \varphi) = C(Q\varphi^{-1} : P\varphi^{-1}). \tag{4.20}$$

We will also use the simplified notation

$$C(Y : X; \varphi) = C(\varphi(Y) : \varphi(X)) \tag{4.21}$$

when P and Q are two distributions of random variables X and Y. Further, we will use the analogous notation for ar :

$$ar(Q : O; \varphi) = ar(C(Q : P; \varphi)), \;\; ar(Y : X; \varphi) = ar(C(Y : X; \varphi)).$$

Order function $\varphi^{\leftarrow}$ (opposite to φ) is defined as

$$\varphi^{\leftarrow}(\omega') \preceq \varphi^{\leftarrow}(\omega) \textit{ iff } \varphi(\omega) \preceq \varphi(\omega')$$
$$\varphi^{\leftarrow}(\omega') = \varphi^{\leftarrow}(\omega) \textit{ iff } \varphi(\omega) = \varphi(\omega').$$

Definition 4.8 *(opposite curve). We say that a Lilliputian curve denoted $C^{\leftarrow}(Q : P; \varphi)$ is opposite to $C(Q : P; \varphi)$ if*

$$C^{\leftarrow}(Q : P; \varphi) \stackrel{df}{=} C(Q : P; \varphi^{\leftarrow}). \tag{4.22}$$

It follows that

$$ar(C^{\leftarrow}(Q : P; \varphi)) = -ar(Q : P; \varphi). \tag{4.23}$$

When Q and P are the distributions of random variables X and Y and when φ is the natural order on the real line, we use a simplified notation $C^{\leftarrow}(Y : X)$. Then, $C^{\leftarrow}(Y : X)$ is equal to $C(f(Y) : f(X))$ for any *decreasing* function $f : \mathbb{R} \to \mathbb{R}$. According to this we have:

Definition 4.9 *Any random variable Z transformed by an arbitrary decreasing function $\mathbb{R} \to \mathbb{R}$ will be called opposite to Z.*

To facilitate the notation, we will write $Z^{\leftarrow}$ to denote variable opposite to Z. We will also write $(X^{\leftarrow}; Y^{\leftarrow})$ to denote a pair of random variables $(f(X); f(Y))$ for an arbitrary (unspecified) decreasing function $f : \mathbb{R} \to \mathbb{R}$. All such pairs have an identical concentration curve

$$C^{\leftarrow}(Y : X) = C(Y^{\leftarrow} : X^{\leftarrow}),$$

which is *opposite to* $C(Y : X)$.

The opposite concentration curve

$$C^{\leftarrow}(Y : X) = C(Y^{\leftarrow} : X^{\leftarrow}),$$

is a Lilliputian curve *symmetric to* $C(X^{\leftarrow} : Y^{\leftarrow})$ *w.r.t. Diag.*
By (4.23)

$$ar(Q : P; \varphi^{\leftarrow}) = -ar(Q : P; \varphi).$$

Illustration (Earnings Data):
We put $M^{\leftarrow} = A - M$, $W^{\leftarrow} = A - W$, where A is the chosen upper bound of male and female earnings. Please remember that any other decreasing function might be selected, in particular threshold A can be chosen arbitrarily. The curves $C(M : W)$ and the opposite curve $C^{\leftarrow}(M : W)$ are shown in Fig. 4.14: the second is symmetric to the first w.r.t. *Diag*, and $ar(M : W) = 0.5574 = -ar(M^{\leftarrow} : W^{\leftarrow})$. The third curve $C(W^{\leftarrow} : M^{\leftarrow})$ (called *dual* to $C(M : W)$) and its ar (equal to $ar(M : W)$) will be discussed in the Illustration in Section 4.4.1.

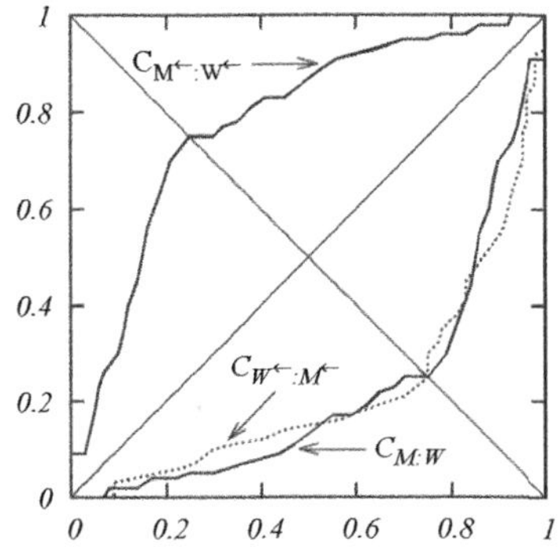

FIGURE 4.14. The curves $C(M : W)$, $C(M^{\leftarrow} : W^{\leftarrow})$ and $C(W^{\leftarrow} : M^{\leftarrow})$

Details: If random variables X and Y take on k values with probabilities $p_1, ..., p_k$, and $q_1, ..., q_k$, respectively, then the concentration curve $C(Y : X)$ is spanned on the set of points

$$\{(S_i^X, S_i^Y);\ i = 0, 1, ..., k\},$$

where $S_i^X = p_1 + \cdots + p_i$, $S_i^Y = q_1 + \cdots + q_k$, $S_0^X = S_0^Y = 0$, while the opposite concentration curve $C^{\leftarrow}(Y : X)$ is spanned on the set

$$\{(1 - S_i^X, 1 - S_i^Y); i = 0, \ldots, k\} = \{(1 - S_{k-i}^X, 1 - S_{k-i}^Y);\ i = 0, 1, ..., k\}.$$

Generally,

$$C^{\leftarrow}(Y : X) = \{(1 - F_X(z), 1 - F_Y(z)); z \in \mathcal{R}\}$$

(completed, if necessary, by points (0,0), (1,1) and by linear interpolation).

For a curve $C \in \mathbb{ULM}$,

$$C^{\leftarrow}(u\pm) = 1 - C((1 - u)\mp).$$

Hence the random variable $\xi^{\leftarrow} = 1 - \xi$ is opposite to $\xi \in \mathbb{ULM}$. The density $c^{\leftarrow}$, called the *opposite concentration density,* is equal to

$$c^{\leftarrow}(u) = c(1 - u).$$

4.3.2 The maximal concentration curve and the maximal concentration index

The likelihood ratio $h_{Y:X}$ defined in Sec. 3.4.3 by formula 3.8 has very important properties:

Given any pair of random variables $(X; Y)$, the concentration curve $C(Y : X; h_{Y:X})$ lies below (not above) $C(Y : X; \varphi)$ for any $\varphi : \mathbb{R} \to \mathbb{R}$, and the concentration curve $C(Y : X; h_{X:Y})$ lies above (not below) $C(Y : X; \varphi)$.

Thus the ordering provided by $h_{Y:X}$ ensures the strongest differentiation in the set of all pairs $(\varphi(X), \varphi(Y))$.

The definition of $h_{Y:X}$ can be extended in the following way:

Definition 4.10 *(likelihood ratio of Q to P). Let P and Q be probability measures on $(\Omega, \mathcal{B}(\Omega))$. The function $h_{Q:P} : \Omega \to \mathbb{R}^+$, defined by*

$$h_{Q:P}(\omega) = \frac{dQ}{d(P+Q)}(\omega) / \frac{dP}{d(P+Q)}(\omega) \quad \text{if } \tfrac{dP}{d(P+Q)}(\omega) > 0,$$

is called the likelihood ratio of Q to P.

Proposition 4.3 *For any pair $(P;Q)$ of distributions defined on $(\Omega_0, \mathcal{B}(\Omega_0))$, and any order function $\varphi : \Omega_0 \to \mathbb{R}$, the curve $C(Q : P; h_{Q:P})$ lies below curve $C(Q : P; \varphi)$ and curve $C(Q : P; h_{P:Q})$ lies above curve $C(Q : P; \varphi)$. This implies that*

$$ar(C(Q : P; h_{P:Q})) \leq ar(C(Q : P; \varphi)) \leq ar(C(Q : P; h_{Q:P})).$$

We note that $h_{P:Q}$ is opposite ordering to $h_{Q:P}$.

Definition 4.11 *(maximal concentration measures). Curve $C(Y : X; h_{Y:X})$ will be called the maximal concentration curve of Y w.r.t. X, and denoted $C_{\max}(Y : X)$. The concentration index of $C_{\max}(Y : X)$ will be called the maximal concentration index of Y w.r.t. X, denoted $ar_{\max}(Y : X)$.*

Illustration (Earnings Data):

We have $C(M_4 : W_4) = C_{\max}(M_4 : W_4)$. For $C(M : W)$ microcategories $z_{4(43)}$, $z_{7(43)}$, $z_{9(43)}$, $z_{11(43)} - z_{16(43)}$, $z_{18(43)}$, $z_{19(43)}$, $z_{21(43)} - z_{26(43)}$, $z_{28(43)} - z_{33(43)}$ have to be reordered to transform this concentration curves into the maximal concentration curve $C_{\max}(M : W)$ (see Fig. 4.15). Similarly, midcategories $z_{6(8)}$, $z_{8(8)}$ have to be reordered to transform $C(M_8 : W_8)$ into $C_{\max}(M_8 : W_8)$.

Theorem 4.3 *(i) $C_{\max}$ is convex (i.e. it has a non-decreasing density);*
(ii) $C_{\max}(Y : X) = C(h_{Y:X}(Y) : h_{Y:X}(X)) = L_{OV_{Y:X}}$;
(iii) $|ar(X, Y)| \leq ar_{\max}(Y : X)$;
(iv) $ar_{\max}(Y : X) = Gini(OV_{Y:X})$.

Proof (i): Lorenz density in point $u \in (0, 1)$ is equal to $z_u/E(Z)$.
(ii): $\frac{dQ^h_Y}{dP^h_X}(z) = z$, where $Q^h_Y = Q_Y h^{-1}$, $P^h_X = P_X h^{-1}$.

Theorem 4.3 (ii) should be given special attention. Due to it, curve $C_{\max}(Y : X)$ and index $ar_{\max}(Y : X)$ measure not only the maximal differentiation of Y w.r.t. X but also the inequality contained in the overrepresentation of Y relative to X.

The constrained maximal concentration curve arises as a special case of the order oriented concentration curve. We restrict this problem here to the example concerning earnings and education level.

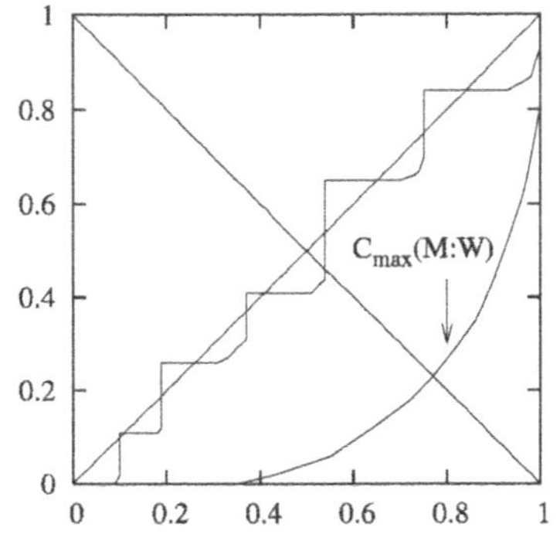

FIGURE 4.15. Concentration curve maximal w.r.t. male-to-female earnings constrained by education level, and maximal concentration curve $C_{\max}(M : W)$ equal to $L_{OV_{M:W}}$

Illustration (Earnings and Education Data):

Let people in Ω_0 be divided into six ordered groups according to education level. Let this be a preliminary partial ordering of people into six groups; *then,* people within each group are ordered according to *the likelihood ratio of male-to-female earnings.* This amounts to a specific ranking: each pearson in Ω is given a rank according to this inner order. Formally, there is a function

$$\widetilde{RANK} : \Omega_0 \to \{1, \ldots , 200\}$$

which plays the role of an order function from $\Omega_0 \to \mathbb{R}$. This function induces two random variables

$$\widetilde{RANK}_M = \widetilde{RANK}|_{\Omega_1}$$

and

$$\widetilde{RANK}_W = \widetilde{RANK}|_{\Omega_2}.$$

The plot of their cdf's is the concentration curve $C(\widetilde{RANK}_M : \widetilde{RANK}_W)$ which is *maximal with respect to male-to-female earnings constrained by the ordered education levels* (see Fig. (4.15)). This curve lies *above* the maximal concentration curve $C_{\max}(M : W)$. The two curves are very different. The constrained curve consists of six parts related to particular education levels. The shape of the constrained curve shows that at each education levels there exists a strong inequality between male and female earnings.

4.3.3 Order oriented Lorenz Curve and inequality (Gini) index

In this section the definition of order oriented concentration curve is applied to the Lorenz curve which is a special case of a concentration curve. The definition and notation from the previous section will be transferred in an obvious way: for $Z \in \overline{\mathbb{RATIO}}_\infty$ we will have $L_{Z;\varphi}$, $L_Z^{\leftarrow}$, $Gini(Z; \varphi)$.

Please note that L_Z is the maximal concentration curve $C_{\max}(Z^{(1)} : Z)$, and hence for any order function $\varphi : \Omega \to \mathbb{R}$ and any $u \in [0,1]$ we have

$$L_Z(u) = L_{Z;Z}(u) \leq L_{Z;\varphi}(u) \leq L_{Z;Z^{\leftarrow}} = L_Z^{\leftarrow}(u) = 1 - L_Z(1-u) \quad (4.24)$$

which implies inequalities

$$-Gini(Z) \leq Gini(Z;\varphi) \leq Gini(Z).$$

Curve $L_Z^{\leftarrow}$ is opposite to L_Z; it is often called the *upper Lorenz Curve.*

A formal definition and some properties of the order-oriented Lorenz Curve are given in the Appendix, Section 4.5.4.

Illustration (Earnings-Education-Gender data):

Let us modify the previous example concerning the order function $\widetilde{RANK}$. We will now restrict ourselves to earnings Z, education T, and $\widetilde{RANK}$. This time, our aim is to calculate *the Lorenz Curve for Z under the constraints introduced by education levels provided by* $\widetilde{RANK}$.

We deal now with set Ω_0 as a whole (i.e. we are *not* interested in dividing Ω_0 into subsets consisting of males and females). The order oriented Lorenz curve $L_{Z;\widetilde{RANK}}$ which is sought is defined as the linearly interpolated set of points

$$\{(Prob(\{\omega \in \Omega_0 : \widetilde{RANK}(\omega) \leq t\}), E(Z(\omega); \widetilde{RANK}(\omega) \leq t)/E(Z); t \in \mathbb{R}^+)\}$$

$$= \{(F_{\widetilde{RANK}}(t), E_P(Z(\omega); \omega \in \Omega_0; \widetilde{RANK}(\omega) \leq t)/E(Z));\ t \in \mathbb{R}^+)\}.$$

This curve is shown in Fig. 4.16, accompanied by L_Z and $L_Z^{\leftarrow}$.

If $\widetilde{RANK}$ is replaced by Z then $L_{Z;Z}$ is equal to L_Z.

In the next subsection order oriented Lorenz Curve and Gini Index will be applied to the density transform of categorical variables.

4.3.4 Order oriented Lorenz Curve and Gini Index for the density transforms of categorical variables

We will now extend the definition of the Lorenz Curve and Gini Index for the density transforms of categorical variables (Section 4.2.6). According to the previous section, instead of $L_{p_X(X)}$ (4.18) we will now have

$$L_{p_X(X);\varphi} = C(F^{(1)}_{p_X(X);\varphi} : F_{p_X(X);\varphi})$$

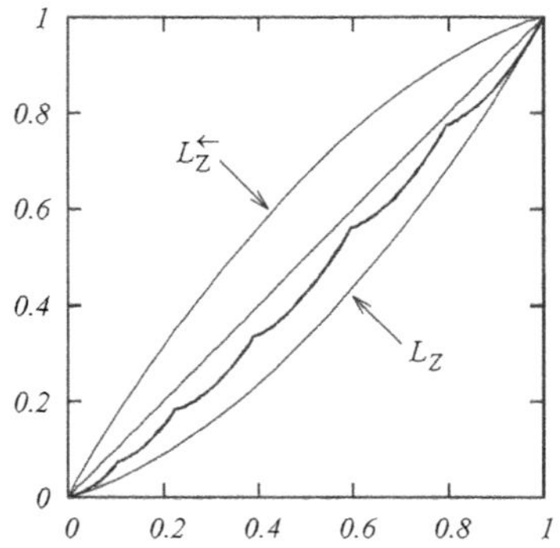

FIGURE 4.16. Order oriented Lorenz Curve $L_{Z;\widetilde{RANK}}$ between L_Z (lower curve) and $L_Z^{\leftarrow}$ (upper curve)

where

$$F_{p_X(X);\varphi}(z) = \int_{\{x:\varphi(x)\le z\}} dF_X(x),$$

$$F^{(1)}_{p_X(X);\varphi}(z) = \int_{\{x:\varphi(x)\le z\}} p_X^2(x)dF_X(x)/E(p_X(X)).$$

Instead of $Gini(p_X(X))$ (formula (4.19)) we will now have

$$Gini(p_X(X);\varphi) = \sum_{i=1}^{m} p_{j_i}^2(S_{j_{i-1}} + S_{j_i} - 1)/E(p_X(X))$$

where $S_{j_i} = p_{j_1} + \cdots + p_{j_i}$, $(j_1, \ldots, j_m)$ is a permutation of numbers $(1, \ldots, m)$ according to ordering $\varphi : \varphi(x_{j_1}) \le \cdots \le \varphi(x_{j_m})$.

Clearly, for any $\varphi : \mathbb{R} \to \mathbb{R}$

$$-Gini(p_X(X)) \le Gini(p_X(X);\varphi) \le Gini(p_X(X)).$$

Moreover,

$$Gini(p_X(X);\varphi^{\leftarrow}) = -Gini(p_X(X);\varphi). \tag{4.25}$$

4.3.5 Link with the two-class discriminant analysis

The order oriented concentration curve is the basic concept of the two-class discriminant analysis (DA). We will illustrate this with the following story.

Illustration (a computer gameplayer):

As a player of a computer game, you are led to an unknown place which you cannot see (your eyes are covered by a scarf). You know however that this place is either a cellar or a room in a palace. You are allowed to make a few measurements which will help you to know where you are: in the cellar or in the palace.

To start with, assume that you can only measure the intensity of light in the quarters (the result observed will be displayed for you on the screen). You know that in this game the intensity of light changes in the cellar according to a probability distribution with cdf F, and in the palace according to a distribution with cdf G; both F and G are known to you, and you also know the maximal concentration curve $C(G:F)$, which is displayed on the screen, also. Since light intensity in the cellar is rather concentrated on smaller values while in the palace it is rather concentrated on larger values, the cdf's satisfy

$$F(x) \geq G(x)$$

for any light intensity x, and their concentration curve $C(G:F)$ is convex and equal to $C_{\max}(G:F)$.

Assume additionally that F and G are both strictly increasing. Then, given the value of observation x, you calculate $F(x)$ and $G(x)$ and locate on curve C the point $(F(x), G(x)) = (u, C(u))$. The concentration density $c(u)$ (equal to the likelihood ratio $g(x)/f(x)$) is smaller or larger than 1. In the first case, it is "more likely" that you are in the cellar; otherwise, it is more likely that you are in the palace.

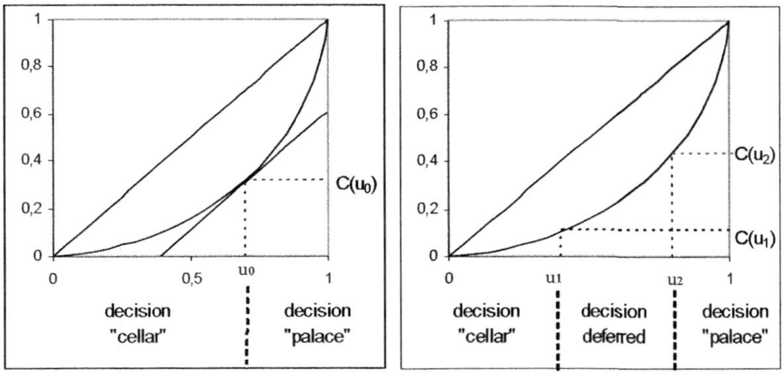

FIGURE 4.17. Visualization of probabilities of wrong decisions in the computer gameplayer example. Horizontal axis: cdf F; vertical axis: cdf G; curve $C = C(G:F)$. Left (forced decision case): $1 - u_0$ is the probability of decision "palace" which is wrong in the cellar, while $C(u_0)$ is the probability of decision "cellar" which is wrong in the palace. Right (deferred decision): $1 - u_2$ is the probability of decision "palace" which is wrong in the cellar, and $C(u_1)$ is the probability of decision "cellar" which is wrong in the palace.

If you intend to play the game just once and you are *forced* to choose ("cellar" or "palace"), it seems reasonable to make your decision according to $c(u) < 1$ or $c(u) > 1$. But what if you intend to play the game many times, repeating your decisions under the same circumstances for differently drawn values of light intensity?

What if you know not only F and G but also π, where π is the probability with which, in each successive game, the game master takes you to the cellar? Moreover, what if you are not forced to make a choice and can defer your decision, but you have to pay for all your mistakes and dodges?

Well, then you are advised to think of your strategy and determine what is known as your optimal decision rule. It consists in fixing two points, u_1 and u_2, with $0 < u_1 < u_2 < 1$, which will serve as reference points for the observed value $u = F(x)$; you will decide "cellar" when $u \leq u_1$, "palace" when $u \geq u_2$, and defer your decision when $u_1 < u < u_2$. The reference points would be established according to the costs connected with your mistakes and dodges, and according to the special requirements set by you and possibly also by the expected game master. A common requirement is to minimize your expected loss (maximize your gain) in an infinite sequence of games. Another possibility is to do the same but controlling the probabilities of one of your possible errors:

- deciding "cellar" when you are in the palace (probability of this error is equal to $C(u_1)$),
- deciding "palace" when you are in the cellar (probability of this error is equal to $1 - u_2$).

Fig. 4.17 visualizes the probabilities of wrong decisions in two cases: when the decision is forced and when it is deferred. In the first case, "cellar" is decided when $c(u) < c(u_0) = 1$ (and "palace" otherwise). In the second case, "cellar" is decided when $c(u) < c(u_1)$, "palace" is decided when $c(u) > c(u_2)$, and decision is deferred when $c(u_1) \leq c(u) \leq c(u_2)$.

4.4 Dual concentration curve

4.4.1 Definition of the dual concentration curve and dual Lorenz curve

Illustration (Earnings-Gender Data):

Recall that $Z^{\leftarrow}$ is the difference between the salary ceiling (for all employees) and the earnings. We will call this variable the deficit. It is a special case of a decreasing function of earnings Z, i.e. of Z opposite. Let $M^{\leftarrow}$ and $W^{\leftarrow}$ denote the deficit for men and for women, respectively.

Whenever male earnings M are more favorable than female earnings W (i.e. male earnings tend to concentrate on higher earnings), female deficits $W^{\leftarrow}$ tend to "overtop" male deficits $M^{\leftarrow}$. Therefore, the result of comparing $W^{\leftarrow}$ with $M^{\leftarrow}$ is in a sense *dual* to the result of comparing M with W.

The comparison of $C(M : W)$ with $C(W^{\leftarrow} : M^{\leftarrow})$ is shown in Fig. 4.14. The two curves are very close one to another, and are symmetric w.r.t. $Diag^{\perp}$ (so that $ar(M : W) = ar(W^{\leftarrow} : M^{\leftarrow})$). It is seen that the curves are a symmetric in a well intepretable way: $C(M : W)$ tends to be below $C(W^{\leftarrow} : M^{\leftarrow})$ for u smaller than the crossing point with $Diag^{\leftarrow}$, and tends to be below $C(W^{\leftarrow} : M^{\leftarrow})$ otherwise. This expresses the fact that male earnings concentrate on earnings higher than females while female deficits concentrate on deficits higher than male deficits. Figure 4.14 shows that extent of this asymmetry is rather small. Analogous figures of dual curves for different sets of objects provide a graphical presentation of the differences between males and females earnings, ready for visual comparisons.

Definition 4.12 *(dual pairs of variables, dual concentration curve and dual Lorenz Curve). (i) For any pair of random variables $(X;Y)$, pair $(Y^{\leftarrow};X^{\leftarrow})$ is called dual to $(X;Y)$, and the concentration curve*

$$C^D(Y : X) \stackrel{df}{=} C(X^{\leftarrow};Y^{\leftarrow})$$

is called dual to $C(Y : X)$.

(ii) if L_Z is the Lorenz Curve of $Z \in \overline{\mathbb{RATIO}}_{\infty}$ then the curve $C^D(Z^{(1)} : Z)$ is called the dual Lorenz Curve of Z and is denoted L_Z^D.

From geometric considerations,

$$C^D(u\pm) = 1 - C^{-1}((1-u)\mp), \quad u \in (0,1). \tag{4.26}$$

Details: (i) Let C be any Lilliputian curve and C^D the curve symmetric to C w.r.t. $Diag^{\perp}$. Then C^D is symmetric to $C^{\leftarrow}$ w.r.t. $Diag$ (recall that the opposite concentration curve $C^{\leftarrow}$ satisfies $C^{\leftarrow}(Y : X) = C(Y^{\leftarrow} : X^{\leftarrow})$), i.e., $C^D = (C^{-1})^{\leftarrow}$.

(ii) Definition 4.12 is easily extended for any pair $(P;Q)$ of probability measures on $(\Omega, \mathcal{B}(\Omega))$, accompanied by an order function $\varphi : \Omega \to \mathbb{R}$:

$$C^D(Q : P;\varphi) = C(P : Q;\varphi^{\leftarrow}). \tag{4.27}$$

It follows immediately from the previous considerations that

$$C^D(Q : P;\varphi)(u\pm) = C(P : Q;\varphi^{\leftarrow})(u\pm) = 1 - C(P : Q;\varphi)((1-u)\mp)$$

and the dual concentration density c^D is equal to

$$c^D(u) = \frac{dF}{dG}(G^{-1}(1-u)) = \frac{1}{c(C(F : G)(1-u))}$$

or, equivalently,

$$c(u) = \frac{1}{c^D(1 - C(G : F)(u))},$$

for all u such that all these densities exist.

The previous considerations justify the following theorem:

Theorem 4.4 *(i) for any Lilliputian curve C,* $ar(C^D) = ar(C)$;
(ii)

$$(C^D(Y:X))^D = C(Y:X). \tag{4.28}$$

(iii) for any $Z \in \overline{\mathbb{RATIO}}_\infty$

$$L_Z^D(u) = 1 - L_Z^{-1}(1-u); \tag{4.29}$$

(iv) for any pair of random variables X and Y,

$$L^D_{OV_{Y:X}} = L_{OV_{X:Y}}; \quad L^D_{1/OV^{(1)}_{Y:X}} = L_{1/OV^{(1)}_{X:Y}}. \tag{4.30}$$

(v)

$$ar(L_Z) = ar(L_Z^D) = -ar(L_Z^{\leftarrow}) = -ar(L_Z^{-1}). \tag{4.31}$$

Examples of Lorenz curves L_Z, $L_Z^{\leftarrow}$ (upper Lorenz curve) L_Z^{-1} (inverse Lorenz curve) and L_Z^D (dual Lorenz curve) are presented in Fig. 4.18 for L_Z given by formula (4.8) and also for $Z = U$. In the latter case, $L_U(u) = u^2$, $L_U^{\leftarrow}(u) = 2u - u^2$, $L_Z^{-1}(u) = \sqrt{u}$, $L_Z^D(u) = 1 - \sqrt{1-u}$.

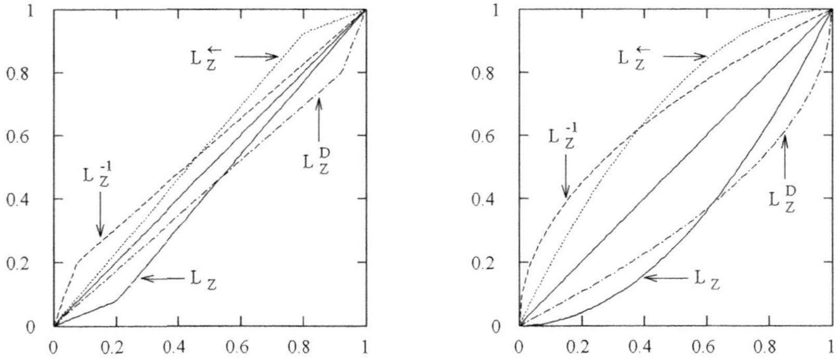

FIGURE 4.18. **Left:** Lorenz curves L_Z, $L_Z^{\leftarrow}$, L_Z^{-1}, and L_Z^D for L_Z given by formula (4.8); **Right:** the same curves for $L_Z(u) = u^2$

4.4.2 Random variable dual to a ratio variable

For any $Z \in \overline{\mathbb{RATIO}}_\infty$, the Lorenz Curve L_Z was introduced in Sec. 4.2.3 as the concentration curve attached to Z :

$$L_Z = C(Z^{(1)} : Z).$$

This suggests the following

Definition 4.13 *(dual ratio variable). For any $Z \in \overline{\mathbb{RATIO}}_\infty$, a variable from $\overline{\mathbb{RATIO}}_\infty$ with the Lorenz curve equal to $(L_Z)^D$ will be called dual to Z and will be denoted Z^D.*

Let us find the dual variable Z^D when $Z \in \overline{\mathbb{RATIO}}_\infty$ and takes on values $0 < z_1 < \cdots < z_k$, with probabilities $p_1, \dots, p_k$.

Let $Z^{(1)}$ be the first moment variable for Z, and let $F_{Z^{(1)}}$ be its cdf. Then $Z^{(1)}$ takes on values $z_1, \dots, z_k$ with probabilities $\frac{z_1 p_1}{EZ}, \dots, \frac{z_k p_k}{EZ}$, and

$$E(Z^{(1)}) = \frac{E(Z^2)}{E(Z)} = \frac{Var(Z)}{E(Z)} + E(Z).$$

Now, note that the random variable $\frac{1}{Z^{(1)}}$ takes on values $(\frac{1}{z_k}, \dots, \frac{1}{z_1})$ with probabilities $(\frac{z_k p_k}{EZ}, ..., \frac{z_1 p_1}{EZ})$. The expected value of $\frac{1}{Z^{(1)}}$ is $\frac{1}{EZ}$:

$$E(\frac{1}{Z^{(1)}}) = \sum_i \frac{z_i p_i}{z_i EZ} = \frac{1}{EZ}.$$

On the other hand, the random variable $\left(\frac{1}{Z^{(1)}}\right)^{(1)}$ takes on values $\frac{1}{z_k}, \cdots, \frac{1}{z_1}$ with probabilities $p_k, \cdots, p_1$, respectively, and hence it is equal to the random variable $\frac{1}{Z}$. These considerations suggest that the random variable Z^D, called dual to Z, is defined as

$$Z^D \stackrel{df}{=} \frac{1}{Z^{(1)}}$$

since

$$\begin{aligned} L_{Z^D} &= C((Z^D)^{(1)} : Z^D) = C(1/Z : 1/Z^{(1)}) \\ &= C(Z^{\leftarrow} : Z^{(1)\leftarrow}) = C^{\leftarrow}(Z : Z^{(1)}) \\ &= C^D(Z^{(1)} : Z). \end{aligned}$$

Remark 4.11 *The reader should be not misled here; one has to keep in mind that Z^D is a random variable with values which are the reciprocal of Z, but with a distribution induced by $Z^{(1)}$.*

Generally we have:

Theorem 4.5 *Let $Z \in \overline{\mathbb{RATIO}}_\infty$ and let $P(Z > 0) = 1$. Then the reciprocal $1/Z^{(1)}$, denoted Z^D, is dual to Z. Moreover,*

(i) $(Z^D)^{(1)} \overset{df}{=} Z^{D(1)} = \frac{1}{Z}$,

(ii) $E(Z^D) = \frac{1}{EZ}$, $E(Z^{(1)}) = \frac{E(Z^2)}{E(Z)} = \frac{Var(Z)}{EZ} + EZ$,

(iii) $E(Z^{D(1)}) = \frac{E((Z^D)^2)}{E(Z^D)} = \frac{Var(Z^D)}{E(Z^D)} + E(Z^D) = E(\frac{1}{Z})$.

(iv) $F^{(1)}_{Z^D}(z) = \overline{F}_Z(z) = 1 - F_Z(z-)$, $F^{(1)}_Z(z) = 1 - F_{Z^D}(\frac{1}{z}-)$.

(v) the densities c, c^D of the order oriented Lorenz curve and the dual order oriented Lorenz curve satisfy, for $u \in (0,1)$

$$c(u) = \frac{1}{c^D(1 - L_{Z;\varphi}(u))}, \quad c^D(u) = \frac{1}{c(L^{-1}_{Z;\varphi}(1-u))}.$$

4.4.3 Dual links between overrepresentation and underrepresentation

It follows from Theorem 4.4(iii) that for any pair of random variables $(X;Y)$ the random variable which is dual to $OV_{Y:X}$ is equal to $OV_{X:Y}$:

$$OV^D_{Y:X} = OV_{X:Y} \tag{4.32}$$

(in particular, $OV^D_{M:W} = OV_{W:M}$). Formula 4.32 and Theorem 4.5 imply that

$$OV_{X:Y}|_{\mathbb{R}} = 1/OV^{(1)}_{Y:X}, \quad OV_{Y:X}|_{\mathbb{R}} = 1/OV^{(1)}_{X:Y}. \tag{4.33}$$

When links refering to duality are incorporated into the diagram in Fig. 4.11 concerning overrepresentation and underrepresentation, we get the extended diagram presented in Fig. 4.19.

The diagram in Fig. 4.19 indicates that the two basic operations:

- derivation of the first moment variable for overrepresentation of Y on X, and

- derivation of the dual variable for overrepresentation of Y on X

are linked: the result of the first operation is the reciprocal of the result of the second operation.

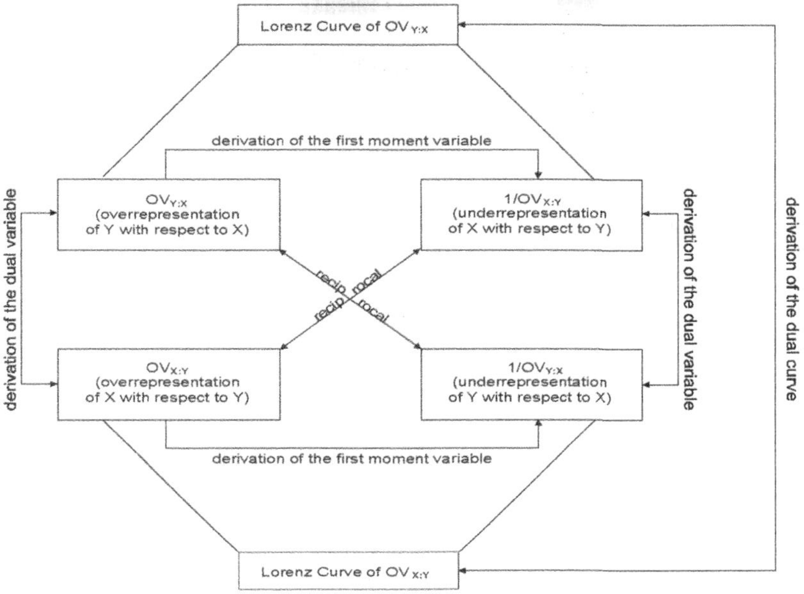

FIGURE 4.19. Diagram linking the concepts of overrepresentation and underrepresentation. Lorenz Curves of $OV_{Y:X}$ and $OV_{X:Y}$ are dual; Gini indices are equal

The link between the two operations is clear when the supports of $OV_{Y:X}$ and its first moment variable are identical, i.e.

$$\Pr(0 < OV_{Y:X} < \infty) = 1).$$

We should remember that $OV_{Y:X}$ represents *departures* from equality of Y and X, and $OV_{X:Y}$ represents also such departures but in a slightly different manner. To pass from $OV_{Y:X}$ to $OV_{X:Y}$, one has to *rotate* values of $OV_{Y:X}$ around point 1, i.e. to find the *reciprocals* of the values of $OV_{Y:X}$, and then *replace* the distribution of $OV_{Y:X}$ by that of its first moment. Then the distributions of $OV_{Y:X}$ and $OV_{Y:X}^{(1)}$ will be interchanged. This is exactly what should be done when pair $(X;Y)$ is replaced by pair $(Y;X)$.

Departures from point 1 of $OV_{Y:X}$ which are demonstrated by its Lorenz Curve are equivalent to departures from point 1 of $OV_{X:Y}$, demonstrated by $L_{OV_{X:Y}}$ which is *dual* to $L_{OV_{Y:X}}$. The shapes of both Lorenz Curves provide exactly the same information on local inequalities, although they are arranged in a reverse way (symmetrically to $Diag^{\perp}$).

It is important to keep in mind that these dual Lorenz Curves are respectively equal to $C_{\max}(Y : X)$ and $C_{\max}(X : Y)$. This means that orders associated with maximal differentiation coincide with orders associated with overrepresentation.

The departures from the value 1 are *on average the same for* $OV_{Y:X}$ *and* $OV_{X:Y}$ (value 1 corresponds to equality of Y and X), which implies that

$$Gini(OV_{Y:X}) = Gini(OV_{X:Y}). \tag{4.34}$$

In the special case, in which $OV_{Y:X} = OV_{X:Y}$ (i.e. $OV_{Y:X} = OV^{D}_{Y:X}$), we have $L_{OV_{Y:X}} = L^{D}_{OV_{Y:X}}$ which means that the Lorenz Curve $L_{OV_{Y:X}}$ (i.e. $C_{\max}(Y : X)$) is symmetric w.r.t. $Diag^{\perp}$.

4.4.4 Towards advantage problems in interpopulation comparisons

It is time now to look at the contents of the present chapter from the point of view of investigators interested in interpopulation comparisons. Suppose that the general problem of interest is male-to-female advantage in access to resources of a certain type (material, intellectual, spiritual etc.). The resources in question might be defined arbitrarily as "wealth", "general state of health", "intellectual efficiency" and so on; in the following we refer to *wealth* as the symbolic resource under consideration, and we treat *poverty* as the *opposing* term.

Please keep in mind what we mean by interpopulation comparisons. It is *not* that we compare males with females: males and females are only involved in the *advantage problem* which serves as *an example* of a decision problem. This problem can be solved by a chosen decision rule which can be applied in *any population consisting of males and females* (more generally, a population consisting of objects of two considered types). So what we mean is to *compare the solutions* of the advantage problem, obtained by the *same decision rule* in two (or more) populations. For instance we might compare the evaluated male to female advantage in wealth in two or more countries, or in a few different periods of time in the same country, etc. This requires that *the solutions should be reasonably comparable.*

The solutions to the advantage problem are Lilliputian curves and we show here how such curves can be compared and how they can be described by various numerical parameters (which are also suitably standardized and hence are comparable). This is a very important reason why grade methods which provide elements of the Univariate Lilliputian Model as solutions are valuable in interpopulation comparisons. However, other essential aspects of the solutions are also worth mentioning, and their suitable behaviour with respect to exchangeable elements of the advantage problem seem particularly important among these aspects. Grade methods based on concentration and inequality possess the required properties of exchangeability.

4.5 Appendix

4.5.1 Measurement scales

The initial example considered in this chapter (which will also appear in the following chapters) refers to a data set of 200 people described by three variables: gender (S), level of education (T), and earnings (Z). The set of people is denoted Ω_0. Variable S maps Ω_0 onto a binary set $\{female,\ male\}$, labeled $\{1,2\}$. Variable T maps Ω_0 into set $\{1,...,6\}$ with label 1 corresponding to the smallest level, and label 6 to the highest level. Variable Z maps Ω_0 into the set $\mathbb{R}^+$ of nonnegative numbers which contain earnings. It follows that each person $\omega \in \Omega_0$ is described by three numbers: $(S(\omega), T(\omega), Z(\omega))$. We can say that $S(\omega)$, $T(\omega)$, $Z(\omega)$ are the results of *measurements* performed on $\omega \in \Omega_0$.

Variables S, T, and Z exemplify three different types of measurements: *nominal, ordinal,* and *ratio.* More formally, one says that there are *measurement scales* called *nominal, ordinal* and *ratio.* The nominal scale is said to be *weaker* than ordinal and ratio, and ordinal to be *weaker* than ratio. The intuitive meaning of this statement is clear:

- for gender, it does not matter which numbers are chosen to label males and females, it is only required that these numbers be *different*;
- for level of education, it is only required that these numbers are different and *increasingly ordered;*
- for earnings, it is required not only that the values of the different earnings are different and increasingly ordered, but also that the ratios $k = Z(\omega_1)/Z(\omega_2)$ are *meaningful* in the following sense: in each of two pairs of people (ω_1, ω_2) and (ω_1', ω_2') for which

$$Z(\omega_1)/Z(\omega_2) = Z(\omega_1')/Z(\omega_2') = k$$

the first person is said to be k times better paid than the second.

These relations can be considered "at the empirical level" (for persons) and "at the numerical level" (for numbers related to persons).

We see that the requirements for the nominal scale form a part of the requirements for ordinal and ratio scales; and the requirements for the ordinal scale form a part of those for the ratio scale. This hierarchy explains what the term "weaker measurement scale" means. However, what is the meaning of the term "measurement scale"?

Clearly, variables S, T, Z in our example relate the "empirical level" to the "numerical level". Generally, at the empirical level, we have set Ω_0 of people and also *relations* (relationships) among these people, among pairs of people, among sets of three people, etc. In our example relations concern the following facts: that males and females are of different sex, that a person with higher degree of education is better educated than a person with a lower degree, that one person is k times better paid than another for any non-negative k.

The structure consisting of set Ω_0 and chosen set of *relations* corresponding to Ω_0 is called an *empirical relational structure.* We have here three different (hierarchically ordered) types of relational structures. Each of them has its numerical counterpart in the form of a relational structure consisting of set $\mathbb{R}$ of real numbers or a subset $\Omega \subset \mathbb{R}$, and of a set of relations corresponding to Ω, $\Omega \times \Omega$, etc. A *measurement scale* is a function from Ω_0 into Ω which maps each empirical relation into its numerical counterpart.

Let $(\Omega_0, \mathbb{REL}_0)$ denote an empirical relational structure, and $(\Omega, \mathbb{REL})$ the corresponding *numerical relational structure.* Further, let $\mathbb{M}$ be the set of all measurement scales

$$X : (\Omega_0, \mathbb{REL}_0) \to (\Omega, \mathbb{REL}).$$

The notation $X : (\Omega_0, \mathbb{REL}_0) \to (\Omega, \mathbb{REL})$ means that X is a function from Ω_0 into Ω such that each relation from $\mathbb{REL}_0$ is mapped into the corresponding relation from $\mathbb{REL}$. Then, clearly, we are interested in set $\boldsymbol{\varphi}$ of all one-to-one transformations which map Ω *onto* Ω in such a way that for any $\varphi \in \{\boldsymbol{\varphi}\}$ and any $X \in \mathbb{M}$, $\varphi(X) \in \mathbb{M}$. A function $\varphi : \Omega \to \Omega$ with this property for a given pair of relational structures (empirical and numerical) is called *admissible* for this pair.

The types of measurement scales mentioned above are identified by specifying sets $\boldsymbol{\varphi}$ of admissible functions. For the nominal scale, $\Omega = \mathbb{R}$, and $\boldsymbol{\varphi}$ is the set of all one-to-one functions $\mathbb{R} \to \mathbb{R}$. For the ordinal scale, $\Omega = \mathbb{R}$ and $\boldsymbol{\varphi}$ is the set of all increasing functions $\mathbb{R} \to \mathbb{R}$. For the ratio scale, $\Omega = \mathbb{R}^+$, and $\boldsymbol{\varphi}$ is the set of all functions of the form $\varphi(x) = ax$ for any $a > 0$. So the set of admissible functions for a *stronger* type of measurement is a *subset* of those for a *weaker* type.

Looking at these three sets of admissible functions, there arises a natural question: is there a type of measurement scale stronger than ordinal, but weaker than ratio? Such a scale exists and is called *interval scale,* with the set of admissible functions $\varphi : \mathbb{R} \to \mathbb{R}$ of the form $\varphi(x) = ax + b$ for $a > 0$ and $b \in \mathbb{R}$. Moreover, there exists a scale *stronger* than the ratio, namely the *absolute scale,* with identity $id(x) = x$ being the unique admissible function.

The absolute scale has a special importance when we are concerned with parameters of grade models. We want to have parameters with values which can be meaningfully compared without any further standardization or modification. However, this is usually possible when the set of models is suitably limited (e.g. to a set of sufficiently regular models).

This topic is important but difficult, and is beyond the scope of the present book.

Finally, let us revisit the data of Example A in Chapter 2, Section 2.2. Table 2.1 provides amount (in millions) spent on five household categories in two countries: country A and country B. This spending is a ratio variable. It is different in country A and country B: say $x_1, ..., x_5$ in A and $y_1, ..., y_5$ in B. So there are two distributions:

$$p = \left(\frac{x_1}{\sum_i x_i}, \cdots, \frac{x_5}{\sum_i x_i}\right) \quad \text{and} \quad q = \left(\frac{y_1}{\sum_i y_i}, \cdots, \frac{y_5}{\sum_i y_i}\right)$$

which can be compared by means of a concentration curve. Such a curve was built and then shown in Figure 2.1 (and also in Figure 2.2 for reordered budget categories). It is the same concentration curve which refers to the distributions of Letter Grades in Geography scores (revisited in Chapter 7).

This example evidently shows that values of a variable can be treated analogously as frequencies.

4.5.2 Supplement to Section 4.2 (the inequality measures)

Proof of Proposition 4.1.

(i) $F_{aZ}(z) = F_Z(\frac{z}{a})$, $F_{aZ}^{(1)}(z) = F_Z^{(1)}(\frac{z}{a})$, hence

$$\{(F_Z(\tfrac{z}{a}), F_Z^{(1)}(\tfrac{z}{a})); z \in \mathbb{R}^+\} = \{(F_Z(z), F_Z^{(1)}(z)); z \in \mathbb{R}^+\}$$

which implies that $C((aZ)^{(1)} : aZ) = C(Z^{(1)} : Z)$.

Now, let $L_Z = L_{Z'}$. Then $F_Z(0) = F_{Z'}(0)$ and for almost all points $u \in (0,1)$ there exist $dL_Z(u)/du = dL_{Z'}(u)/du$, i.e. $F_Z^{-1}(u)/E(Z) = F_{Z'}^{-1}(u)/E(Z')$, and hence $F_Z^{-1}(u) = E(Z)F_{Z'}^{-1}(u)/E(Z')$, i.e. $(Z|Z > 0) = a(Z'|Z' > 0)$ for $a = E(Z)/E(Z')$.

(ii)

$$
\begin{aligned}
Gini(Z) &= \frac{1}{2E(Z)} \int_0^\infty \int_0^\infty |z' - z| dF_Z(z) dF_Z(z') \\
&= \frac{1}{E(Z)} \int_0^\infty z' F_Z(z') dF_Z(z') - \int_0^\infty F_Z^{(1)}(z') dF_Z(z') \\
&= ar(Z^{(1)} : Z) = ar(L_Z).
\end{aligned}
$$

4.5.3 Supplement to Section 4.3.2 (the maximal concentration measures)

$$
\begin{aligned}
C_{\max}(Q:P) &\le C(Q\varphi^{-1} : P\varphi^{-1}; h_{Q\varphi^{-1}:P\varphi^{-1}}) \le C(Q\varphi^{-1} : P\varphi^{-1}; \psi) \\
&\le C(Q\varphi^{-1} : P\varphi^{-1}; h_{P\varphi^{-1}:Q\varphi^{-1}}) \le C(Q : P; h_{P:Q}).
\end{aligned}
$$

Also

$$C_{\max}(P:Q) = C(P:Q; h_{P:Q}) = C^D_{\max}(Q:P),$$

$$C^{\leftarrow}_{\max}(Q:P) \stackrel{df}{=} (C_{\max}(Q:P))^{\leftarrow} = C(Q:P; h_{P:Q}).$$

4.5.4 Supplement to Section 4.3.3 (the ordered Lorenz Curve and Gini Index)

Let cdf $F^{(1)}_{Z;\varphi}$ be the cdf corresponding to the measure $P^{(1)}_{Z;\varphi} = P^{(1)}_Z \varphi^{-1}$:

$$F^{(1)}_{Z;\varphi}(t) = \frac{1}{E(Z)} \int_{\{\omega:\varphi(\omega)\le t\}} Z(\omega) P(d\omega) = \frac{E(Z(\omega); \varphi(\omega) \le t)}{E(Z)}.$$

Then, the φ-ordered Lorenz Curve is defined by

$$L_{Z;\varphi} = C(F^{(1)}_{Z;\varphi} : F_\varphi) \tag{4.35}$$

If cdf F_φ is continuous, then

$$L_{Z;\varphi}(u) = F^{(1)}_{Z;\varphi}(\varphi_u) = \frac{E(Z(\omega); \varphi(\omega) \le \varphi_u)}{E(Z)}.$$

In general

$$L_{Z;\varphi}(u) = \frac{E(Z(\omega); \varphi(\omega) < \varphi_u)}{E(Z)} + \frac{E(Z; \varphi = \varphi_u)(u - F_\varphi(\varphi_u -))}{E(Z)(F_\varphi(\varphi_u) - F_\varphi(\varphi_u -))}$$

where the second term is equal to zero if $F_\varphi(\varphi_u) = F_\varphi(\varphi_u -)$. As previously stated, the definition of an order oriented concentration curve can be adapted to the case when measure Q is equal to $P^{(1)}_Z$ and $P = P_Z$ where P are measures defined $(\Omega_0, \mathcal{B}(\Omega_0))$ and $Z \in \overline{\mathbb{RATIO}}_\infty$.

4.5.5 Supplement to Section 4.4.2 (the random variable dual to a ratio variable)

In practice, it is often convenient to adjust the dual variable $\frac{1}{X^{(1)}}$ to have the same expectation as X. The variable which is dual to X and has the same expectation is given by

$$(EX)^2 X^D = \frac{(EX)^2}{X^{(1)}}. \tag{4.36}$$

The Lorenz Curve of $(EX)^2 X^D$ is still equal L_X^D.

4.5.6 Bibliographical remarks to Chapter 3 and 4

The main idea of the present book is to develop statistical models and methods based on one general concept called *concentration.* This unique "soft" concept serves to describe models as well as to compare one with another, starting from very simple and proceeding to very complex.

Concentration curves measure and describe visually concentration of one univariate distribution in respect to another. The set of all such curves forms the Univariate Lilliputian Model, which is equivalently the set of the cdfs of the Lilliputian variables (i.e. random variables defined on interval (0,1)). Linearly transformed expectation of a Lilliputian variable is the concentration index which measures and describes visually divergence of one univariate distribution from another. Suitable grade transformations applied to both distributions transform them into their concentration curve. In the general case, grade transformations are the probability transition functions and therefore all discrete-continuous variables are "softly" approached in the same way.

The authors of the present book published several papers which tackle this topic (although the term "Lilliputian" is introduced by them only now in the present book). In particular, we will mention here [59], [26], [39], [100], [13].

The problem of decomposition of the concentration index ar and of the maximal concentration index $ar_{\max}$ was not incorporated to Chapter 3 and 4 although the authors of the present chapter investigated it before in several papers ([65], [55], [84]).

The links between discriminant analysis and grade models and methods (just mentioned in Chapter 4) were presented in detail in [100] and tackled in [83] and in [31].

An early approach to the grade inequality measures and to the link between inequality and concentration can be found in [28] and in [39]; this initial link has been widely developed in Chapter 4.

Dual concentration curve which plays an important role in Chapter 4 was introduced in [28] and reintroduced in [17] in a novel form; it will be extensively used in Chapter 5 of the present book.

The statistical literature concerning concentration and its measurement is enormous. Therefore, we will mention only a few positions which were to us especially valuable landmarks of this research area: [37], [8], [2], [15], [91], [34], [5], [38], [96], [109], [88].

5

Asymmetry and the inverse concentration set

Alicja Ciok

5.1 Introduction

The concentration index ar and the maximal concentration index $ar_{\max}$ discriminate curves in the Univariate Lilliputian Model. However these indices often lack sufficient precision. It is due to the fact that there are many curves which have the same index value. Problems related to how greatly such curves may differ are discussed in Sec. 5.2. In particular, the boundaries of the smallest region containing all maximal concentration curves with a common value of $ar_{\max}$ are shown. Comparisons of curves from this region reveal that asymmetry (skewness) could be the appropriate discriminative tool.

To this purpose relative asymmetry curves are introduced in this chapter. These curves measure relative asymmetry, that is how one curve (random variable) is skewed with respect to another. If these pairs of random variables are suitably chosen, the relative asymmetry reduces to the common asymmetry of a single variable. It should be emphasized that in the $\mathbb{ULM}$ the skewness of random variables and the skewness of Lilliputian curves can be treated equivalently - thanks to the equivalence between cdfs and Lilliputian curves.

There are two main ideas underlying the proposed concepts of skewness. The first is the classical approach based on comparisons of the tails of distributions. The second, which seems very natural in the considered context, is that the concentration curves should somehow be involved. Thanks to the used formalism, which exploits concentration curves and opposite variables, many problems concerning a variety of measurement scales are avoided. Links between asymmetry and opposite variables are discussed in Sec. 5.3.

A function-valued measure of asymmetry (i.e., asymmetry curve) of a single random variable from the $\mathbb{ULM}$ and - based on this curve - a numerical measure (i.e., asymmetry index) are introduced in Sec. 5.4. In this section a generalization of these indices for the case of curves (which are not necessarily Lilliputian) is introduced.

A further generalization of the asymmetry measures to the case of any pairs of random variables is introduced in Sec. 5.5. These generalized indices enable the measurement of the relative asymmetry of one variable with respect to another.

Both these sections concentrate on presentation and interpretation of the basic ideas. Therefore all concepts are explained starting from the simplest case and using many illustrative examples. Formal proofs and additional properties of asymmetry indices are included in Sec. 5.6.2.

Special asymmetry curves in the form of circle or ellipsis segments and parabolas are considered in Sec. 5.4.3. There are two main reasons for our interest. The first is their relation to the first moment distribution of a variable uniform on $[0,1]$. Parabolas are the cdfs of the mixtures of two components: the uniform variable and its first moment distribution. Circles are the asymmetry curves of parabolas. The second reason is regularity. All segments of circles with constant length have similar shapes; each of them can be transformed into the other by only shift and rotation. This makes it possible to restore the whole curve if one segment, even infinitesimally small, is given. The specific properties of these curves give rise to the formulation of a regularity concept introduced in Chapter 6 and applied to simultaneous discretization of pairs of variables.

5.2 Concentration curves with a common value of the concentration index

Let us consider the family of all concentration curves with a common value of the concentration index ar.

Definition 5.1 *The class of concentration curves with α ($\alpha \in [0,1]$) as the common value of ar will be called the inverse concentration set and will be denoted by $ar^{-1}(\alpha)$.*

As described in Chapter 3, the $\mathbb{ULM}$ cdfs of random variables and concentration curves are equivalent (each cdf is uniquely determined by the respective curve and vice versa). Therefore the inverse concentration set $ar^{-1}(\alpha)$ contains cdfs of all random variables defined on the unit square which differ equally from the uniform variable (as the difference is measured by ar). Due to the equivalence between the concentration index of a random variable and the variable mean (formula (3.7)), all variables with cdfs belonging to $ar^{-1}(\alpha)$ have identical means. Are these the only features which they have in common? How greatly may they differ among themselves? What are the properties which differentiate them?

Figure 5.1 contains three examples of curves from $ar^{-1}(0.5)$. Curve C_3 does not contain linear segments, curves C_1 and C_2 contain such segments. Curve C_1 is particularly distinct. It turns sharply at one point and then crosses the diagonal. Two areas (situated between the diagonal and the curve) enter into the formula of ar with different signs (cf Chapter 3), consequently the value of ar is much lower than it would seem to be. The two remaining curves are comparatively smooth, but one has both a horizontal and a vertical segment. Generally, it may be stated that the curves shown in Fig. 5.1 differ to such an extent that it is difficult to detect similar features among them.

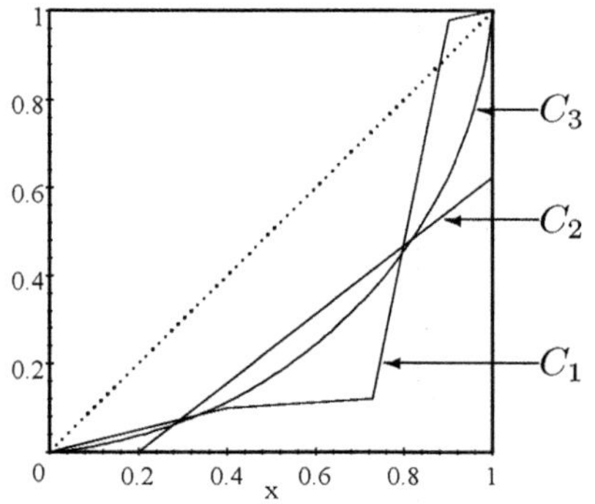

FIGURE 5.1. Three curves from the inverse concentration set $ar^{-1}(0.5)$

The situation changes when we restrict our considerations to the more regular concentration curves, namely the maximal concentration curves.

Definition 5.2 *The set of maximal concentration curves with* α *(*$\alpha \in [0,1]$*) as the common value of the concentration index will be called the inverse maximal concentration set and denoted by* $ar_{\max}^{-1}(\alpha)$.

As was shown in Sec. 4.3.2, the maximal concentration curves are convex. Hence the corresponding derivatives are non-decreasing. This property has a great impact on the variability of curves from $ar_{\max}^{-1}(\alpha)$. The next figure (Fig. 5.2) shows three curves from $ar_{\max}^{-1}(0.5)$. One of them (C_4) contain linear segments, the others (C_5 and C_6) do not. Although their diversity is still great, these curves are much more similar then curves C_1, C_2, C_3 and lie in a smaller area of the unit square.

The question arises: is it possible to find the smallest region containing all curves from the inverse maximal concentration set? The response is positive and the formal description of this region is presented in Sec. 5.6.1.

Two curves from $ar_{\max}^{-1}(0.5)$ visible in Fig. 5.3 have special shapes. Linked with the diagonal of the unit square, they form triangles or polygons similar to trapezoids. Therefore these curves will be called triangular or trapezoidal. They are particularly interesting because they form the boundaries of the smallest region containing all curves belonging to $ar_{\max}^{-1}(\alpha)$.

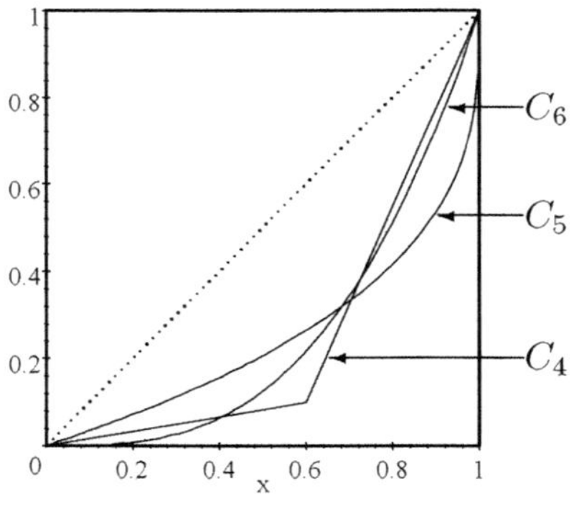

FIGURE 5.2 Three curves from the inverse maximal concentration set $ar^{-1}_{\max}(0.5)$

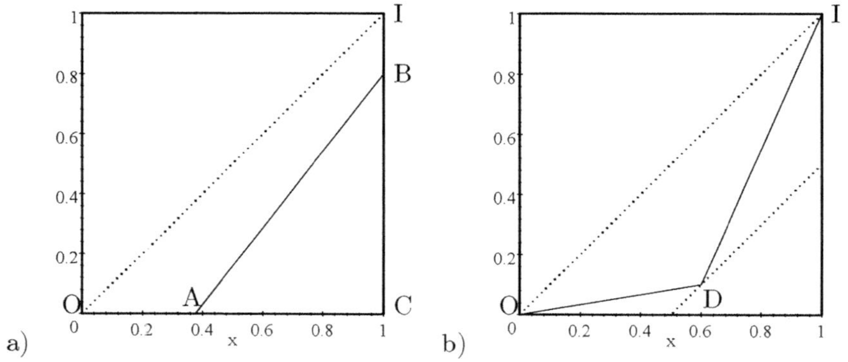

FIGURE 5.3. Trapezoidal (a) and triangular (b) curves from $ar^{-1}_{\max}(0.5)$

Let us consider a trapezoidal curve defined by points $OABI$ (Fig. 5.3a). Such a curve belongs to $ar^{-1}_{\max}(\alpha)$ if

$$|AC||BC| = 1 - \alpha$$

(where $|.|$ is the length of the segment), which is equivalent to the conditions:

$$|AC| = \sqrt{(1-\alpha)\beta} \text{ and } |BC| = \sqrt{\frac{1-\alpha}{\beta}}$$

for some β such that $0 \leq (1-\alpha)\beta \leq 1$ and $0 \leq \frac{1-\alpha}{\beta} \leq 1$. Since $\alpha < 1$, these inequalities are equivalent to the condition:

$$1 - \alpha \leq \beta \leq \frac{1}{1-\alpha}.$$

The parameter β can be expressed as the ratio $\frac{|AC|}{|BC|}$.

Remark 5.1 *Note that in this and the following chapters notations α and β have different meanings than those used in the previous chapters.*

Trapezoidal curves from $ar_{\max}^{-1}(\alpha)$ parametrized by (α,β) are defined as:

$$M_\beta^\alpha(t) = \begin{cases} 0 & \text{for } 0 \le t \le 1-\sqrt{(1-\alpha)\beta}, \\ \frac{t-1}{\beta} + \sqrt{\frac{1-\alpha}{\beta}} & \text{for } 1-\sqrt{(1-\alpha)\beta} < t < 1, \\ \left[\sqrt{\frac{1-\alpha}{\beta}}, 1\right] & \text{for } t = 1. \end{cases} \tag{5.1}$$

One may observe that curve $M_{1/\beta}^\alpha$ is symmetric to M_β^α with respect to the diagonal $y = 1-x$ (cf Fig. 5.4), so it belongs to $ar_{\max}^{-1}(\alpha)$. If $\beta = 1-\alpha$, the pair of curves $M_{1-\alpha}^\alpha$ and $M_{1/(1-\alpha)}^\alpha$ assumes a triangular shape with one segment parallel to one of the axis. One of the curves has the vertical segment, the second has the horizontal, therefore they are extreme in $ar_{\max}^{-1}(\alpha)$ (cf Fig. 5.4).

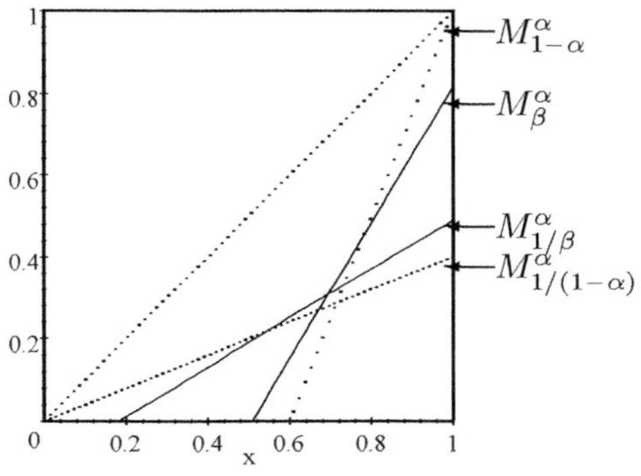

FIGURE 5.4. Curves M_β^α, $M_{1/\beta}^\alpha$, $M_{1-\alpha}^\alpha$, $M_{1/(1-\alpha)}^\alpha$ for $\alpha = \beta = 0.6$

The curve K_{min}^α defined as the maximum of M_β^α in the set of all admissible β:

$$K_{min}^\alpha(t) = \max_\beta\{M_\beta^\alpha(t): \ \beta \in [1-\alpha, \frac{1}{1-\alpha}]\}, \ t \in [0,1], \tag{5.2}$$

forms the upper boundary of the region covered by the family $\{M_\beta^\alpha\}$ (the proof is included in Sec. 5.6.1).

$$K_{min}^\alpha(t) = \begin{cases} t(1-\alpha) & \text{for } t \le 0.5, \\ \frac{1-\alpha}{4(1-t)} & \text{for } 0.5 < t < 0.5(1+\alpha), \\ \frac{t-\alpha}{1-\alpha} & \text{for } 0.5(1+\alpha) \le t \le 1. \end{cases} \tag{5.3}$$

The lower part of the boundary of the region covered by the family $\{M_\beta^\alpha\}$ (denoted here by K_{max}^α) is a piecewise curve determined by curves $M_{1-\alpha}^\alpha$

and $M^{\alpha}_{1/(1-\alpha)}$. K^{α}_{max} has a trapezoid shape with one segment parallel to the diagonal of the unit square. It can be proven (cf Sec. 5.6.1) that all triangular curves from $ar^{-1}_{\max}(\alpha)$ have such a property that their vertex points (point D in Fig. 5.3b) lie on the segment of K^{α}_{max} which is parallel to the diagonal. Formally

$$K^{\alpha}_{max}(t) = \begin{cases} 0 & \text{for } 0 \le t < \alpha, \\ t - \alpha & \text{for } \alpha \le t < 1, \\ [1-\alpha, 1] & \text{for } t = 1. \end{cases} \tag{5.4}$$

All curves from $ar^{-1}_{\max}(\alpha)$ lie in the region bounded by the K^{α}_{min} and K^{α}_{max} .

Figure 5.5 shows both boundary curves for $\alpha = 0.5$.

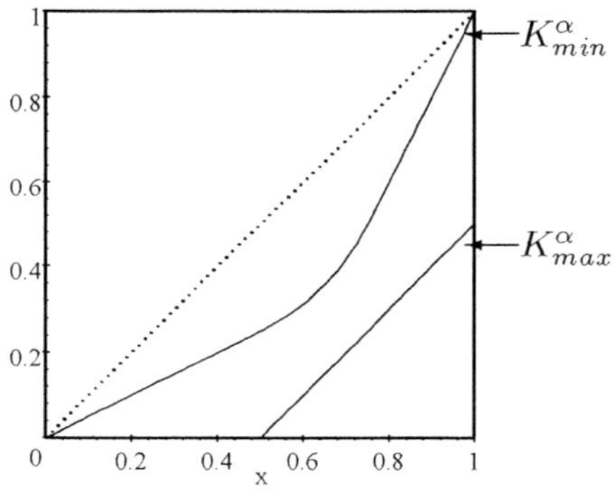

FIGURE 5.5. The boundary curves: K^{α}_{min} and K^{α}_{max} for $\alpha = 0.5$

For any $\alpha > 0$, the boundary curves K^{α}_{min} and K^{α}_{max} do not belong to $ar^{-1}_{\max}(\alpha)$.

Since $K^{\alpha}_{\min}$ and $K^{\alpha}_{\max}$ are nondecreasing and convex, they may be considered as maximal concentration curves, but they belong to different inverse concentration sets.

The region bounded by K^{α}_{min} and K^{α}_{max} has a very important property: it is completely covered by curves from $ar^{-1}_{\max}(\alpha)$ (the proof in Sec. 5.6.1). More precisely, every point of the region lies on a trapezoidal or a triangular curve from $ar^{-1}_{\max}(\alpha)$.

The area of this region may be large (cf Fig. 5.5). Obviously, if α is going to 0 or 1, the respective areas converge to 0. For α from the middle of the interval $(0, 1)$, the region is great enough to contain a variety of curves. In such a case, it would be very useful to detect some features, which would enable us to discriminate among curves from $ar_{\max}^{-1}(\alpha)$. For this purpose an asymmetry concept can be used.

5.3 Links between asymmetry and opposite orderings

In the concentration framework the symmetry term can be used in various contexts. Usually a random variable is regarded as symmetric if its tails are identical (that is the respective probabilities from both sides of the symmetry point are identical). Comparison of these tails induces a transformation of the initial variable into one with exchanged tails. Such a transfomation depends on the variable measurement scale and its point of symmetry.

The symmetry term is also used when a variable and its counterpart having the value scale reversed are identical. This reversal also depends on the measurement scale. A scale reversal and a rotation around the symmetry point often lead to the same results, but there are variables (for instance normal variables) where the scale reversal is equivalent to a shift transformation.

The shift transformation suggests that the symmetry term can be also used in connection with the behaviour of one random variable with respect to the other. For example, to say that the behaviour of Y relative to X is identical to the behaviour of X relative to Y.

Relative asymmetry curves, which will be introduced in this chapter, can express all forms of symmetry mentioned above. The proposed formalism is based on concentration curves and opposite variables. Thanks to these tools many problems concerning a variety of measurement scales can be avoided.

We start to introduce the formalism by recalling the definition of opposite variables. The opposite variable measures the same feature as the original variable, but starting from the other side of the scale, it goes in the opposite direction (it reverses the value ordering). Such variables are well known and frequently used. An example would be the reversed variable scales used in socio-psychological research to measure the reliabilty of respondent's answer. Let us remind the formal definition (see Sec. 4.3.1).

The random variable $X^{\leftarrow}$ is opposite to variable X, if there exists a decreasing function $\varphi : R \to R$ such that $X^{\leftarrow} = \varphi(X)$. φ will be called an opposite order function.

If variable X is continuous and F_X is its cdf, then the cdf $F_X^{\leftarrow}$ of a opposite variable $X^{\leftarrow}$ is equal to

$$F_X^{\leftarrow}(t) = 1 - F_X(\varphi^{-1}(t)). \tag{5.5}$$

Taking φ^{-1} when $X^{\leftarrow} = \varphi(X)$, we can obtain: $(X^{\leftarrow})^{\leftarrow} = \varphi^{-1}(X^{\leftarrow}) = X$. In other words, reversal done twice (with the same order function φ) does not change the initial variable.

5.4 Asymmetry in the Univariate Lilliputian Model

5.4.1 Asymmetry curves

In this section the concept of an asymmetry curve will be introduced in the simplest case of single Lilliputian variables and of one opposite order function. Let ξ be a continuous Lilliputian variable measured on an interval scale with f_ξ and F_ξ as its density function and its cdf, respectively. To simplify the explanations, we assume that the inverse F_ξ^{-1} is a function. A more general case will be disccussed later in this section.

If ξ is symmetric then 0.5 must be the symmetry point. Therefore the symmetry of ξ can be expressed by the condition: $f_\xi(0.5-t) = f_\xi(0.5+t)$, $t \in [0,1]$. The equivalent conditions for symmetry are:

$$f_\xi(t) = f_\xi(1-t),\ t \in [0,1] \tag{5.6}$$

and

$$F_\xi(t) = 1 - F_\xi(1-t) = F_{1-\xi}(t),\ t \in [0,1] \tag{5.7}$$

where $F_{1-\xi}$ is the cdf of the opposite variable $1-\xi$. The formula (5.7) shows that the opposite variable $1-\xi$ retains the symmetry of ξ.

The cdfs of symmetric Lilliputian variables are not convex (their densities are not increasing), therefore they cannot belong to any inverse maximal concentration set $ar_{\max}^{-1}(\alpha)$ for $\alpha > 0$.

Fig. 5.6 shows examples of the density function and of the cdf for a symmetric variable ξ. The cdf F_ξ is convex up to point 0.5, then it crosses the diagonal *Diag* and goes over it. The cdf $F_{1-\xi}$ of the opposite variable is identical. Fig. 5.6 shows also the inverse (i.e. quantile) functions $F_\xi^{-1} = F_{1-\xi}^{-1}$. Fig. 5.7 shows the density f_ξ and the cdf F_ξ for a non-symmetric Lilliputian variable ξ and their counterparts ($f_{1-\xi}$ and $F_{1-\xi}$) for the opposite variable $1-\xi$. The cdf is convex ($F_\xi(x) = x^5$), therefore it belongs to the set $ar_{\max}^{-1}(\alpha)$ where $\alpha = 0.66$. ξ is strongly skewed because the difference between F_ξ and $F_{1-\xi}$ is great.

In the concentration framework an asymmetry index should measure the deviations of opposite variables from the initial variables using concentration curves. However, this can be done in various ways. In the proposed measure the quantile functions: $F_\xi^{-1}(t)$ and $(F_{1-\xi})^{-1}(t) = 1 - F_\xi^{-1}(1-t)$, $t \in [0,1]$ are exploited. Let us note that in the ULM these quantile functions are also cdfs of some Lilliputian variables.

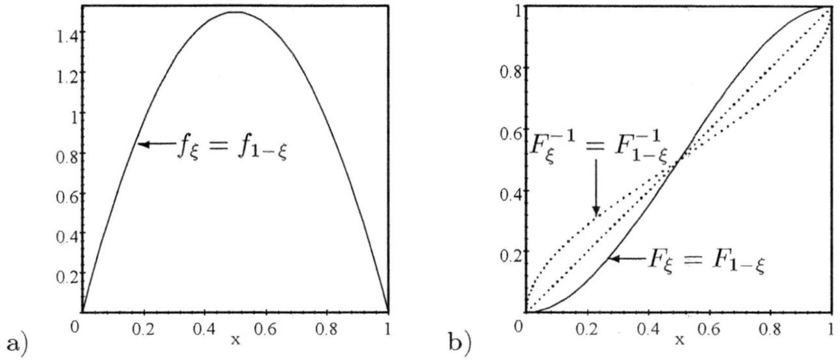

FIGURE 5.6. The densities (a), the cdfs and their inverses (b) for a symmetric variable ξ and its opposite $1-\xi$

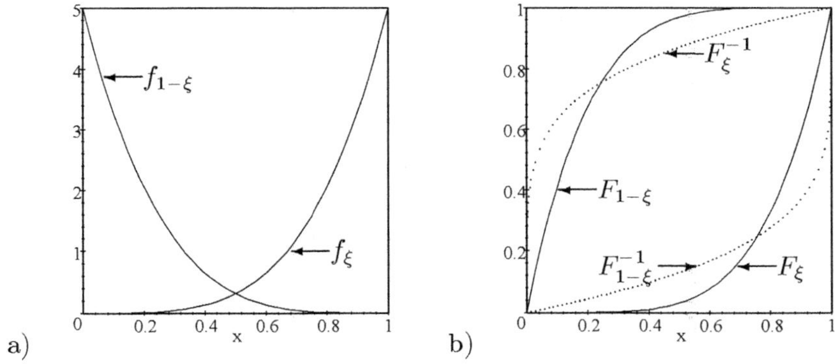

FIGURE 5.7. The densities (a), the cdfs and their inverses (b) for an asymmetric variable ξ and its opposite $1-\xi$

Definition 5.3 *(asymmetry curve in* $\mathbb{ULM}$*) Let* $\xi \in \mathbb{ULM}$ *be a continuous variable measured on an interval scale and let* F_ξ *be its cdf. If* F_ξ^{-1} *is a function then the asymmetry curve* $A(\xi)$ *of variable* ξ *is the composition of two functions:*

$$A(\xi) = F_{1-\xi}^{-1} \circ F_\xi \tag{5.8}$$

As both components of the composition are nondecreasing, the asymmetry curve $A(\xi)$ is a Lilliputian curve. Other implications of such a choice will be seen in the following sections.

Fig. 5.8a shows the asymmetry curve $A(\xi)$ of variable ξ as in Fig. 5.7 and its components: F_ξ and $(F_{1-\xi})^{-1}$. Let us note that $(F_{1-\xi})^{-1}$ is symmetric to F_ξ with respect to the diagonal $y = 1 - x$ ($Diag^\perp$).

The curves visible in Fig. 5.8b correspond to the opposite variable $1-\xi$. In each case the cdf and the asymmetry curve lie in the same half of the unit square (below or above the diagonal), which makes it easier to compare the respective curves. (It was this reason that such a definition of the asymmetry curve was chosen.)

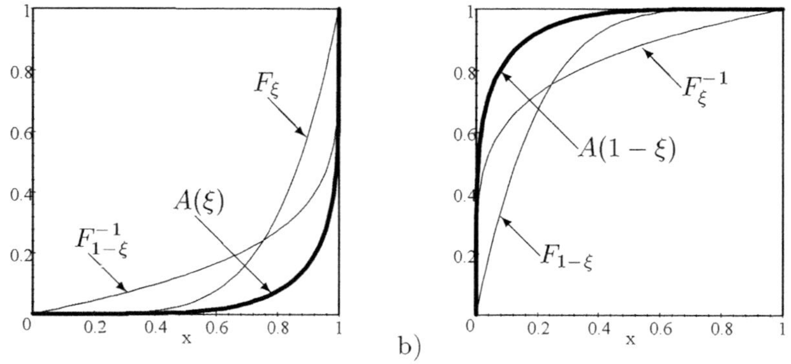

a) b)

FIGURE 5.8. The asymmetry curves and their components for variable ξ (a) and its opposite $1-\xi$ (b), when $F_\xi(x) = x^5$

Our example brings us closer to an appropriate interpretation of the asymmetry curves. If greater values of the density of ξ coincide with greater values of ξ (cf Fig. 5.7a), the asymmetry curve lies in the lower part of the unit square (below the diagonal $Diag$). We then say that ξ is skewed to the right. For the opposite variable the greater densities concentrate at the beginning of the variable scale, therefore we say that this variable is skewed to the left. Its asymmetry curve lies in the upper half of the square (above the diagonal).

A natural requirement is that the asymmetry curve of the opposite variable $1-\xi$ should be an inversion of the asymmetry curve for ξ. This requirement is fulfilled because:

$$A(\xi) = A^{-1}(1-\xi).$$

(the proof is included in Sec. 5.6.2).

If ξ and η are continuous Lilliputian variables with the respective cdfs F_ξ and F_η and the inverse F_ξ^{-1} is function then their concentration curve $C(\eta:\xi)$ is the composition of functions: $F_\eta \circ F_\xi^{-1}$. Therefore the asymmetry curve $A(\xi)$ is also the concentration curve of two Lilliputian variables: one correspond to the cdf F_ξ^{-1}, the second to the cdf $F_{1-\xi}^{-1}$.

Let us also note that F_ξ^{-1} and $F_{1-\xi}^{-1}$ (both from the $\mathbb{ULM}$) can be expressed as the concentration curves:

$$F_\xi^{-1} = C(U:\xi)$$

and

$$F_{1-\xi}^{-1} = C(1 - U : 1 - \xi),$$

where U is a variable uniform on $[0, 1]$.

Replacing all components in the definition of an asymmetry curve by the respective concentration curves provides a generalization of this definition.

Definition 5.4 *(asymmetry curve in* $\mathbb{ULM}$*) The asymmetry curve* $A(\xi)$ *of a Lilliputian variable* ξ *measured on an interval scale is defined by:*

$$A(X) = C(C(1 - U : 1 - \xi) : C(U : \xi)). \tag{5.9}$$

Before we proceed, particular attention should be paid to the concentration curves with vertical segments. If $C(U : \xi)$ or $C(1 - U : 1 - \xi)$ contains vertical segments, then the calculation of $A(\xi)$ needs the transformation of its component curves into cdfs (according to the definition of a concentration curve). As shown in Sec. 3.2, if vertical segments are removed from the concentration curves, these curves become the plots of nondecreasing functions These functions can be treated as cdfs of some Lilliputian variables and consequently their concentration curves can be calculated. Therefore we may use the notation: $A(K) = A(\xi_K)$, where K is a Lilliputian curve and ξ_K is the corresponding Lilliputian variable such that the plot of its cdf is equal to K minus the vertical segments. This approach will be called the variable approach.

Another approach (the curve approach) is also possible. In this case the definition of asymmetry curves as the composition of two curves (not necessary cdfs) is exploited. The same notation $A(K)$, where K is a curve in the unit square (not necessary the Lilliputian), can be also used but there is no necessity to consider the corresponding random variable.

We will consider only curves which lie in finite rectangles and which are plots of monotone functions (not necessary cdfs) extended by necessary vertical segments. (Further generalizations are possible, but they are not considered in this book.) Let $Cr^{V \times W}$ denote the set of such curves which lie in $V \times W$, where V and W are intervals: $V = [v_1, v_2] \subset R$ and $W = [w_1, w_2] \subset R$. Curve $K \in Cr^{V \times W}$ can be formally defined as subsets of points $\{(v, w)\} \subset V \times W$. Since continuity is the main characteristic of any curve, curves from $Cr^{V \times W}$ must satisfy the following condition:

$$\forall_{v \in V}\ \exists_{w \in W}\ (v, w) \in K \text{ and } \forall_{w \in W}\ \exists_{v \in V}\ (v, w) \in K; \text{ for } K \in Cr^{V \times W}$$

The composition of curves can be defined analogously to the composition of functions and it reduces to the composition of functions when curves are plots of functions.

Definition 5.5 *(composition of curves) If curves $K_1 \in Cr^{V\times Z}$ and $K_2 \in Cr^{Z\times W}$ then their composition $K_2 \circ K_1$ is the curve from $Cr^{V\times W}$ such that:*

$$(v,w) \in K_2 \circ K_1 \text{ iff } \exists_{z\in Z} \left((v,z) \in K_1 \text{ and } (z,w) \in K_2\right) \text{ and } (v - v_1^z)(w_2^z - w_1^z) = (v_2^z - v_1^z)(w - w_1^z)$$

where $v_1^z = \min_{(v,z)\in K_1} v$, $v_2^z = \max_{(v,z)\in K_1} v$, $w_1^z = \min_{(z,w)\in K_2} w$, $w_2^z = \max_{(z,w)\in K_2} w$.

If K is a curve and H is a function then $H(K)$ denotes the composition of K and the plot of H. Thanks to this $1-K$ is also properly defined. If $K \in Cr^{V\times W}$ then $K(v)$ is defined as the subset $\{w \in W : (v,w) \in K\}$.

In this chapter only curves from $Cr^{[0,1]\times[0,1]}$ are taken into account. Let $K \in Cr^{[0,1]\times[0,1]}$, then

- K^{-1} is the curve symmetric to K w.r.t. the diagonal $y = x$ ($Diag$) and $K^{-1} \circ K = K \circ K^{-1} = Diag$. Therefore we can write

$$C^{-1}(U : \xi) = C(\xi : U).$$

- K_{Sym} defined as follows:

$$K_{Sym}(t) = 1 - K^{-1}(1-t), t \in [0,1]$$

  is symmetric to K w.r.t. the diagonal $y = 1 - x$ ($Diag^{\perp}$). If $K = C(X : U)$, then K_{Sym} is the dual concentration curve (cf. Chapter 4)

$$K_{Sym} = C(1-U : 1-\xi) = C^{-1}(1-\xi : 1-U).$$

In consequence the asymmetry curve for a curve K in the unit square is the composition of this curve and of the curve symmetric to it w.r.t. the diagonal $Diag^{\perp}$.

$$A(K) = K_{Sym} \circ K. \tag{5.10}$$

For Lilliputian curves both approaches are equivalent, however outside $\mathbb{ULM}$ the curve approach is more general - it makes possible to investigate curves other than the nondecreasing. Moreover this approach is very intuitive, easy to apply and to interpret. In particular it helps to simplify the proofs of various properties of asymmetry curves.

> The asymmetry curves can be expressed in a more convenient form (common to both approaches) :
>
> $$A(K_\xi)(t) = 1 - K_\xi^{-1}(1 - K_\xi(t)) \quad \text{for } t \in [0,1] \tag{5.11}$$
>
> For a Lilliputian variable ξ the asymmetry curve $A(\xi) = A(K_\xi)$ where $K_\xi = C(\xi : U)$.

It is interesting that the asymmetry curves are symmetric w.r.t. the diagonal $Diag^{\perp}$:

$$A(K)(t) = 1 - A^{-1}(K)(1-t),\ t \in [0,1]$$

or equivalently

$$(A(K))_{Sym} = A(K).$$

Now let us consider the Lilliputian curve $M^{\alpha}_{1-\alpha}$ introduced in Sec. 5.2. It is a triangular curve with one horizontal segment; hence the symmetric curve $(M^{\alpha}_{1-\alpha})_{Sym} = M^{\alpha}_{1/(1-\alpha)}$ has the vertical segment. Whichever of the two approaches is used, the resulting asymmetry curve $A(M^{\alpha}_{1-\alpha})$ is a trapezoid curve identical to the curve $K^{\alpha}_{\max}$ introduced also in Sec. 5.2 as the part of the boundary of the inverse maximal concentration set $ar^{-1}_{\max}(\alpha)$.

$$A(M^{\alpha}_{1-\alpha}) = K^{\alpha}_{max}$$

Fig. 5.9 shows the curve $M^{\alpha}_{1-\alpha}$, the curve symmetric to it $(M^{\alpha}_{1-\alpha})_{Sym}$, and the asymmetry curve $A(M^{\alpha}_{1-\alpha})$ for $\alpha = 0.5$.

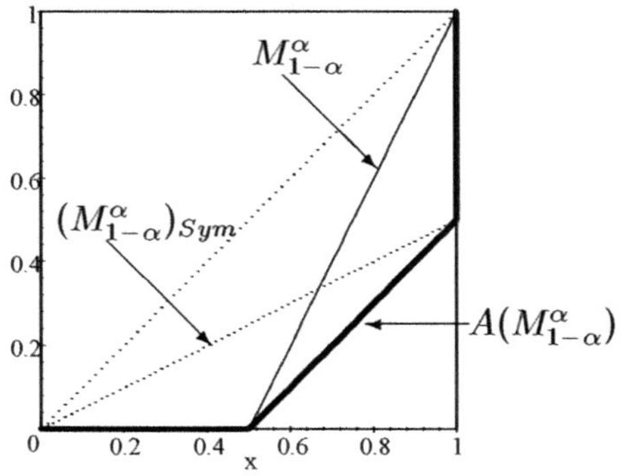

FIGURE 5.9. The curves $M^{\alpha}_{1-\alpha}$, $(M^{\alpha}_{1-\alpha})_{Sym}$ and the asymmetry curve $A(M^{\alpha}_{1-\alpha})$ for $\alpha = 0.5$

The asymmetry curve for the curve $(M^{\alpha}_{1-\alpha})_{Sym} = M^{\alpha}_{1/(1-\alpha)}$ (which also forms the boundary of $ar^{-1}_{\max}(\alpha)$) is a trapezoid curve similar to $K^{\alpha}_{\max}$. However it lies below $K^{\alpha}_{\max}$ and for $\alpha \geq 0.5$ it reduces to the triangular curve with one horizontal and one vertical segment.

The Lilliputian variable corresponding to the curve $M^{\alpha}_{1-\alpha}$ is wholly concentrated on the interval $[\alpha, 1]$ (with a constant density). The variable corresponding to $M^{\alpha}_{1/(1-\alpha)}$ has also a constant density in $[0, 1)$ but the remaining probability α is concentrated in point 1. Therefore we can say that both curves are skewed to the right.

The observations provided by the presented examples as well as the properties of asymmetry curves described in this section lead to the simple rules of interpretation of the asymmetry curves:

> If the asymmetry curves lie on the diagonal then the corresponding curves (variables) are symmetric. If asymmetry curves lie below the diagonal *Diag*, the corresponding curves (variables) are skewed to the right. If they lie above this diagonal, the respective curves (variables) are skewed to the left.

Several additional but interesting properties of the asymmetry curves and all proofs are included in Sec. 5.6.2.

5.4.2 Asymmetry index

On the basis of the function-valued measure of asymmetry a numeric measure is defined:

Definition 5.6 *(asymmetry index in* $\mathbb{ULM}$*) The asymmetry index* $AI(\xi)$ *for a Lilliputian variable* ξ *measured on an interval scale is defined as the concentration index* ar *calculated for the asymmetry curve* $A(\xi)$*:*

$$AI(\xi) = ar(A(\xi))$$

The asymmetry index for a curve K *in the unit square is defined analogously:* $AI(K) = ar(A(K))$.

The index is equal to twice the area between the diagonal and the asymmetry curve:

$$AI(\xi) = 2\int_0^1 (t - A(\xi)(t))dt$$

As the asymmetry curve for the opposite variable is the inversion of the initial asymmetry curve, the corresponding asymmetry index changes its sign while retaining the same absolute value:

$$AI(1-\xi) = -AI(\xi).$$

The properties and the interpretation of the asymmetry index follow the respective properties and interpretation of the asymmetry curves.

> If variables (curves) are symmetric then $AI = 0$. If asymmetry curves lie below the diagonal *Diag* (the corresponding variables (curves) are skewed to the right) then $AI > 0$. When they lie above the diagonal (the respective variables (curves) are skewed to the left) then $AI < 0$.

The deviation of asymmetry curves from the diagonal *Diag* as well as the deviation of the asymmetry index from zero measure strength of skewness. Therefore both indices (asymmetry curve and asymmetry index) generate orderings in the set of Lilliputian variables and make them appropriate tools for the discrimination of curves in the inverse maximal concentration sets. It is particularly important that the analysis of special shapes of asymmetry curves gives rise to the formulation of the ideal regularity concept considered in Chapters 6 and 11.

5.4.3 Families of curves with special properties

Analysis of the asymmetry curves enables us to distinguish very special families of Lilliputian curves. It is a common practice to evaluate diversity of variables by comparing them with a uniform variable with the same value range (its cdf with the diagonal *Diag* of the unit square), but if one wishes to compare two variables with the cdfs from the same $ar_{\max}^{-1}(\alpha)$ such a reference point is not appropriate. It seems that in $ar_{\max}^{-1}(\alpha)$ parabolas play a similar role to this diagonal as an appropriate reference point. Moreover, curves derived from them by means of asymmetry curves have very special properties.

Let us consider three families of continuous variables from the $\mathbb{ULM}$ with cdfs defined by the following formulas:

1. Par_δ: parabolas

$$\delta x^2 + (1-\delta)x, 0 < \delta \leq 1$$

2. Crc_δ: circles

$$1 + \delta - \sqrt{(1+\delta)^2 + \delta^2 - (x+\delta)^2}, \delta \geq 0$$

3. Ell_Δ: ellipses

$$\delta_1 - \delta_2\sqrt{\delta_3^2 - (x-\delta_4)^2},$$

where $\Delta = (\delta_1, \delta_2, \delta_3, \delta_4)$ and $\delta_1 = \delta_2\sqrt{\delta_3^2 - \delta_4{}^2}$, $\delta_2\sqrt{\delta_3^2 - (1-\delta_4)^2} = \delta_1 - 1$.

Parabola Par_δ is a mixture of two curves: the cdf of the variable uniform on $[0, 1]$ and the Lorenz curve of the same variable (cf Chapter 4). So it reflects changes in the first moment of the uniform variable smoothed by the uniform factor x.

The circles and ellipses may be derived from the parabolas by means of asymmetry or concentration curves. Let F_ξ denote the cdf of a continuous variable ξ defined on $[0, 1]$ and $(F_\xi)_I$ - the cdf of variable ξ_I which is the restriction of ξ to the interval $I = [B_1, B_2] \subset [0, 1]$. U_I is uniform on $[0, 1]$ and also restricted to the same interval. The concentration curves for restricted variables are given by the formula:

$$C(\xi_I : U_I)(t) = \frac{F_\xi(t(B_2 - B_1) + B_1) - F_\xi(B_1)}{F_\xi(B_2) - F_\xi(B_1)}$$

Such restricted concentration curves will be considered in the next chapter in a more general form.

Connections between the considered families are the following:

- The asymmetry curves calculated for parabolas are circles ($A(Par_\delta) \in \{Crc_\delta\}$). Moreover, the concentration curves for restricted parabolas are also parabolas ($C((Par_\delta)_I : U_I) \in \{Par_\delta\}$). Consequently, their asymmetry curves are also circles.

- The concentration curves for restricted circles are ellipses ($C((Crc_\delta)_I : U_I) \in \{Ell_\Delta\}$). It is also possible to obtain ellipses by another way but this will be discussed in the next chapter.

The specific properties of these three families of curves give rise to formalization of a regularity concept, which is discussed in Chapter 6.

5.5 Relative asymmetry

5.5.1 *Links with measurement scales*

In this section we will not restrict our consideration to Lilliputian variables. Also, all opposite order transformations will be considered.

A very natural requirement is that opposite order transformations should not change the types of measurement scales - a variable X and its opposite $X^\leftarrow$ should be measured on the same scale. But this requirement implies that the measurement scale determines the admissible opposite order transformations of variables. For example $X^\leftarrow = -aX$ is suitable for variables measured on an interval scale, $X^\leftarrow = \frac{a}{X}$ is appropriate for ratio variables ($a \in R$, $a > 0$).

Unfortunately, sometimes the choice of direction (i.e., initial variable or its opposite) is just a matter of chance, and the possible influence of the direction on the obtained results is not considered.

The comparison of a random variable with its opposite can measure asymmetry of this variable when the opposite order function does not distort the symmetry of the symmetric variables. Usually, whether an opposite variable retains symmetry or not, depends on the measurement scale and the symmetry point.

When we say that an ordinal variable X with n categories $c_1, c_2, ..., c_n$ is symmetric, then the condition $\Pr(X = c_i) = \Pr(X = c_{n-i+1})$, $i = 1, ..., n$ is expected. This would not be appropriate for continuous variables on an interval or on a ratio scale. If Y is on an interval scale and symmetric with respect to the point m, then we expect that $\Pr(X < m-x) = \Pr(X > m+x)$ for $x \in R$. In the first case the order function $\varphi(c_i) = c_{n-i+1}$ preserves the symmetry of X. In the latter $\varphi(x) = -x$ retains the symmetry of Y only when the symmetry point is equal to 0.

The situation is different when relative asymmetry of one variable with respect to another (called a reference variable) is considered. The relative asymmetry measures introduced below are generalizations of the asymmetry indices for single variables, and reduces to them when reference variables are suitably chosen. The relative asymmetry indices are based on the concepts of opposite variables and of concentration curves (like the asymmetry measures for single variables). However, the relative measures are invariant with respect to the chosen opposite order functions. Also the interpretation is broader and depends on the chosen reference variable.

5.5.2 Relative asymmetry measures

Let $(X; Y)$ be a pair of random variables having an unrestricted value range. One may ask how much is Y skewed with respect to X. In this case X is a reference to Y. (In the previous section a uniform variable U was considered as the reference variable.) For example, possession of "goods" by a chosen group of people can be analysed against a background of the group sizes. This kind of problem is often met in practice, for example in poverty research. Economists and sociologists discuss how to define and compare poverty or wealth in different economic systems or different cultures. Sometimes seemingly identical features are almost not comparable among different countries, and can by no means be expressed in a unified form. In such cases it is reasonable to analyse the discrimination of groups in a given country against a background of the average distribution in this country.

Formally, in the definition of asymmetry curve the reference variable U is replaced now by the variable X. In consequence we obtain the generalized definition of the relative asymmetry curve:

Definition 5.7 *(relative asymmetry curve) If variables X and Y are at least ordinal then the asymmetry curve $A(Y:X)$ of Y with respect to X is defined as:*

$$A(Y:X) = C(C(X^{\leftarrow}:Y^{\leftarrow}):C(X:Y)), \tag{5.12}$$

where $Y^{\leftarrow}$ and $X^{\leftarrow}$ are any variables opposite to Y and X respectively, determined by the same decreasing function. X will be called the reference variable.

Definition 5.8 *(relative asymmetry index) Asymmetry index $AI(Y:X)$ of Y with respect to X is the concentration index of $A(Y:X)$:*

$$AI(Y:X) = ar(A(Y:X) = 2\int_0^1 (u - A(Y:X)(u))du.$$

If X and Y are ordinal with the same support and X is uniform then both relative asymmetry indices measure asymmetry of Y in the sense appropriate for ordinal variables.

The conclusion is that the interpretation of the relative asymmetry indices depends on the chosen reference variable.

Remark 5.2 *Let us note that the discussion concerning the definition of asymmetry curves where the respective concentration curves are not identical with cdfs (presented in the previous section) can be repeated for the relative asymmetry curves; the uniform variable U is replaced now by a variable X and both variables (X and Y) may not belong to the $\mathbb{ULM}$. As before we can use one of the two approaches: the random variable approach or the curve approach; both formalisms are suitable. However we will use the latter as the more universal when possible (that is when a reference to random variables is not necessary).*

Now to simplify the following explanations, let us assume that F_Y, F_X are increasing cdfs of the variables Y and X respectively and both variables are defined on $[0,1]$. Let $F_Y^{\leftarrow}$, $F_X^{\leftarrow}$ correspond to the opposite variables $Y^{\leftarrow}$ and $X^{\leftarrow}$. If $H = C(Y:X) = F_Y \circ F_X^{-1}$ then $C(X:Y) = H^{-1}$ and from (5.5) $C(X^{\leftarrow}:Y^{\leftarrow})(t) = F_X^{\leftarrow} \circ (F_Y^{\leftarrow})^{-1}(t) = 1 - F_X \circ F_Y^{-1}(1-t) = 1 - H^{-1}(1-t)$. Hence $C(Y^{\leftarrow}:X^{\leftarrow})$ is the opposite curve and when treated as a cdf corresponds to the opposite variable. The variable to which this variable is opposite has the cdf equal to H. Consequently

$$A(Y:X) = 1 - H^{-1}(1-H) = 1 - F_X \circ F_Y^{-1} \circ (1 - F_Y \circ F_X^{-1}) \tag{5.13}$$

Generally, for any random variables X and Y

$$A(Y:X) = H_{Sym} \circ H,$$

where $H = C(Y:X)$ and H_{Sym} is symmetric to H w.r.t. the diagonal $Diag^{\perp}$.

Formula (5.13) shows why the concentration curve is invariant with respect to the opposite order functions. The proof in the general case when the concentration curves are not plots of functions is presented in Sec. 5.6.2. Also many properties of the asymmetry indices (and their proofs) are presented in that section. This section concentrates on interpretation. Therefore a few examples, which help to interpret the introduced formalism will now be presented.

5.5.3 Examples

Example 1

Suppose that in an analyzed population some "resources" or "goods" were distributed. It may be income, education or any feature describing socio-economical status of population members. The population consists of two groups discriminated by some factor (education, sex, place of residence, etc.). Suppose next, that we omit the diversity inside the particular groups; we wish to analyse the group as a whole. Hence, a random variable Y with only two values 1, 2 corresponding to the groups will be taken into consideration. The respective probabilities are b_0 and $1 - b_0$, where the first value is equal to the percent of all "goods" possesed by members of the first group, the second corresponds to the other group. Another considered variable X has also two values, identical to the values of Y. The corresponding probabilities a_0 and $1 - a_0$ characterize the size of the groups. Their concentration curve $C(Y : X) = C_0$ (shown in Fig. 5.10a, where $a_0 = 0.5$ and $b_0 = 0.25$) is the triangular curve described in Sec. 5.2. This curve treated as a cdf of some variable has a density shown in the same figure (Fig. 5.10b). This new variable will be denoted by Y_X. Its density is constant in the range $[0, a_0]$ with value b_0/a_0. In the remaining interval $(a_0, 1]$ it is also constant but with value $(1-b_0)/(1-a_0)$. If value b_0 decreases for a fixed a_0, the density also decreases in the first subinterval and consequently must increase in the second. This means that not only the concentration index (equal to $a_0 - b_0$) increases but also that the distribution becomes more skewed to the right. So, in this case skewness can measure the socio-economical inequality.

The main question in poverty research is: does the "goods" distribution differ from uniformity and in what manner. That is, which classes are discriminated against and to what extent. If all population members have an equal portion of "goods", C_0 lies on the diagonal and the corresponding concentration index is equal to 0. A similar situation occurs when the cumulative percent of "goods" in every group is proportional to group size. Then, it can be said that none of the group is discriminated against. Usually, distributions of "goods" among particular groups are far from uniform. This total deviation is well characterized by the value of the concentration index of C_0 but the structure of these deviations (local deviations, deviation trend) is not.

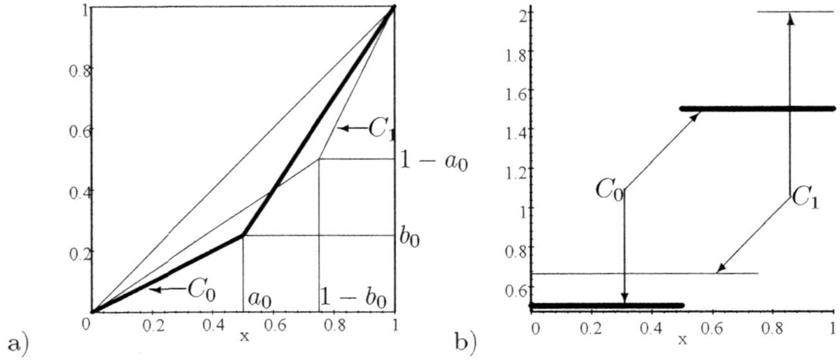

FIGURE 5.10. Two triangular curves (a) and the corresponding densities (b)

Suppose now that comparisons of the distributions of the same "goods" in various populations are needed. For example, one wishes to compare group discrimination in various countries or its variability across time. If respective concentration curves are ordered according to strength and type of skewness (symmetric, skewed to the right, skewed to the left), one will have a clear answer as to where (or when) the groups are discriminated against most and where least.

Fig. 5.10 contains also another concentration curve (C_1) of the same shape as C_0 (precisely, it is the curve symmetric to C_0 w.r.t. the diagonal $Diag^{\perp}$) and its corresponding density. This concentration curve belongs to the same inverse set $ar_{\max}^{-1}$, therefore the concentration index has no discriminating power here. Comparison of both densities provides a clear although intuitive result. The density of C_1 increases strongly in the rather short interval $(1-b_0, 1]$, which means that on the average every member of the second group has a much greater percent of "goods" than any member of the first. The difference between these percents is much greater for C_1 than for C_0. Therefore, our intuition say that the second distribution is more skewed to the right. Do the intuitive and the formal approach agree? Figure 5.11 shows the asymmetry curves for both distributions. The second asymmetry curve (A_1 corresponding to C_1) covers a greater area than the first (A_0), hence the respective value of the asymmetry index AI for C_1 is greater than that for C_0 which confirms this agreement.

This property can be generalized. Section 5.6.2 contains the proof, that for a fixed α the asymmetry index of triangular curves from $ar_{\max}^{-1}(\alpha)$ is a monotone increasing function of value a_0 (which reflects a shift to the right). This fact provides an easy interpretation of the orderings of the asymmetry curves and the asymmetry indices. The lower the asymmetry curve lies (the asymmetry index is positive and large), the more skewed to the right is the analyzed

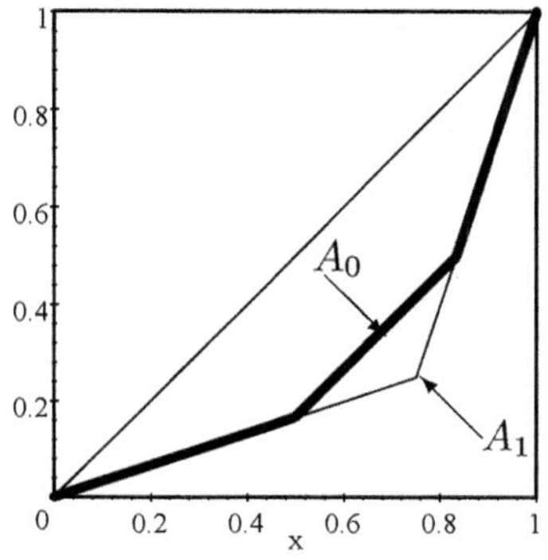

FIGURE 5.11. The asymmetry curves corresponding to the triangular curves from Fig. 5.10

distribution. The interpretation will be even easier when other properties of the asymmetry measures are taken into account.

Example 2

The cdfs F_X and F_Y of two continuous variables X and Y defined on $[0,1]$ are shown in Fig. 5.12a. Their asymmetry curves are presented in Fig. 5.12b. It is clearly visible that both variables are skewed to the right but Y is much more skewed than X. More precisely, asymmetry of Y with respect to the variable U uniform on $[0,1]$ is much greater than the analogous asymmetry of X ($A(Y:U) < A(X:U)$). The relative asymmetry $A(Y:X)$ is comparatively weaker than any of the others (see Fig. 10b). That completely agrees with the expectations, because Y and X are much more similar than Y and U (or X and U).

Example 3

A similar effect is observable in some families of interval variables generated by means of the shift transformation. Let us consider a family of continuous variables X_α ($\alpha \in A \subset R$) defined on R and measured on an interval scale with cdfs F_α. These variables are obtained by shift transformations from variable X_0, that is their cdf's are equal to

$$F_\alpha(x) = F_0(x - \alpha).$$

Let us assume that F_0 is increasing, and that X_0 is symmetric. That is, there exists $m \in R$ such that for every pair of points x and $2m - x$ (which is symmetric with respect to m) holds:

$$1 - F_0(x) = F_0(2m - x), x \in R \tag{5.14}$$

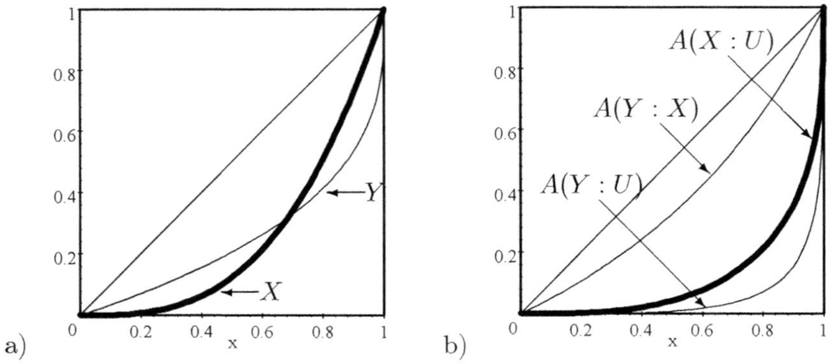

FIGURE 5.12. The cdfs of two variables (a) and the respective asymmetry curves: $A(Y : X)$, $A(X : U)$, $A(Y : U)$ (b)

(or equivalently: $F_0^{-1}(1-u) - m = m - F_0^{-1}(u)$ for $u \in [0,1]$). Obviously, X_α for $\alpha \neq 0$ are also symmetric with symmetry point $m + \alpha$. Well known examples of such a family are the normal variables with common standard deviation (e.g. equal to 1) and various means.

It is easy to check that the concentration curve of X_β w.r.t. X_α is equal to:

$$H_{\beta:\alpha}(u) = C(X_\beta : X_\alpha)(u) = F_0(F_0^{-1}(u) - \beta + \alpha) = F_0(F_0^{-1}(u) - \mu),$$

where $\mu = \beta - \alpha$. Since $H_{\beta:\alpha}^{-1}(u) = C^{-1}(X_\beta : X_\alpha)(u) = F_0(F_0^{-1}(u) + \mu)$, then

$$1 - H_{\beta:\alpha}^{-1}(1-u) = 1 - F_0(F_0^{-1}(1-u) + \mu).$$

From the symmetry (5.14)

$$\begin{aligned} 1 - H_{\beta:\alpha}^{-1}(1-u) &= F_0(2m - \mu - F_0^{-1}(1-u)) = \\ &= F_0(2m - \mu - (2m - F_0^{-1}(u))) = H_{\beta:\alpha}(u). \end{aligned}$$

This means that the concentration curves $H_{\beta:\alpha}$ are symmetric with respect to the diagonal $y = 1 - x$. In such a case $A(X_\beta : X_\alpha)$ is equal to the composition of $C(X_\beta : X_\alpha)$ with itself and consequently

$$A(X_\beta : X_\alpha)(u) = F_0(F_0^{-1}(u) - 2\mu).$$

The last equation shows that for this family the asymmetry curves are identical with the concentration curves. Let us note that if $\mu = \beta - \alpha \geq 0$ then $A(X_\beta : X_\alpha)(u) \leq u$, therefore X_β is skewed to the right with respect to X_α. It is interesting that such a popular family as the normal distributions possess this property. In Fig. 5.13, several examples of asymmetry curves for the normal family are shown.

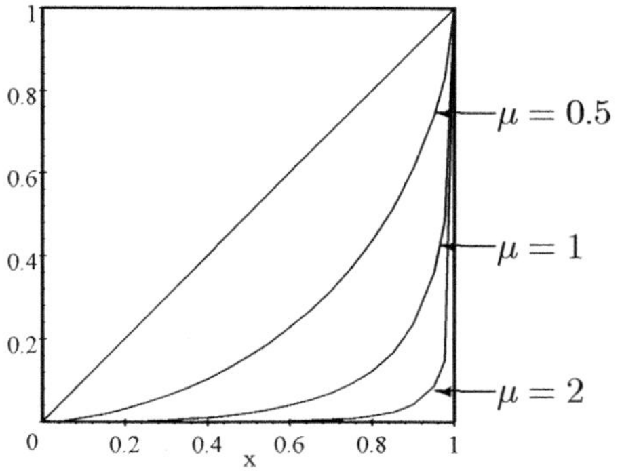

FIGURE 5.13. The asymmetry curves for the normal variables

Example 4

Let us consider two random variables X and Y. Let $h_{Y:X}$ and $h_{X:Y}$ denote the likelihood ratios (Def. 3.8) corresponding to this pair of variables. These ratios determine four random variables (overrepresentation and underrepresentation variables also introduced in Sec. 4.2.4): $h_{Y:X}(Y)$, $h_{Y:X}(X)$ and $h_{X:Y}(Y)$, $h_{X:Y}(X)$. Let us note that $\frac{1}{h_{Y:X}(Y)} = h_{X:Y}(Y)$ and $\frac{1}{h_{Y:X}(X)} = h_{X:Y}(X)$. Therefore we have two pairs of mutually opposite variables $((h_{Y:X}(Y), \frac{1}{h_{Y:X}(Y)})$ and $(h_{Y:X}(X), \frac{1}{h_{Y:X}(X)}))$ and the relative asymmetry curve of $h_{Y:X}(Y)$ w.r.t. $h_{Y:X}(X)$ is given by the following formula:

$$\begin{aligned} A(h_{Y:X}(Y) \ : \ h_{Y:X}(X)) &= \\ &= C(C(\frac{1}{h_{Y:X}(X)} : \frac{1}{h_{Y:X}(Y)}) : C(h_{Y:X}(X) : h_{Y:X}(Y))) = \\ &= C(C(h_{X:Y}(X) : h_{X:Y}(Y)) : C(h_{Y:X}(X) : h_{Y:X}(Y))) \end{aligned}$$

According to the Theorem 4.3 (Sec. 4.3.2)

$$\begin{aligned} C_{max}(X : Y) &= C(h_{X:Y}(X) : h_{X:Y}(Y)) \\ C_{max}(Y : X) &= C(h_{Y:X}(Y) : h_{Y:X}(X)) = C^{-1}(h_{Y:X}(X) : h_{Y:X}(Y)). \end{aligned}$$

Therefore

$$\begin{aligned} A(h_{Y:X}(Y) : h_{Y:X}(X)) &= C(C_{max}(X : Y) : C^{-1}_{max}(Y : X)) = \\ &= C_{max}(X : Y) \circ C_{max}(Y : X). \end{aligned}$$

The immediate conclusion is that this asymmetry curve is equal to the diagonal *Diag* only when $C^{-1}_{max}(Y : X)$ and $C_{max}(X : Y)$ are equal. In other words it measures the difference between $C_{max}(X : Y)$ and $C^{-1}_{max}(Y : X)$. (Let us note that always $C(X : Y) = C^{-1}(Y : X)$.)

5.6 Appendix

5.6.1 *The inverse concentration set*

Theorem 5.1 *(a) For any curve* $Z \in ar_{\max}^{-1}(\alpha)$ *and* $t \in [0,1]$ $K_{max}^{\alpha}(t) \le Z(t) \le K_{min}^{\alpha}(t)$.
(b) Each point of the region bounded by K_{min}^{α} *and* K_{max}^{α} *lies on a curve from* $ar_{\max}^{-1}(\alpha)$.

Proof:

(a)

Suppose Z crosses K_{max}^{α} say at a point $D = (d_x, d_y)$. Consider $Z_1 = ODI$ (Fig. 5.3b). It is easy to calculate that $ar(Z_1) = d_x - d_y$. Hence, by (5.4), $ar(Z_1) = \alpha$. Since Z is convex and crosses K_{max}^{α} at D, it must lie strictly below Z_1. Hence $ar(Z) > ar(Z_1) = \alpha$, which contradicts $Z \in ar_{\max}^{-1}(\alpha)$.

Suppose now that Z crosses K_{min}^{α}. Near $t = 0$ Z must lie below K_{min}^{α}, otherwise it would be strictly above $M_{1/(1-\alpha)}^{\alpha}$. Therefore it must first cross K_{min}^{α} from below and if P is the first point of crossing, the (right) derivative of Z at P, Z'_P, is greater than the derivative of K_{min}^{α} at P, $(K_{min}^{\alpha})'_P$. Since both curves are convex their derivatives are increasing. Therefore either there is a point on K_{min}^{α}, say P', right of the point P, such that $(K_{min}^{\alpha})'_{P'} = Z'_P$, or Z'_P is greater than $(K_{min}^{\alpha})'_I$ (see Fig. 5.14). In the first case Z lies strictly above the curve M_{β}^{α} which is tangent to K_{min}^{α} at P', so that $ar(Z) < \alpha$. In the second case Z could never reach the value 1 which again contradicts $Z \in ar_{\max}^{-1}(\alpha)$.

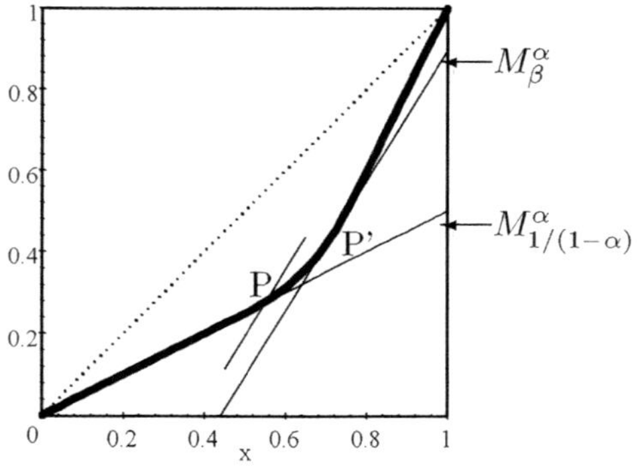

FIGURE 5.14. Curves illustrating construction of K_{min}^{α}

(b)

Each point $P \in ar_{\max}^{-1}(\alpha)$ lies either on (at least) one trapezoidal curve M_{β}^{α} or on a triangular curve of the type ODI (Fig. 5.3b) constructed in the first part of the proof (a).

5.6.2 Asymmetry indices

Let us consider the most general case, when pairs of considered variables have an unrestricted value range. Moreover, their concentration curve may contain vertical or horizontal segments. While the language of curves introduced in Sec. 5.4 (inverse curves and composition of curves) and the language of variables are equivalent in the ULM, we usually use the first as more universal and convenient, except when referring to random variables.

Theorem 5.2 *The relative asymmetry curve $A(Y : X)$ does not depend on the selected opposite order function and can be expressed in the form:*

$$A(Y : X)(u) = 1 - C^{-1}(Y : X) \circ (1 - C(Y : X)(u)) = \tag{5.15}$$

$$= 1 - C(X : Y) \circ (1 - C^{-1}(X : Y)(u)); \; u \in [0,1]. \tag{5.16}$$

Proof:
It follows immediately from the definitions that

$$\begin{aligned} A(Y : X) &= C(C(X^{\leftarrow} : Y^{\leftarrow}) : C(X : Y)) = \\ &= \{(C(X : Y)(t), 1 - C(X : Y)(1-t)); t \in [0,1]\} = \\ &= \{(C^{-1}(Y : X)(t), 1 - C^{-1}(Y : X)(1-t)); t \in [0,1]\}, \end{aligned}$$

so

$$\begin{aligned} A(Y : X)(u) &= 1 - C(X : Y) \circ (1 - C^{-1}(X : Y)(u)) = \\ &= 1 - C^{-1}(Y : X) \circ (1 - C(Y : X)(u)). \end{aligned}$$

The properties of the asymmetry indices (A and AI) for opposite variables are described by the following lemma.

Lemma 5.3 *If opposite order functions for variables $X^{\leftarrow}$ and $Y^{\leftarrow}$ are the same, then*
a) $A(Y^{\leftarrow} : X^{\leftarrow}) = A^{-1}(Y : X)$,
b) $AI(Y^{\leftarrow} : X^{\leftarrow}) = -AI(Y : X)$.

Proof.
Part (a) follows immediately from the definitions of concentration curves and asymmetry curves:
$A(Y^{\leftarrow} : X^{\leftarrow}) = C(C(X : Y) : C(X^{\leftarrow} : Y^{\leftarrow})) = C^{-1}(C(X^{\leftarrow} : Y^{\leftarrow}) : C(X : Y)) = A^{-1}(Y : X)$
Part (b) follows from two facts. The first is that curve A^{-1} is symmetric to A with respect to diagonal $y = x$. The second is that AI is proportional to the area between A and diagonal $y = x$.

Asymmetry measures are invariant w.r.t. increasing transformations, that is $A(Y : X) = A(g(Y) : g(X))$, where g is an increasing transformation $R \to R$. It follows from the similar invariance of concentration curves.

Lemma 5.4 *For any X, Y, the curve $A(Y:X)$ is symmetric with respect to diagonal $y=1-x$, i.e.*

$$A(t)=1-A^{-1}(1-t)$$

Proof.
From (5.15) $A^{-1}(Y:X)(u)=H^{-1}\circ(1-H(1-u))$, where $H=C(Y:X)$, therefore $1-A^{-1}(Y:X)(1-u)=1-H^{-1}\circ(1-H(u))=A(Y:X)(u)$.

Let Z be a continuous variable on $[0,1]$ with cdf F_Z, which is symmetric w.r.t. $y=1-x$. Let U be the uniform variable on $[0,1]$. It is easy to check that $A(Z:U)=F_Z(F_Z)$. Thus, if a variable has symmetric (in this sense) cdf then this cdf is also an adequate asymmetry measure.

Lemma 5.5 *Let X,Y be continuous variables on $[0,1]$ such that their cdf's F_X,F_Y are increasing; let Z be such that $F_Z=F_Y\circ F_X$; let U be uniform on $[0,1]$. If Y is symmetric, that is if $F_Y(u)=1-F_Y(1-u)$, $u\in[0.1]$ then*

$$A(Z:U)=A(X:U).$$

Proof.
From (5.15) $A(Z:U)(u)=1-F_Z^{-1}(1-F_Z(u))=1-F_X^{-1}\circ F_Y^{-1}(1-F_Y\circ F_X(u))$. But $F_Y^{-1}(1-F_Y)(u)=1-u$, so $A(Z:U)(u)=1-F_X^{-1}(1-F_X(u))=A(X:U)(u)$.

Lemma 5.6 *Let X,Y be random variables on $[0,1]$ and let U be uniform on $[0,1]$. If X is stochastically smaller than Y ($X\prec_{st}Y$) then $A(X:U)>A(Y:U)$.*

Proof.
Let F and G be respective cdfs of X and Y. If $X\prec_{st}Y$ then $F>G$. The curves symmetric to them w.r.t. the diagonal $y=1-x$ (F_{Sym} and G_{Sym} respectively) satisfy a similar inequality $F_{Sym}>G_{Sym}$. Therefore $A(X:U)=F_{Sym}\circ F>G_{Sym}\circ G=A(Y:U)$.

Let X be a continuous random variable on $[0,1]$ with increasing cdf F. Let F_{Sym} denote the curve symmetric to F w.r.t. the diagonal $y=1-x$. It is also a cdf of some variable X_{Sym} continuous on $[0,1]$. If U is the uniform variable on $[0,1]$, then according to the definition $A(X_{Sym}:U)=(F_{Sym})_{Sym}\circ F_{Sym}$. As $(F_{Sym})_{Sym}$ denotes the curve symmetric to F_{Sym} then it is equal to F, so $A(X_{Sym}:U)=F\circ F_{Sym}=(F_{Sym}^{-1}\circ F^{-1})^{-1}$. On the other hand $A(U:X)=F_{Sym}^{-1}\circ F^{-1}$. Finally

$$A(X_{Sym}:U)=A^{-1}(U:X).$$

Generally, this formula does not enable to determine which variable X or X_{Sym} is more skewed and in what direction. However, if X belongs to a very special family such an evaluation is possible.

Let us consider a triangular curve TR defined on $[0,1]$ and described in Sec. 5.2. It is given by the formula:

$$TR(t) = \begin{cases} \frac{v_y}{v_x} t & \text{for } t \le v_x \\ \frac{1-v_y}{1-v_x} t + \frac{v_y - v_x}{1-v_x} & \text{for } t > v_x \end{cases}$$

where (v_x, v_y) is the vertex point which determines it. Its asymmetry curve is equal to:

$$A(TR)(t) = \begin{cases} \frac{(1-v_x)v_y}{(1-v_y)v_x} t & \text{for } t \le v_x \text{ and } t \le \frac{(1-v_y)v_x}{v_y} \\ t - \frac{v_x - v_y}{1-v_y} & \text{for } t > v_x \text{ and } t \le 1 - \frac{v_y(1-v_x)}{1-v_y} \\ & \text{or } t \le v_x \text{ and } t > \frac{(1-v_y)v_x}{v_y} \\ \frac{(1-v_y)v_x}{(1-v_x)v_y} t - \frac{v_x - v_y}{(1-v_x)v_y} & \text{for } t > v_x \text{ and } t > 1 - \frac{v_y(1-v_x)}{1-v_y} \end{cases}$$

The respective formula for the asymmetry index is the following:

$$AI(TR) = \begin{cases} 1 - \left(\frac{1-v_x}{1-v_y}\right)^2 (1 - 2v_y) - 2v_y \frac{1-v_x}{1-v_y} & \text{for } v_y \le 0.5 \\ (v_x - v_y)\left(\frac{(v_x - v_y)(2v_y - 1)}{v_y^2} + 2\right) & \text{for } v_y > 0.5 \end{cases}$$

If a triangular curve K belongs to $ar_{\max}^{-1}(\alpha)$, where $\alpha > 0$, then $ar(K) = v_x - v_y = \alpha$ and consequently

$$AI(TR) = \begin{cases} 1 - \left(1 - \frac{\alpha}{1-v_y}\right)^2 (1 - 2v_y) - 2v_y \left(1 - \frac{\alpha}{1-v_y}\right) & \text{for } v_y \le 0.5 \\ \alpha\left(\frac{\alpha(2v_y - 1)}{v_y^2} + 2\right) & \text{for } v_y > 0.5 \end{cases}$$

As $1 - \frac{\alpha}{1-v_y}$ is a decreasing function of v_y, the first part of AI is an increasing function of v_y. The same property holds for the second part of the formula. This enable us to formulate the following corollary:

Corollary 5.7 *Let $\{TR_V\}$ be a family of triangular curves defined by vertex points $V = (v_x, v_y)$ which belong to $ar_{\max}^{-1}(\alpha)$, where $\alpha > 0$ In this family the asymmetry index $AI(X_V)$ is an increasing function of v_x (as well as v_y).*

In other words in the inverse maximal concentration set, if vertex points move to the right, the asymmetry index increases, hence the skewness to the right of the corresponding variables (curves) increases.

5.6.3 Bibliographical remarks

Concepts of skewness of the univariate distribution have been extensively studied in the literature. For example, a very short summary of theoretical developments in this area with a reference list may be found in [6].

The asymmetry measures discussed in this chapter are based on the classical approach (see, for example [71] or [32]) which compares the tails of distributions.

Asymmetry measures (simple and relative) were introduced by Ciok in [17]. In the same paper it was shown that relative asymmetry measures are invariant with respect to the chosen reverse functions (in this book the reverse functions are introduced under the name "opposite order functions"). Dual curves were first introduced by the authors in [28] as the Lorenz curve of a ratio variable which is symmetric w.r.t. the diagonal *Diag*. In [17] this curve was expressed in a new way as a concentrate curve for opposite variables. Thanks to this formula, the dual curve was used for the construction of relative asymmetry curves.

The form of the relative asymmetry curve (and its being composed of curves, and not functions) brings to the notice of the author of this chapter the problem of concentration curves which include vertical segments. To deal with this problem, operations on curves (composition, inversion) were introduced in [17]. This formalism make possible a generalization of asymmetry curves for the case of curves which are not Lilliputian.

Concentration curves in the context of single distribution skewness, under the name of orbits, were used by [71].

The boundaries of the smallest region containing all maximal concentration curves with a common value of $ar_{\max}$ were found by Ciok in [23].

6

Discretization and regularity

Alicja Ciok

6.1 Introduction

Discretizations of continuous attributes are very common and very important in data analysis. In this chapter, the problem of simultaneous discretization of two variables is considered. Due to the formalism used (i.e., grade framework) which is based on concentration indices, the discretization of one random variable can be treated as a particular example of this problem. The grade framework is described in Sec. 6.2.

Next, it will be shown, by a carefully chosen example, that the loss of information about the mutual behaviour of both variables can be great. Assuming that the number of categories is known, the simultaneous discretization which minimizes this loss is presented (Sec. 6.3). Thanks to the language of concentration curves, it is possible to distinguish between the loss of information due to the specific discretization and the loss associated with a chosen number of categories. It is important to pay particular attention to the quality of the discretization used, because the error due to bad discretizations can be greater than that caused by the wrong number of categories.

Sections 6.4 - 6.6 are devoted to the determination of the proper number of categories. First, the concept of a "natural" category is discussed. Next, the concept of *ideally regular* concentration curves is introduced. These concepts enable us to classify a pair of variables into one of the three following classes: no category structure is present (i.e., the respective concentration curves are ideally regular); a natural category structure is present; an intermediate structure between the first two.

It will be shown that in ideally regular concentration curves there are no natural categories. In such a case, *any number of categories is fine provided that the generated error* (i.e., the difference between the initial curve and its counterpart for the discretized variables) *is acceptable.* Therefore, the first step in the determination of the proper number of categories is to check whether the curve is, or is not, regular.

However, to determine whether a curve is ideally regular may be difficult. To make the search easier a family of ideally regular curves from the $\mathbb{ULM}$ is presented. Its members are parametrized by two indices: the concentration index and the asymmetry index. Knowing the values of these parameters for a given curve, we can determine the ideally regular curve with the same (or at least very similar) values of these parameters, or we can state that there is no such curve. If the similarity between the curve and the respective regular curve is great, the curve will be recognized as almost ideally regular (i.e., without a category structure) and the number of categories is then determined only by the required precision.

6.2 Discretization framework

As stated previously, the problem of the discretization of one random variable and that of the simultaneous discretization of two varables are usually considered separately. However, in the Univariate Lilliputian Model the first task is an example of the second. Due to the equivalence between cdfs and concentration curves (see Chapter 3), the discretization of a pair: a variable from the $\mathbb{ULM}$ and a variable uniform on $[0, 1]$, is equivalent to the discretization of the first variable. That is, it is the common value range of both variabls which is partitioned into non-overlapping subsets, and therefore does not matter whether it is a value range of one variable or a range of two variables. These subsets determine the categories of the newly formed discrete variable(s). Simultaneous discretization of pairs is more general, therefore we will use the language of concentration curves rather than the language of cdfs. However, the reader should bear in mind that all conclusions are also valid for the case of one variable.

Let us assume that n is the number of new categories. Let C be a concentration curve of variable Y on X, their respective cdfs are F_Y and F_X. We will consider only *convex concentration curves* because this assumption is necessary for the construction of a discretization procedure. It is not a strong restriction: we can always take into account the maximal concentration curves which are convex. This usually transforms the initial variables, but a discretization of the maximal concentration curve always determines a categorization of the initial variables

If vector $w = (w_0 = 0, w_1, ..., w_{n-1}, w_n = 1)$, where $w_{i-1} \leq w_i$, defines the partition of interval $[0, 1]$ into n consecutive intervals $(w_{i-1}, w_i]$, then the intervals $(F_X^{-1}(w_{i-1}), F_X^{-1}(w_i)]$ define the partition of the common range S into non-overlapping subsets. These subsets constitute n categories of the newly formed discrete variables.

Let us note that if the initial variables are discrete (we wish to aggregate the original categories), only these partition points are considered which correspond to the initial categories. Also if the concentration curve includes a horizontal segment then the inversion F_X^{-1} should be extended in two ways: $F_X^{-1}(w) = \max\{v : F_X(v) = w\}$ or $F_X^{-1}(w) = \min\{v : F_X(v) = w\}$. The choice of minimum or maximum depends on which end of the interval is considered. For the interval $(F_X^{-1}(w_{i-1}), F_X^{-1}(w_i)]$ the minimum is appropriate for $F_X^{-1}(w_{i-1})$, the maximum for $F_X^{-1}(w_i)$.

According to the definition, the concentration curves (or their maximal counterparts) corresponding to these new discrete variables, are piecewise curves which have points $(w_i, F_Y(w_i))$ $(i = 0, ..., n)$ in common with the respective initial curves.

Figure 6.1 presents two concentration curves: one corresponds to the pair of continuous variables (X, U), where U is uniformly distributed. Interval $[0, 1]$ is the common support of both variables, and $F_X(x) = 1 - \sqrt[4]{1-x}$ is the cdf of X. The second curve corresponds to the discretized counterparts $(\widehat{X}, \widehat{U})$. Discretization points for both variables are $w_i = \frac{i}{n}$, where $i = 0, ..., n$ and $n = 4$. In this case, support S is equal to interval $[0, 1]$, hence the intervals $(\frac{i-1}{n}, \frac{i}{n}]$ immediately determine the new categories of variable X (and of U as well).

In the general case, if the inversion F_X^{-1} exists then the discretization points w_i for curve $C(Y : X)$ are the quantiles of order $F_X(w_i)$ of variable X. The discretization determined by $w_i = \frac{i}{n}$ will be called the uniform quantile discretization.

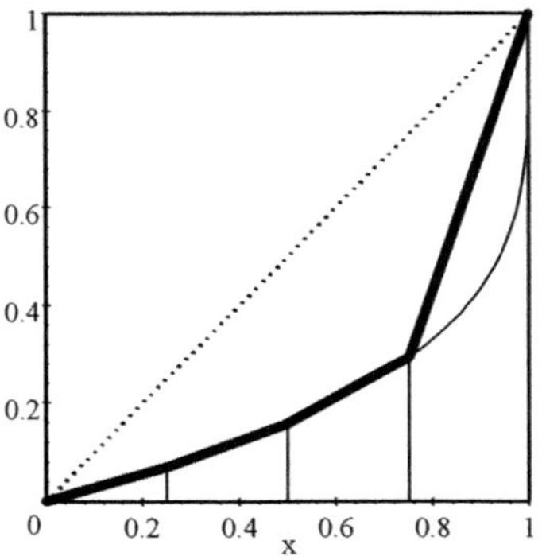

FIGURE 6.1. Two concentration curves: one for a pair of continuous variables and the other for their discretized counterparts (bold)

Looking at Fig. 6.1 we can see that in the first three intervals both concentration curves are very similar. In the last interval ($[0.75, 1]$) the difference between the curves is greater, hence much information about the initial distributions is lost when this discretization is used. This loss can distort the evaluation of mutual relations between the analyzed variables.

6.3 Optimal discretization for a given number of categories

A discretization should provide for the minimum possible loss of information. In our framework, this means that the maximal concentration curves for the initial variables and for their discretized counterparts must be as similar as possible. This similarity is measured by the difference between the respective values of the maximal concentration index ar, or equivalently by the area between the original curve and its discretized counterpart (thanks to the interpretation of the concentration index (see Sec. 3.2.2)). Therefore the minimization of this difference is equivalent to the maximization of the ar for the discretized variable.

Definition 6.1 *(optimal discretization) For a given number of categories the simultaneous discretization of two random variables is called optimal when the respective value of the maximal concentration index is maximal.*

Let $w = (w_0 = 0, w_1, ..., w_{n-1}, w_n = 1)$ be the partition vector which determines the discretization and let n be the number of categories. The procedure which generates an optimal partition is based on the necessary conditions for optimality provided by the following theorem (the proof is given in Sec. 6.7.1).

Theorem 6.1 *Let C be a convex concentration curve $C(Y : X)$. If partition vector w determines the optimal simultaneous discretization of variables (X, Y) into n categories, then the following condition must be satisfied:*

$$\frac{dC}{dx}(w_i - 0) \leq \frac{C(w_{i+1}) - C(w_{i-1})}{w_{i+1} - w_{i-1}} \leq \frac{dC}{dx}(w_i + 0) \tag{6.1}$$

If C has a continuous derivative (denoted by c) then condition (6.1) is transformed into the equality:

$$c(w_i) = \frac{C(w_{i+1}) - C(w_{i-1})}{w_{i+1} - w_{i-1}} \tag{6.2}$$

> In many practical cases, the necessary condition is also sufficient. For example, if the derivative c of C is continuous and increasing then the solution of condition (6.1) is unique. Moreover, even if more solutions exist, they are equivalent in the sense that they have identical values of the concentration index. (The proofs are included in Sec. 6.7.1.)

Fig. 6.2 shows the pair of concentration curves similar to those visible in Fig. 6.1. One curve corresponds to the initial continuous pair of variables (hence it is identical to the curve visible in Fig. 6.1), the second corresponds to their optimally discretized counterparts.

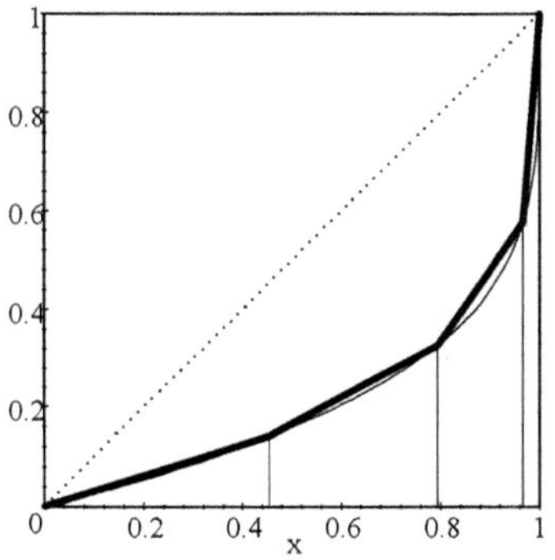

FIGURE 6.2. Two concentration curves: one for a pair of continuous variables and the other for their optimally discretized counterpart (bold)

The simultaneous discretizations of two variables are frequently used in the estimation of the concentration index ar. The difference between the results of the uniform quantile and the optimal discretizations (compare Figs 6.1 and 6.2) clearly shows how great the influence of a chosen discretization on this estimation can be. In the optimal discretization the partition points w_i for $i = 1, 2, 3$ are 0.454, 0.79447, 0.96763 respectively. The analogous points in the uniform discretization are $0.25, 0.5, 0.75$. The value of ar for the initial continuous variables is 0.6, when the respective values after discretizations are: 0.57 for the optimal discretization and 0.49 for the uniform quantile discretization. Looking at Fig. 6.2, one can see that the more skewed to the right one variable is with respect to the other (cf Chapter 5), the worse the uniform quantile discretization. Perhaps, this is the reason for the dispersion produced by the estimator of ar, which is based on the quantile discretization (described in [23]).

Discretizations of two variables which do not take the shape of the concentration curve into account, can cause great dispersion in the estimated values of ar

In a simulation study presented in the same paper ([23]), the influence of the number of categories on the efficiency of the estimation of the concentration index is also shown. For any discretization into n categories, if each category is divided into k subcategories, then the latter discretization is better. Unfortunately, one usually works on data samples whose sizes restrict, sometimes drastically, the quality of cdf approximation and consequently the possible choice. It seems that in such cases, the optimal discretization is the best remedy.

The loss of information caused by a bad discretization can be much greater than the loss caused by too few categories.

Let n_0 is the chosen number of categories for the curve C. The following difference measures the quality of one discretization of C into n categories compared with the optimal discretization into n_0 categories. This quality can be expressed by the sum of two differences:

$$ar(C(n, D_n)) - ar(C(n_0, D_{n_0}^{opt})) = [ar(C(n, D_n)) - ar(C(n, D_n^{opt}))] + [ar(C(n, D_n^{opt})) - ar(C(n_0, D_{n_0}^{opt}))]$$

where $C(i, D_i)$ denotes the discretized counterpart of curve C for i categories and a discretization D_i $(i = n, n_0)$. D_i^{opt} denotes the optimal discretization for i categories.

The first difference in this formula measures the quality of discretization D_n, the second measures whether the number of clusters n was properly chosen.

6.4 Ideally regular concentration curves

There are pairs of variables which have convex concentration curves with extraordinary properties. Let us consider the optimal discretization of variables (X, U) from the $\mathbb{ULM}$ into n categories. Let variable U be uniform on $[0, 1]$ and let the cdf of X be the parabola: $F_X(t) = t^2$, $t \in [0, 1]$. In this case the optimal discretization is equal to the uniform quantile discretization, that is, the optimal discretization points are equal to $\frac{i}{n}$, $i = 0, ..., n$. As this property is valid for every n, it is clear that the optimal discretization into kn categories only adds new partition points to those defining the discretization into n categories. This property gives rise to the following definition.

Definition 6.2 *(ideally regular concentration curve) Let vector* $w = (w_0 = 0, w_1, ..., w_{n-1}, w_n = 1)$ *define the optimal discretization of the convex concentration curve* C *into* n *categories. Vectors* $w^i = (w_0^i = w_{i-1}, w_1^i, ..., w_{k-1}^i, w_k^i = w_i)$ *determine the partitions of the subintervals* $[w_{i-1}, w_i]$ $(i = 1, ..., n)$ *according to condition (6.1). Moreover, let* $w_j^i \neq w_{j+1}^i$ *for all* $i = 0, ..., n$ *and* $j = 0, ..., k-1$. *If for every* n *and* k *the discretization into* nk *categories defined by* w^i $(i = 1, ..., n)$ *is optimal, then the curve* C *is called ideally regular.*

Figure 6.3 presents two ideally regular curves: a parabola (a) and a circle (b) with their two optimally discretized counterparts divided into 2 and 4 categories. The four subcategories are derived from separate discretizations (according to condition (6.1)) of the previously obtained two categories.
Circles possess a particularly interesting property in that the optimal discretization always divides them into segments of identical shape. This property provides immediate proof of their regularity without the necessity of calculations determined by the regularity definition.

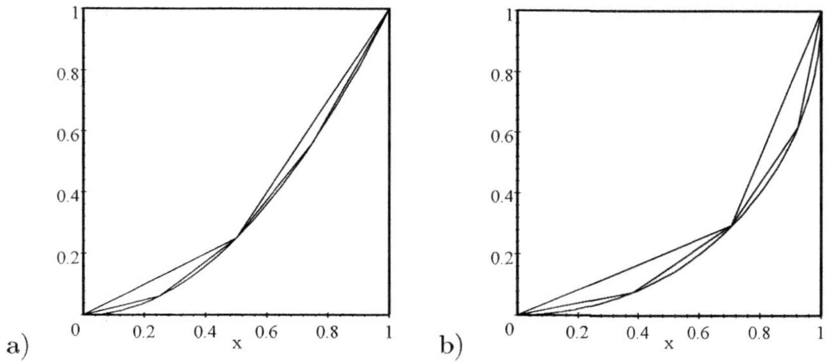

FIGURE 6.3. Two ideally regular curves: parabola (a) and circle (b) each with their two optimally discretized counterparts divided into 2 and 4 categories

Fig. 6.3 is a good illustration of the ideal regularity concept. Each category can be partitioned separately into further subcategories according to condition (6.1) and the discretization still remains optimal. Moreover, these partitions can proceed endlessly. Therefore we can say that *ideal regularity means a property of infinite optimal divisibility.*

Let us consider variables X and Y with the same value range S and respective differentiable cdfs: F_X and F_Y. Let X^A and Y^A be the same variables, but restricted to subinterval $A = [a, b] \subset S$. Their cdfs are denoted respectively by F_X^A and F_Y^A. Assuming that F_X^{-1} exists, it is easy to calculate a so called restricted concentration curve.

Definition 6.3 *(restricted concentration curve) The restricted concentration curve $C_A(Y : X)$ is equal to $C(Y^A : X^A)$.*

Proposition 6.1 *The optimal simultaneous discretization of the variables restricted to a chosen interval is determined by the optimal partition of the respective interval of the initial concentration curve.*

Proof:

If F_X^{-1} exists then $C(u) = C(Y : X)(u) = F_Y(F_X^{-1}(u))$ $(u \in [0,1])$. Since $F_Z^A(v) = (F_Z(v) - F_Z(a))/(F_Z(b) - F_Z(a))$ where $Z \in \{X, Y\}$ and $v \in A$. Consequently $(F_X^A)^{-1}(u) = F_X^{-1}[u(F_X(b) - F_X(a)) + F_X(a)]$ for $u \in [0,1]$ and the restricted concentration curve is given by the formula:

$$\begin{aligned} C_A(u) &= C(Y^A : X^A)(u) = \\ &= \frac{F_Y[F_X^{-1}[u(F_X(b) - F_X(a)) + F_X(a)]] - F_Y(a)}{F_Y(b) - F_Y(a)} = \\ &= \frac{C[u(F_X(b) - F_X(a)) + F_X(a)] - F_Y(a)}{F_Y(b) - F_Y(a)}, \quad u \in [0,1]. \end{aligned}$$

Hence its derivative satisfies

$$\frac{dC_A}{du}(u) = \frac{dC}{du}(u(F_X(b) - F_X(a)) + F_X(a))\frac{F_X(b) - F_X(a)}{F_Y(b) - F_Y(a)}$$

Let $w_i^A \in [0, 1]$ $(i = 0, ..., k)$ be the partition points of C_A according to condition (6.4). Therefore

$$\frac{dC_A}{du}(w_i^A) = (C_A(w_{i+1}^A) - C_A(w_{i-1}^A))/(w_{i+1}^A - w_{i-1}^A)$$

and consequently

$$\begin{aligned}\frac{dC}{du}[w_i^A(F_X(b) - F_X(a)) + F_X(a)](F_X(b) - F_X(a)) = \\ = C(w_{i+1}^A(F_X(b) - F_X(a)) + F_X(a))/(w_{i+1}^A - w_{i-1}^A) \\ - C(w_{i-1}^A(F_X(b) - F_X(a)) + F_X(a))/(w_{i+1}^A - w_{i-1}^A)\end{aligned}$$

Substituting $w_j^A(F_X(b) - F_X(a)) + F_X(a) = w_j$ and changing the differentiation variable u into $v = u(F_X(b) - F_X(a)) + F_X(a)$ $(v \in [F_X(a), F_X(b)])$ we obtain

$$\frac{dC}{du}(w_i) = (C(w_{i+1}) - C(w_{i-1}))/(w_{i+1} - w_{i-1})$$

This formula is nothing other than the equality (6.2), therefore w_i are the optimal partition points for curve C.

The immediate conclusion of this proposition is that if a concentration curve is ideally regular, then any restricted concentration curve related to it is also ideally regular.

As the ideal regularity property repeats at any optimal partition level, the ideally regular curves can be regard as fractals.

There are many definitions of fractals. The most appropriate in this context is the following: *a fractal is a fragmented pattern that can be subdivided in parts, each of which is similar to the whole.* Let us note that in this context, a pattern has a more general meaning here than just a geometrical shape. This *fractal property* of ideally regular curves provides an additional argument for the interpretation of ideal regularity as infinite divisibility.

Currently, two families of ideally regular curves are known: parabolas and ellipses introduced in Sec. 5.3.4. The regularity of ellipses follows from the fact that they are restricted concentration curves for curves in the form of circles.

It can be proven (see Sec. 6.7.1) that:

> Curves which are symmetric to ideally regular curves w.r.t. the diagonal $Diag^{\perp}$ are also ideally regular. Therefore a concentration curve dual to an ideally regular curve is also ideally regular.

Consequently, we have two additional families of ideally regular curves: symmetric to parabolas, and to ellipses (w.r.t. the diagonal $Diag^{\perp}$).

6.5 On the determination of the number of categories

Two possible approaches which concern the problem of category identification can be considered:

- The first (natural category approach) starts from a formalisation of natural category concept and next tries to reveal categories in the data.
- The second (discretization approach) is an optimization task where one tries to replace continuous variables by their discretized conterparts with a minimal (or at least acceptable) loss of information.

Of course, both approaches overlap. Up to now we concentrated on the optimization approach, now we will turn to the other approach.

The grade discretization framework provides a natural definition of categories. According to the definition of concentration curves, they include linear segments if the variables are discrete. On the other hand, continuous variables can also produce concentration curves which include linear segments. This occurs when the common range of the variables is partitioned into consecutive intervals where the density ratios for the corresponding variables are constant (in particular when both densities are constant). This case will be used as a category model. The idea of inner uniformity (constant density) as a category (or more generally a cluster) distinguishing feature is not new (for example, in cluster context it was discussed in [41]). However under the discretization framework this idea is a straightforward generalization of the natural categories of discrete variables.

Now the problem arises: how to find categories in data or how to check that there are no clusters at all. The concept of ideal regularity can provide an appropriate solution.

Let us consider a convex concentration curve which consists of several linear segments. To simplify the explanation, let us assume that this curve ($C(X:U)$) corresponds to variables from the $\mathbb{ULM}$ and that U is uniform.

Figure 6.4 shows an example of such a curve which consists of four linear segments. These segments correspond to the intervals at which the density of X is constant. According to the definition, the variable $X \in \mathbb{ULM}$ has four natural categories.

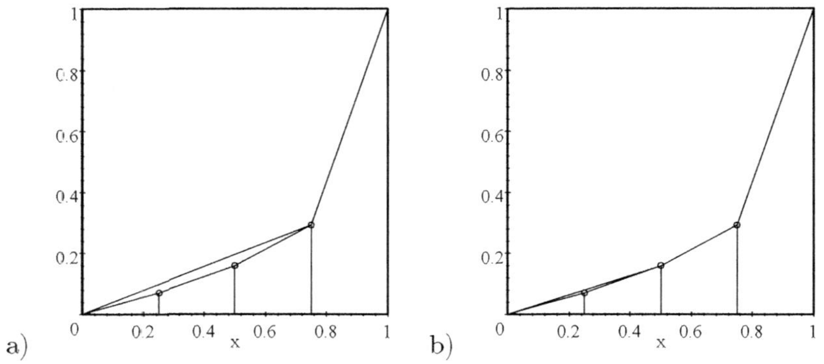

FIGURE 6.4. A concentration curve with its two discretized counterparts: a) into 2 optimal categories, b) into 4 categories where each of the two intervals determined previously was optimally partitioned into 2 subcategories

The curve is defined by points $(w_i, C(w_i))$, where $w_1 = 0.25$, $w_2 = 0.5$, $w_3 = 0.75$ (they are marked by the tiny circles in Fig. 6.4). Of course the optimal discretization into 4 categories reconstructs the original curve. The discretized curve after the optimal grouping into 2 categories is shown in Fig. 6.4a. Let us note that the partition point is w_3 and not w_2, therefore both curves (the initial and the discretized) are identical in the interval $[w_3, 1]$ and this interval (which determines a category) is not partitioned further. In the next step only the interval $[0, w_3]$ is partitioned into 2 subcategories according to formula (6.1). This yields w_2 as the next partition point (see Fig. 6.4b). Thus the curve is not ideally regular. The optimal partitions into 2 subcategories of each category obtained previously do not provide the optimal partition into 4 categories. There is also another reason. *The definition of ideally regular curve requires that consecutive partition points should not be identical* - this requirement is not fulfilled in our example. Let us note that the consecutive optimal partitions of categories provides finally the optimal solution if the number of categories used in the optimization procedure is sufficiently great.

These observations lead to the following statements which are true for all concentration curves which include only linear segments.

- the partition process which checks for ideal regularity must end when some consecutive partition points are equal,

- thanks to the same partition process, it is always possible to reconstruct the categories present in the data, however, to obtain such results, sometimes more partitions must be used than the real number of categories.

As shown above, the errors caused by optimal discretizations is measured by the difference $ar(C(Y : X)) - ar(C(\widehat{Y} : \widehat{X}))$, where $\widehat{X}$ and $\widehat{Y}$ are the discretized counterparts of X and Y. Comparison of these errors for ideally regular curves and for the curves which include only linear segments provides additional characteristics which can differentiate these two kinds of curves. It is obvious that the error decreases when the number of categories increases. However, for ideally regular curves the error is always relatively high (greater than zero) contrary to the error for the latter curves, where for the appropriate number of categories it reduces to zero. Summarizing the above, the following classification of possible data structures can be introduced:

- Ideal regularity implies a lack of natural categories in data and produces discretization error (measured by the concentration index) that is relatively high for any number of categories. In this case, any number of categories can be chosen, provided that this error is acceptable.
- When a curve is not regular, but the discretizations provide reconstruction of the initial curve, it indicates that the real category structure has been found. Then the discretization error is equal to zero.
- The intermediate case, where the curve is not ideally regular, but the optimal discretization (when an appropriate number of categories is given) provides an approximation of natural categories with a relatively small error but greater than zero.

A practical conclusion is that the determination as to whether a category structure exists in data (at least approximately) or not, should be the first step in the determination of the proper number of categories. In other words we should know at the beginning whether the concentration curve is regular or not. However, this is sometimes difficult to determine. If a regular curve sufficiently similar to the curve of interest can be found, then one may say that there is no category structure in the data (at least approximately). To this purpose a family of ideally regular curves, appropriately parametrized is needed. Such a family will be presented in the following section.

6.6 A parametric family of ideally regular Lilliputian curves

It was shown in Chapter 5, that the concentration indices and the asymmetry indices can discriminate well curves from the $\mathbb{ULM}$. The task formulated in the previous section is to find a regular curve - as similar as possible - to a given convex concentration curve. The required similarity of curves implies that they should have identical values of the concentration index ar and of the asymmetry index AI. In other words, we must find regular curves in the set of curves with the given values of both indices. For this reason, the family of ideally regular curves should be characterized by these two parameters - the values of the indices: ar and AI.

As before, we restrict ourselves to variables belonging to the $\mathbb{ULM}$ and the pairs (X, U), where U is uniform. Recall that $C(X:U)$ is identical with the cdf of X (strictly, the graph of the cdf). Let Par denote the family of variables X^α ($\alpha \in [0,1]$) with cdfs F^α in the form of parabolas:

$$F^\alpha(x) = \alpha x^2 + (1-\alpha)x, \ 0 < \alpha \leq 1$$

This family of variables is discussed in Chapters 4 and 5. Values of the concentration index in this family are:

$$ar(X^\alpha : U) = \frac{\alpha}{3}$$

It is easy to confirm that the mixtures $X^{\alpha\beta}$ having the cdf $F^{\alpha\beta}(x) = \beta F^\alpha(x) + (1-\beta)x$, where $\beta \in [0,1]$, also belong to the family Par.

Let $Par_{Sym} = \{X^\alpha_{Sym}\}$ denote the family of variables for which the plots of corresponding cdfs F^α_{Sym} are symmetric w.r.t. the diagonal $y = 1 - x$ to F^α (cf Chapter 5). The asymmetry curve for X^α was defined (cf Chapter 5) as the plot of $F^\alpha_{Sym} \circ F^\alpha$. For the Par family such a combination produces a circle.

Now, let us consider a generalization of this composition in the form:

$$T(\alpha,\beta,\gamma) = F^{\alpha\gamma}_{Sym} \circ F^{\alpha\beta} \tag{6.3}$$

When $\gamma = 0$, $T(\alpha,\beta,\gamma) = F^{\alpha\beta}$. Also if $\beta = 0$, $T(\alpha,\beta,\gamma) = F^{\alpha\gamma}_{Sym}$.

Substituting: $\delta_1 = \alpha\beta$, $\delta_2 = \alpha\gamma$ when $\beta \neq 0$ and $\gamma \neq 0$, we obtain:

$$T(\alpha,\beta,\gamma)(x) = \frac{1+\delta_2}{2\delta_2} - \sqrt{\frac{\delta_1}{\delta_2}\left[\frac{(1-\delta_2)^2}{4\delta_1\delta_2} + \frac{(1+\delta_1)^2}{4\delta_1^2} - (x + \frac{1-\delta_1}{2\delta_1})^2\right]},$$

where $x \in [0,1]$. This formula defines an ellipse, which is an ideally regular curve.

If $T(\alpha,\beta,\gamma)$ is treated as a cdf of some variable $Y^{\alpha,\beta,\gamma}$ from the $\mathbb{ULM}$, the value of the concentration index for this variable is given by the formula:

$$ar(Y^{\alpha,\beta,\gamma} : U) = \frac{\delta_1 - \delta_2}{2\delta_1\delta_2} - \frac{1}{\delta_2} + \\ + \sqrt{\frac{\delta_1}{\delta_2}} M^2 \left(\arcsin \frac{1+\delta_1}{2M\delta_1} - \arcsin \frac{1-\delta_1}{2M\delta_1} \right),$$

$$\text{where } M = \frac{(1-\delta_2)^2}{4\delta_1\delta_2} + \frac{(1+\delta_1)^2}{4\delta_1^2}$$

Since $ar(X^\alpha : U)$ and $ar(Y^{\alpha,\beta,\gamma} : U)$ should be identical, one of the parameters (say γ) can be expressed as a function of the remaining two. Fig. 6.5 presents the plot of γ as a function of β when α is constant and equal to 0.9.

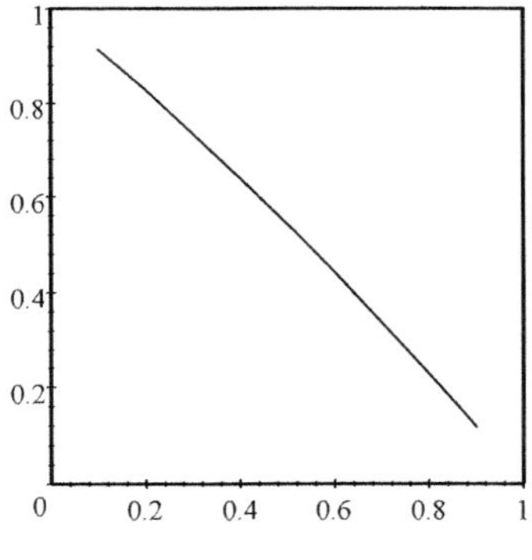

FIGURE 6.5. Plot of the parameter β as a function of γ and α when $\alpha = 0.9$

The function shown is almost linear. Calculations made for other values of α confirm that this property holds irrespective of the values of the concentration index $ar(X^\alpha : U)$.

Figure 6.6 presents several curves produced by the transformation $T(\alpha, \beta, \gamma(\alpha,\beta))$, when α is constant and equal to 0.9. Bold lines mark the extremal curves: the parabola, and the curve symmetric to it w.r.t. $Diag^{\perp}$.
It is clearly visible that the ellipses cover all the area between the extremal curves. Their ar is constant and equal to 0.3. Their asymmetry index AI (cf. Chapter 5) also changes monotonically (see the following table).

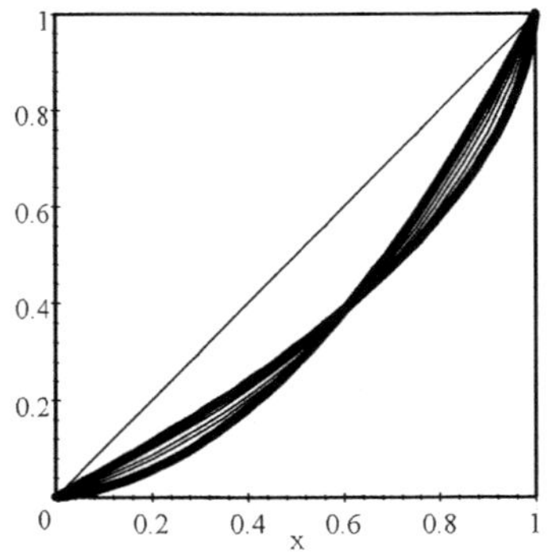

FIGURE 6.6. Family of ideally regular curves with constant ar equal to 0.3

TABLE 6.1. Values of asymmetry index for the chosen curves generated by T transformation when $ar = 0.3$

β	$\gamma(\alpha,\beta)$	AI
0.2	0.826	0.582
0.521	0.521	0.556
0.8	0.229	0.538

Let $TC(\alpha)$ denote the family of curves generated by the transformation $T(\alpha, \beta, \gamma(\alpha,\beta))$. This family includes members of all currently known families of ideally regular curves. Any member of $TC(\alpha)$ is determined by two parameters: ar and AI. These parameters are functions of parameters α and β. As mentioned earlier, the pair of indices: ar and AI has a high discriminative power in the set of convex Lilliputian curves. Therefore this pair helps to characterize the shape of a given curve.

If this curve is very similar to an ideally regular curve, then it can be recognized as almost ideally regular and a lack of natural categories can be concluded (at least approximately) Consequently any number of categories can be chosen provided that it provides an acceptable degree of precision for approximation. It should be emphasised that we really want to know whether the curve has a specific shape. To this purpose the chosen similarity measure should consider also the concentration and the asymmetry indices. In other words, we should compare the given curve with the ideally regular curve which has the same value of ar and of AI. The discrepancy between them (measured, for instance, by the concentration curve) can provide additional information about possible natural categories.

The idea of the ideal regularity of curves can be generalized on two-way data tables and exploited for determination of the proper number of clusters. This generalization will be discussed in Chapter 11.

6.7 Appendix

6.7.1 Optimal discretization

Theorem 6.2 *Let C be the convex concentration curve $C(Y:X)$. If the partition vector $w = (w_0, w_1, ..., w_{n-1}, w_n)$ determines the optimal simultaneous discretization of variables (X,Y) into n categories, then the following condition must be satisfied:*

$$\frac{dC}{dx}(w_i - 0) \leq \frac{C(w_{i+1}) - C(w_{i-1})}{w_{i+1} - w_{i-1}} \leq \frac{dC}{dx}(w_i + 0) \tag{6.4}$$

If C has a continuous derivative (denoted by c) then condition (6.1) is transformed into the equality:

$$c(w_i) = \frac{C(w_{i+1}) - C(w_{i-1})}{w_{i+1} - w_{i-1}} \tag{6.5}$$

Proof:

The optimality condition may be replaced by an equivalent one, minimalization of the area below the curve $C(Y_w : X_w)$ in the set of all discretization of adjacent points, where X_w and Y_w are discretized according the partition vector w. As this curve is a piece-wise curve linking points: $(0,0)$, $(w_i, C(w_i))$ $(i = 1, ..., n-1)$, $(1,1)$ this area is equal to

$$\frac{1}{2}\sum_{i=0}^{n-1}(C(w_i) + C(w_{i+1}))(w_{i+1} - w_i)$$

Let us consider another discretization determined by $\widetilde{w} = (w_0 = 0, w_1, ..., w_{i-1}, \widetilde{w}_i, w_{i+1}, ..., w_n = 1)$. $\widetilde{w}$ differs from w only in the i-th point w_i. As this discretization is not optimal, the area below the respective curve is greater, which leads to the following inequality:

$$\begin{aligned}&(C(w_{i-1}) + C(w_i))(w_i - w_{i-1}) + (C(w_i) + C(w_{i+1}))(w_{i+1} - w_i)\\ \leq\ &(C(w_{i-1}) + C(\widetilde{w}_i))(w_i - \widetilde{w}_{i-1}) + (C(\widetilde{w}_i) + C(w_{i+1}))(\widetilde{w}_{i+1} - w_i)\end{aligned}$$

and consequently:

$$\frac{C(w_{i+1}) - C(w_{i-1})}{w_{i+1} - w_{i-1}}(w_i - \widetilde{w}_i) \geq C(w_i) - C(\widetilde{w}_i)$$

Convergence of $\widetilde{w}_i$ to w_i from the left and the right provides both sides of the inequality (6.4). If the left side and the right side derivatives are equal, we obtain condition (6.5).

Condition (6.4) does not provide a unique solution. Any point w_i whose value $C(w_i)$ lies on the line parallel to that linking points $(w_{i-1}, C(w_{i-1}))$

and $(w_{i+1}, C(w_{i+1}))$ is equally good. We will call these points *C- parallel to* (w_{i-1}, w_{i+1}). All of them satisfy condition (6.4), moreover the selection of any of them does not change $ar(Y_w : X_w)$.

Proposition 6.2 *If derivative c is continuous and increasing then the solution of condition (6.4) is unique.*

Proof:

In the continuous case, condition (6.4) reduces to condition (6.5), which implies that all points w_i for $i = 2, ..., n$ are increasing functions of chosen w_1 (given w_j for $j = 0, ..., i$ we may always compute w_{i+1} using formula (6.5)). If points w_i $(i = 1, ..., n)$ define the optimal discretization then w_n must be equal to 1. The immediate conclusion is that they must be unique.

Proposition 6.3 *Let* $w = (0, w_1, ..., w_{n-1}, 1)$ *and* $w' = (0, w'_1, ..., w'_{n-1}, 1)$ *be partition vectors which determine two discretizations of the convex curve C. If these vectors satisfy the necessary condition for optimality and differ in points* $w_j \neq w'_j$, $j = j_1, ..., j_r$ *then* w_j *and* w'_j *are C-parallel to* (w_{j-1}, w_{j+1}). *Therefore both discretized curves have the same value of ar.*

Proof:

Without a loss of generality, we may assume that there exists such k and m $(k < m)$ that $w_i = w'_i$ for all $i < k$ or $i > m$, moreover $w_k < w'_k$. It also follows from (6.1) that $w_j \leq w'_j$ for $j > k+1$. The only possibility to reconcile this condition with $w_{m+1} = w'_{m+1}$ is that w_k and w'_k are C-parrallel to some interval (w_{k-1}, w_{k+1}). The same reasoning may be repeated for further points w_j.

Proposition 6.4 *If a convex concentration curve C is ideally regular then the curve* C_{Sym} *symmetric to it (w.r.t. the diagonal* $y = 1 - x$*) is also ideally regular.*

Proof:

Let points $w_i \in (0, 1)$ define the optimal discretization of C_{Sym}, then:

$$\frac{C_{Sym}(w_{i+1}) - C_{Sym}(w_{i-1})}{w_{i+1} - w_{i-1}} = \frac{C^{-1}(1 - w_{i-1}) - C^{-1}(1 - w_{i+1})}{(1 - w_{i-1}) - (1 - w_{i+1})}$$

and if the derivative exists

$$\frac{dC_{Sym}(w_i \pm)}{dx} = \frac{1}{\frac{dC}{dx}(C^{-1}(1 - w_i \mp))}$$

Substituting $v_{n-i} = C^{-1}(1 - w_i)$, where n is the number of categories, and comparing both equations according to (6.4), we obtain the analogous optimality condition, but for points v_{n-i}. Therefore, for any vector of optimal partition points for C, we have the corresponding optimal vector for C_{Sym}. Consequently, if C is regular, C_{Sym} must be also regular.

6.7.2 *Algorithm of optimal discretization*

The starting point of the algorithm is the choice of the number of classes (say n).

1. Next, the vector $w^0 = (w_0^0 = 0, w_1^0, ..., w_{n-1}^0, w_n^0 = 1)$, which defines the initial configuration of the discretization points must be chosen.
2. Given the configuration vector $w^i = (w_0^i = 0, w_1^i, ..., w_{n-1}^i, w_n^i = 1)$ calculated in the previous step, the next configuration $w^{i+1} = (w_0^{i+1} = 0, w_1^{i+1}, ..., w_{n-1}^{i+1}, w_n^{i+1} = 1)$ is generated using formula (6.4) for discrete variables, or formula (6.5) for continuous variables. In these formulas w_j^{i+1} plays the role of w_j. Points w_{j-1} and w_{j+1} are replaced by w_{j-1}^{i+1} and w_{j+1}^{i}, respectively.
3. Step 2 is consecutively repeated for $j = 1, ..., n-1$.
4. Step 3 is repeated for each consecutive i until the difference between vectors w^i and w^{i+1} is smaller than the chosen level.

Proposition 6.5 *The procedure for optimal discretization always converges.*

Proof:

Each partition vector w^i generated in step 3 of the algorithm provides a better discretization than the vector w^{i-1}. That is, the values of ar corresponding to the consecutive discretizations generated by the procedure form a nondecreasing sequence. If the curve includes only linear segments then the number of possible discretizations is finite (Note: only vertices of the curve are taken into account) and consequently the optimization process must end in a finite number of steps.

Any curve can be approximated by a curve which consists of linear segments. If the procedure starts from the same configuration for both curves (the approximation and the initial), then the result (i.e., a partition vector) obtained for the first curve (say in the m-th iteration) is an approximation of the result from the m-th iteration for the initial curve. Taking a sufficiently good approximation curve we can obtain an approximation of the optimal partition vector for the initial curve with the required precision which completes the proof.

Previous experience shows that the procedure converges very quickly. The chosen measure of vector differences has no influence on the convergence process. The influence of the initial configuration is very small on the convergence speed but as the number of classes increases, the convergence is slower.

6.7.3 Bibliographical remarks

An enormous volume of literature is devoted to the problem of "real" cluster detection (cf [12]) which makes even a short review impossible.

The optimal discretization method described above was developed by Ciok for grade cluster analysis (cf [22]). This clustering method is discussed in Chapter 10.

7

Preliminary concepts of bivariate dependence

Teresa Kowalczyk, Wiesław Szczesny

7.1 Introduction

This chapter is devoted to the dependence of one variable with respect to another. It presents the infrastructure of parameters describing bivariate distributions. Traditional infrastructure (in particular, regression and correlation meaningful for at least interval variables) is completed with parameters based on concentration measures (applied e.g., to pairs of conditional distributions). In this way concepts belonging to the Univariate Lilliputian Model ($\mathbb{ULM}$) supplement and help one to visualize the traditional model of bivariate dependence.

We will assume that basic terms like *correlation coefficient, correlation ratio, binormal distribution, regression function,* etc. are known by the reader. However, these terms will be seen now in a new context.

The next chapter deals with the extension of the $\mathbb{ULM}$ (i.e. with the set $\mathbb{BLM} = \mathbb{ULM}^2$) and presents the *grade approach* to bivariate dependence.

7.2 Contingency tables with m rows and k columns

Illustration (Earnings-Education-Gender data):

We are going to explore the earnings dataset, with particular attention directed towards two variables: earnings Z and level of education T. The third variable, gender S, will appear in this and also the next chapter, but will play a secondary role only.

So we deal with the pair (Z,T). Here, a comma is put between Z and T to indicate that we are interested in the *joint distribution* of these two variables, while the notation $(Z;T)$ refers to a pair of *univariate* distributions with cdf's F_Z and F_T. Distributions of Z and T are called *marginals* (row marginal and column marginal) of the joint distribution of (Z,T).

TABLE 7.1. Contingency table for earnings Z_4 and education T

macrocategory	$T=1$	$T=2$	$T=3$	$T=4$	$T=5$	$T=6$	sum
$z_{1(4)} = 1084$	8	6	8	4	5	4	35
$z_{2(4)} = 1932$	7	6	9	10	13	15	60
$z_{3(4)} = 2942$	5	9	10	17	12	12	65
$z_{4(4)} = 3963$	1	3	6	10	10	10	40
sum	21	24	33	41	40	41	200

TABLE 7.2. Probability table for earnings Z_4 and education T

macrocategory	$T=1$	$T=2$	$T=3$	$T=4$	$T=5$	$T=6$	sum
$z_{1(4)} = 1084$	0.040	0.030	0.040	0.020	0.025	0.020	0.175
$z_{2(4)} = 1932$	0.035	0.030	0.045	0.050	0.065	0.075	0.300
$z_{3(4)} = 2942$	0.025	0.045	0.050	0.085	0.060	0.060	0.325
$z_{4(4)} = 3963$	0.005	0.015	0.030	0.050	0.050	0.050	0.200
sum	0.105	0.120	0.165	0.205	0.200	0.205	1.000

The cdf F_Z was presented in Fig. 4.1, cdf F_T in Table 4.3. There were $m = 43$ values of Z (microcategories) and $k = 6$ values of T (levels of education). The corresponding contingency table

$$[N_{ij},\ i = 1, ..., m,\ j = 1, ..., k] \tag{7.1}$$

is too large to be shown here; instead, we show aggregated tables for (Z_4, T) (Table 7.1) and for (Z_8, T) (Table 7.3). Table 7.1 is transformed into its respective "probability" table $P = [p_{ij}]$ (cf table 7.2), where

$$p_{ij} = \frac{N_{ij}}{N},\quad N = \sum_i \sum_j N_{ij} = ...,\quad i = 1, \ldots, m;\ j = 1, \ldots, k. \tag{7.2}$$

The sums of rows and columns in (7.2) are denoted

$$p_{i+},\ \ i = 1, \ldots, m, \quad \text{and } p_{+j},\ \ j = 1, \ldots k,$$

where p_{i+}, p_{+j}, are the respective marginal "probabilities" (of the earnings macrocategories and of the education levels).

The cdf of (Z_4, T), is calculated as

$$F_{(Z_4,T)}(i_0, j_0) = \sum_{i=1}^{i_0} \sum_{j=1}^{j_0} p_{ij},\quad i_0 = 1, \ldots, 4,\ j_0 = 1, \ldots, 6. \tag{7.3}$$

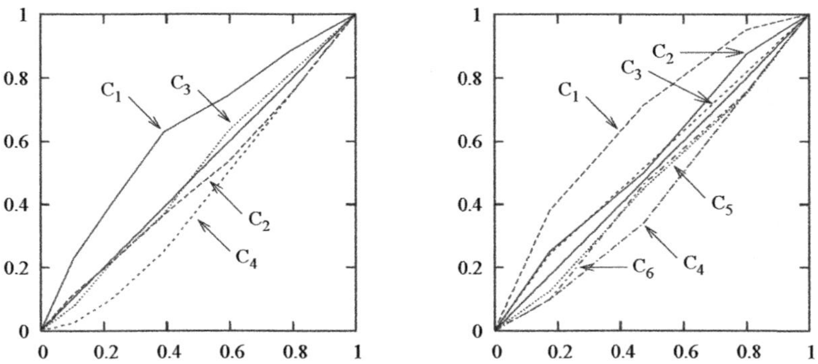

FIGURE 7.1. Concentration curves $C_i = C((T|Z_4 = z_{i(4)}) : T)$ $(i = 1, \ldots, 4)$ (for row profiles - **left** figure) and $C_j = C((Z_4|T = j) : Z_4)$ $(j = 1, \ldots, 6)$ (for column profiles - **right** figure)

TABLE 7.3. Contingency table for earnings Z_8 and education T

midcategory	$T=1$	$T=2$	$T=3$	$T=4$	$T=5$	$T=6$	sum
$z_{1(8)} = 758$	4	1	3	2	1	1	12
$z_{2(8)} = 1254$	4	5	5	2	4	3	23
$z_{3(8)} = 1736$	4	3	5	6	10	9	37
$z_{4(8)} = 2246$	3	3	4	4	3	6	23
$z_{5(8)} = 2765$	4	6	7	11	6	6	40
$z_{6(8)} = 3226$	1	3	3	6	6	6	25
$z_{7(8)} = 3711$	1	2	4	6	5	5	23
$z_{8(8)} = 4303$	0	1	2	4	5	5	17
sum	21	24	33	41	40	41	200

We will consider now the profiles of rows and columns. Let the probabilities in each row of the probability table 7.2 be divided by the row sum. Each row is transformed in this way into a probability distribution with $k = 6$ categories. Such a row distribution is called *the conditional distribution* or shortly *row profile* of T given $Z_4 = z_{i(4)}$, $i = 1, \ldots, 4$. The variable $(T|Z_4 = z_{i(4)})$ takes on values $1, \ldots, 6$, with probabilities p_{ij}/p_{i+}, $i = 1, \ldots, 4$, shown in Table 7.4. Similarly, for columns we have the profiles p_{ij}/p_{+j}, $i = 1, \ldots, 4$, shown in Table 7.5. The conditional random variable corresponding to such a profile is denoted $(Z_4|T = j)$.

TABLE 7.4. Contingency table for earnings Z_4 and education T : row profiles

macrocategory	$T=1$	$T=2$	$T=3$	$T=4$	$T=5$	$T=6$	sum
$z_{1(4)} = 1084$	0.2286	0.1714	0.2286	0.1143	0.1429	0.1143	1
$z_{2(4)} = 1932$	0.1167	0.10	0.15	0.1667	0.2167	0.25	1
$z_{3(4)} = 2942$	0.0769	0.1385	0.1538	0.2615	0.1846	0.1846	1
$z_{4(4)} = 3963$	0.025	0.075	0.15	0.25	0.25	0.25	1

TABLE 7.5. Contingency table for earnings Z_4 and education T : column profiles

macrocategory	$T=1$	$T=2$	$T=3$	$T=4$	$T=5$	$T=6$
$z_{1(4)}=1084$	0.381	0.250	0.242	0.098	0.125	0.098
$z_{2(4)}=1932$	0.333	0.250	0.273	0.244	0.325	0.367
$z_{3(4)}=2942$	0.238	0.375	0.303	0.415	0.300	0.293
$z_{4(4)}=3963$	0.048	0.125	0.182	0.244	0.250	0.244
sum	1	1	1	1	1	1

TABLE 7.6. Contingency table for earnings Z_4 and education T_3 : column profiles

macrocategory	$T_3=1$	$T_3=2$	$T_3=3$
$z_{1(4)}$	0.311	0.162	0.111
$z_{2(4)}$	0.289	0.257	0.346
$z_{3(4)}$	0.311	0.365	0.296
$z_{4(4)}$	0.089	0.216	0.247
sum	1	1	1

Note that no information is lost when the probability table 7.2 is expressed jointly by two elements: the table of row profiles (for $(T|Z_4=i)$) and the distribution of Z_4 (called in this context *the marginal distribution of* Z_4); equivalently, we can express the probability table 7.2 by the table of column profiles (for $(Z_4|T=j)$) and the marginal distribution of T.

However, the contents of tables containing profiles are not easy to perceive when one merely looks at all these numbers; these profiles are best presented visually. In Fig. 7.1 (left figure) each row is represented by the *concentration curve* $C((T|Z_4=i):T)$ of its profile w.r.t. the *marginal column distribution* $(p_{+j}, j=1,\ldots,6)$. Similarly, the concentration curves for the column profiles $(Z_4|T=j)$ are shown in Fig. 7.1 (right figure).

Details: Recall that for any pair $(X;Y)$ such that the concentration curve $C(Y:X)$ is continuous, the concentration curve $C(Y:X)$ and the cdf F_X suffice to calculate F_Y. Therefore, row and column profiles can be calculated from the respective concentration curves $C((Y|X=i):Y)$, $C((X|Y=j):X)$ and cdf's F_Y, F_X.

We will comment on only the right figure of Fig. 7.1 leaving the interpretation of the left figure of Fig. 7.1 to the reader. Line *Diag* in Fig. 7.1 represents the marginal distribution of Z_4. The upper concentration curve (corresponding to the least educated people) lies high above the *Diag* which means that very low education distinctly handicaps earnings. The curve corresponding to the best educated people lies below the *Diag*, but its distance from the *Diag* is much smaller than in the case

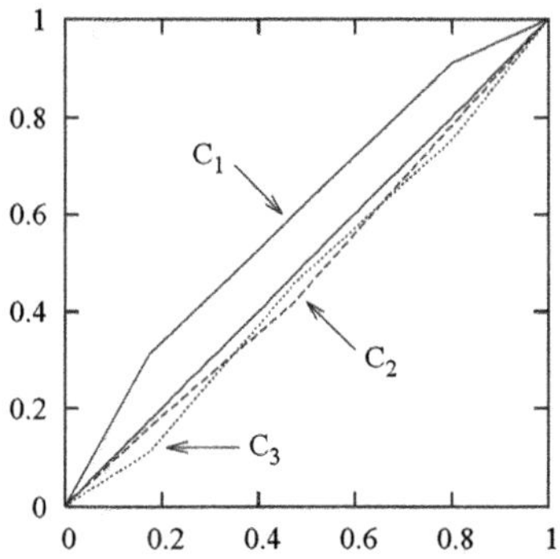

FIGURE 7.2. Concentration curves $C((Z_4|T_3 = j) : Z_4)$ $(j = 1, \ldots, 3)$ (for column profiles)

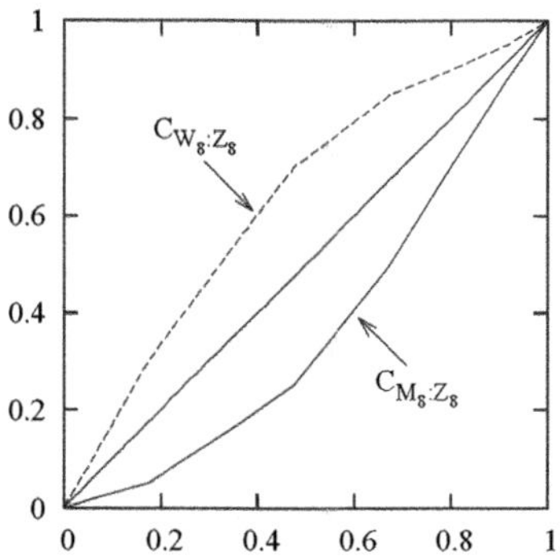

FIGURE 7.3. Concentration curves $C(W_8 : Z_8)$ and $C(M_8 : Z_8)$, where $W_8 = (Z_8|S = 1)$ and $M_8 = (Z_8|S = 2)$

of the least educated people. The remaining curves lie between these two extreme curves. Education levels 2 and 3 provide two very similar curves lying slightly above the *Diag*; levels 4 and 5 provide two very similar curves lying slightly below the *Diag*. So, generally, high education tends to increase earnings, but medium levels 2,3,4,5 also give relatively similar effects.

Remark 7.1 *Please note that the relative position of the curves in Fig. 7.1 can be expressed using stochastic ordering of curves (Sec. 3.4).*

The relations between the profiles and the *Diag* can also be evaluated numerically by the respective concentration indices. In Fig. 7.1 (right), these indices are negative for $j = 1, 2, 3$ and positive for $j = 4, 5, 6$ and their absolute values are equal to the respective areas between each particular curve and the *Diag*, multiplied by 2.

The effect of haphazard aggregation of education levels will now be shown when we replace variable T by its aggregation assuming three values only. The aggregated variable will be denoted T_3; it assigns value "1" to people with former levels of education 1 or 2, value "2" to people with former levels 3 or 4, and value "3" to people with former levels 5 or 6. The column profiles of $(Z_4|T_3 = i)$, $i = 1, 2, 3$ are easily calculated from the probabilities provided in table 7.2, and the respective concentration curves are presented in Fig. 7.2. Now let us compare curves in Figs. 7.1 and 7.2. The conclusions based on Fig. 7.2 will be different than those made previously on the basis of Fig. 7.1. Namely, we see now that the first education level of T_3 confirms the negative effect of low education on earnings, but the two higher levels of T_3 produce almost indistinguishable curves. The aggregation of T which produced T_3 masks the better earnings of people with the highest education (i.e., with $T = 6$) and it ceases to be visible. This demonstrates the importance of the aggregation.

We turn now to consideration of pair (Z_8, S). The relation between male and female earnings was considered in Chapter 3 and 4; here we want to show that we may also start from the related contingency table (with two columns), and then proceed in the same way as before in the case of (Z_4, T) or (Z_4, T_3).

So we deal now with a 8×2 table with two column profiles:
$W_8 = (Z_8|S = 1)$ (female earnings), $M_8 = (Z_8|S = 2)$ (male earnings).

The marginal row distribution is now that for Z_8. Thus we consider two concentration curves $C(W_8 : Z_8)$ and $C(M_8 : Z_8)$, shown in Fig. 7.3.

7.3 Quadrant dependence

The results of the previous section can be applied to any bivariate distribution of random variables (X, Y) which are both continuous (more generally, discrete-continuous). As an example we will consider the binormal distribution, denoted

$$N_2(\rho; m_1, m_2, \sigma_1, \sigma_2) \quad \text{or} \quad N_2(\rho),$$

where ρ denotes the correlation coefficient, and m_1, m_2, σ_1, σ_2 are the means and standard deviations of the row and column variables. Symbol $N_2(\rho)$ will be used whenever the values of the means and standard deviations have no impact on the considerations.

We will consider conditional variables related to quantiles of the marginal distributions. So, for $(X, Y) \sim N_2(\rho)$, we have row profiles for variables $(Y|X = x_p)$ and column profiles for variables $(X|Y = y_p)$. In this case we will obtain, for any $p \in (0,1)$ and any $u \in (0,1)$

$$C((Y|X = x_p) : Y)(u) = F_{(Y|X=x_p)}(F_Y^{-1}(u)) = \Phi\left(\frac{\Phi^{-1}(u) - \rho\,\Phi^{-1}(p)}{\sqrt{1-\rho^2}}\right).$$

The resulting concentration curves are shown in Figs. 7.4.

Binormal densities possess an elegant symmetry and smoothness which then leads to the symmetry and smoothness of the concentration curves in Figs. 7.4 and 7.5. In particular, there exists a symmetry between the respective curves for ρ and $-\rho$. Pairs of variables with positive (respectively, negative) correlation are called positively (respectively, negatively) correlated. From the equalities $\Phi^{-1}(u) = -\Phi^{-1}(1-u)$ for any $u \in (0,1)$ and $\Phi(-x) = 1 - \Phi(x)$ for any $x \in \mathcal{R}$, for $C_p^{(\rho)}(u) \stackrel{df}{=} C((Y|X = x_p) : Y)(u)$ in the case of cdf $N_2(\rho)$ we obtain

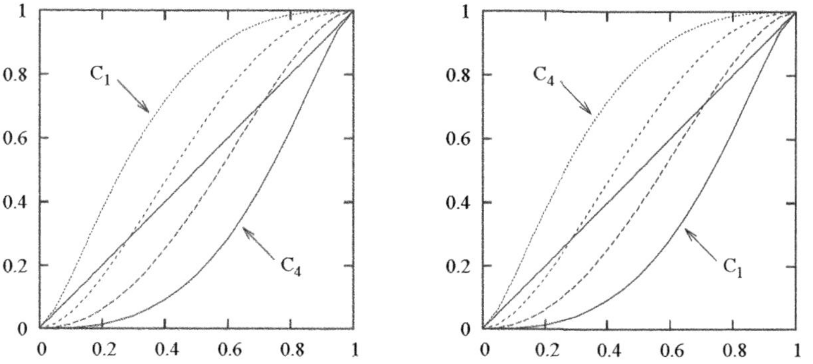

FIGURE 7.4. Concentration curves $C_i = C(Y|X = x_{p_i} : Y)$ for $p_1 = 0.2$, $p_2 = 0.4$, $p_3 = 0.6$, $p_4 = 0.8$, in the case of the binormal distribution $N_2(0.75)$ (left) and in the case of the binormal distribution $N_2(-0.75)$ (right); x_p is the quantile of X of order p

$$C_p^{(\rho)}(u) = C_{1-p}^{\rho\leftarrow}(u) = C_{(1-p)}^{(-\rho)}(u) = 1 - C_p^{(-\rho)}(1-u) = C_p^{(-\rho)\leftarrow}(u).$$

Remark 7.2 *The value of the correlation coefficient ρ of $N_2(\rho; \mathbf{m}, \boldsymbol{\sigma})$ where $\mathbf{m} = (m_1, m_2)$, $\boldsymbol{\sigma} = (\sigma_1, \sigma_2)$, contains the complete information about the positive and the negative dependence, while other parameters of the distribution N_2 contain information about the marginal distributions only.*

In intuitive terms, positive dependence means that smaller (resp. larger) values of one variable tend to appear together with smaller (resp. larger) values of another variable.

The shape of the concentration curves in Fig. 7.4 left visualize positive dependence existing in $N_2(0.75)$. When p increases from 0.2 to 0.8 (i.e., when the curves are changing from C_1 to C_4) and when $u < 0.5$, the densities $c_1(u), ..., c_4(u)$ of the respective curves decrease. This is one way of formalizing the intuitive concept of positive dependence.

Generally, for $\rho > 0$ curves $C(Y|X = x_p : Y)$ in $N_2(\rho)$ are changing in a very regular way. The differences between densities c_i and c_{i+1} for $p_i = 0.2i$, $i = 1, 2, 3$, will diminish with ρ and turn to zero for $\rho = 0$ (all curves lie then on the diagonal). Fig. 7.4 right shows that for $\rho < 0$ the situation is reversed: smaller (resp. larger) values of X tend to appear together with larger (resp. smaller) values of Y and therefore the concentration curves $C_1, ..., C_4$ are reversely ordered (although their shape is preserved due to the symmetry of the densities).

Further comments on duality of shapes of the curves in left and right parts of Fig. 7.4 will be given in a forthcoming Remark 7.5 concerning the distributions, which possess the feature of so called total positive and negative dependence of order two.

An even more significant effect of the positive dependence revealed in $N_2(0.75)$ is evident when the probability mass is accumulated for values of X not exceeding the quantile x_p. For $N_2(0.75)$, the respective concentration curves all lie *above* the *Diag* (Fig. 7.5 left); for $N_2(-0.75)$ they all lie *below* the *Diag* (Fig. 7.5 right).

This accumulation of probability mass appears in the following definition of positive and negative dependence

Definition 7.1 *(quadrant dependence). A pair of variables (X, Y) is positively quadrant dependent if for any $p \in (0,1)$, curve $C((Y|X \le x_p) : Y)$ lies above the Diag (property QD^+), and is negatively quadrant dependent if curve $C((Y|X \le x_p) : Y)$ lies below the Diag (property QD^-). The family of bivariate distributions positively (resp. negatively) quadrant dependent will be denoted $\mathbb{QD}^+$ (resp. $\mathbb{QD}^-$); we will also use the notation*

$$\mathbb{QD} = \mathbb{QD}^+ \cup \mathbb{QD}^-.$$

Due to Figs. 7.5,

$$N_2(\rho) \in \begin{cases} \mathbb{QD}^+ \text{ for } \quad 0 \le \rho \le 1, \\ \mathbb{QD}^- \text{ for } -1 \le \rho \le 0. \end{cases}$$

Also the discretization (either *full*, which provides discrete variables, or *partial*, in which at least one variable is partly continuous) of the binormal distribution preserves this property.

We will use the notation

$$N_2(\rho; \mathbf{p}, \mathbf{q})$$

for a discretized binormal distribution obtained from $N_2(\rho)$, which has the marginal probability vectors $\mathbf{p} = (p_1, \dots, p_m)$ (for a row variable) and $\mathbf{q} = (q_1, \dots, q_k)$ (for a column variable). Also, we write

$$N_2(\rho; m, k) \stackrel{df}{=} N_2\left(\rho; \left(\frac{1}{m}, \dots, \frac{1}{m}\right), \left(\frac{1}{k}, \dots, \frac{1}{k}\right)\right).$$

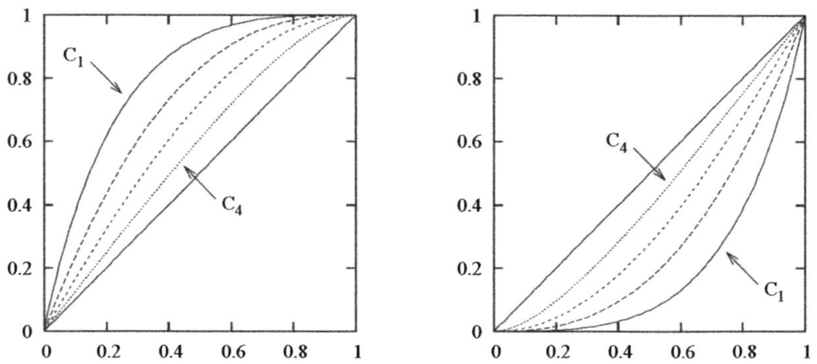

FIGURE 7.5. Concentration curves $C_i = C((Y|X \leq x_{p_i}) : Y)$ for $p_1 = 0.2$, $p_2 = 0.4$, $p_3 = 0.6$, $p_4 = 0.8$ in the case of the binormal distribution $N_2(0.75)$ (left) and $N_2(-0.75)$ (right); x_p is the quantile of X of order p

Proposition 7.1 *(i)* $N_2(\rho)$ *and* $N_2(\rho; \mathbf{p}, \mathbf{q})$ *belong to* $\mathbb{QD}^+$ *when* $\rho \geq 0$ *and to* $\mathbb{QD}^-$ *when* $\rho \leq 0$.

(ii) $(X,Y) \in \mathbb{QD}^+(\mathbb{QD}^-)$ *iff for every* $(x,y) \in \mathcal{R} \times \mathcal{R}$

$$F_{(X,Y)}(x,y) \geq F_X(x)F_Y(y) \;\; (\leq).$$

(iii) $(X,Y) \in \mathbb{QD}^+$ $(\mathbb{QD}^-)$ *iff* $(Y,X) \in \mathbb{QD}^+$ $(\mathbb{QD}^-)$.

(iv) If $(X,Y) \in \mathbb{QD}^+$ $(\mathbb{QD}^-)$ *then for any functions* f, $g : \mathcal{R} \to \mathcal{R}$ *which are both non-decreasing or both non-increasing, the pair* $(f(X), g(Y))$ *belongs to* $\mathbb{QD}^+$ $(\mathbb{QD}^-)$. *If one function is non-decreasing, and one non-increasing, then the pair* $(f(X), g(Y)) \in \mathbb{QD}^-$ $(\mathbb{QD}^+)$.

(v) If $(X,Y) \in \mathbb{QD}$ *then*

$$(X,Y) \in \mathbb{QD}^+ \text{ iff } corr(X,Y) \geq 0,$$

$$(X,Y) \in \mathbb{QD}^- \text{ iff } corr(X,Y) \leq 0.$$

Due to proposition 7.1 (iv), the property QD^+ (resp. QD^-) is preserved when X and/or Y are discretized (fully or partly).

Illustration (Earnings-Education data):
To check whether the pair (Z_4, T_3) is quadrant dependent (as could be expected from just looking at Fig. 7.2), we calculated in Table 7.7 the cdf's of $(Z_4|T_3 \leq j)$, $j = 1, 2, 3$.

The last column (for $j = 3$) provides the cdf of Z_4. According to Prop. 7.1 (ii), we have to compare the first and second column with the third column. Since all

TABLE 7.7. Cdf's of $(Z_4|T_3 \leq j)$, $j = 1, 2, 3$

macrocategory	j=1	j=2	j=3
$z_{1(4)}$	0.31	0.22	0.175
$z_{2(4)}$	0.6	0.49	0.475
$z_{3(4)}$	0.91	0.83	0.8
$z_{4(4)}$	1	1	1

TABLE 7.8. Cdf's of $(Z_4|T \leq j)$, $j = 1, ..., 6$

macrocategory	$j=1$	$j=2$	$j=3$	$j=4$	$j=5$	$j=6$
$z_{1(4)}$	0.38	0.31	0.282	0.22	0.195	0.175
$z_{2(4)}$	0.714	0.6	0.564	0. 49	0.48	0.475
$z_{3(3)}$	0.95	0.91	0.87	0.8.3	0.81	0.8
$z_{4(4)}$	1	1	1	1	1	1

differences are negative (equal to -0.09,-0.11, -0.08, etc.) we conclude that Z_4 and T_3 are positively quadrant dependent. A counterpart of Table 7.7 is also provided for (Z_4, T) (as Table 7.8), and shows that (Z_4, T) is also positively quadrant dependent (although the distribution of $(Z_4|T \leq 5)$ is very close to Z_4 given in the last column of Table 7.8).

Similar checks of property QD^+ were done for less aggregated earnings Z_8 and Z. Pair $(Z_8, T) \in \mathbb{QD}^+$ while pair (Z, T) is not in $\mathbb{QD}$.

Note that the set $\mathbb{QD}$ contains bivariate distributions with dependence more regular than those outside $\mathbb{QD}$ because in $\mathbb{QD}^+$ the accumulation of probability mass for small values of X and small values of Y is non-decreasing w.r.t. threshold x_p (and in $\mathbb{QD}^-$ this monotonicity is reversed). In the next section we will define a family of distributions which are *totally monotone dependent.* They form a subset of $\mathbb{QD}$ which is more regular than $\mathbb{QD}$.

Definition 7.1 leads to the following intuitive partial ordering of positive dependence:

Definition 7.2 *(quadrant ordering). (X, Y) is less positively (negatively) quadrant dependent than (X', Y') if for any $p \in (0, 1)$, $C((Y|X \leq x_p) : Y) \leq C((Y'|X' \leq x'_p) : Y')$ $(\geq)$. Then the following notation is used*

$$(X, Y) \leq_{QD^+} (X', Y') \quad (\leq_{QD^-}).$$

This partial ordering plays a similar role for pairs of bivariate distributions as does stochastic ordering for pairs of univariate distributions. For binormal distributions,

$$N_2(\rho) \leq_{QD^+} N_2(\rho') \quad \text{iff} \quad \rho \leq \rho'.$$

Note that $(X,Y) \leq_{QD^+} (X',Y')$ iff $(X',Y') \leq_{QD^-} (X,Y)$.

Now we will consider the functional relationship between two variables. We start with a relatively easy situation, illustrated by the singular binormal distributions $N_2(1)$ and $N_2(-1)$.

Let $(X,Y) \sim N_2(\rho; m_1, m_2, \sigma_1, \sigma_2)$. Please note that for $|\rho| < 1$, $N_2(\rho)$ is a continuous distribution such that its density exists everywhere, while for $|\rho| = 1$ we deal with two binormal distributions called *the upper Frechet bound of* (X,Y) (for $\rho = 1$) and *the lower Frechet bound of* (X,Y) (for $\rho = -1$), which have probability masses concentrated on *lines*. The upper bound concentrates on

$$\frac{y - m_2}{\sigma_2} = \frac{x - m_1}{\sigma_1} \tag{7.4}$$

(which means that for this degenerated distribution one variable is an *increasing linear function* of the other); the lower Frechet bound of (X,Y) concentrates on the line

$$\frac{y - m_2}{\sigma_2} = -\frac{x - m_1}{\sigma_1} \tag{7.5}$$

(which means that for this degenerated distribution one variable is a *decreasing linear function* of the other). Their cdf's in point $(x,y) \in \mathcal{R} \times \mathcal{R}$ are defined by

$$Fr^+_{(X,Y)}(x,y) = \min\left(\Phi\left(\frac{x - m_1}{\sigma_1}\right), \Phi\left(\frac{y - m_2}{\sigma_2}\right)\right), \tag{7.6}$$

$$Fr^-_{(X,Y)}(x,y) = \max\left(\Phi\left(\frac{x - m_1}{\sigma_1}\right) + \Phi\left(\frac{y - m_2}{\sigma_2}\right) - 1, 0\right), \tag{7.7}$$

where Φ is the cdf of a standardized normal random variable $N(0,1)$. Clearly, for any vector of parameters $(\rho; \mathbf{m} = (m_1, m_2), \boldsymbol{\sigma} = (\sigma_1, \sigma_2))$

$$N_2(-1; \mathbf{m}, \boldsymbol{\sigma}) \leq_{QD^+} N_2(\rho; \mathbf{m}, \boldsymbol{\sigma}) \leq_{QD^+} N_2(1; \mathbf{m}, \boldsymbol{\sigma}). \tag{7.8}$$

Property (7.8) holds for every bivariate distribution of (X,Y) and Frechet distributions corresponding to it. Namely, if we keep the notation $Fr^+_{(X,Y)}$ and $Fr^-_{(X,Y)}$ for the cdf's of the Frechet upper and lower bound of (X,Y) and define them as in formulas 7.6 and 7.7 with $\Phi\left(\frac{x-m_1}{\sigma_1}\right)$ replaced by $F_X(x)$ and $\Phi\left(\frac{y-m_2}{\sigma_2}\right)$ replaced by $F_Y(y)$, then

$$Fr^-_{(X,Y)} \leq_{QD+} F_{(X,Y)} \leq_{QD+} Fr^+_{(X,Y)}.$$

This means that any bivariate distribution is more (less) quadrant dependent than its lower (upper) Frechet bound.

For any univariate cdf F, let

$$x_t = F^{-1}(t) = \begin{cases} inf\{x : F(x) \geq t\}, & \text{if } 0 < t < 1, \\ sup\{x : F(x) = 0\}, & \text{if } t = 0, \\ inf\{x : F(x) = 1\}, & \text{if } t = 1. \end{cases}$$

Remark 7.3 *We note that Frechet distributions are defined using marginal distributions only. Thus the Frechets depend solely on the transformation which relates the cdf of X to the cdf of Y. The upper and lower Frechets related to distribution (X,Y) have the marginal distributions equal respectively, F_X and F_Y. When variables X and Y are continuous, the upper Frechet is concentrated on the set $\{(x,y); F_X(x) = F_Y(y)\}$, i.e. on the graph of the function $y = F_Y^{-1}(F_X(x))$ (or equivalently on the graph of function $x = F_X^{-1}(F_Y(y))$); the lower Frechet is concentrated on the set $\{(x,y); F_X(x) = 1 - F_Y(y)\}$, i.e. on the graph of the function $y = F_Y^{-1}(1-F_X(x))$ (or equivalently on the graph of function $x = F_X^{-1}(1-F_Y(y))$). In general, the upper Frechet is concentrated on the set*

$$\{(x_u, y_u); u \in [0,1]\}$$

and the lower Frechet is concentrated on the set

$$\{(x_u, y_{1-u}); u \in [0,1]\} = \{(x_{1-u}, y_u); u \in [0,1]\}.$$

These sets form in general a nondecreasing or noincreasing relation on $\mathcal{R}^2$, respectively.

It is clear that the functional monotonic relationship between variables always corresponds to Frechet distributions, but Frechets are not necessarily concentrated on the graph of some monotonic functions. Generally they can be concentrated on a *a non-decreasing or non-increasing relation* as happens in the following 2×2 probability table:

$$\begin{bmatrix} 0.3 & 0.2 \\ 0.1 & 0.4 \end{bmatrix}. \tag{7.9}$$

In this example the probability tables corresponding to the upper and lower Frechet bounds are respectively of the form

$$\begin{bmatrix} 0.4 & 0.1 \\ 0 & 0.5 \end{bmatrix}, \begin{bmatrix} 0 & 0.5 \\ 0.4 & 0.1 \end{bmatrix}, \tag{7.10}$$

i.e. for (X, Y) distributed according to (7.9)

$$Fr^{+}_{(X,Y)}(x_i, y_j) = \begin{cases} 0.4 \text{ if } i = j = 1 \text{ or } i = 2, j = 1, \\ 0.5 \text{ if } i = 1, j = 2, \\ 1 \quad \text{if } i = j = 2, \end{cases}$$

$$Fr^{-}_{(X,Y)}(x_i, y_j) = \begin{cases} 0 \quad \text{if } i = j = 1, \\ 0.4 \text{ if } i = 2, j = 1, \\ 0.5 \text{ if } i = 1, j = 2, \\ 1 \quad \text{if } i = j = 2. \end{cases}$$

So the table given by (7.9) is less positively (negatively) quadrant dependent than the left (right) table given by (7.10).

Illustration (Earnings-Education data):
The upper and lower Frechet bounds for the pair (Z_4, T_3) are respectively equal to

$$\begin{bmatrix} 0.175 & 0 & 0 \\ 0.05 & 0.25 & 0 \\ 0 & 0.12 & 0.205 \\ 0 & 0 & 0.2 \end{bmatrix}, \begin{bmatrix} 0 & 0 & 0.175 \\ 0 & 0.07 & 0.23 \\ 0.025 & 0.3 & 0 \\ 0.2 & 0 & 0 \end{bmatrix}.$$

7.4 Matrices of *ar*'s for pairs of profiles. Total positivity of order two

Sets of row profiles (i.e. the profile corresponding to $(Y|X = i)$) and of column profiles (i.e. corresponding to $(X|Y = j)$) will now be described by means of concentration measures applied to *pairs of profiles.* We take into account all pairs in which a column profile is paired with a profile for one of the columns which succeed it.

Thus, in this section, we will deal with concentration curves

$$C_{(q:p;row)} = C((Y|X = x_q) : (Y|X = x_p))$$

instead of curves

$$C_{(q;row)} = C((Y|X = x_q) : Y)$$

as considered in Section 7.2.

The concentration indices related to curves $C_{(q:p;row)}$ will be denoted $ar_{(q:p;row)}$. Special notation will also be used for $m \times k$ tables:

$$C_{(s:i;row)},\quad C_{(t:j;col)},\quad ar_{(s:i;row)},\quad ar_{(t:j;col)}.$$

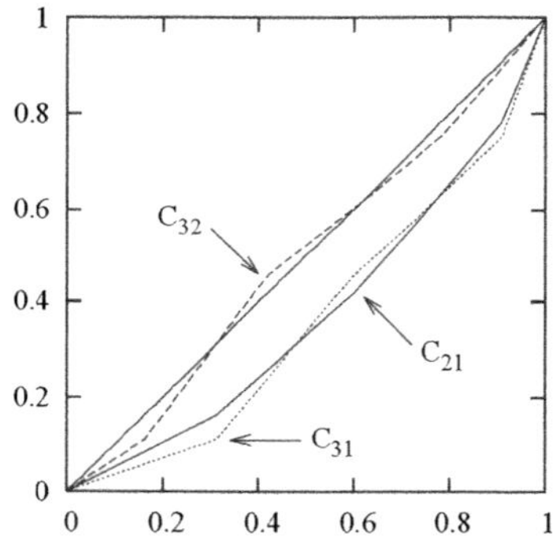

FIGURE 7.6. Concentration curves for pairs of column profiles for (Z_4, T_3) : $C_{ij} = C_{ji;col} = C((Z_4|T_3 = j) : (Z_4|T_3 = i))$, $i = 1, 2$, $j = 2, 3$.

Illustration (Earnings-Education data):

For pair (Z_4, T_3) we have pairs of earning profiles w.r.t. education levels (j, t), where $j < t$, $t = 1, 2, 3$. The number of such pairs is 2+1=3. For each of them we calculate the values of ar and of $ar_{\max}$; they form the matrix:

$$\begin{bmatrix} 0 & 0.2488 & 0.269 \\ 0.2488 & 0 & 0.0373 \\ 0.3131 & 0.1387 & 0 \end{bmatrix} \tag{7.11}$$

in which the values of ar appear above the diagonal, and the values of $ar_{\max}$ below the diagonal. We see that there are very small differences between these two parts of the table. This is illustrated by the set of concentration curves $C_{(t:j;col)} = C((Z_4|T_3 = t) : (Z_4|T_3 = j))$ presented in Fig. 7.6.

We see from Fig. 7.6 that

(i) the second and third column profiles are almost identical (curve $C_{(3:2;col)}$ almost coincides with the *Diag*),

(ii) curves $C_{(2:1;col)}$ and $C_{(3:1;col)}$ are very close to one another and each of them is almost the same distance from the *Diag* (and also from $C_{(3:2;col)}$) as the other,

(iii) curve $C_{(2:1;col)}$ is convex (so that $ar_{\max}(2:1;col) = ar(2:1;col)$),

(iv) curve $C_{(3:1;col)}$ is almost convex (so that $ar_{\max}(3:1;col)$ is slightly larger than $ar(3:1;col)$),

(v) curve $C_{(3:2;col)}$ is not convex and therefore $ar_{\max}(3:2;col)$ is slightly larger than $ar(3:2;col)$.

Similar matrices of indices $ar_{\max}$ and ar can be calculated for pairs of *row* profiles.

The comparison of matrices of ar's and $ar_{\max}$'s provides more information than the comparison of the vectors of ar's of row (column) profiles w.r.t. the respective marginal profile.

Definition 7.3 *(total positivity of order two). Pair (X, Y) is totally positive dependent of order two (abbreviated TP_2) if*

$$ar_{\max}(q:p;col) = ar(q:p;col) \quad \text{for } p < q,\ p,q \in (0,1).$$

Pair (X, Y) is totally negative dependent of order two (abbreviated TN_2) if

$$ar_{\max}(q:p;col) = -ar(q:p;col) = ar(p:q;col) \quad \text{for } p < q,\ p,q \in (0,1).$$

> The set of TP_2 distributions will be denoted $\mathbb{TP}_2$, the set of TN_2 distributions wil be denoted $\mathbb{TN}_2$, while $\mathbb{TM}_2 = \mathbb{TP}_2 \cup \mathbb{TN}_2$. The set $\mathbb{TM}_2$ contains *totally monotone dependent distributions* which have regular monotone trends between row and column variables.

Illustration (Earnings-Education data):

A regular positive almost TP_2 trend exists between Z_4 and T_3 : the contingency table related to (Z_4, T_3), equal to

$$\begin{bmatrix} 14 & 12 & 9 \\ 13 & 19 & 28 \\ 14 & 27 & 24 \\ 4 & 16 & 20 \end{bmatrix}, \tag{7.12}$$

becomes TP_2 if $n_{21} = 13$ is replaced by 17 and $n_{22} = 19$ is replaced by 32.

For the pairs consisting of less aggregated earnings and education, the monotone trend between earnings and education will be less and less regular (please note that a similar observation was made for quadrant dependence).

Remark 7.4 *It is easy to construct $m \times k$ probability tables which are not TP_2 but becomes TP_2 after a suitable aggregation. For instance, the probability table*

$$\begin{bmatrix} 0.07 & 0.34 & 0.17 \\ 0.02 & 0.16 & 0.08 \\ 0 & 0.02 & 0.04 \\ 0 & 0.06 & 0.04 \end{bmatrix}, \tag{7.13}$$

is not TP_2 (see Proposition 7.2(iv)) since $0.02 \times 0.04 - 0.06 \times 0.04 = -0.0016 < 0$ *but becomes TP_2 after an aggregation of the second and third row:*

$$\begin{bmatrix} 0.07 & 0.34 & 0.17 \\ 0.02 & 0.18 & 0.12 \\ 0 & 0.06 & 0.04 \end{bmatrix}. \tag{7.14}$$

Proposition 7.2 *Properties of $\mathbb{TM}_2$:*

(i)

$$(X,Y) \in \mathbb{TP}_2 \Longrightarrow (X,Y) \in \mathbb{QD}^+,$$

$$(X,Y) \in \mathbb{TN}_2 \Longrightarrow (X,Y) \in \mathbb{QD}^-,$$

(ii) the set of equalities

$$ar_{\max}(q:p;col) = ar(q:p;col), \quad \textit{for all}\, p < q,\ p,q \in (0,1)$$

implies the set of equalities

$$ar_{\max}(q:p;row) = ar(q:p;row) \quad \textit{for all}\ \ p < q,\ p,q \in (0,1).$$

(iii) For a contingency table $P_{m \times k}$ which belongs to $\mathbb{TP}_2$ ($\mathbb{TN}_2$) any table which arises from $P_{m \times k}$ by deleting some rows and/or some columns belongs to $\mathbb{TP}_2$ ($\mathbb{TN}_2$).

(iv) (X,Y) distributed according to $[p_{ij}, i = 1, \ldots, m, j = 1, \ldots, k]$ belongs to $\mathbb{TP}_2$ ($\mathbb{TN}_2$) iff

$$det_{isjt} = p_{ij}p_{st} - p_{it}p_{sj} \geq 0 \quad (\leq 0)$$

for $i < s$, $j < t$, $s = 2, \ldots, m$, $t = 2, \ldots, k$;

The pair (X,Y) with density $f(x,y)$ belongs to $\mathbb{TP}_2$ ($\mathbb{TN}_2$) iff

$$\begin{vmatrix} f(x,y) & f(x,y') \\ f(x',y) & f(x',y') \end{vmatrix} \geq 0 \quad (\leq 0) \tag{7.15}$$

for $x < x'$ and $y < y'$, $x,y \in \mathcal{R}$.

(v) If $(X,Y) \in \mathbb{TP}_2$ $(\mathbb{TN}_2)$ then concentration curves $C_{(q:p;row)}$ and $C_{(q:p;col)}$ are convex (concave) for $p < q$, $p, q \in (0,1)$.

(vi) If $(X,Y) \in \mathbb{TP}_2$ $(\mathbb{TN}_2)$ then $(Y|X = x_p) \leq_{st} (Y|X = x_q)$ for $p < q$ $(q < p)$ (stochastic ordering of row profiles). An analogous property holds for variables $(X|Y = y_p)$ (stochastic ordering of column profiles).

(vii) If $(X,Y) \in \mathbb{TP}_2$ $(\mathbb{TN}_2)$ then for any functions $f,\ g : \mathcal{R} \to \mathcal{R}$ which are both non-decreasing or both non-increasing, the pair $(f(X), g(Y)) \in \mathbb{TP}_2$ $(\mathbb{TN}_2)$. If one function is non-decreasing and one non-increasing then the pair $(f(X), g(Y)) \in \mathbb{TN}_2$ $(\mathbb{TP}_2)$.

Remark 7.5 *Formula (7.15) can be rewritten for the quantiles x_p and $x_{p'}$ corresponding to $p \leq p'$: if $\rho > 0$, for any $0 \leq u \leq u' \leq 1$,*

$$f_{Y|X=x_p}(y_{u'})f_{Y|X=x_{p'}}(y_u) \leq f_{Y|X=x_p}(y_u)f_{Y|X=x_{p'}}(y_{u'}).$$

This means that the densities of the curves $C((Y|X = x_p) : Y)$ $(p \in (0,1))$ decrease with p for $\rho > 0$ and, analogously, increase with p for $\rho < 0$. Figure 7.4 illustrates this for the binormal distributions $N_2(0.75)$ and $N_2(-0.75)$.

Properties which describe the family $\mathbb{TM}_2$ provide a refinement of properties which describe the family $\mathbb{QD}$. For instance, in $\mathbb{QD}^+$ it is necessary to aggregate the probability table up to a given threshold to obtain stochastically ordered aggregated profiles for increasing thresholds. In $\mathbb{TP}_2$, not only aggregated profiles, but also the profiles themselves, are stochastically ordered. Moreover, $\mathbb{TP}_2$ is contained in the set of distributions which "have a TP_2 property for cdf's", namely

$$\begin{vmatrix} F_{(X,Y)}(x,y) & F_{(X,Y)}(x,y') \\ F_{(X,Y)}(x',y) & F_{(X,Y)}(x',y') \end{vmatrix} \geq 0 \tag{7.16}$$

for any $x < x'$, $y < y'$.

Stochastic ordering of profiles is very important. Intuitively it shows that in $\mathbb{TM}_2$ the gain in regularity of positive or negative dependence cannot be increased by reordering rows and/or columns. This statement will be considered further in Section 8.6.1 and Section 8.6.3

Let us turn now to $N_2(\rho)$. It was not mentioned in Proposition 7.2 that $N_2(\rho) \in \mathbb{TP}_2$ for $\rho \geq 0$ and $N_2(\rho) \in \mathbb{TN}_2$ for $\rho \leq 0$ because a much stronger property holds, which concerns bivariate distributions *totally monotone dependent of order $s \geq 2$*. Their definition refers to point (iv) in Prop. 7.2. The extension is straightforward: instead of determinants of 2×2 matrices as in the case of $\mathbb{TM}_2$, determinants of $s \times s$ matrices are considered.

Note: $N_2(\rho)$ belongs to $\mathbb{TM}_s$ for any finite s, while

$$N_2(\rho; (p_1, \ldots, p_m), (q_1, \ldots, q_k))$$

belongs to $\mathbb{TM}_s$ for $s = \min(m, k)$.

Property (vi) in Prop. 7.2 determines that in $\mathbb{TP}_2$ pairs of more distant profiles have stochastically ordered concentration curves. This is illustrated in Fig. 7.7 by four profiles of $(Y|X = x_{S_i})$, when $(X,Y) \sim N_2(0.75)$ and $N_2(-0.75)$. We have chosen $S_1 = 0.2$, $S_2 = 0.4$, $S_3 = 0.6$, $S_4 = 0.8$. Clearly, profiles corresponding to 0.2 and 0.8 are the most distant, profiles corresponding to 0.2 and 0.6 are more distant than profiles corresponding to 0.2 and 0.4, etc. So we deal with four profiles of $(Y|X = x_{S_i})$ where $S_1 = 0.2$, $S_2 = 0.4$, $S_3 = 0.6$, $S_4 = 0.8$, and with six pairs of profiles.

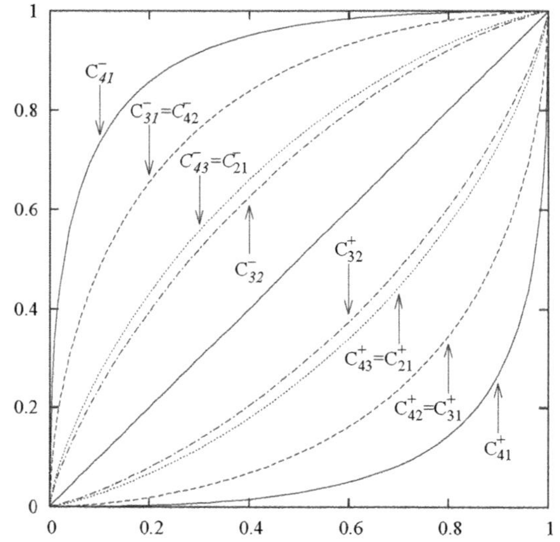

FIGURE 7.7. Concentration curves $C_{S_j:S_i}$ for pairs of row profiles denoted C^+_{ji} for $N_2(0.75)$, and (C^-_{ji}) for $N_2(-0.75)$, when $(S_1, ..., S_4) = (0.2, 0.4, 0.6, 0.8)$

Since for $N_2(\rho)$ the concentration curves $C^{(\rho)}_{(q:p)}$ are as follows:

$$C^{(\rho)}_{(q:p)}(u) = \Phi\left(\Phi^{-1}(u) - \frac{\rho\,(\Phi^{-1}(q) - \Phi^{-1}(p))}{\sqrt{1-\rho^2}}\right) \qquad (7.17)$$

for any $p, q \in (0,1)$ and any $u \in (0,1)$, we obtain

$$C^{(-\rho)}_{(q:p)}(u) = C^{(\rho)}_{(p:q)}(u) = 1 - C^{(\rho)}_{(q:p)}(1-u), \qquad (7.18)$$
$$C^{(\rho)}_{(q:p)}(u) = C^{(\rho)}_{(1-p:1-q)}(u).$$

So there are only four different Lilliputian curves which are stochastically ordered:

$$Diag(u) \geq C_{(0.6:0.4)}(u) \geq C_{(0.8:0.6)}(u) \geq C_{(0.8:0.4)}(u) \geq C_{(0.8:0.1)}(u)$$

for $\rho = 0.75$ and any $u \in (0,1)$, and

$$C_{(0.8:0.1)}(u) \geq C_{(0.8:0.4)}(u) \geq C_{(0.8:0.6)}(u) \geq C_{(0.6:0.4)}(u) \geq Diag(u)$$

for $\rho = -0.75$ and any $u \in (0,1)$.

The distances between the curves are measured by the differences of the respective areas $ar_{\max}$. For $N_2(\rho)$ we have for any $p, q \in (0,1)$

$$ar^{(\rho)}(q:p) = ar^{(\rho)}_{\max}(q:p) = 1 - 2\Phi\left(\frac{\rho\left(\Phi^{-1}(p) - \Phi^{-1}(q)\right)}{\sqrt{2(1-\rho^2)}}\right). \tag{7.19}$$

It is important to note that bivariate distributions are TP_2 if and only if $ar's$ and $ar_{\max}'s$ for all pairs of quantiles are equal. This will be exploited in Section 9.5 and 14.5 where outliers from total positivity of order two will be detected and used to decompose probability tables.

Columns of probability tables $2 \times k$ can be always reordered to form a TP_2 table. The ordering coincides with this rearrangement of columns for which the concentration curve for the distributions in the first and second row becomes convex. In the rearranged table, the determinant in each 2×2 subtable is non-negative and then all $ar's$ are equal to $ar_{\max}'s$. The corresponding concentration curves are triangular and are partially stochastically ordered so that the matrix $[ar(j:i;col)]$ is Robinson (i.e., its elements increasing monotonically when moving away from the matrix diagonal consisting of nulls).

Illustration (revisit of Example B from Chapter 2, Section 2.2, Geography scores):

In Example B we have a 2×5 contingency table. It follows that there is just one concentration curve $C_{2:1;row}$ and 10 concentration curves $C_{j:i;col}$ for $i, j = 1, ..., 5$, $j > i$. Each of these curves has its counterpart $C_{i:j;col}$, which is symmetric to $C_{j:i;col}$ with respect to *Diag* and provides equivalent information.

The curve $C_{1:2;col}$ is equal to the curve presented in Fig. 2.1, which was constructed in Chapter 2 for the budget comparison in two different countries (the data in Example A (Table 2.1) and Example B (Table 2.2) are identical, although the labels are different: *countries* instead of *schools* and *budget categories* instead of *letter grades.*) The interpretation of this curve in Fig. 2.1 for the letter grades is analogous to that for budget categories.

Each column distribution is two point. For instance, the distribution for grade "D" is defined by the proportion of school children from the East City School who got grade "D" among all school children with grade letter "D" : $30/(30+15) = 0.67$ (rounded). So we have

A	B	C	D	F
0.5	0.45	0.25	0.67	0.308

The grades can be ordered C, F, B, A, D (from the smallest proportion p_C through the largest proportion p_D); this is the ordering looked for. Please note that this is the ordering under which the concentration curve for East City School w.r.t. West City School is convex (see Figure 2.2). Indeed, the concentration indices ar in Table 7.9 form a anti-Robinson matrix, i.e.,

TABLE 7.9. The concentration indices $ar(S_j : S_i; col)$, $j < i$, $i, j = 1, ..., 5$ in the Example B

	C (S_1)	F (S_2)	B (S_3)	A (S_4)	D (S_5)
C	0	0.0577	0.2046	0.25	0.4167
F		0	0.1468	0.1923	0.3589
B			0	0.0455	0.2121
A				0	0.1667
D				0	0

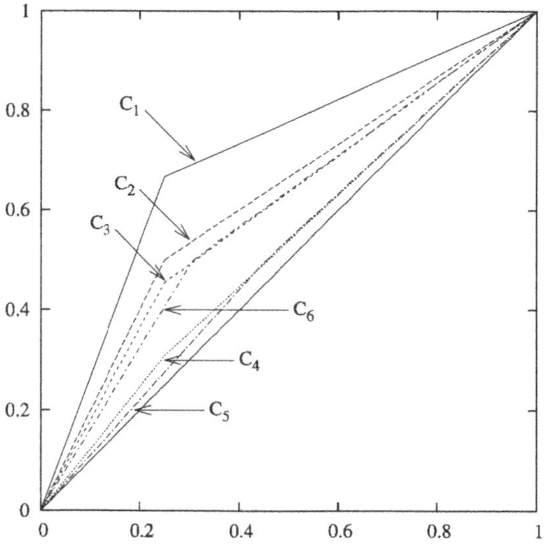

FIGURE 7.8. Concentration curves in example B: $C_1 = C(D : C)$, $C_2 = (A : C)$, $C_3 = C(B : C)$, $C_4 = C(F : C)$, $C_5 = C(A : B)$, $C_6 = C(A : F)$.

Moreover, for any $i = 1, ..., 4$, the concentration curves $C_{S_j:S_i;col}$, $j = i+1, \dots, 5$ are stochastically ordered, and so are for any j the concentration curves $C_{S_j:S_i;col}$, $j = i + 1, \dots, 5$. The curves $C(D : C)$, $C(A : C)$, $C(B : C)$, $C(F : C)$, $C(A : B)$, $C(A : F)$ presented in Figure 7.8 illustrate this property.

The ordering C, F, B, A, D helps to find out where are the main differences between the distributions of particular grade letters in both schools. The crucial difference is between D and C. The difference between adjacent distributions are small and they are *ordered.*

Note: Exactly the same results will be obtained for Example A concerning budget categories. This will be considered in Chapter 12 since in this case the 2×5 table is not a contingency table but a 2×5 data matrix. The partial stochastic ordering of the distributions of "expenses on food" and "expenses on transportation" in country A and country B seems then worth consideration.

Note: Let us stress the difference between the examples A, B, C, D in Chapter 2, Section 2.2 (supplemented by the example on Winners and Losers in Section 2.6) and the examples concerning cylinder with gas (Section 2.6). In all examples from the first group it is sensible to compare one row with another as well as one column with another. In the examples refering to the cylinder with gas, we have the set of data concerning *cylinder space* and the set of data concerning amount of gas (where cylinder space and amount of gas are ratio variables). Each set of data corresponds to cyllinder cells, since it is assumed that in each cell gas is uniformly distributed and space is uniformly increasing with the position in cylinder. The two distributions are continuous with density which is constant in each cell. Then, it is possible to compare the cumulative distribution of space and the cumulative distribution of amount of gas (i.e., to compare both row distributions). But it is evident that the columns do not correspond to variables and therefore their comparison will not have a sensible interpretation.

7.5 The regression function

We will illustrate this section using pairs of variables consisting of earnings and education.

Please note that in general in the previous sections of this chapter it was not necessary to take into account the measurement scales attached to variables X and Y, since most of the concepts considered up to now, such as probability tables, row and column profiles, and concentration measures related to the profiles, are *meaningful for arbitrary pairs of measurement scales of the considered variables.*

Now, however, we will consider such parameters of bivariate distributions for which measurement scales of particular variables *are* important. Regression

TABLE 7.10. Row/column regression for pairs (Z_4, T) and (Z_8, T)

	T=1	T=2	T=3	T=4	T=5	T=6
$E(Z_4\|T)$	1937	2347	2396	2761	2637	2638
$E(Z_8\|T)$	1892	2374	2385	2748	2647	2674

functions (i.e. sequences of expectations of conditional variables) are among them.

Regression functions $E(Y|X = x)$ and $E(X|Y = y)$ indicate roughly the trend of dependence. $E(Y|X = x)$ is *meaningful* if Y is at least interval, and *calculable* if Y has a finite expectation.

Illustration (Earnings-Education data):

Both regressions are meaningful (and also calculable) for *earnings w.r.t. education*, but they are not meaningful for *education w.r.t. earnings*. Recall that education T is ordinal (which is a measurement scale weaker than interval), while earnings Z, Z_8 and Z_4 are *ratio* (which is a measurement scale stronger than interval).

Definition 7.4 *Let Y have a finite expectation $E(Y)$. The regression of variable Y on variable X is a function $r_{Y|X} : \mathcal{R} \to \mathcal{R}$ such that for any x belonging to the support of X*

$$r_{Y|X}(x) \overset{df}{=} E(Y|X = x) = \int_{-\infty}^{+\infty} y dF_{(Y|X=x)}(y).$$

We will use the convention that in the distribution of (X, Y), X corresponds to the row variable and Y to the column variable. Therefore, in a considered probability or contingency table we will sometimes write $r_{col|row}$ instead of $r_{Y|X}$, and $r_{row|col}$ instead of $r_{X|Y}$.

Illustration (Earnings-Education data):

The (meaningful) regression function of Z_4 on T is equal to the sequence of expectations of column profiles in Table 7.10:

$$E(Z_4|T = j) = \sum_{i=1}^{4} z_{i(4)} p_{ij} / p_{+j}, \quad j = 1, \ldots, 6$$

These expectations are presented in Table 7.10 and shown graphically in Fig. 7.9 (the regression functions of Z_8 on T and of Z_4 on T are also presented).

TABLE 7.11. Regression functions for (Z_4, T) and (Z_8, T)

Z_4	$z_{1(4)}$		$z_{2(4)}$		$z_{3(4)}$		$z_{4(4)}$	
$E(T\|Z_4)$	3.12		4.02		3.89		4.38	
Z_8	$z_{1(8)}$	$z_{2(8)}$	$z_{3(8)}$	$z_{4(8)}$	$z_{5(8)}$	$z_{6(8)}$	$z_{7(8)}$	$z_{8(8)}$
$E(T\|Z_8)$	2.83	3.26	4.14	3.83	3.68	4.24	4.17	4.65

The (not meaningful) regression of T on Z_4 (sequence $E(T|Z_4 = z_{i(4)})$, $i = 1, ..., 4$) is given in Table 7.11. We recall that this regression depends on the choice of education levels; if the six levels 1,2,3,4,5,6 were replaced by, say, 1,2,3,5,7,10, the shape of the regression function will be strongly affected by this fact.

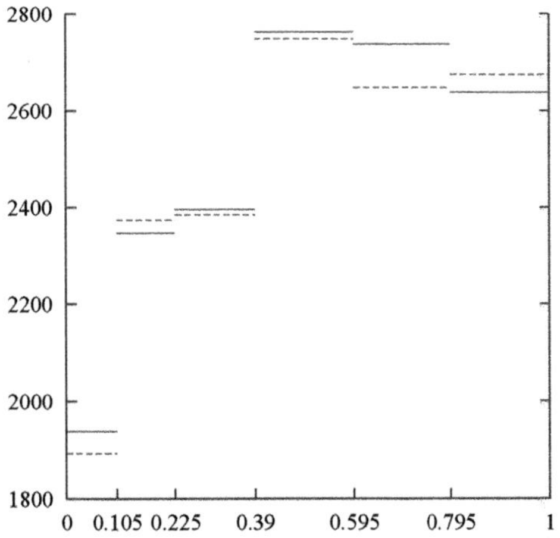

FIGURE 7.9. Regression functions $E(Z_4|T = t_u)$ (solid line) and $E(Z_8|T = t_u)$ (dashed line); t_u is $u - th$ quantile of T for $u \in (0, 1)$

For $N_2(\rho; (m_1, m_2), (\sigma_1, \sigma_2))$, the regression functions are simultaneously linear. Indeed, when both regressions are presented as functions on the real line, the graph of $r_{Y|X}$ lies on the line $y = \rho(\sigma_2/\sigma_1)(x - m_1) + m_2$, and the graph of $r_{X|Y}$ lies on the line $y = (\sigma_2/\rho\sigma_1)(x - m_1) + m_2$. For $m_1 = m_2 = 0$ and $\sigma_1 = \sigma_2 = 1$, we have the lines $y = \rho x$ and $y = (1/\rho)x$. If $\rho = 0$, the graphs lie on the lines $y = m_2$ and $x = m_1$.

It is sometimes convenient to write the formulas for both binormal regressions as

$$\begin{aligned} E(Y|X = x_u) &= \rho(\sigma_2/\sigma_1)(x_u - m_1) + m_2 \\ &= \rho\sigma_2\Phi^{-1}(u) + m_2 = \rho y_u + (1-\rho)m_2 \\ E(X|Y = y_v) &= \rho(\sigma_1/\sigma_2)(y_v - m_2) + m_1 \\ &= \rho\sigma_1\Phi^{-1}(v) + m_1 = \rho x_v + (1-\rho)m_1 \end{aligned} \tag{7.20}$$

where x_u and y_v are the respective quantiles of the marginal distributions (we can then use the standardized normal cdf Φ).

Both binormal regressions $r_{Y|X}$ and $r_{X|Y}$ are equal iff $\rho = 1$.

Example. For the 2×2 probability table given in Table 7.9 the regression functions are

$$E(Y|X = x_1) = \frac{3}{5}y_1 + \frac{2}{5}y_2, \qquad E(Y|X = x_2) = \frac{1}{5}y_1 + \frac{4}{5}y_2,$$

$$E(X|Y = y_1) = \frac{3}{4}x_1 + \frac{1}{4}x_2, \qquad E(X|Y = y_2) = \frac{1}{3}x_1 + \frac{2}{3}x_2,$$

(x_1, x_2 are any specified values of X, and y_1, y_2 are any specified values of Y).

Regression functions represent (numerically and graphically) *the trend* between the marginal variables, exhibited by the joint distribution. We want to know *how far (i.e. how distant) is that trend – represented by the regression function* $r_{Y|X}$ *– from the functional relationship of* Y *on* X, which is expressed by the upper Frechet. Analogously, we want to know how distant is that trend represented by $r_{X|Y}$ from the functional relationship of X on Y.

These distances are measured numerically by the indices considered in the next section.

A well established practice is to represent trends not only by the regressions $r_{Y|X}$ and $r_{X|Y}$ but also by the *second type regression functions* which are the best *linear* approximations of the functional relationship of Y on X and of X on Y. As is well known, the second type of regression functions are defined as

$$r^{(2)}_{Y|X}(x) = corr(X,Y)\frac{\sigma_2}{\sigma_1}(x - m_1) + m_2,$$

$$r^{(2)}_{X|Y}(x) = \frac{1}{corr(X,Y)}\frac{\sigma_2}{\sigma_1}(x - m_1) + m_2;$$

$r^{(2)}_{Y|X}$ is increasing when $corr(X,Y) > 0$ and $r^{(2)}_{Y|X}$ is decreasing when $corr(X,Y) < 0$. For $N_2(\rho; m_1, m_2, \sigma_1, \sigma_2)$, $r^{(2)}_{Y|X} = r_{Y|X}$, and $r^{(2)}_{X|Y} = r_{X|Y}$.

Regression functions are expectations corresponding to profiles of rows or of columns. It is also important to consider the expectations corresponding to *cumulated profiles.* This leads to functions which will be called here *the cumulated expectations.*

Definition 7.5 *Let Y be a variable with finite expectation. The cumulated expectation of Y on X is a function denoted $E(Y; X \leq)$, and defined as*

$$E(Y; X \leq x) = \int_{-\infty}^{x} r_{Y|X}(z) dF_X(z)$$

(according to tradition, we write $E(Y; X \leq x)$ instead of $E(Y; X \leq)(x)$).

The following modification of the cumulated expectation denoted $C_{cor(Y:X)}$ will be called the correlation function of Y on X :

$$C_{cor(Y:X)}(u) = \int_{0}^{u} r_{Y|X}(F_X^{-1}(t)) dt, \qquad u \in [0, 1], \tag{7.21}$$

where

$$r_{Y|X}(F_X^{-1}(t)) = E(Y|X = x_t) = \int_{-\infty}^{\infty} y dF_{(Y|X=x_t)}(y),$$

Definition 7.5 looks complicated, but is quite simple for continuous distributions of (X, Y) as well as for $m \times k$ probability tables (with categories valued $x_1 \leq \cdots \leq x_m$ for X, and $y_1 \leq \cdots \leq y_k$ for Y). In the first case,

$$C_{cor(Y:X)}(u) = E(Y; X \leq x_u) = \int_{-\infty}^{x_u} r_{Y|X}(z) dF_X(z) = \int_{0}^{u} r_{Y|X}(x_t) dt,$$

and thus for $(X, Y) \sim N_2(\rho; m_1, m_2, \sigma_1, \sigma_2)$, C_{cor} is expressed by the density φ and cdf Φ of $N(0, 1)$ as follows:

$$C_{cor(Y:X)}^{(\rho)}(u) = m_2 \, u - \rho \, \sigma_2 \varphi(\Phi^{-1}(u)), \quad u \in (0, 1).$$

For (X, Y) distributed according to a table $P_{m \times k}$,

$$r_{Y|X}(F_X^{-1}(t)) = r_{Y|X}(x_i) = \sum_{j=1}^{k} y_j \frac{p_{ij}}{p_{i+}}, \qquad \text{for } t \in (F_X(x_i-), F_X(x_i)),$$

so that for any $u \in (F_X(x_i-), F_X(x_i)]$

$$C_{cor(Y:X)}(u) = \sum_{s=1}^{i-1} r_{Y|X}(x_s) p_{s+} + (u - F_X(x_i-)) r_{Y|X}(x_i).$$

Note that $C_{cor(Y:X)}(u) = C_{cor(r_{Y|X}(X):X)}(u)$; moreover, for $Y \geq 0$

$$C(E(Y/E(Y):X\leq):F_X)(u)=C_{cor(Y:X)}(u)/E(Y).$$

For any variable Y with $|E(Y)|<\infty$ and any $u\in(0,1)$ we define the function $C_{cor(Y:Y)}$:

$$C_{cor(Y:Y)}(u)=E(Y;Y<y_u)+y_u(u-F_Y(y_u-)).$$

Further considerations concerning C_{cor} and the so called *monotone dependence function* related to C_{cor} are presented in Section 7.6. An important special case of the correlation function for a pair of uniform Lilliputian variables will be considered in Section 8.5.

7.6 The monotone dependence function and the Gini Index

We start with the properties of the correlation function $C_{cor(Y:X)}$ defined by formula (7.21) and the opposite correlation function $C^{\leftarrow}_{cor(Y:X)}$ defined by

$$C^{\leftarrow}_{cor(Y:X)}(u)\stackrel{df}{=}E(Y)-C_{cor(Y:X)}(1-u).$$

Proposition 7.3 *Let $f^{\nearrow}$ and $f^{\searrow}$ denote respectively, an increasing and decreasing function and let f be either $f^{\nearrow}$ or $f^{\searrow}$. We have*

$$C_{cor(Y:X)}(u)=\begin{cases}C_{cor(Y:f(X))}(u) & \text{iff } f \text{ is increasing},\\ E(Y)-C_{cor(Y:f(X))}(1-u) & \text{iff } f \text{ is decreasing},\\ uE(Y) & \text{iff } r_{Y|X}\equiv E(Y),\\ C_{cor(Y:Y)}(u) & \text{iff } X=f(Y),\ f^{\nearrow},\\ E(Y)-C_{cor(Y:Y)}(1-u) & \text{iff } X=f(Y),\ f^{\searrow},\\ C_{cor(r_{Y|X}:r_{Y|X})}(u) & \text{iff } r_{Y|X} \text{ is non-decreasing},\\ E(Y)-C_{cor(r_{Y|X}:r_{Y|X})}(1-u) & \text{iff } r_{Y|X} \text{ is non-increasing}.\end{cases}$$

It follows that for $Y\in\overline{\mathbb{RATIO}}$ (i.e. for $Y\geq 0$ and $E(Y)<+\infty$),

$$C_{cor(Y:Y)}(u)=E(Y)L_Y(u),\quad C_{cor(Y:r_{Y|X})}=E(Y)L_{r_{Y|X}},\tag{7.22}$$

$$C^{\leftarrow}_{cor(Y:Y)}(u)=E(Y)L^{\leftarrow}_Y(u)=E(Y)(1-L_Y(1-u))\tag{7.23}$$

where L_Y is the Lorenz curve of Y.

Fig. 7.10 illustrates the fact that whenever the regression function $r_{Y|X}$ is nearly constant, the graph of the related correlation function divided by $E(Y)$ almost lies on the diagonal. Moreover, in this particular case $C_{cor(Y:X)}$ and $C_{cor(X:Y)}$ are almost equal.

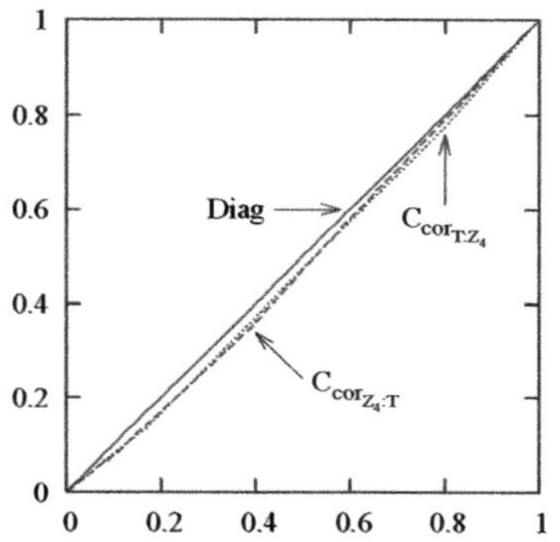

FIGURE 7.10. Graphs of the correlation functions $C_{cor(Z_4:T)}$ and $C_{cor(T:Z_4)}$

Now, we introduce the "normalized" correlation function named the *monotone dependence function* $\mu_{Y:X} : [0,1] \to [-1,1]$.

Definition 7.6 *For any pair (X,Y) of non-degenerate random variables such that $E(Y)$ is finite,*

$$\mu_{Y:X}(u) = \begin{cases} \mu^+_{Y:X}(u) & \textit{if } \mu^+_{Y:X}(u) \geq \mu^+_{-Y:X}(u), \\ \mu^-_{Y:X} & \textit{otherwise}, \end{cases} \tag{7.24}$$

where

$$\mu^+_{Y:X}(u) = \frac{uE(Y) - C_{cor(Y:X)}(u)}{uE(Y) - C_{cor(Y:Y)}(u)},$$

$$\mu^-_{Y:X}(u) = -\mu^+_{-Y:X}(u) = \frac{uE(Y) - C_{cor(Y:X)}(u)}{(1-u)E(Y) - C_{cor(Y:Y)}(1-u)},$$

Remark 7.6 *The condition that $E(Y)$ is finite under which the monotone dependence function $\mu_{Y:X}$ is defined is only a formal requirement (analogous e.g. to the condition that random variable Z belongs to set $\mathbb{RATIO}$ in the definition 4.2 of the Lorenz Curve). If one wants to refer to the measurement scales, the monotone dependence function will be meaningful if it is assumed that variable X is (at least) ordinal, and variable Y is (at least) interval (analogously the assumption that Z is a ratio variable makes the Lorenz Curve meaningful).*

We note, that at fixed point u, either $\mu^+_{Y:X}(u) = \mu^-_{Y:X}(u) = 0$ or only one function: $\mu^+_{Y:X}(u)$ or $\mu^+_{-Y:X}(u)$ is positive. So, if $\mu^+_{Y:X}(u)$ is positive then $\mu_{Y:X}(u) = \mu^+_{Y:X}(u)$. If $\mu^+_{Y:X}(u)$ is negative then $\mu_{Y:X}(u) = \mu^-_{Y:X}(u) = -\mu^+_{-Y:X}(u)$.

Proposition 7.4 *(i) If* $(X,Y) \in \mathbb{QD}^+$ $(\mathbb{QD}^-)$ *then* $\mu_{Y:X}(u) = \mu^+_{Y:X}(u)$ $(\mu^-_{Y:X}(u))$ *for every* $u \in (0,1)$.

(ii)

$$\mu_{aY+b:f(X)}(u) = \begin{cases} (sgn\, a)\mu_{Y:X}(u) & \text{if } f \text{ is increasing,} \\ (-sgn\, a)\mu_{Y:X}(1-u) & \text{if } f \text{ is decreasing.} \end{cases}$$

So,

$$\mu_{Y:X}(u) = \begin{cases} \mu^+_{Y:r_{Y|X}}(u) & \text{if } r_{Y|X} \text{ is non-decreasing,} \\ -\mu^+_{Y:r_{Y|X}}(1-u) & \text{if } r_{Y|X} \text{ is non-increasing.} \end{cases}$$

By immediate calculations, for any $u \in (0,1)$ we obtain: if $\mu_{Y:X}(u) = \mu^+_{Y:X}(u)$ then

$$C_{cor(Y:X)}(u) = \mu^+_{Y:X}(u)C_{cor(Y:Y)}(u) + (1 - \mu^+_{Y:X}(u))uE(Y); \tag{7.25}$$

if $\mu_{Y:X}(u) = \mu^-_{Y:X}(u)$ then

$$C_{cor(Y:X)}(u) = -\mu^-_{Y:X}(u)C_{\overleftarrow{cor}(Y:Y)}(u) + uE(Y)(1 + \mu^-_{Y:X}(u)). \tag{7.26}$$

Thus, the correlation function $C_{cor(Y:X)}$ for any $u \in (0,1)$ is a convex combination of two correlation functions: that for a functional monotone relationship $(f(Y), Y)$ where f is strictly monotone, and that in the case where random variables X and Y are independent.

When $\mu_{Y:X}(u) \equiv const$, we obtain a convex combination with coefficients non dependent on u.

The graph of $\mu_{Z_8:T}$ is shown in Fig. 7.11. According to the figure, dependence of earnings Z_8 w.r.t. education T is positive but relatively high only for low education; then it decreases practically to null.

The horizontal line $v = 0.174$ in Fig. 7.11 corresponds to the value of the so called correlation ratio $\eta_{Y:X}$ defined as follows.

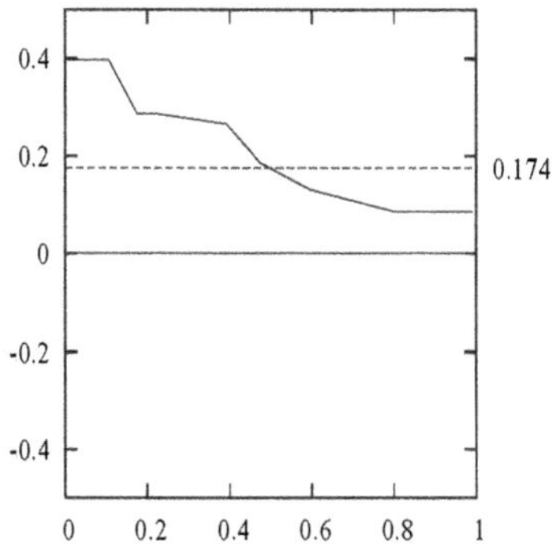

FIGURE 7.11. Graph of the monotone dependence function for Z_8 w.r.t. T, contrasted with the dotted horizontal line corresponding to the correlation ratio η of Z_8 w.r.t. T (equal 0.174)

Definition 7.7 *(the correlation ratio $\eta_{Y:X}$). Let (X,Y) be a pair of non-degenerated random variables (i.e. non concentrated at one point) with $E(Y^2) < \infty$ and with cdf H. Let (Y',Y) be the distribution which arises from (X,Y) when the marginal distribution of X is transformed to that of Y by the probability transition function $\Pi(y;x)$:*

$$\Pi(y;x) = \begin{cases} 1 & \text{if } F_X(x) \leq F_Y(y), \\ \frac{F_Y(y)-F_X(x-)}{F_X(x)-F_X(x-)} & \text{if } F_X(x-) \leq F_Y(y) < F_X(x), \\ 0 & \text{if } F_X(x-) > F_Y(y). \end{cases}$$

We have

$$F_{(Y',Y)}(y',y) = \int_{-\infty}^{\infty} \Pi(y';x)H(dx,y) \tag{7.27}$$
$$= \int_{-\infty}^{\infty} \Pi(y';x)F_{Y|X=x}(y)dF_X(x).$$

The correlation coefficient $corr(Y,Y')$ will be called the correlation ratio of Y on X and denoted $\eta_{Y:X}$.

Theorem 7.1 *(properties of $\eta_{Y:X}$). If Y is not concentrated in one point and if (Y',Y) is a pair with cdf given by (7.27) then*

(i) if $E(Y)$ is finite then $\eta_{Y:X} = \eta_{Y:Y'}$,

(ii) if $E(Y^2)$ is finite then $\eta^2_{Y:X} = Var(r_{Y|Y'})/Var(Y)$,

(iii) $\eta_{Y:X} = corr(X,Y)$ iff $r_{Y|X}$ is linear,

(iv) $\mu_{Y:X} \equiv \eta_{Y:X}$ iff $r_{Y|Y'}$ is linear with the slope equal to $\eta_{Y:X}$ (i.e. $\mu_{Y:X}(u) \equiv corr(Y',Y)$),

(v) $\mu_{Y:X} \equiv corr(X,Y)$ iff $r_{Y|X}$ is linear and there exists a linear function f such that

$$(Y', Y) \sim (f(X), Y).$$

We say that Y *is strongly linearly dependent on* X if $r_{Y|X}$ is linear and there exists a linear function f such that $f(X) \sim Y$ i.e. $\mu_{Y:X} \equiv corr(X, Y)$.

The horizontal line $v = \eta_{Y:X}$ can be compared with the graph of $\mu_{Y:X}$ as seen in Fig. 7.11. This shows how distant is the trend of monotone dependence of Y on X represented by $\mu_{Y:X}$ from the linear trend between Y and Y' which reduces the graph of the monotone dependence function to the line $v = \eta_{Y:X}$.

Linearity of the regression of Y on Y' holds if and only if

$$C_{cor(Y:X)}(u) = \eta_{Y:X}\, C_{cor(Y:Y)}(u) + (1 - \eta_{Y:X})E(Y)u, \text{ (for } \eta_{Y:X} > 0),$$

$$C_{cor(Y:X)}(u) = -\eta_{Y:X}\, C^{\leftarrow}_{cor(Y:Y)}(u) + (1 + \eta_{Y:X})uE(Y), \text{ (for } \eta_{Y:X} < 0).$$

Strong linear dependence of Y on X holds if and only if $\eta_{Y:X}$ in the above equalities is replaced by $corr(X, Y)$.

7.7 Appendix - Bibliographical remarks

As mentioned in the introduction (Sec. 7.1), we assume that the reader is acquainted with the classic descriptive infrastructure used in statistics. Therefore, no references to such infrastructure are given.

This chapter concerns orderings in the set of bivariate distributions and parameters concordant with these orderings. Among thousands of papers and books in this research area, the key ones were written by Lehmann. Lehmann built a hierarchical system of formal concepts of positive dependence in [70], which were then extended by Yanagimoto [114], Sampson [49], Shaked [95] and Schriever [93]. Orders of positive dependence were considered in [111] in a way consistent with the approach introduced here.

In the present chapter, only two elements of the hierarchical family of regular distributions were introduced and described: quadrant dependence and total positivity of order two. The essence of this approach was to develop the concept of *regularity* of monotone dependence, although the term "regularity" was not mentioned. Certain concepts in the present book were inspired by the ideas presented in [59].

The description takes into account regularity and strength of dependence. Both these main aspects of stochastic dependence should be expressed using an infrastructure based on concentration, which is the main goal of the present chapter.

The correlation function was considered in [105] by Taguchi who recognized the importance of this notion and other related notions to it (e.g. the ratio curve). It is worth mentioning that the same name (correlation function) has been given to another concept introduced in [33].

The monotone dependence function, which is closely linked with the correlation function, was introduced in [66] for continuous distributions and in [61] for the general case (see also [67], [89], [62], [60] [7]).

8

Dependence Lilliputian Model

Teresa Kowalczyk, Wiesław Szczesny, Włodzimierz Wysocki

8.1 Introduction

We introduce here the extension of $\mathbb{ULM}$, called *the Bivariate Lilliputian Model* and denoted $\mathbb{BLM}$.

> The $\mathbb{BLM}$ consists of the set of *pairs* of Lilliputian variables. The $\mathbb{BLM}$ may also be equivalently specified as the set of cdf's of bivariate distributions on the unit square $[0,1]^2$. As in the $\mathbb{ULM}$, any bivariate Lilliputian distribution represents a whole set of appropriately selected *pairs* of bivariate distributions.
>
> Our attention will be focused on a special *subset* of the $\mathbb{BLM}$, called *the Dependence Lilliputian Model* and denoted $\mathbb{DLM}$. Each Lilliputian pair $(\xi, \eta) \in \mathbb{DLM}$ differs from the Lilliputian pairs outside $\mathbb{DLM}$ in that the marginal distributions of (ξ, η) are *uniform.*
>
> Any Lilliputian pair from $\mathbb{DLM}$ represents the whole set of *pairs* of bivariate distributions, namely that of (X, Y) and that of the *independent counterpart* of (X, Y), denoted $(X, Y)^{\perp}$ (i.e. such that its components are independent).

Due to the independence of the components of $(X, Y)^{\perp}$, the cdf of this pair is the product of the cdf's of X and of Y. The distribution of (X, Y) is transformed into the distribution of (X^*, Y^*) which belongs to $\mathbb{DLM}$. This is done by the relevant uniform grade transformations applied jointly to X and to Y, with $X^* = F_X^*(X)$, $Y^* = F_Y^*(Y)$.

> The cdf of (X^*, Y^*) is a function defined on the unit square, with values in [0,1]. It is called the *copula* of (X, Y) and denoted $Cop_{(X,Y)}$; the distribution of (X^*, Y^*) is called the *grade distribution* of (X, Y). The set $\mathbb{DLM}$ consists of copulas or, equivalently, of grade distributions.

If X and Y are categorical variables with, respectively, m and k categories, then their joint distribution can be given by the respective probability table with m rows and k columns. The grade density of the distribution of (X, Y) is then constant on each of the corresponding $m \times k$ rectangles which are contained in the unit square; the areas of these rectangles refer to independent components in $(X, Y)^{\perp}$.

The grade density of a bivariate pair (X, Y) can be visualized as a so called *overrepresentation map* (described in Section 8.2). This map is the basic element of visualization connected with grade methodology. Maps will be seen in this and the following chapters as a useful tool to visualize trends of dependence and of maximal dependence (see Section 8.6.1).

Sections 8.2 and 8.6 are specifically addressed to practitioners. These sections show via many examples how grade probability tables can be ordered, displayed and described. Section 8.3 is more theoretical; it refers to the copulas and their Lilliputian surfaces.

Special effort is made in this chapter to present the Lilliputian surfaces and their parameters as being analogs to the Lilliputian curves and their parameters presented in Chapters 3 and 4. In particular, the grade correlation ρ^* is defined in Section 8.4 analogously as the concentration index ar, and the maximal grade correlation $\rho^*_{\max}$ is defined in Section 8.6 analogously as the maximal concentration index $ar_{\max}$. The only change is that areas are now replaced by volumes. The grade correlation ρ^* is known as Spearman's *rho*. The links between ρ^* and another measure of dependence called Kendall's *tau* are investigated (in particular the links between $\rho^*_{\max}$ and $\tau_{\max}$). Parameter τ serves to introduce a measure of departure from total positivity of order two.

In Section 8.6.1 grade distributions of $m \times k$ tables are compared with their grade binormal analogs. This is done exactly according to the lines introduced in Chapter 4 (Section 4.2) for any two Lilliputian variables in $\mathbb{ULM}$.

The last section (8.7)provides an interesting insight into the structure of a probability table due to the so called hinged sequences of rows (columns), in which some elements are replaced by their opposites.

8.2 Grade bivariate distributions and overrepresentation maps for probability tables

This section presents an important tool of visualization: the overrepresentation map of a probability table.

Let $P_{m\times k}$ be a bivariate probability table

$$P_{m\times k} = [p_{ij}], \quad i = 1,\dots,m, \;\; j = 1,\dots,k.$$

This table will be transformed first into a table in which each p_{ij} is divided by the product $p_{i+}p_{+j}$. Such a transformation of Tables 8.1 and 8.2 which concern the pair (Z_4, T_3) results in Table 8.3. It can be visualized as a *map* on the unit square where the square is divided into $m \times k$ rectangles, each of them coloured by one of five shades of gray.

The rectangles are formed by a network of vertical and horizontal lines. Horizontal lines join the left and right sides of the square at the heights $\sum_{i\le s} p_{i+}$, $s = 1,\dots,m$ (heights 0.175,0.475,0.8,1)). Vertical lines join the bottom and top sides of the square starting from point $\sum_{j\le t} p_{+j}$, $t = 1,\dots,k$ on the bottom side (coordinates 0.225,0.595,1).

> The network of rectangles corresponding to the pair (Z_4, T_3) is shown in Fig. 8.1. The area of the rectangle $\mathcal{R}_{ij}$ lying in the ith row and j^{th} column is equal to the product $p_{i+}p_{+j}$.

TABLE 8.1. Probability table for (Z_4, T_3)

	1	2	3	p_{i+}	$\sum_{i\le j} p_{i+}$
1	0.07	0.06	0.045	0.175	0.175
2	0.065	0.095	0.14	0.3	0.475
3	0.07	0.135	0.12	0.325	0.8
4	0.02	0.08	0.1	0.2	1
p_{+j}	0.225	0.37	0.405		
$\sum_{j\le i} p_{+j}$	0.225	0.595	1		

TABLE 8.2. Products $[p_{i+}p_{+j}]$ for the probability table of (Z_4, T_3)

	1	2	3	p_{i+}	$\sum_{i\le j} p_{i+}$
1	0.0393	0.0648	0.0709	0.175	0.175
2	0.0675	0.111	0.1215	0.3	0.475
3	0.0731	0.1203	0.1316	0.325	0.8
4	0.045	0.074	0.081	0.2	1
p_{+j}	0.225	0.37	0.405		
$\sum_{j\le i} p_{+j}$	0.225	0.595	1		

TABLE 8.3. Overrepresentation table for (Z_4, T_3)

	1	2	3
1	1.781	0.9259	0.6347
2	0.9630	0.8559	1.1523
3	0.9576	1.122	0.9118
4	0.4444	1.081	1.2346

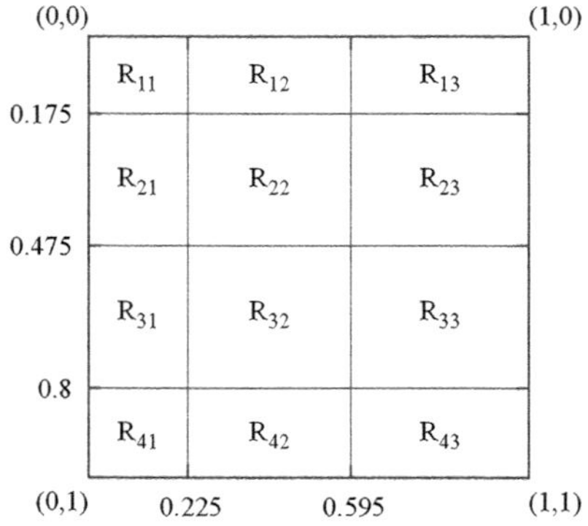

FIGURE 8.1. Rectangles $\mathcal{R}_{ij}$ in the unit square, corresponding to (Z_4, T_3)

We will consider now the function which assigns to each point of a rectangle which lies in the ith row and j^{th} column, the ratio

$$c(u,v) = \frac{p_{ij}}{p_{i+}p_{+j}} \quad \text{for } (u,v) \in \mathcal{R}_{ij}, \quad i = 1,\dots,m, \ j = 1,\dots,k \qquad (8.1)$$

which will be called the *overrepresentation index* of points in $\mathcal{R}_{ij}$. (Recall that the values of the overrepresentation indices for (Z_4, T_3) are given in Table 8.3). Note that

$$\int\int_{\mathcal{R}_{ij}} c(u,v)dudv = p_{ij}, \quad i = 1,\dots,m, \ j = 1,\dots,k, \qquad (8.2)$$

$$\int_0^1 \int_0^1 c(u,v)dudv = 1,$$

$$\int_0^1 c(u,v)du = \sum_i \frac{p_{ij}}{p_{+j}} = 1 \ \text{ for } v \in (T_{j-1}, T_j), \ \ T_j = \sum_{s=1}^{j} p_{+s}, \ j = 1, \dots, k$$

$$\int_0^1 c(u,v)dv = \sum_j \frac{p_{ij}}{p_{i+}} = 1 \ \text{ for } u \in (S_{i-1}, S_i), \ \ S_i = \sum_{s=1}^{i} p_{s+}, \ i = 1, \dots, m.$$

Thus, the overrepresentation indices (which are constant for each rectangle) form a density function of the joint distribution of a pair of uniform Lilliputian variables.

In the case of (Z_4, T_3), the row Lilliputian variable is the *uniform grade transform of* Z_4, denoted Z_4^*; the column Lilliputian variable is the *uniform grade transform of* T, denoted T^*. Suitably discretized values of the density function can be visualized as a map over the unit square.

Values of overrepresentation indices are discretized with the use of five intervals:

$$(0, 2/3),\ [2/3, 0.99),\ [0.99, 1.01],\ (1.01, 3/2],\ (3/2, +\infty).$$

The shading used in the maps is shown in Fig. 8.2.

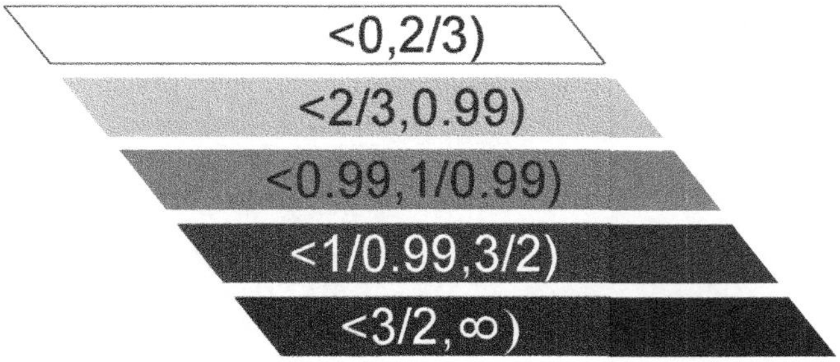

FIGURE 8.2. Categories of shading used in discretization of overrepresentation indices

The overrepresentation map for (Z_4, T_3) is shown in Fig. 8.3 (right). It represents the density of the distribution of (Z_4^*, T_3^*).

Please note that point (0,0) is placed at the top left corner of the unit square. In addition, the vertical axis for Z_4^* is directed down, which fits with the customary way of representing frequencies in a contingency table (see for instance Table 7.1 in Chapter 7 corresponding to the pair (Z_4, T)).

Overrepresentation maps serve to give quick insight into the inner structure of a bivariate distribution. A series of such maps is displayed in Figures 8.3 to 8.8.

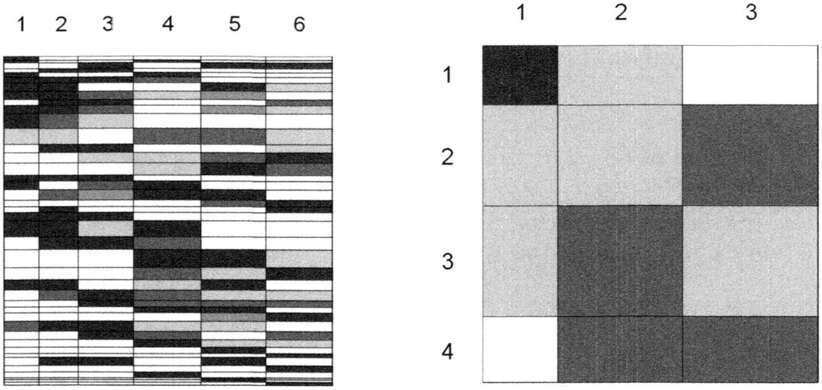

FIGURE 8.3. Overrepresentation maps for (Z, T) (**left**) and for (Z_4, T_3) (**right**). Rows in the left map are labelled 1,...,41.

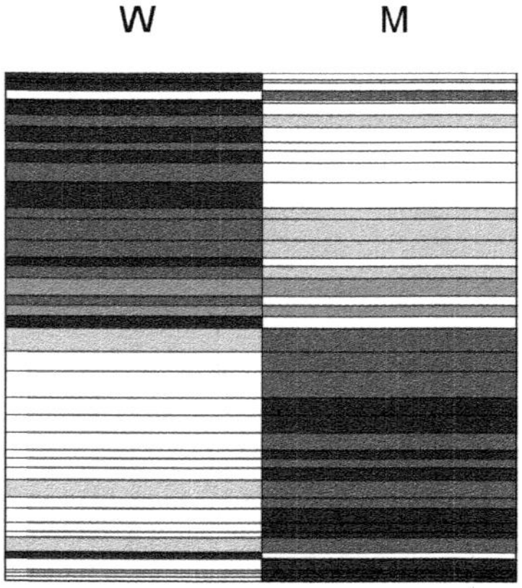

FIGURE 8.4. Overrepresentation map for (Z, S)

Illustration (Earnings-Education-Gender Data):

The first two maps represent the pairs (Z, T) and (Z_4, T_3) (Fig. 8.3 left and right). They show the effect of aggregation (from Z to Z_4 and from T to T_3). The left map contains 41×6 rectangles (two rows with zero frequencies are not visible on the figure), the right map contains 12 rectangles. The white rectangle in the right upper corner of the right map indicates that the frequency for highly educated but poorly paid individuals is much less than what would be expected in the case of independence between education and earnings; the white rectangle in the opposite corner (i.e., bottom left) corresponds to the opposite, i.e. poorly educated, but highly paid. The unique black rectangle (i.e., upper left) represents individuals who are lowly paid and poorly educated; such persons are overrepresented in the data.

The next map (Fig. 8.4) represents the pair (Z, S). It has two columns since gender has two values: so there are female earnings in the left column (concentrated on lower earnings) and male earnings in the right column (concentrated on higher earnings). A few exceptions are visible: black stripes in the light parts and white stripes in the dark parts of the map. One can find in the frequency table that these stripes correspond to frequencies exceptionally low or high.

The columns are of equal width (100 females and 100 males).

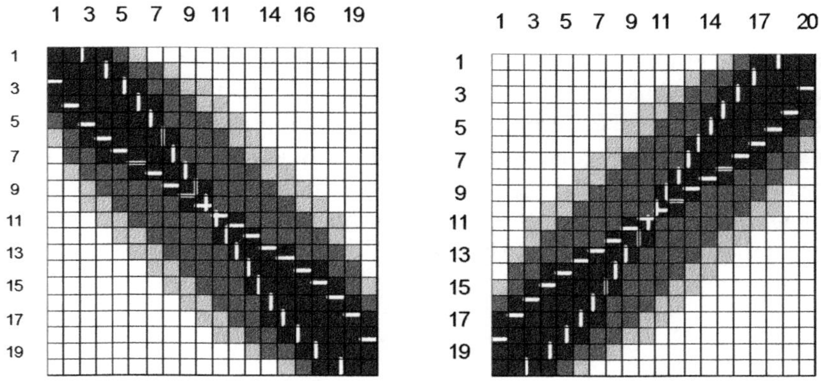

FIGURE 8.5. Overrepresentation maps showing both grade regression functions for the discretized binormal $N_2(0.75; 20, 20)$ (**left**) and for the discretized binormal $N_2(-0.75; 20, 20)$ (**right**). Discretized intervals are numbered from 1 to 20 (not all labels are shown)

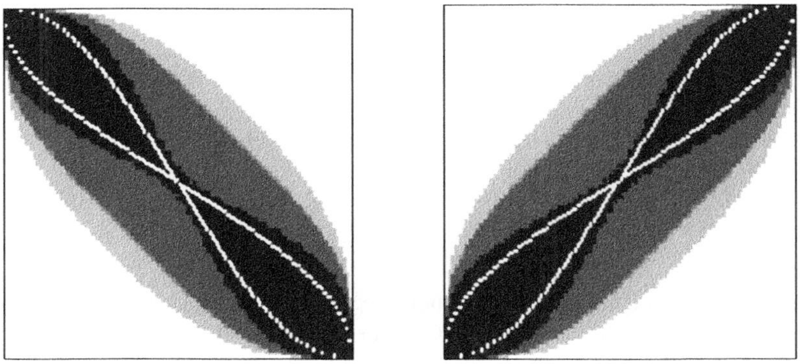

FIGURE 8.6. Overrepresentation maps showing both grade regression functions for the discretized binormal $N_2(0.75; 100, 100)$ (**left**) and for the discretized binormal $N_2(-0.75; 100, 100)$ (**right**).

The next four maps (Figs. 8.5-8.6) correspond to discretized grade binormal distributions. There are two levels of uniform discretization and two correlation coefficients ρ (0.75 or -0.75) in the distribution before discretization. The four maps (two of size 20×20 and two of size 100×100) exibit high regularity (i.e. white neighboring with light grey, then a border line between light grey, and dark grey and finally a border line between dark gray and black. These border lines are very regular (especially when the discretized distribution is very close to a continuous one).

There are small white segments, vertical and horizontal, which show, respectively, the expected values of the row profiles (vertical segments) and of the column profiles (horizontal segments). The whole set of vertical profiles forms the graph of the *column-on-row regression function* of the grade distribution; the whole set of horizontal segments forms the graph of the *row-on-column regression function* of the grade distribution. The distributions with positive correlation concentrate around the diagonal from the left upper vertex (point (0,0)) to the right lower vertex (point (1,1)), i.e. they concentrate along the diagonal $v = u$ (the upper Frechet bound of this distribution); when the correlation is negative, the map is "rotated 90° clockwise".

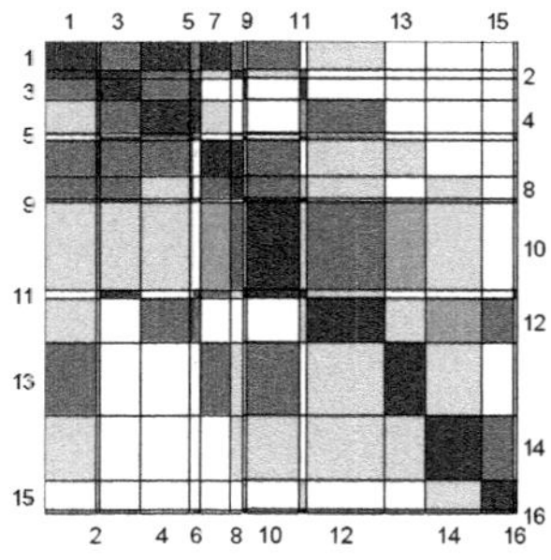

FIGURE 8.7. Overrepresentation map for $CARS_{16\times 16}$ (rows correspond to old cars, columns correspond to new cars. Rows and columns 2,5,6,11, and 16 are so narrow that not all of them are labeled)

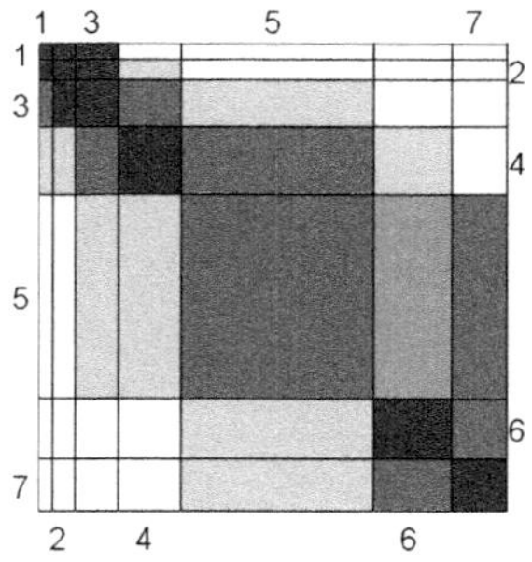

FIGURE 8.8. Overrepresentation map for $BRIT_{7\times 7}$ (rows correspond to fathers, columns correspond to sons)

Details: The diagonal $v = u$ and both regressions are "going down" on the map because point (0,0) is at the top left corner and the vertical axis is directed down. The diagonal $v = 1 - u$ and both regressions are "going up" on the map.

The regularity of binormal distributions is in contrast to the irregularity appearing in the map representing the dataset $CARS_{16\times 16}$ (Fig. 8.7).

Illustration ($CARS_{16\times16}$ data):

This dataset was introduced in Chapter 2 where the concentration curve of one marginal to another was shown (Fig. 2.12). It was stated there that the complete "inner" table of frequencies would be analyzed in Section 8.6. From now on, we use the name $CARS_{16\times16}$ for the complete datset. We recall that this table contains frequencies for Car Models Purchased by Repeat Car Buyers (see Tables 2.3a and 2.4a in Chapter 2) with columns corresponding to a previous purchase and rows corresponding to the present purchase.

The widths of rows as well as columns in the marginals of Figure 8.7 are very differentiated. The rectangles lying on the diagonal $v = u$ are all black which corresponds to the tendency to buy the same type of car as owned before. The owners of cars of types 1 through 5 tend to buy a car in this range of types, and the owners of cars of types 14 and 15 tend to buy one of these types. But otherwise, the table is shaded very irregularly. The 16 categories of cars are:

1. Subcompact/domestic	9. Importcompact
2. Subcompact/captiveimports	10. Midsizedomestic
3. Subcompact/imports	11. Midsize imports
4. Small specialty/domestic	12. Midsize specialty
5. Small specialty/captive imports	13. Low price standard
6. Small specialty/imports	14. Medium price standard
7. Low price compact	15. Luxury domestic
8. Medium price compact	16. Luxury import

Illustration ($BRIT_{7\times7}$ data):

The next map (Fig. 8.8) presents the data contained in Table 8.4, called $BRIT_{7\times7}$. The name refers to the fact that the data were gathered in Britain (see Section 8.8). The objects in $BRIT_{7\times7}$ are pairs (father, son); the row and column variables are father's and son's occupation, with the same occupational categories in the case of fathers and in the case of sons; these categories are the following:

1. PROF - professionals
2. EXEC - managerial and executive
3. HSUP - higher supervisory
4. LSUP - lower supervisory
5. SKILL - skilled manual and routine nonmanual
6. SEMI - semiskilled manual
7. UNSK - unskilled manual.

Since the categories were subjectively ordered according to the original investigator's plan, they might be classified as nominal or ordinal. The particular order in Table 8.4 is often treated by investigators as a "social prestige" order (but there is no general agreement among sociologists as to how such an order should be established experimentally).

TABLE 8.4. Contingency and probability table for Father-Son occupation categories (according to the dataset $BRIT_{7\times7}$). Rows and columns denote occupation categories (PROF, EXEC, HSUP, LSUP, SKIL, SEMI and UNSK). The upper numbers denotes frequencies, the lower relative frequencies

	1	2	3	4	5	6	7	sum
1	50	19	26	8	18	6	2	129
	.0143	.0054	.0074	.0023	.0051	.0017	.0006	.0369
2	16	40	34	18	31	8	3	150
	.0046	.0114	.0097	.0052	.0089	.0023	.0009	.0429
3	12	35	65	66	123	23	21	345
	.0034	.0100	.0186	.0189	.0352	.0066	.0060	.0987
4	11	20	58	110	223	64	32	518
	.0031	.0057	.0166	.0315	.0638	.0183	.0092	.1481
5	14	36	114	185	714	258	189	1510
	.0040	.0103	.0326	.0529	.2042	.0738	.0541	.4318
6	0	6	19	40	179	143	71	458
	0	.0017	.0054	.0114	.0512	.0409	.0203	.1310
7	0	3	14	32	141	91	106	387
	0	.0009	.0040	.0091	.0403	.0260	.0303	.1107
sum	103	159	330	459	1429	593	424	3497
	.0295	.0455	.0944	.1313	.4086	.1696	.1212	1

The map visualization of the $BRIT_{7\times7}$ data is quite regular: the related probability table is almost TP_2. (These probabilities are shown in Table 8.4 along with frequencies.) The determinants of the 2×2 subtables are almost all positive (among the few negative exceptions are $p_{43}p_{54}-p_{44}p_{53} = -0.00015$, $p_{65}p_{76} - p_{75}p_{66} = -0.0000065$, $p_{64}p_{75} - p_{74}p_{65} = -0.00032$). This shows that closeness to TP_2 coincides with the regular representation visible in the overrepresentation map.

8.3 Lilliputian surfaces with uniform marginal distributions

The reader now has an idea of *grade bivariate distributions* from the examples given in the previous section. More specifically, we dealt in Section 8.2 with *grade densities* on the unit square. Now, we turn to *grade cdf's*. However, in contrast to the grade densities introduced in Section 8.2 for $m \times k$ tables, we will now discuss the cdf's in a quite general context. This will show how concepts in $\mathbb{DLM}$ are analogous to concepts in $\mathbb{ULM}$ which were introduced in the first part of Chapter 3 (particularly in Section 3.2).

We hope that the reader remembers the Lilliputian curves, i.e. the graphs of the cdf's of Lilliputian variables (see e.g. Figs. 3.1-3.2), supplemented by vertical segments when needed to produce a continuous curve. Instead of Lilliputian curves defined on the unit interval we will investigate now *Lilliputian surfaces defined on the unit square.* Similarly as in the case of Lilliputian curves, each Lilliputian surface can be equivalently identified with the graph of a *bivariate function which is non-decreasing in each coordinate.* The graph is supplemented by vertical walls (quadrangles) whenever needed to turn the graph into a continuous surface. Each non-decreasing function H which fulfils $H(u_1, v_1) + H(u_2, v_2) \geq H(u_1, v_2) + H(u_2, v_1)$ for any $u_1 \leq u_2$, $v_1 \leq v_2$, $u_1, u_2, v_1, v_2 \in [0, 1]$ can be treated as the cumulative distribution function (cdf) in the joint distribution of a Lilliputian pair.

The set of all bivariate Lilliputian surfaces will be called the Bivariate Lilliputian Model and denoted $\mathbb{BLM}$. As in $\mathbb{ULM}$, the symbol $\mathbb{BLM}$ is used to denote the set of pairs of Lilliputian variables $\{(\xi, \eta)\}$ and also cdf's on the unit square.

In $\mathbb{ULM}$, each Lilliputian curve joins the opposite vertices of the unit square: (0,0) and (1,1). In $\mathbb{BLM}$, each Lilliputian surface joins the opposite vertices of the unit cube: (0,0,0) and (1,1,1).

In $\mathbb{ULM}$, the cdf of the uniformly distributed Lilliputian variable U is denoted *Diag* and is a straight line joining (0,0) and (1,1). In $\mathbb{BLM}$, the cdf of the Lilliputian pair (U, V) with uniform marginals is a paraboloida denoted $Diag_2$:

$$Diag_2(u, v) = uv = Diag(u) \times Diag(v), \quad (u, v) \in [0, 1] \times [0, 1],$$

which is spread over the unit square and joins vertices (0,0,0) and (1,1,1).

It was shown in Chapter 3 that the pair $(X; Y)$ can be transformed to a pair $(F_X^*(X); F_X^*(Y))$, where $F_X^*(X)$ (denoted X^*) is uniform on (0,1). Generally, there is no easy way to introduce analogous transformations when pairs of univariate distributions are replaced by pairs of bivariate distributions. However it becomes easy when we are concerned with *the dependence* of two Lilliputian variables.

Remark 8.1 *It will be shown in Sec. 8.4 that it is also easy in the case of the differentiation of one bivariate distribution w.r.t. another when both distributions are specifically chosen (see also Section 8.6.1).*

So, our approach to dependence is to treat it as (ordered) differentiation of components in *a special pair* of bivariate distributions: the first component is the considered joint distribution of (X, Y) and the second component is the respective distribution with independent marginal variables. Therefore, we are interested in pairs of bivariate distributions, consisting of the joint distribution of (X,Y) and the distribution of its *independent counterpart* denoted $(X,Y)^{\perp}$, where the cdf of $(X,Y)^{\perp}$ is the product of F_X and F_Y. Thus any pair $((X,Y),(X,Y)^{\perp})$ will be transformed to the pair

$$((X^*,Y^*),(X^*,Y^*)^{\perp}),$$

where $(X^*,Y^*)^{\perp} = (U,V)^{\perp} \sim Diag_2$.

When the distribution of (X,Y) is continuous (e.g., binormal), variables $F_X(X)$ and $F_Y(Y)$ are uniformly distributed and hence equal to X^* and Y^*. If the distribution of (X,Y) is not continuous, the respective *uniform grade transformations* F^*_X and F^*_Y have to be applied to each marginal distribution. This was exemplified in the previous section for various pairs of categorical variables X and Y (e.g. (Z_4,T_3)).

Thus for a continuous bivariate distribution of (X,Y), the cdf of (X^*,Y^*) is

$$Cop_{(X,Y)}(u,v) = F_{(X,Y)}(x_u,y_v).$$

For (X,Y) distributed according to $m\times k$ probability tables $[p_{ij}]$, where $p_{ij} = \Pr(X=x_i, Y=y_j)$, we have

$$\begin{aligned} Cop_{(X,Y)}(u,v) &= F_{(X,Y)}(x_{i-1},y_{j-1}) + (u - F_X(x_{i-1}))F_{Y|X=x_i}(y_{j-1}) \\ &+ (v - F_Y(y_{j-1}))F_{X|Y=y_j}(x_{i-1}) \\ &+ (u - F_X(x_{i-1}))(v - F_Y(y_{j-1}))\frac{p_{ij}}{p_{i+}p_{+j}} \end{aligned} \tag{8.3}$$

for $u \in (F_X(x_{i-1}),F_X(x_i)]$, $v \in (F_Y(y_{j-1}),F_Y(y_j)]$, $i = 1,\dots,m$, $j = 1,\dots,k$.

Details: The symbol "*Cop*" and the term "copula" are traditionally used with respect to the cdf of a bivariate distribution with uniform marginals.

Since a copula is a bivariate continuous cdf, its graph coincides with its Lilliputian surface. For any copula *Cop* the derivative $\partial^2 Cop_{(X,Y)}(u,v)/\partial u\partial v$ exsists almost everywhere in $[0,1]^2$; further,

$$Cop(u,v) = A_C(u,v) + S_C(u,v)$$

where A_C is an absolutely continuous component with density

$$cop_{(X,Y)}(u,v) = \partial^2 Cop_{(X,Y)}(u,v)/\partial u\partial v$$

and S_C is a singular component for which $\partial^2 Cop_{(X,Y)}(u,v)/\partial u\partial v = 0$ almost everywhere in $[0,1]^2$ (note that except $Cop(u,v) = A_C(u,v)$ or $Cop(u,v) = S_C(u,v)$, neither A_C nor S_C is a copula, since neither has uniform marginals). A singular component corresponds to a functional relationship between X^* and Y^* (to be considered later on).

A uniform grade transformation applied respectively to each marginal of a bivariate distribution transforms bivariate distribution into an element of $\mathbb{BLM}$ *with uniform marginals.* This subset of $\mathbb{BLM}$ will be called the Dependence Lilliputian Model and denoted $\mathbb{DLM}$:

$$\begin{aligned} \mathbb{DLM} &= \{(\xi,\eta) \in \mathbb{BLM} : \; F_\xi = F_\eta = Diag\} \qquad (8.4) \\ &= \{(X^*,Y^*) : (X,Y) \text{ is any pair of random variables}\}. \end{aligned}$$

The cdf of an element of $\mathbb{DLM}$ is called *a copula.* The symbol $\mathbb{DLM}$ will also be used to denote the set of copulas.

Definition 8.1 *(Copula and grade distribution). The cdf of* $(X^*,Y^*) = (F_X^*(X), F_Y^*(Y))$ *is called the copula of* (X,Y) *and denoted* $Cop_{(X,Y)}$*; the distribution of* (X^*,Y^*) *is called the grade distribution of* (X,Y).

The measurement of *grade dependence* consists in comparing $Cop_{(X,Y)}$ with $Diag_2$. In other words, *the differentiation* between the grade distribution of (X,Y) and the uniform distribution on the unit square is measured. By analogy with the Univariate Lilliputian Model, $Cop_{(X,Y)}$ might be treated as the *concentration surface,* denoted

$$\mathbf{C}((X,Y) : (X,Y)^{\perp}).$$

Definition 8.2 *(grade equivalence). Two bivariate distributions* $P_{(X,Y)}$ *and* $P'_{(X',Y')}$ *are grade equivalent if* $Cop_{(X,Y)} = Cop_{(X',Y')}$.

Grade equivalence is preserved when the order of the variables is changed, since

$$Cop_{(X,Y)}(u,v) = Cop_{(Y,X)}(v,u)$$

Set $\mathbb{DLM}$ contains distributions from $\mathbb{QD}$ (but is not contained in $\mathbb{QD}$). Those distributions which belong to $\mathbb{QD}^+$ have Lilliputian surfaces *above* the paraboloida $Diag_2$, while those which belong to $\mathbb{QD}^-$ have the Lilliputian surfaces *below* $Diag_2$, and

$$(X^*, Y^*) \in \mathbb{QD}^+ \ (\mathbb{QD}^-) \ \text{ iff } \begin{cases} (1-X^*, 1-Y^*) \in \mathbb{QD}^+ \ (\mathbb{QD}^-), \text{ or} \\ (1-X^*, Y^*) \in \mathbb{QD}^- \ (\mathbb{QD}^+), \text{ or} \\ (X^*, 1-Y^*) \in \mathbb{QD}^- \ (\mathbb{QD}^+). \end{cases}$$

Definition 8.3 *A copula is called opposite to $Cop_{(X,Y)}$ (and denoted $Cop^{\leftarrow}_{(X,Y)}$) if either*

$$Cop^{\leftarrow}_{(X,Y)}(u,v) = v - Cop_{(X,Y)}(1-u, v) \tag{8.5}$$

or

$$Cop^{\leftarrow}_{(X,Y)}(u,v) = u - Cop_{(X,Y)}(u, 1-v). \tag{8.6}$$

Formula (8.5) corresponds to the copula of $(f(X), g(Y))$ for any decreasing f and any increasing g; formula (8.6) corresponds to the copula of $(f(X), g(Y))$ for any increasing f and any decreasing g. Formulas (8.5) and (8.6) indicate that the distance from $Cop^{\leftarrow}_{(X,Y)}$ to the $Diag_2$ at point (u,v) is equal to the distance from the $Diag_2$ to $Cop_{(X,Y)}$ at point $(1-u, v)$ or point $(u, 1-v)$:

$$Cop^{\leftarrow}_{(X,Y)}(u,v) - uv = \begin{cases} (1-u)v - Cop_{(X,Y)}(1-u, v) & \text{in the first case,} \\ u(1-v) - Cop_{(X,Y)}(u, 1-v) & \text{in the second case.} \end{cases}$$

Set $\mathbb{DLM} \cap \mathbb{QD}^+$ consists of all elements opposite to the elements of $\mathbb{DLM} \cap \mathbb{QD}^-$. Geometrically, a copula in $\mathbb{DLM} \cap \mathbb{QD}^+$ is transformed into the opposite copula if either: for all points (u,v) each vertical segment at point (u,v) which joins the $Cop_{(X,Y)}(u,v)$ with the $Diag_2$ is transferred to $(1-u, v)$ and placed vertically under the $Diag_2(1-u, v)$; or, for all points (u,v), the same thing happens when $(1-u, v)$ is replaced by $(u, 1-v)$.

Definition 8.4 *A copula is called the survival copula of (X,Y) (and denoted $Cop^{\Leftarrow}_{(X,Y)}$) if*

$$Cop^{\Leftarrow}_{(X,Y)}(u,v) = u + v - 1 + Cop_{(X,Y)}(1-u, 1-v) \tag{8.7}$$

(note that symbol $Cop^{\Leftarrow}$ is different from $Cop^{\leftarrow}$).

Formula (8.7) indicates, that the distance from $Cop^{\Leftarrow}_{(X,Y)}$ to the $Diag_2$ at point (u, v) is equal to the distance from the $Cop_{(X,Y)}$ to the $Diag_2$ at point $(1-u, 1-v)$:

$$Cop^{\Leftarrow}_{(X,Y)}(u,v) - uv = Cop_{(X,Y)}(1-u, 1-v) - (1-u)(1-v).$$

Note that $Cop^{\Leftarrow}_{(X,Y)}$ corresponds to the copula of $(f(X), g(Y))$ where f, g are decreasing functions.

The extremal singular copulas in $\mathbb{DLM}$ correspond to the functional relationship between X^* and Y^*. The first extreme belongs to $\mathbb{QD}^+$ and corresponds to the relationship $X^* = Y^*$, the second belongs to $\mathbb{QD}^-$ and corresponds to the relationship $Y^* = 1 - X^*$. The probability mass of the first is concentrated on the line $v = u$; the probability mass of the second is concentrated on the line $v = 1 - u$. The surface corresponding to the first is called the upper Frechet Lilliputian surface and (according to the Frechet definition introduced in Section 7.3 is defined by

$$Fr^+_{(X^*,Y^*)}(u,v) = \min(u,v).$$

The surface corresponding to the second extreme is called the lower Frechet Lilliputian surface and is defined by

$$Fr^-_{(X^*,Y^*)}(u,v) = \max(u + v - 1, 0).$$

Thus, the upper Frechet Lilliputian surface is opposite to the lower Frechet Lilliputian surface, because

$$\begin{aligned} Fr^+_{(X^*,Y^*)}(u,v) &= v - Fr^-_{(X^*,Y^*)}(1-u, v) \\ &= u - Fr^-_{(X^*,Y^*)}(u, 1-v). \end{aligned}$$

Any Lilliputian surface from $\mathbb{DLM}$ lies between the lower and the upper Lilliputian Frechet surfaces.

A more precise location of the copula for a specified $(X, Y) \in \mathbb{DLM}$ is possible when the copulas for the Frechet bounds of (X, Y) are calculated. Let (X^+, Y^+) and (X^-, Y^-) be pairs of random variables with joint cdf's $Fr^+_{(X,Y)}$ and $Fr^-_{(X,Y)}$, respectively (marginals of X^+, X^- are equal to F_X, marginals of Y^+, Y^- are equal to F_Y.) Then,

$$\begin{aligned} Fr^-_{(X^*,Y^*)}(u,v) &\le Cop_{(X^-,Y^-)}(u,v) \le Cop_{(X,Y)}(u,v) \\ &\le Cop_{(X^+,Y^+)}(u,v) \le Fr^+_{(X^*,Y^*)}(u,v). \end{aligned} \tag{8.8}$$

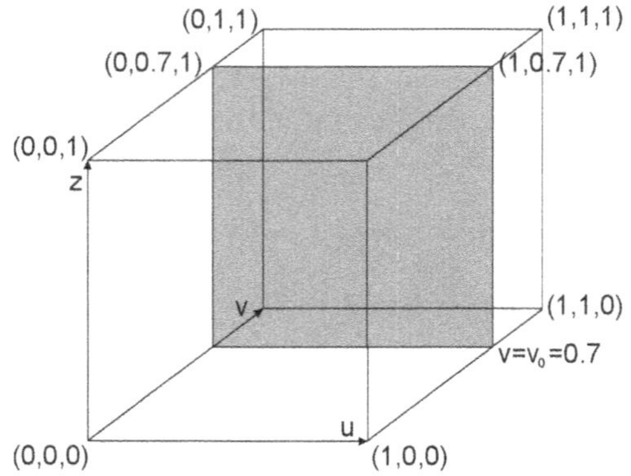

FIGURE 8.9. The intersection of the unit cube with the plane $v = v_0$ (for $v_0 = 0.7$)

It is easy to verify that $Cop_{(X^-,Y^-)}$ belongs to $\mathbb{QD}^-$ and $Cop_{(X^+,Y^+)}$ belongs to $\mathbb{QD}^+$.

Illustration ((X, Y) distributed according to table 2×2 in formula (7.9)):

The values of $Cop_{(X,Y)}$ and related copulas at points $(u, 0.7)$ are given by:

$$Cop_{(X,Y)}(u, 0.7) = \begin{cases} \frac{4}{5}u & \text{if } 0 \le u \le 0.5, \\ \frac{3}{5}u + \frac{1}{10} & \text{if } 0.5 \le u \le 1. \end{cases} \tag{8.9}$$

$$Cop_{(\overleftarrow{X},Y)}(u, 0.7) = \begin{cases} \frac{3}{5}u & \text{if } 0 \le u \le 0.5, \\ \frac{4}{5}u - \frac{1}{10} & \text{if } 0.5 \le u \le 1. \end{cases}$$

$$Cop_{(X^-,Y^-)}(u, 0.7) = \begin{cases} \frac{1}{2}u & \text{if } 0 \le u \le 0.5, \\ \frac{9}{10}u - \frac{2}{10} & \text{if } 0.5 \le u \le 1. \end{cases}$$

$$Cop_{(X^+,Y^+)}(u, 0.7) = \begin{cases} \frac{9}{10}u & \text{if } 0 \le u \le 0.5, \\ \frac{1}{2}u + \frac{1}{5} & \text{if } 0.5 \le u \le 1. \end{cases}$$

A Lilliputian surface from $\mathbb{DLM}$ in the unit cube is difficult to visualize. Therefore, we will consider instead the intersections with the plane $v = v_0$, $v_0 \in (0, 1)$. Examples are presented in Fig. 8.10 (for $v_0 = 0.7$) and Fig. 8.11 (for $v_0 = 0.8$).

The intersection of the surface $Diag_2$ with the plane $v = v_0$ is the line $Diag_2(u, v) = uv_0$.

The Lilliputian surfaces from $\mathbb{QD}^+$ are represented by non-decreasing curves joining point $(0, v_0, 0)$ with point $(1, v_0, v_0)$, these curves lie *above* the line uv_0. The Lilliputian surfaces from $\mathbb{QD}^-$ are also represented by non-decreasing curves as joining point $(0, v_0, 0)$ with $(1, v_0, v_0)$; these curves lie *below* the line uv_0.

Each Lilliputian surface in $\mathbb{DLM}$ might be presented as the set of all corresponding curves lying at the intersections of the unit cube with the plane $v = v_0$, $v_0 \in [0, 1]$.

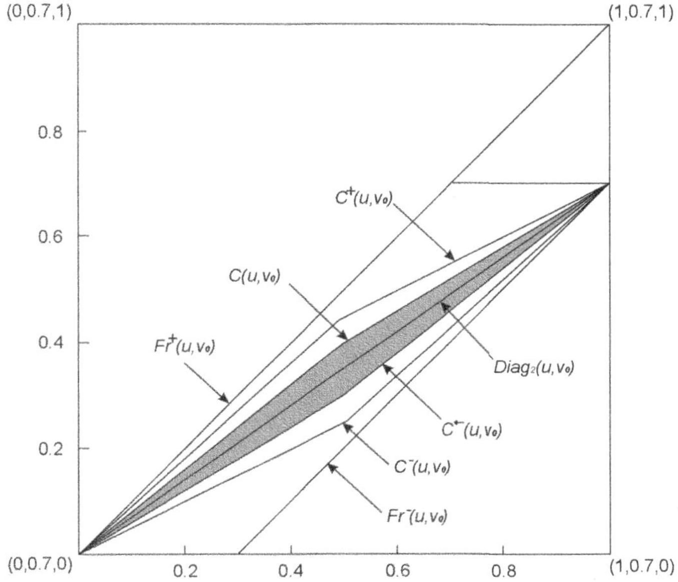

FIGURE 8.10. Curves $C(u, 0.7)$ and $C^{\leftarrow}(u, 0.7)$ arising (at the intersection of the unit cube with the plane $v_0 = 0.7$) from the Lilliputian surfaces for (X, Y) distributed according to formula (7.9). Other curves: curve $C^+(u, 0.7)$ denotes $Cop_{(X^+,Y^+)}(u, 0.7)$, curve $C^-(u, 0.7)$ denotes $Cop_{(X^-,Y^-)}(u, 0.7)$, and curves $Fr^+(u, 0.7)$, $Fr^-(u, 0.7)$ correspond to the Frechet Lilliputian surfaces.

Figures 8.10 and 8.11 present the intersections corresponding to $v_0 = 0.7$ and $v_0 = 0.8$, respectively. At each figure (i.e. at each intersection), there are the extremal curves corresponding to the upper and lower Frechet Lilliputian surfaces. The upper Frechet Lilliputian curve $Fr^+(u, v_0)$ consists of two segments: one from point $(0, v_0, 0)$ to point (v_0, v_0, v_0), the second from (v_0, v_0, v_0) to $(1, v_0, v_0)$. The lower Frechet Lilliputian curve $Fr^-(u, v_0)$ is opposite to the upper, and both curves form a *parallelogram*. The diagonal of this parallelogram is the line $Diag_2(u, v_0)$. The set of all parallelograms bound the space in the unit cube in which lie all Lilliputian surfaces from $\mathbb{DLM}$; any point (u, v, z) of the unit cube lies on some Lilliputian surface from $\mathbb{DLM}$.

Moreover, the set of the upper triangles of the parallelograms bound the space in which lie the Lilliputian surfaces from $\mathbb{QD}^+ \cap \mathbb{DLM}$, and the set of the lower triangles bound the space for $\mathbb{QD}^- \cap \mathbb{DLM}$.

Figure 8.10 presents curves $C(u, 0.7)$ and $C^{\leftarrow}(u, 0.7)$, where $C(u, 0.7)$ is calculated for (X^*, Y^*) when (X, Y) is distributed according to table (7.9); it also presents curves $C^+(u, 0.7)$ and $C^-(u, 0.7)$ corresponding to cdf's $Cop_{(X^+,Y^+)}$ and $Cop_{(X^-,Y^-)}$.

In Fig. 8.11 we present curves for $v_0 = 0.8$ in the case of (X, Y) distributed according to $N_2(0.75)$. Since the distribution $N_2(\rho)$ is continuous, then $Cop_{(X^\pm,Y^\pm)} = Fr^{\pm}_{(\xi,\eta)}$ and therefore there are just two curves $C(u, 0.8) = Cop_{(X,Y)}(u, 0.8)$ and $C^{\leftarrow}(u, 0.8) = Cop^{\leftarrow}(u, 0.8)$ inside the parallelogram lying on both sides of $Diag_2(u, 0.8)$. Due to the high value of ρ, these curves are very close to the sides of the parallelogram.

It is often convenient to present the curves $C(u, v_0)$ – for a few chosen values of v_0 – on *one* figure. This is illustrated in Figure 8.11 (down) for the curves $C(u, v_0)$ corresponding to the Lilliputian surfaces for (X^*, Y^*) when (X, Y) is distributed according to $N_2(0.75)$.

There are five curves $C(u, v_0)$ for $u \in [0, 1]$ when $v_0 = 0.2, 0.5, 0.8, 0.95, 1$. Each curve corresponds to the intersection with another plane. The curve $C(u, 0.8)$ appears in both figures: in Figure 8.11 (upper) which concerns the intersection with the plane $v = 0.8$, and in Figure 8.11 (down) where are presented the curves from *five* intersections. It is easy to see that the curve (ending in 1,0.8,0.8) is identical in both curves. Figure 8.11 (down) contains the lines $v = uv_0$. The curves at the intersections of $Cop_{(X,Y)}$ with $v = v_0$ when $(X, Y) \sim N_2(0.75)$ lie above the respective lines $v = uv_0$, since $corr(X, Y)$ is positive and $(X, Y) \in \mathbb{QD}^+$.

The areas between lines $v = uv_0$ and the curves are shaded. This helps to get the impression of intersections situated one after another in the unit cube. The last intersection, for $v_0 = 1$, consists of the diagonal of the rear side of the unit cube (the diagonal and the curve coincide).

8.4 Spearman's *rho* and Kendall's *tau* expressed by volumes and masses in the unit cube

The set of Lilliputian surfaces in $\mathbb{DLM}$ was discussed in the previous section in the same manner as the set of Lilliputian curves in $\mathbb{ULM}$ in Section 3.2.2. In the case of $\mathbb{ULM}$, the next step was to consider the *expectation* of a Lilliputian variable, expressed as some area of the unit square (more precisely, the area above the Lilliputian curve (see formula (3.7)). The concentration index ar was then defined as a suitably standardized difference of the expectations $E(\xi)$ and $E(U)$. So, in $\mathbb{ULM}$ we have the formula

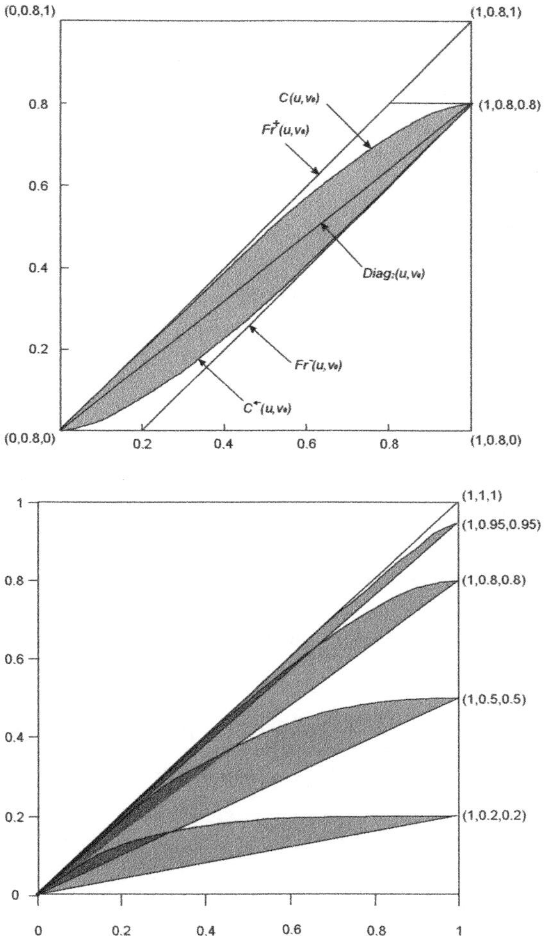

FIGURE 8.11. **Upper:** Curves $C(u, 0.8)$ and $C^{\leftarrow}(u, 0.8)$ at the intersection of the unit cube with the plane $v = 0.8$, arising from the Lilliputian surfaces for (X^*, Y^*) when (X, Y) is distributed according to $N_2(0.75)$. Other curves: $Fr^+(u, 0.8)$ and $Fr^-(u, 0.8)$ arising from the Frechet Lilliputian surfaces. **Down:** Curves $C(u, 0.2)$, $C(u, 0.5)$, $C(u, 0.8)$ and $C(u, 0.95)$ and the lines $v = uv_0$ for $v_0 = 0.2, 0.5, 0.8, 0.95$

$$ar(\xi) = \frac{E(\xi) - E(U)}{E(\xi^+) - E(U)} = \frac{E(\xi) - E(U)}{1 - E(U)} = 2E(\xi) - 1 \tag{8.10}$$

where ξ^+ is the Lilliputian variable with the largest possible expectation equal to 1; (the Lilliputian curve of ξ^+ consists of the bottom side and right side of the unit square, and therefore the area above this curve is the area of the unit square, equal to 1).

Instead of ξ^+, it is possible to use the Lilliputian variable ξ^- which is opposite to ξ^-; its Lilliputian curve consists of the left side and top side of the unit square, and therefore $E(\xi^-) = 0$. So we have

$$ar(\xi) = \frac{E(\xi^{\leftarrow}) - E(U)}{E(\xi^-) - E(U)} = \frac{E(\xi^{\leftarrow}) - E(U)}{-E(U)} = 1 - 2E(\xi^{\leftarrow}). \tag{8.11}$$

Both equivalent expressions for the concentration index lead to the difference of two areas: the area between the *Diag* and the Lilliputian curve above the *Diag*, and the area between the *Diag* and the part of the Lilliputian curve under the *Diag* (see Fig. 3.5 in Sec. 3.2.2).

The same procedure is adopted to define the *concentration index* in $\mathbb{DLM}$, provided that: the unit square is replaced by the unit cube; the curve *Diag* is replaced by the surface $Diag_2$; the Lilliputian curve is replaced by the Lilliputian surface; cdf of ξ^+ is replaced by $Fr^+(X^*, Y^*)$; cdf of ξ^- is replaced by $Fr^-(X^*, Y^*)$, and the areas are replaced by their respective volumes.

The bivariate concentration index related to (8.10) will be denoted ρ^* (Spearman's *rho*):

$$\begin{aligned}\rho^*(X,Y) &= \frac{\int_0^1\int_0^1 Cop_{(X,Y)}(u,v)dudv - \int_0^1\int_0^1 uvdudv}{\int_0^1\int_0^1 Fr^+_{(X^*,Y^*)}(u,v)dudv - \int_0^1\int_0^1 uvdudv} \\ &= 12\int_0^1\int_0^1 (Cop_{(X,Y)}(u,v) - uv)dudv\end{aligned} \tag{8.12}$$

or (the version related to (8.11)):

$$\begin{aligned}\rho^*(X,Y) &= \frac{\int_0^1\int_0^1 Cop^{\leftarrow}_{(X,Y)}(u,v)dudv - \int_0^1\int_0^1 uvdudv}{\int_0^1\int_0^1 Fr^-_{(X^*,Y^*)}(u,v)dudv - \int_0^1\int_0^1 uvdudv} \\ &= 12\int_0^1\int_0^1 (uv - Cop^{\leftarrow}_{(X,Y)}(u,v))dudv \\ &= 6\int_0^1\int_0^1 (Cop_{(X,Y)}(u,v) - Cop^{\leftarrow}_{(X,Y)}(u,v))dudv.\end{aligned}$$

Both expressions for ρ^* are equal and can be presented as the difference of two volumes: one between the $Diag_2$ and that part of the Lilliputian surface which lies above the $Diag_2$, and the other between the $Diag_2$ and that part of the Lilliputian surface which lies under the $Diag_2$.

The shaded areas in Fig. 8.10 and Fig. 8.11 are the areas on the intersections of the space contained between $Cop(u,v)$ and $Cop^{\leftarrow}(u,v)$ by the planes $v = v_0$. The volume of this space is equal to

$$\int_0^1 \int_0^1 (Cop(u,v) - Cop^{\leftarrow}(u,v))dudv.$$

Thus ρ^* is the normalized integral of the shaded areas.

An important property of ρ^* follows from formula (8.8): ρ^* can take the values -1 or 1 only when $Cop_{(X,Y)} = Fr^-_{(X^*,Y^*)}$ or $Cop_{(X,Y)} = Fr^+_{(X^*,Y^*)}$, respectively. These conditions are fulfilled only if the marginals are continuous and there exists a functional monotonic relationship between them.

Proposition 8.1 $\rho^*(X,Y) = corr(X^*,Y^*) \in [-1,1]$.

The proof follows from the equality

$$\int_0^1 \int_0^1 Cop_{(X,Y)}(u,v)dudv = \int_0^1 \int_0^1 uvdCop_{(X,Y)}(u,v);$$

the numerator of the expression in (8.12) is equal to $cov(X^*,Y^*)$; the denominator of the expression in (8.12) is equal to $cov(X^*,X^*) = var(X^*) = var(Y^*) = 1/12$.

Using formula (8.3) we obtain

$$\rho^*(X,Y) = 12 \int_0^1 \int_0^1 (\tilde{F}_{(X,Y)}(x_u,y_v) - \tilde{F}_X(x_u)\tilde{F}_Y(y_v))dudv$$

where

$$\tilde{F}_{(X,Y)}(x,y) = \frac{1}{4}(F_{(X,Y)}(x,y) + F_{(X,Y)}(x-,y) + F_{(X,Y)}(x,y-) + F_{(X,Y)}(x-,y-)),$$

or equivalently

$$\rho^*(X,Y) = 12cov(\tilde{F}_X(X), \tilde{F}_Y(Y)).$$

Proposition 8.2 *Properties of Spearman's rho* (ρ^*)

(i) For (X,Y) *distributed according to a* $m \times k$ *probability table* $[p_{ij}]$

$$\rho^*(X,Y) = 3\sum_{j=1}^{k}\sum_{i=1}^{m}(S_{i-1}+S_i-1)(T_{j-1}+T_j-1)p_{ij} \tag{8.13}$$

$$= \frac{3}{2}\sum_{r=1}^{m}\sum_{i=1}^{m}(S_r+S_{r-1}-S_i-S_{i-1})p_{i+}p_{r+}ar(r:i;row)$$

$$= \frac{3}{2}\sum_{r=1}^{m}\sum_{i=1}^{m}(S_r+S_{r-1}-S_i-S_{i-1})\sum_{t=2}^{k}\sum_{j=1}^{t-1}(p_{ij}p_{rt}-p_{it}p_{rj})$$

$$= \frac{3}{2}\sum_{t=1}^{k}\sum_{j=1}^{k}(T_t+T_{t-1}-T_j-T_{j-1})p_{+j}p_{+t}ar(t:j;col)$$

$$= \frac{3}{2}\sum_{t=1}^{k}\sum_{j=1}^{k}(T_t+T_{t-1}-T_j-T_{j-1})\sum_{r=2}^{m}\sum_{i=1}^{r-1}(p_{ij}p_{rt}-p_{it}p_{rj})$$

where $S_i = p_{1+}+\cdots+p_{i+}$, $T_j = p_{+1}+\cdots+p_{+j}$, $S_0 = T_0 = 0$, $i = 1,\ldots,m$, $j = 1,\ldots,k$. *Hence*

$$\rho^*(P_{m\times k}) = \frac{1}{2}\sum_{r=1}^{m}\sum_{i=1}^{m}(S_r+S_{r-1}-S_i-S_{i-1})(p_{i+}+p_{r+})^2\,\rho^*_{ir;row} \tag{8.14}$$

where $\rho^*_{ir;rou}$ *is the index* ρ^* *calculated for the table consisting of only two row profiles:* i^{th} *and* r^{th}. *Similarly we can present* ρ^* *as a convex combination of* $\rho^*_{jt;col}$ *for pairs of column profiles.*

For a $m\times 2$ *probability table* $P_{m\times 2}$

$$\rho^*(P_{m\times 2}) = 3\sum_{r=2}^{m}\sum_{i=1}^{r-1}(p_{i1}p_{r2}-p_{r1}p_{i2}). \tag{8.15}$$

Each expression for ρ^* *defined above for table* $[p_{ij}]$ *can be generalized for any pair of discrete-continuous random variables. For instance, the expression in the second line of (8.13) can be given by*

$$\rho^*(X,Y) = 3\int_0^1\int_0^1 (u-s)ar((Y|X=x_u):(Y|X=x_s))dsdu$$

$$= 3\int_0^1\int_0^1 (\tilde{F}_X(x_u)-\tilde{F}_X(x_s))ar((Y|X=x_u):(Y|X=x_s))dsdu.$$

(ii) $\rho^*(X^{\leftarrow},Y) = \rho^*(X,Y^{\leftarrow}) = -\rho^*(X,Y)$.

(iii) If $(X,Y)\in\mathbb{QD}^+$ *then* $\rho^*(X,Y)\geq 0$. *If* $(X,Y)\in\mathbb{QD}^-$ *then* $\rho^*(X,Y)\leq 0$. *If* $(X,Y)\in\mathbb{QD}$ *then* $\rho^*(X,Y)=0$ *iff* (X,Y) *are independent.*

(iv) If $Y=f(X)$ *for a non-decreasing function* f *then*

$$\rho^*(X,Y) = \rho^*(Y,Y) = 1 - \int_{-\infty}^{\infty} (F_Y(y_u) - F_Y(y_u-))^2 du. \tag{8.16}$$

If $X = f(Y)$ for a non-decreasing function f then

$$\rho^*(X,Y) = \rho^*(X,X) = 1 - \int_{-\infty}^{\infty} (F_X(x_u) - F_X(x_u-))^2 du. \tag{8.17}$$

If $Y = f(X)$ or $X = f(Y)$, respectively, for a non-increasing function f then

$$\rho^*(X,Y) = -\rho^*(Y,Y) \quad \text{or} \quad \rho^*(X,Y) = -\rho^*(X,X), \tag{8.18}$$

respectively.

Hence $\rho^*(X,Y) = 1$ iff (X,Y) are continuous variables and $Y = f(X)$ where f is increasing; $\rho^*(X,Y) = -1$ iff (X,Y) are continuous variables and $Y = f(X)$ where f is decreasing.

(v) For $(X,Y) \sim N_2(\rho)$,

$$\rho^*(X,Y) = \frac{6\, arcsin(\rho/2)}{\pi} \approx \rho. \tag{8.19}$$

Due to Prop. 8.2 (ii)

$$\rho^*(X^{\leftarrow}, Y^{\leftarrow}) = \rho^*(X,Y). \tag{8.20}$$

There exist many formulas for ρ^*. Two of them, namely (8.12) and the second line of (8.13), have a special interpretational meanings.

Formula (8.12) shows that $\rho^*(X,Y)$ measures *strength of dependence* between X and Y (since the denominator in (8.12) refers to functional dependence between X^* and Y^* compared with independence).

The second line of (8.13) is a linear combination of concentration indices of two conditional row distributions and therefore measures *total differentiation of rows* which is also equal to *global differentiation of columns* (due to the fourth line of (8.13)).

This means that the strength of grade dependence can be identified with total differentiation of rows profiles or of column profiles.

***

Another parameter in $\mathbb{DLM}$ which is a twin to ρ^* is defined analogously as ρ^* with the essential difference that *volumes are replaced by masses* (so that the uniform density used in formula (8.12) to integrate $Cop_{(X,Y)}$ over the unit square is replaced by density $cop_{(X,Y)}$ whenever it exists. Generally, we consider the integrals $\int\int Cop(u,v)dCop(u,v)$. For instance, in the case of the Frechet distributions the densities does not exist but then we integrate $\int\int F^+(u,v)dF^+(u,v)$).

This new parameter will be denoted τ and called Kendall's *tau*.

So, instead of formula (8.12) for ρ^*, we have for τ :

$$\tau(X,Y) = \frac{\int_0^1 \int_0^1 Cop_{(X,Y)}(u,v)dCop_{(X,Y)}(u,v) - \int_0^1 \int_0^1 uvdudv}{\int_0^1 \int_0^1 Fr^+_{(X^*,Y^*)}(u,v)dFr^+_{(X^*,Y^*)}(u,v) - \int_0^1 \int_0^1 uvdudv}. \quad (8.21)$$

Since

$$\int_0^1 \int_0^1 Fr^+_{(X^*,Y^*)}(u,v)dFr^+_{(X^*,Y^*)}(u,v) = \int_0^1 udu = \frac{1}{2},$$

we obtain

$$\tau(X,Y) = 4\int_0^1 \int_0^1 Cop_{(X,Y)}(u,v)dCop_{(X,Y)}(u,v) - 1 \in [-1,1]. \quad (8.22)$$

The *univariate* counterpart of $\tau(X,Y)$ as given by formula (8.21) is useless: it is constantly equal to 0, since for any $\xi \in \mathbb{ULM}$ with cdf C

$$\tau(\xi) = 2\left(\int_0^1 \tilde{C}(u)dC(u) - \int_0^1 udu\right) = 0,$$

where $\tilde{C}(u) = (C(u-) + C(u))/2$.

Proposition 8.3 *Properties of Kendall's tau.*

(i) For (X,Y) distributed according to a probability table $P_{m\times k} = [p_{ij}]$,

$$\begin{aligned}\tau(X,Y) = \tau(P_{m\times k}) &= 2\sum_{r=2}^{m}\sum_{i=1}^{r-1} p_{i+}p_{r+}ar(r:i;row) \quad (8.23)\\ &= 2\sum_{r=2}^{m}\sum_{i=1}^{r-1}\sum_{t=2}^{k}\sum_{j=1}^{t-1}(p_{ij}p_{rt} - p_{it}p_{rj}),\\ &= 2\sum_{t=2}^{k}\sum_{j=1}^{t-1} p_{+j}p_{+t}ar(t:j;col).\end{aligned}$$

If $m = 2$,

$$\begin{aligned}\tau(P_{2\times k}) &= 2\sum_{t=2}^{k}\sum_{j=1}^{t-1} p_{1+}p_{2+}ar(2:1;row) \quad (8.24)\\ &= 2\sum_{t=2}^{k}\sum_{j=1}^{t-1}(p_{1j}p_{2t} - p_{1t}p_{2j}) = \frac{2}{3}\rho^*(P_{2\times k}).\end{aligned}$$

Similarly as for ρ^*, each expression for τ defined above for $[p_{ij}]$ can be generalized for any pair of discrete-continuous random variables. For instance, the expressions in (8.24) are given by

$$\tau(X,Y) = 2\int_{u=0}^{u=1}\int_{s=0}^{s=u} ar((Y|X=x_u):(Y|X=x_s))dsdu \tag{8.25}$$

$$= 2\int_{x'=-\infty}^{x'=\infty}\int_{x=-\infty}^{x=x'} ar((Y|X=x'):(Y|X=x))dF_X(x)dF_X(x'). \tag{8.26}$$

Using formula (8.3) we obtain

$$\tau(X,Y) = 4\int_{-\infty}^{\infty}\int_{-\infty}^{\infty} \tilde{F}_{(X,Y)}(x,y)dF_{(X,Y)}(x,y)$$
$$-4\int_{-\infty}^{\infty}\int_{-\infty}^{\infty} \tilde{F}_X(x)\tilde{F}_Y(y)dF_X(x)dF_Y(y).$$

(ii) $\tau(X^{\leftarrow},Y) = \tau(X,Y^{\leftarrow}) = -\tau(X,Y)$.

(iii) If $(X,Y) \in \mathbb{QD}^+$ *then* $\tau(X,Y) \geq 0$. *If* $(X,Y) \in \mathbb{QD}^-$ *then* $\tau(X,Y) \leq 0$.

(iv) If $Y = f(X)$ *for a non-decreasing function* f *then*

$$\tau(X,Y) = \tau(Y,Y) = 1 - \int_0^1 (F_Y(y_u) - F_Y(y_u-))du. \tag{8.27}$$

If $X = f(Y)$ *for a non-decreasing function* f *then*

$$\tau(X,Y) = \tau(X,X) = 1 - \int_0^1 (F_X(x_u) - F_X(x_u-))du. \tag{8.28}$$

If $Y = f(X)$ *or* $X = f(Y)$ *for a non-increasing function* f *then, respectively,*

$$\tau(X,Y) = -\tau(Y,Y) \quad \text{or} \quad \tau(X,Y) = -\tau(X,X).$$

Hence $\tau(X,Y) = 1$ *iff* X *and* Y *are continuous variables and* $Y = f(X)$ *where* f *is increasing;* $\tau(X,Y) = -1$ *iff* X *and* Y *are continuous variables and* $Y = f(X)$ *where* f *is decreasing.*

(v) For $(X,Y) \sim N_2(\rho)$:

$$\tau(X,Y) = \frac{2\,arcsin(\rho)}{\pi}, \qquad \frac{\rho^*(X,Y)}{\tau(X,Y)} = \frac{6\,arc\,sin(\rho/2)}{2\,arc\,sin(\rho)}; \tag{8.29}$$

$$\frac{\rho^*(X,Y)}{\tau(X,Y)} = \begin{cases} \frac{3}{2} & \text{if } \rho \to 0, \\ 1 & \text{if } \rho \to 1. \end{cases}$$

(vi) According to property 8.3 (ii)

$$\tau(X^{\leftarrow}, Y^{\leftarrow}) = \tau(X,Y).$$

Due to formulas (8.21) and (8.24) (first line) parameter τ measures *strength of dependence* as well as *total differentation in the set of rows as well as in the set of columns.* Then it is clear that τ is twin to ρ^*; they differ only due to different weights in the formulas concerning strength of dependence and global differentiation (compare (8.21) with (8.12), and (8.24) with (8.13)).

Remark 8.2 *The symbols ρ^* and τ are introduced here because, in the case of continuous distributions, ρ^* and τ are equal to the parameter rho introduced by Spearman and the parameter tau introduced by Kendall.*

8.5 Grade regression functions and related measures

In this section we consider various parameters of the bivariate distribution of (X,Y) applied to the grade distribution of (X,Y). Such parameters are called *grade*: e.g., grade regression functions, grade correlation functions, grade monotone dependence functions, grade Gini Indicies. Note that the grade regression function of Y on X is not equal to the regression function of Y on X (which means that regression function of bivariate distributions is not invariant on grade transformations which transform (X,Y) to (X^*,Y^*)).

Definition 8.5 *The grade regression function of Y on X is a function from $[0,1]$ to $[0,1]$ defined by*

$$r^*_{Y|X}(u) \stackrel{df}{=} r_{Y^*|X^*}(u) = \int_0^1 v dF_{Y^*|X^*=u}(v) = \int_0^1 (1 - Cop_{(X,Y)}(v|u))dv,$$

where $Cop_{(X,Y)}(v|u) = P(Y^* \le v | X^* = u)$, $u \in (0,1)$.

Proposition 8.4 *(i) Let $\tilde{F}_Y(y) = \frac{1}{2}(F_Y(y-) + F_Y(y))$. Then*

$$r^*_{Y|X}(u) = \int_{-\infty}^{\infty} \tilde{F}_Y(y) dF_{(Y|X=x_u)}(y). \tag{8.30}$$

For the $m \times k$ *probability table we obtain*

$$r^*_{Y|X}(u) = \sum_{j=1}^{k} \frac{(T_{j-1} + \frac{1}{2}p_{+j})p_{ij}}{p_{i+}}, \quad \text{for } u \in (S_{i-1}, S_i) \tag{8.31}$$

where $S_i = p_{1+} + \cdots + p_{i+}$, $T_j = p_{+1} + \cdots + p_{+j}$, $S_0 = T_0 = 0$, $i = 1, \ldots, m$, $j = 1, \ldots, k$.

(ii) We have

$$\begin{aligned} r^*_{Y|X}(u) &= \frac{1}{2} + \frac{1}{2} ar((Y^*|X^* = u) : Y^*) \\ &= \frac{1}{2} + \frac{1}{2} ar((Y|X = x_u) : Y). \end{aligned} \tag{8.32}$$

(iii) The maximal value of the grade regression is equal to 1, and the minimal value is equal to 0. In the case of independence (X, Y), $r^*_{Y|X} = \frac{1}{2}$ *(but not conversely).*

(iv) For (X, Y) *with cdf* $N_2(\rho)$

$$r^*_{Y|X}(u) = \Phi\left(\frac{\rho\,\Phi^{-1}(u)}{\sqrt{2 - \rho^2}}\right), \quad u \in (0, 1). \tag{8.33}$$

Both grade regression functions in $N_2(\rho)$ ($r^*_{Y|X}$ *and* $r^*_{X|Y}$) *are the same. Moreover, if* $(X, Y) \sim N_2(\rho)$ *and* $(X', Y') \sim N_2(-\rho)$, *then*

$$r^*_{Y'|X'}(u) = 1 - r^*_{Y|X}(u) = r^*_{Y|X}(1 - u) \tag{8.34}$$

(see Fig. 8.12).

For small $|\rho|$ *the graph of* $r^*_{Y|X}$ *consists of approximately three segments: the first joins points (0,0) and* $(u_0, r^*_{Y|X}(u_0))$, *the second joins points* $(u_0, r^*_{Y|X}(u_0))$ *and* $(u_1, r^*_{Y|X}(u_1))$, *and the third joins points* $(u_1, r^*_{Y|X}(u_1))$ *and (1,1), where*

$$u_- = \Phi\left(-\sqrt{ln\left(\frac{2-\rho^2}{\rho^2}\right)}\right), \quad u_+ = \Phi\left(\sqrt{ln\left(\frac{2-\rho^2}{\rho^2}\right)}\right).$$

Moreover, for sufficiently small values of $\frac{\rho\,(\Phi^{-1}(q) - \Phi^{-1}(p))}{\sqrt{2(1-\rho^2)}}$ *and for any portion* $p < p_1 < \cdots < p_n < q$ *of the interval* (p, q),

$$ar^{\rho}(q : p) = ar^{\rho}(p_1 : p) + ar^{\rho}(p_2 : p_1) + \cdots + ar^{\rho}(q : p_n)$$

where $ar^{\rho}(q : p)$ *is given by formula (7.19) in Chapter 7.*

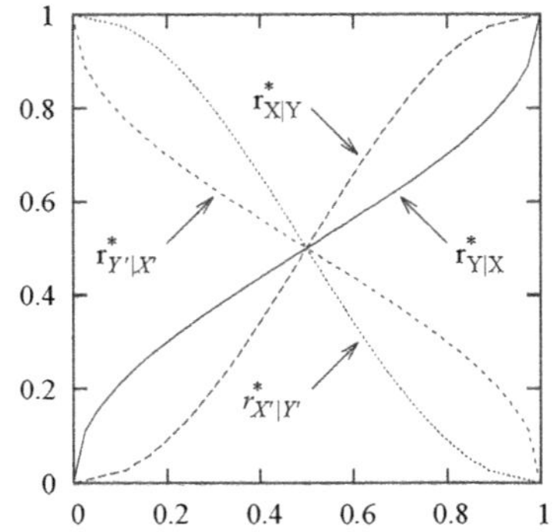

FIGURE 8.12. The graphs of the grade regression functions $r^*_{X|Y}$, $r^*_{Y|X}$ for $N_2(0.75)$ and $r^*_{X'|Y'}$, $r^*_{Y'|X'}$ for $N_2(-0.75)$ (compare Fig. 8.5).

Definition 8.6 *The grade correlation function of* Y *on* X *is the cumulated expectation of* Y^* *on* X^*, *denoted* $C^*_{cor(Y:X)}$:

$$C^*_{cor(Y:X)}(u) \overset{df}{=} E(Y^*; X^* \leq u).$$

The grade correlation function is equal to the cumulated regression:

$$C^*_{cor(Y:X)}(u) = E(r_{Y^*|X^*}(X^*); X^* \leq u).$$

Let $Parab(u) = u^2$, $Parab^{\leftarrow}(u) = 1-(1-u)^2 = 2u-u^2$, $u \in (0,1)$.

Proposition 8.5 *For* $u \in (0,1)$:

$$2C^*_{cor(Y:X)}(u) = \begin{cases} u & \textit{iff } r^*_{Y|X}(u) \equiv \frac{1}{2}, \\ Parab(u) & \textit{iff } Cop_{(X,Y)} = Fr^+_{(X^*,Y^*)}, \\ Parab^{\leftarrow}(u) & \textit{iff } Cop_{(X,Y)} = Fr^-_{(X^*,Y^*)}, \\ L_{r^*_{Y|X}}(u) & \textit{iff } r^*_{Y|X} \textit{ is non-decreasing}, \\ L^{\leftarrow}_{r^*_{Y|X}}(u) & \textit{iff } r^*_{Y|X} \textit{ is non-increasing}. \end{cases}$$

Due to the first equivalence in Proposition 8.5, $2C^*_{cor(Y:X)}$ coincides with *Diag* for independent X and Y.

We will now standardize each value of the grade correlation function according to the same scenario used earlier for parameter $ar(\xi)$ (formula 8.10) or for parameter ρ^* (formula (8.12)). The resulting standardized grade correlation, which is a function defined on (0,1) and taking values from [-1,1], is called *the grade monotone dependence function of* Y on X; and is denoted $\mu^*_{Y:X}(p)$.

For any $p \in [0,1]$ we have

$$\mu^*_{Y:X}(p) = \frac{C^*_{cor(Y:X)}(p) - pE(Y^*)}{C^*_{cor(Y:Y)}(p) - pE(Y^*)} = \frac{p - 2C^*_{cor(Y:X)}(p)}{p(1-p)}. \qquad (8.35)$$

Thus the grade monotone dependence function $\mu^*_{Y:X}$ is a continuous function on [0,1]. It takes a positive value at point p iff the curve $2C^*_{cor(Y:X)}(p)$ lies under the $Diag(p)$, and a negative value iff the curve $2C^*_{cor(Y:X)}(p)$ lies above the $Diag(p)$.

Remark 8.3 *(visualization of the grade monotone dependence function). Let us come back to Figure 8.11 (down). Suppose that all areas between $Cop(u,v)$ and uv are integrated with respect to v :*

$$\int_0^1 (Cop(u,v) - uv)dv = \frac{1}{2}u - E(Y^*; X^* \leq u) = \frac{1}{2}u - C^*_{cor(Y:X)}(u)$$

which is equal to doubled numerator of (8.35); the denominator of (8.35) is equal the double maximal value of integrated intersections.

It follows from (8.35) that the grade correlation function is a mixture of the values of the $Parab$ and the $Diag$, and the grade monotone dependence function is the mixer:

$$2C^*_{cor(Y:X)}(p) = \mu^*_{Y:X}(p)Parab(p) + (1 - \mu^*_{Y:X}(p))Diag(p). \qquad (8.36)$$

This mixer is constant and equal to ρ^* if and only if the regression function is linear (see (8.38)).

Proposition 8.6 *Expressions for ρ^* based on the grade regression and grade correlation curve.*

(i)

$$\begin{aligned}
\rho^*(X,Y) &= 3ar(2C^*_{cor(Y:X)}) = 6\int_0^1 (u - 2C^*_{cor(Y:X)}(u))du \\
&= 6\int_0^1 (2C^{*\leftarrow}_{cor(Y:X)}(u) - u)du \\
&= 6\int_0^1 (C^{*\leftarrow}_{cor(Y:X)}(u) - C^*_{cor(Y:X)}(u))du \\
&= -3 + 12\int_0^1 ur^*_{Y|X}(u)du \\
&= 6\int_0^1 \tilde{F}_X(x_u))\, ar((Y|X = x_u) : Y)du \qquad (8.37) \\
&= 6\int_{u=0}^{u=1}\int_{s=0}^{s=u} (r^*_{Y|X}(u) - r^*_{Y|X}(s))dsdu.
\end{aligned}$$

(ii)

$$\mu^*_{Y:X}(u) \equiv \rho^*(X,Y) \quad \textit{iff} \quad r^*_{Y|X}(u) = \rho^*\,u + \tfrac{1}{2}(1-\rho^*),\ u \in (0,1). \tag{8.38}$$

Proof of (i) follows from the equalities:

$$\begin{aligned}\int_0^1 Cop_{(X,Y)}(u,v)dv &= \int_0^1 \int_0^u F_{Y^*|X^*=t}(v)dtdv \\ &= \int_0^u (1 - E(Y^*|X^* = t))dt = u - E(Y^*; X^* \leq u).\end{aligned}$$

The proof of (ii) follows from the property of the monotone dependence function defined in the Section 7.6.

We note that all graphs of the grade correlation functions multiplied by 2 lie in the unit square between $Parab$ and $Parab^{\leftarrow}$:

$$\begin{aligned}2E(X^*; X^* \leq u) = u^2 \leq 2C^*_{cor(Y:X)}(u) \leq \\ \leq 2E(X^*; X^* \geq 1-u) = 2u - u^2.\end{aligned} \tag{8.39}$$

8.6 On permuting rows and columns of $m \times k$ probability tables

8.6.1 *Maximal grade correlation*

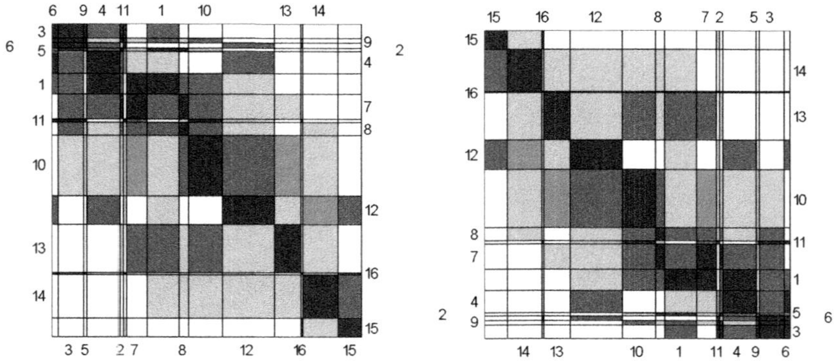

FIGURE 8.13. Overrepresentation maps for $CARS_{\max}$ (**left**) and $CARS^{\Leftarrow}_{\max}$ (**right**). Rows (car owned previously) and columns (car owned presently) are ordered so that ρ^* is maximized ($\rho^* = 0.4239$); the orderings in the right table are opposite to those in the left table

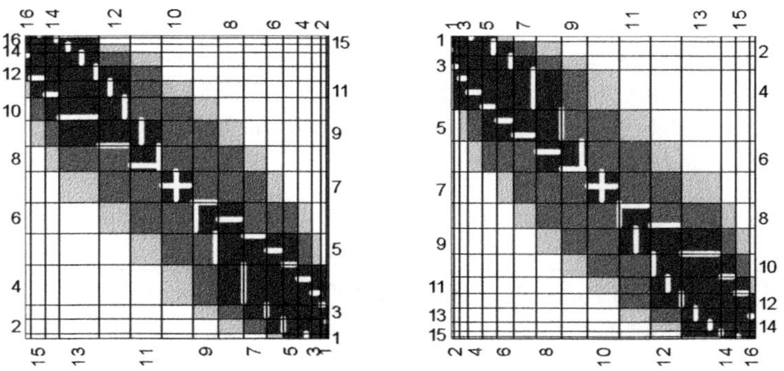

FIGURE 8.14. Two overrepresentation maps for discretized bivariate normal distributions $NORM_{16\times16}(0.75;\mathbf{p},\mathbf{q})$ and $NORM^{\Leftarrow}_{16\times16}(0.75;\mathbf{p},\mathbf{q})$ where $\mathbf{q}=\mathbf{p}^{\leftarrow}$ (i.e. $p_1=q_{16}, p_2=q_{15}$ etc). Graphs of the grade regression functions are visible on both maps (i.e., the white lines)

This section provides an introduction to the general problems of order oriented copulas (however, this material is particularly important for practical applications). It concerns categorical variables.

We will reconsider two probability tables: $CARS_{16\times16}$ (Fig. 8.7) and $BRIT_{7\times7}$ (Table 8.4, Fig. 8.8), when their rows and columns are permuted in various ways.

We will also consider the table $NORM_{16\times16}$ which is a discretized binormal table $N(0.75;\mathbf{p},\mathbf{q})$ (Fig. 8.14 left).

First, we permute each table so that the grade correlation ρ^* attains its maximal value (the relevant procedure will be given in Chapter 9). The datasets $BRIT_{7\times7}$ and $NORM_{16\times16}$ are already ordered (i.e., permuted) in this way, while for $CARS_{16\times16}$ the initial ordering has to be changed (see Fig. 8.13).

The maximal value of the grade correlation of (X,Y) is denoted $\rho^*_{\max}(X,Y)$ and called the *maximal grade correlation.* A table with $\rho^*=\rho^*_{\max}$ exhibits the *strongest monotone dependence between row variable and column variable* and therefore also the *strongest possible global differentiation (called maximal global differentiation) between rows and also between columns.*

A specific feature of each distribution with $\rho^*=\rho^*_{\max}$ is that *both grade regression functions are non-decreasing* (see their graphs in Fig. 8.14).

Remark 8.4 *If $\rho^* = \rho^*_{\max}$ then both grade correlation curves: $C^*_{cor(Y:X)}$ and $C^*_{cor(X:Y)}$ are convex (if not they could be turned to convex by suitable permutations of rows and columns which would increase the value of ρ^*; this will be considered in detail in Section 9.2.1). Therefore, the derivatives of $C^*_{cor(Y:X)}$ and $C^*_{cor(X:Y)}$, are equal respectively to $r^*_{Y|X}$ and $r^*_{X|Y}$, and are non-decreasing.*

If $(X,Y) \in TM_2$ then $\rho^*_{\max}(X,Y) = |\rho^*(X,Y)|$. In view of the properties of TP_2 (proposition 7.2) and the formula in the second line of (8.13)), in any TP_2-table the value of $\rho^*_{\max}$ is a linear combination of $ar_{\max}(r : i; row)$ for all pairs consisting of an arbitrary row i and a successive row r, and also a linear combination of $ar_{\max}(t : j; col)$ for all pairs consisting of an arbitrary column j and a successive column t :

$$\rho^*_{\max}(X,Y) = \tag{8.40}$$
$$= 3\sum_{r=2}^{m}\sum_{i=1}^{r-1}(S_r + S_{r-1} - S_i - S_{i-1})p_{i+}p_{r+}ar_{\max}(r : i; row)$$
$$= 3\sum_{t=2}^{k}\sum_{j=1}^{t-1}(T_t + T_{t-1} - T_j - T_{j-1})p_{+j}p_{+t}ar_{\max}(t : j; col)$$

where $S_i = p_{1+} + \cdots + p_{i+}$, $T_j = p_{+1} + \cdots + p_{+j}$, $i = 1, \ldots, m$, $j = 1, \ldots, k$.

Therefore, for $(X,Y) \in TM_2$, $\rho^*_{\max}$ is equal to the *maximal global differentiation* between rows, and also to the *maximal global differentiation* between columns.

Remark 8.5 *Please note that formula (8.40) is true if and only if pair (X,Y) is totally monotone dependent of order two. If $\rho^*_{\max}(X,Y) = \rho^*(X,Y)$ but (X,Y) does not belong to TP_2, then at least some of $ar_{\max}$'s in formula (8.40) have to be replaced by the respective values of ar's (so that $\rho^*_{\max}$ is smaller than the right-hand side of (8.40)). However, in a table with $\rho^* = \rho^*_{\max}$) it is rare to have a negative value of $ar(r : i; row)$ or $ar(t : j; col)$ for $r > i$ or $t > j$.*

Example. Let (X,Y) be distributed according to the probability table

$$M_{m\times k} = [m_{ij}] = \begin{bmatrix} P & \emptyset \\ \emptyset & Q \end{bmatrix} \tag{8.41}$$

where P and Q are tables $P_{i_1 \times j_1}$ and $Q_{(m-i_1)\times(k-j_1)}$, respectively, such that $\sum_{i=1}^{i_1}\sum_{j=1}^{j_1} m_{ij} = \alpha$, $\sum_{i=i_1+1}^{m}\sum_{j=j_1+1}^{k} m_{ij} = 1-\alpha$.

Then we have the following decomposition of $\rho^*_{\max}(M)$:

$$\rho^*_{\max}(M) = 3\alpha(1-\alpha) + \alpha^3\,\rho^*_{\max}(P/\alpha) + (1-\alpha)^3\rho^*_{\max}(Q/(1-\alpha)). \quad (8.42)$$

We note that the first term in (8.42) corresponds to the value of ρ^* calculated for the aggregated table

$$\begin{bmatrix} \alpha & 0 \\ 0 & 1-\alpha \end{bmatrix}.$$

It is clear that $M_{\max}$ belongs to TP_2 iff P/α and $Q/(1-\alpha)$ belong to the family of total monotonicity of order two (TM_2).

If P/α and $Q/(1-\alpha)$ belong to TM_2, then this decomposition of $\rho^*_{\max}$ corresponds to a partition of the sum in formula (8.40) onto three sums: the first for $i \in \{1,\dots,i_1\}$, $r \in \{i_1+1,\dots,m\}$ (the "between term"), the second for $i, r \in \{1,\dots,i_1\}$ (the first component of the "within term") and the third for $i, r \in \{i_1+1,\dots,m\}$ (the second component of the "within term") in the decomposition of $\rho^*_{\max}$.

* * *

The row concentration curves and column concentration curves of a probability table $P_{m\times k}$ are the same as the concentration curves for respective pairs of conditional distributions of the grade distribution of $P_{m\times k}$. Let us illustrate it in the case of a $P_{2\times k}$ distribution.

TABLE 8.5. The overrepresentation values for optimally permuted geography scores

	D	A	B	F	C
East	1.667	1.25	1.136	0.769	0.625
West	0.556	0.833	0.909	1.154	1.25

Illustration (Optimally permuted Geography Scores):

Figure 8.15 (left) presents the overrepresentation map for 2×5 table with optimally permuted geography scores given in Table 2.2 in Chapter 2, Section 2.3. The overrepresentation values are given in Table 8.5. The ratios of overrepresentations of the upper row to the lower row are: 3, 1.5, 1.25, 0.67, 0.5, which are exactly the same as indicated in Fig. 2.2 for the concentration curve of East City School w.r.t. West City School.

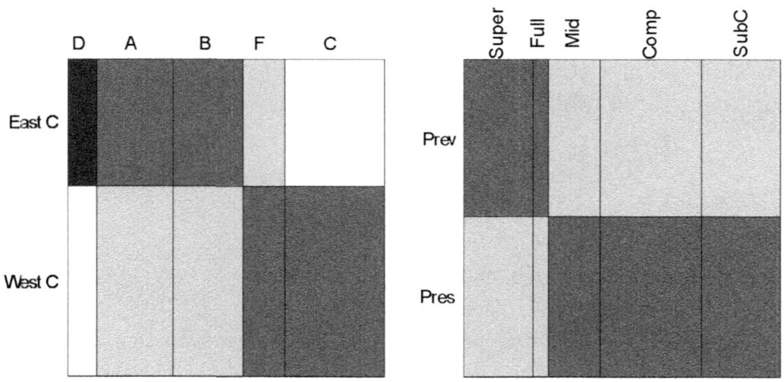

FIGURE 8.15. Overrepresentation maps for optimally permuted geography scores (**left**) and optimally permuted car models (**right**)

Illustration (Optimally permuted car models - data from Table 2.7):
Figure 8.15(right) presents the overrepresentation map for optimally permuted car models given in Table 2.7. The respective concentration curve is shown in Fig. 2.14.

8.6.2 Ordered Gini indices for marginal density transforms

Since $\rho^*(X,Y) = \rho^*(X^{\leftarrow}, Y^{\leftarrow})$ (formula (8.20)), each table with maximal value of ρ^* is accompanied by another one which has rows and columns ordered in opposite directions (e.g. Fig. 8.13 left and right, Fig. 8.14 left and right).

Let us now find such a parameter of tables $[p_{ij}]$ which will allow to distinguish one of the two twin tables $[p_{ij}]$ and $[p_{ij}]^{\Leftarrow}$. This parameter will be specially useful when $\rho^*([p_{ij}]) = \rho^*_{\max}([p_{ij}])$.

Remark 8.6 *Note that there can be many tables with the same value of* $\rho^*_{\max}$. *For instance, for the table*

$$[p_{ij}] = \begin{bmatrix} 0.15 & 0 & 0 & 0.05 \\ 0 & 0.1 & 0.15 & 0 \\ 0 & 0.05 & 0.2 & 0 \\ 0.05 & 0 & 0 & 0.25 \end{bmatrix}$$

with $\rho^* = 0.52125$, *there are 8 pairs of permutations of rows and columns such that* $\rho^* = \rho^*_{\max} = 0.82125$, *for instance*

$$[q_{ij}] = \begin{bmatrix} 0.1 & 0.15 & 0 & 0 \\ 0.05 & 0.2 & 0 & 0 \\ 0 & 0 & 0.25 & 0.05 \\ 0 & 0 & 0.05 & 0.15 \end{bmatrix}.$$

Each such table, say $[q_{ij}]$, *is accompanied by* $[q_{ij}]^{\Leftarrow}$. *The parameter which we are going to introduce is intended to distinguish between* $[q_{ij}]$ *and* $[q_{ij}]^{\Leftarrow}$. *This parameter may be applied to rows or to columns and is connected with the sequences of widths (of rows or of columns, respectively).*

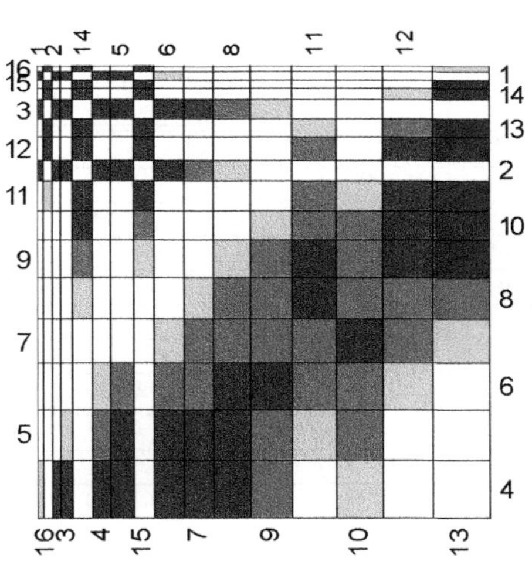

FIGURE 8.16. Overrepresentation map for $NORM_{16\times16}$ with rows and columns ordered increasingly in size

First of all, we recall (Section 4.3.4) the Gini Index for the density transform $p_X(X)$ of a categorical variable X (i.e. $Gini(p_X(X))$ where $p_X(x_i) = p_i$). This index can be calculated both for row variable (X) and for column variable (Y) of any $m \times k$ table. Its values for $BRIT_{7\times7}$, $CARS_{16\times16}$, and $NORM_{16\times16}$ are given in Table 8.6 in its last column (when rows and columns are respectively permuted according to their nondecreasing sizes). We see that $BRIT_{7\times7}$ has the Gini Index for the density transform equal to 0.335 for rows and 0.321 for columns; in the case of $CARS_{16\times16}$ it is equal to 0.275 for rows and 0.199 for columns; in the case of $NORM_{16\times16}(0.75)$, it is equal to 0.222 both for rows and for columns.

TABLE 8.6. Selected parameters describing variously permuted contingency tables ($BRIT_{7\times 7}$, $CARS_{16\times 16}$, $NORM_{16\times 16}$). The symbols *Gini r* and *Gini c* denote $Gini(p_X(X);\varphi)$ and $Gini(p_Y(Y);\psi)$ where φ and ψ indicate the respective orderings of rows and columns in the current table (i.e. according to $\rho^* = \rho^*_{\max}$, to $\tau = \tau_{\max}$, or to increasing sizes of rows and columns). The left and right columns under the headings $\rho^* = \rho^*_{\max}$ and $\tau = \tau_{\max}$ refer to the twin opposite tables

		Tables permuted according to:				
		$\rho^* = \rho^*_{\max}$		$\tau = \tau_{\max}$		increasing sizes of rows and columns
BRIT	Fig. No	8.8	-	-	-	8.17 (l)
	ρ^*	0.3720	0.3720	0.3720	0.3720	0.2009
	τ	0.2566	0.2566	0.2566	0.2566	0.1369
	τ/ρ^*	0.6898	0.6898	0.6898	0.6898	0.6814
	Gini r	0.0670	-0.0670	0.0670	-0.0670	0.3350
	Gini c	0.0540	-0.0540	0.0540	-0.0540	0.3210
	τ_{abs}	0.264	0.264	0.264	0.264	0.264
	τ/τ_{abs}	0.9720	0.9720	0.9720	0.9720	0.5186
CARS	Fig. No	8.13 (l)	8.13 (r)	-	-	8.17 (r)
	ρ^*	0.4239	0.4239	0.4231	0.4231	0.1650
	τ	0.3007	0.3007	0.3013	0.3013	0.1086
	τ/ρ^*	0.7094	0.7094	0.7121	0.7121	0.6582
	Gini r	0.104	-0.104	0.1209	-0.1289	0.275
	Gini c	0.035	-0.035	-0.1289	0.1289	0.199
	τ_{abs}	0.4137	0.4137	0.4137	0.4137	0.4137
	τ/τ_{abs}	0.7269	0.7269	0.7283	0.7283	0.2625
NORM	Fig. No	8.14 (l)	8.14 (r)	8.14 (l)	8.14 (r)	8.16
	ρ^*	0.7226	0.7226	0.7226	0.7226	0.0107
	τ	0.5287	0.5287	0.5287	0.5287	-0.0133
	τ/ρ^*	0.7317	0.7317	0.7317	0.7317	-1.2430
	Gini r	-0.107	0.107	-0.107	0.107	0.222
	Gini c	0.107	-0.107	0.107	-0.107	0.222
	τ_{abs}	0.5287	0.5287	0.5287	0.5287	0.5287
	τ/τ_{abs}	1	1	1	1	-0.0252

However, the parameters which we will use to distinguish between $[p_{ij}]$ and $[p_{ij}]^{\Leftarrow}$ *do not* refer to the Gini Indices mentioned above but to the *ordered Gini Indices* for *the density transform* introduced in Section 4.3.4 The orderings of rows and columns which will be taken into account will refer to the case when $\rho^* = \rho^*_{\max}$ (i.e. to the orders introduced by the non-decreasing grade regression functions $r^*_{Y|X}$ and $r^*_{X|Y}$).

The marginal variables in the $m \times k$ table are categorical variables with m categories in the case of the row variable denoted X and with k categories in the case of the column variable Y. The probabilities of X and Y are respectively $\mathbf{p}_{\bullet+} = (p_{1+}, \dots, p_{m+})$ and $\mathbf{p}_{+\bullet} = (p_{+1}, \dots, p_{+k})$; the related functions p_X and p_Y are given by:

$$p_X(x_i) = p_{i+}, \quad i = 1, \dots, m; \qquad p_Y(y_j) = p_{+j}, \quad j = 1, \dots, k.$$

So we deal with two variables $p_X(X)$ and $p_Y(Y)$, each measured on the absolute scale. When rows or/and columns of a considered table are permuted according to φ and ψ, respectively, we get the Lorenz Curves of the density transforms $L_{p_X(X);\varphi}$, $L_{p_Y(Y);\psi}$, and the respective concentration indices $Gini(p_X(X);\varphi)$, $Gini(p_Y(Y);\psi)$.

Due to formula (4.24) in Section 4.3.3

$$Gini(p_X(X);\varphi^{\leftarrow}) = -Gini(p_X(X);\varphi),$$
$$Gini(p_Y(Y);\psi^{\leftarrow}) = -Gini(p_Y(Y);\psi).$$

It follows that in the twin tables $[p_{ij}]$ and $[p_{ij}]^{\Leftarrow}$ the value of ρ^* is the same while $Gini(p_X(X);\varphi)$ and $Gini(p_X(X);\varphi^{\leftarrow})$ have equal absolute values but opposite signs, and the same holds for $Gini(p_Y(Y);\psi)$ and $Gini(p_Y(Y);\psi^{\leftarrow})$. This holds in particular for the two twin tables with $\rho^* = \rho^*_{\max}$.

It is convenient (for reasons which will be given in Chapter 12) to distinguish this table for which the ordered Gini Index for the density transform of the *row* variable is *positive*. This is possible whenever this index is not equal to zero (which happens in particular when all probabilities p_{i+} are equal). If the value of this index is zero, but the respective ordered Gini Index for columns is not equal to zero, we will distinguish this one of the twin tables for which the latter parameter is positive.

The orderings of rows and columns in the table permuted in the way mentioned above will be denoted $(\varphi_{\max}, \psi_{\max})$; this table will be denoted $P_{\max}$ and called *the maximal table.*

If both ordered Gini Indices for the density transform of row and column variables are null, table $P_{\max}$ will be distinguished in a way to be described in Section 9.2.1.

For $BRIT_{7\times7}$, $CARS_{16\times16}$ and $NORM_{16\times16}$ the parameters of the twin tables with $\rho^* = \rho^*_{\max}$ are considered in the first two columns of Table 8.6; the left corresponds to the pair of orderings $(\varphi_{\max}, \psi_{\max})$, the right to $(\varphi^{\leftarrow}_{\max}, \psi^{\leftarrow}_{\max})$. So we have:

- for $BRIT_{7\times7}$ in Fig. 8.8, the ordered Gini Index for the density transform is equal to 0.067 for rows and 0.054 for columns; for $BRIT^{\Leftarrow}_{7\times7}$, these indices have the same absolute values but different signs (-0.067 and -0.054);
- for $CARS_{16\times16}$ the ordered Gini Index for the density transform is equal to 0.104 for rows and 0.035 for columns; for $CARS^{\Leftarrow}_{16\times16}$ we have, respectively, -0.104 and -0.035;
- for $NORM_{16\times16}$ the ordered Gini Index for the density transform is equal 0.107 for rows and -0.107 for columns; for $NORM^{\Leftarrow}_{16\times16}$ we obtain -0.107 and 0.107, respectively (the absolute values are the same since the *sets* of marginal probabilities (i.e. sizes) of rows and columns are identical).

When $Gini(p_X(X); \varphi_{\max}) > 0$, the sizes (widths) of the subsequent rows in the table tend to increase more than in the twin table $[p_{ij}]^{\Leftarrow}$. This tendency can be observed in Figs. 8.13 and 8.14.

A practical interpretation of a table with rows and columns ordered according to their non-decreasing sizes might be interesting e.g. for $CARS_{16\times16}$. In this case proportions of different types of cars which are previously owned and presently owned are certainly important. The overrepresentation map related to such ordering is shown in Fig. 8.17 right. We can observe in this figure the small differences appearing in both marginals.

For $BRIT_{7\times7}$ it is seen from Fig. 8.17 left that the proportions of particular occupational categories for fathers and for sons are ordered almost identically and the respective widths are only slightly different.

> The global differentiation of rows or of columns (equal to ρ^*) in a table with rows and columns ordered according to their non-decreasing sizes is much smaller than the maximal global differentiation. As shown in Table 8.6, it is equal to only 0.201 for $BRIT_{7\times7}$, 0.165 for $CARS_{16\times16}$, and 0.011 for $NORM_{16\times16}$ (while the values of $\rho^*_{\max}$ are 0.372, 0.4239, and 0.723).

***

It is interesting just how different the grade distributions are for tables which have the same marginals, the same value of ρ^*, and moreover, where for both tables $\rho^* = \rho^*_{\max}$.

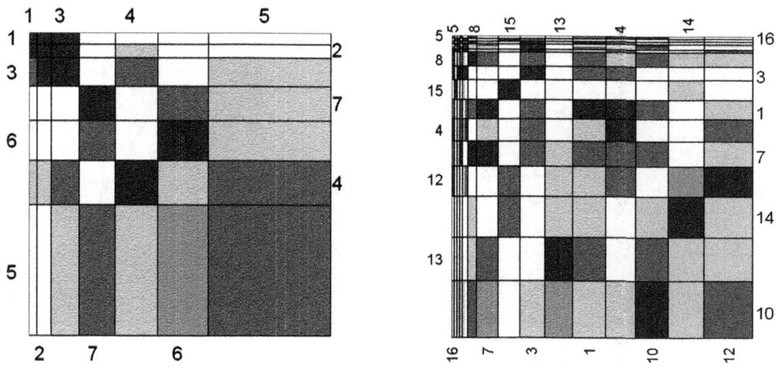

FIGURE 8.17. Overrepresentation maps for $BRIT_{7\times7}$ (**left**) and $CARS_{16\times16}$ (**right**) with rows and columns ordered by increasing size

The respective measures of departure, i.e., the maximal concentration curve $C_{\max}$ and the maximal concentration index $ar_{\max}$ were considered in Chapter 4. The general definition of $C_{\max}$ and $ar_{\max}$ can be applied, in particular, to pairs of grade distributions. So we start with two tables $Q = [q_{ij}]_{\max}$ and $P = [p_{ij}]_{\max}$ and calculate the likelihood ratio (see formula (3.53)).

$$h_{Q:P}(u,v) = \begin{cases} \frac{q_{ij}/q_{i+}q_{+j}}{p_{ij}/p_{i+}p_{+j}} = \frac{q_{ij}}{p_{ij}} & \text{if } p_{ij} > 0,\ \ u \in (S_{i-1}, S_i),\ v \in (T_{j-1}, T_j), \\ \infty & \text{if } p_{ij} = 0,\ q_{ij} > 0, \\ 1 & \text{if } p_{ij} = q_{ij} = 0, \end{cases}$$

where $S_i = p_{1+} + \cdots + p_{i+}$, $T_j = p_{+1} + \cdots + p_{+j}$, $S_0 = T_0 = 0$, $i = 1, \ldots, m$ $j = 1, \ldots, k$. Further calculations of $C_{\max}$ and $ar_{\max}$ are made as in Section 4.3.2.

Remark 8.7 *If we are interested in the departure of P from Q, we use $h_{P:Q}$ and proceed as before. Then $C_{\max}$ will turn into its dual, and $ar_{\max}$ will be the same (see Section 4.4.1).*

We will apply the maximal concentration measures to find the distance between any table $P = [p_{ij}]$ and its binormal analog.

Definition 8.7 *A binormal analog of a table $P_{m\times k}$ with $\rho^* = \rho^*_{\max}$ is a discretized binormal distribution with marginal distributions the same as in $P_{m\times k}$ and with the same value of ρ^*. It will be denoted $P^{NORM}_{m\times k}$.*

The overrepresentation maps of the binormal analogs of $BRIT_{7\times7}$ and $CARS_{16\times16}$ are shown in Fig. 8.18 left and right. The values of $ar_{\max}(P_{m\times k}$:

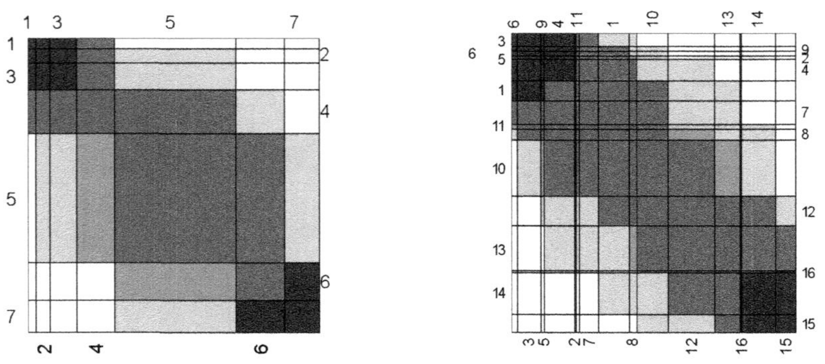

FIGURE 8.18. Overrepresentation maps of the binormal analogs of $BRIT_{7\times7}$ (**left**) and $CARS_{16\times16}$ (**right**). They refer to the maps of $BRIT_{7\times7}$ (Fig. 8.8) and to $CARS_{\max}$ (Fig. 8.13 left)

$P_{m\times k}^{NORM}$) calculated for $BRIT_{7\times7}$, $CARS_{16\times16}$, and $NORM_{16\times16}$ are 0.1099, 0.2661, and 0, respectively. Indeed, the map for $BRIT_{7\times7}^{NORM}$ is quite similar to that for $BRIT_{7\times7}$ (Fig. 8.8), however, the map for $CARS_{16\times16}^{NORM}$ strongly differs from $CARS_{\max}$ (Fig. 8.13 left).

8.6.3 Maximal Kendall's tau

Analogously to Section 8.6.1, we will consider the maximal value of parameter τ, obtained in the set of all possible permutations of rows and columns. In the case of $BRIT_{7\times7}$ the maximal value is already obtained by the original table (see the columns corresponding to $\tau = \tau_{\max}$ in table 8.6). The same holds for $NORM_{16\times16}$. For $CARS_{16\times16}$ the permutations leading to the maximal value of τ are slightly different than those leading to the maximal value of ρ^* : for rows, we get now the permutation

$$[3, 9, 6, 2, 11, 4, 1, 5, 7, 8, 10, 12, 13, 14, 16, 15]$$

instead of the permutation obtained for $\rho^*_{\max}$

$$[3, 9, 6, 2, 5, 14, 4, 1, 7, 11, 8, 10, 12, 13, 16, 14, 15]$$

for columns we get

$$[6, 9, 3, 11, 5, 4, 2, 7, 1, 8, 10, 12, 13, 14, 16, 15]$$

instead of the permutation obtained for $\rho^*_{\max}$

$$[6, 3, 9, 5, 4, 2, 11, 7, 1, 8, 10, 12, 13, 16, 14, 15].$$

Note that these changes correspond only to rows and columns with the smallest widths (2,5,6,8,9,11,16) which have a very small influence on the values of ρ^* and τ. For this reason, the value of $\tau_{\max}$ shown in Table 8.6 is almost the same as the value of τ when $\rho^* = \rho^*_{\max}$ (0.3013 instead of 0.3007).

If $(X,Y) \in TM_2$ then

$$\tau_{\max}(X,Y) = |\tau(X,Y)| \tag{8.43}$$

but not vice versa.

Table 8.7 illustrates the following property of tables in which $\tau = \tau_{\max}$:

If $\tau([p_{ij}]) = \tau_{\max}([p_{ij}])$ then

$$ar(i+1:i;row) \geq 0 \quad \text{for any} \quad i = 1, \dots, m-1, \tag{8.44}$$

and

$$ar(j+1:j;col) \geq 0 \quad \text{for any} \quad j = 1, \dots, k-1. \tag{8.45}$$

Remark 8.8 *Formulas (8.44) and (8.45) do not hold in general in tables satisfying $\rho^* = \rho^*_{\max}$ (as illustrated for $CARS_{16\times 16}$ in the right-hand side of Table 8.7). In such cases, $\tau(P_{\max}) < \tau_{\max}(P)$.*

Proposition 8.7 *Properties of $\tau_{\max}$.*

(i) If (X,Y) are categorical variables distributed according to a probability table $P_{m\times k}$ and $\tau(X,Y) = \tau_{\max}(X,Y)$ then for any $1 \leq s < t \leq m$,

$$\sum_{s \leq l < t} p_{l+} ar(t:l;row) \geq 0, \quad \sum_{s < l \leq t} p_{l+} ar(l:s;row) \geq 0, \tag{8.46}$$

and for any $1 \leq s < t \leq k$

$$\sum_{s \leq l < t} p_{+l} ar(t:l;col) \geq 0, \quad \sum_{s < l \leq t} p_{+l} ar(l:s;col) \geq 0. \tag{8.47}$$

(ii) If $(X,Y) \in TM_2$ then

$$\begin{aligned}\tau_{\max}(X,Y) &= \int_0^1 \int_0^1 ar_{\max}((Y|X=x_u):(Y|X=x_s))dsdu \\ &= \int_0^1 \int_0^1 ar_{\max}((X|Y=y_u):(X|Y=y_s))dsdu.\end{aligned}$$

TABLE 8.7. The values of ar for neighbouring rows in $CARS_{16\times 16}$ permuted so that $\tau = \tau_{\max}$ **(left)** and $\rho^* = \rho^*_{\max}$ **(right)**

i	labels of rows permuted according to $\tau = \tau_{\max}$	$ar(i+1 : i; row)$	labels of rows permuted according to $\rho^* = \rho^*_{\max}$	$ar(i+1 : i; row)$
1	3	0.0057	3	0.0400
2	9	0.0368	9	0.0307
3	6	0.1186	6	0.1008
4	2	0.0456	2	0.0889
5	11	0.0196	5	-0.0332
6	4	0.0533	4	0.0521
7	1	0.0038	1	0.0491
8	5	0.0631	7	-0.0252
9	7	0.1198	11	0.1115
10	8	0.1407	8	0.1409
11	10	0.1255	10	0.1252
12	12	0.0313	12	0.0315
13	13	0.2338	13	0.1811
14	14	0.1286	16	0.1061
15	16	0.3815	14	0.4060
16	15		15	

For the distribution of (X, Y) given by a probability table $P_{m\times k}$, we obtain

$$\begin{aligned}\tau_{\max}(P_{m\times k}) &= \sum_{i=1}^{m}\sum_{j=1}^{m} p_{i+}p_{+j}ar_{\max}(j : i; row) \qquad (8.48)\\ &= \sum_{s=1}^{k}\sum_{r=1}^{k} p_{+s}p_{+r}ar_{\max}(r : s; col)\\ &= \sum_{i=1}^{m}\sum_{j=1}^{k}\sum_{r=1}^{m}\sum_{s=1}^{k} |p_{ij}p_{rs} - p_{is}p_{rj}|.\end{aligned}$$

(iii) If $P = P_{2\times k}$ or $P = P_{m\times 2}$ then

$$\tau_{\max}(P)/\rho^*_{\max}(P) = 2/3$$

and the same permutation provides the table maximal w.r.t. ρ^* and τ.

(iv) $\tau_{\max}$ is invariant on the aggregation of the identical row profiles as well as of identical columns profiles.

***

Example. For the probability table given by (8.41) we have the following decomposition of $\tau_{\max}(M)$ (analogous to the decomposition of $\rho^*_{\max}(M)$) :

$$\tau_{\max}(M) = 2\alpha(1-\alpha) + \alpha^2\tau_{\max}(P/\alpha) + (1-\alpha)^2\tau_{\max}(Q/(1-\alpha)).$$

Let τ_{abs} be a parameter called *absolute tau* given by

$$\tau_{abs}(X,Y) = \int_0^1\int_0^1 ar_{\max}((Y|X=x_u):(Y|X=x_s))dsdu. \tag{8.49}$$

Hence for a probability table $[p_{ij}]$,

$$\tau_{abs}([p_{ij}]) = \sum_{i=1}^{m}\sum_{j=1}^{k}\sum_{r=1}^{m}\sum_{s=1}^{k}|p_{ij}p_{rs} - p_{is}p_{rj}| \tag{8.50}$$

$$= \sum_{i=1}^{m}\sum_{j=1}^{k}\sum_{r=1}^{m}\sum_{s=1}^{k}|det_{ijrs}|.$$

Absolute *tau* is sometimes called *absolute global differentiation of rows (or of columns)*. It is attained only in sufficiently regular tables as stated in the following proposition:

Proposition 8.8 *(i) For any* (X,Y)

$$|\tau(X,Y)| \leq \tau_{\max}(X,Y) \leq \tau_{abs}(X,Y).$$

(ii) $(\tau_{\max}(X,Y) = \tau_{abs}(X,Y)) \Leftrightarrow ((X,Y) \in TM_2)$.
(iii) $\tau_{abs}(X,Y) = 0 \Leftrightarrow (X,Y)$ *are independent.*

The proofs of proposition 8.7 and 8.8 (i), (ii) are obvious.

To prove property (iii) in proposition 8.8 we note that $\tau_{abs}(X,Y) = 0$ if and only if for any $u \in (0,1)$ cdf's of variables $(Y|X=x_u)$ are equal to the cdf of Y.

The values of τ_{abs} and of τ/τ_{abs} for variously permuted tables *BRIT*, *CARS* and *NORM* are calculated in Table 8.6.

***

The parameter

$$\delta(X,Y) = \begin{cases} 1 - \frac{\tau_{\max}(X,Y)}{\tau_{abs}(X,Y)} & \text{if } \tau_{abs}(X,Y) > 0, \\ 0 & \text{if } (X,Y) \text{ are independent} \end{cases} \tag{8.51}$$

may be treated as the distance of the distribution of (X,Y) from family TP_2 (if $\tau(X,Y) \geq 0$) and from the family TN_2 (if $\tau(X,Y) \leq 0$).

The expression (8.51) is equal to 0.028 for $BRIT_{7\times 7}$, 0.27 for $CARS_{16\times 16}$, and 0 for $NORM_{16\times 16}$. According to these results we are ensured that $BRIT_{7\times 7}$ is "almost TP_2" while $CARS_{16\times 16}$ is much more distant from TP_2, and that $NORM_{16\times 16} \in TP_2$.

***

Finally, we shall introduce a parameter defined for tables $P_{m\times k}$, related to Kendall's *tau* and called *tau signum*. We recall (formula (8.24)) that

$$\begin{aligned} \tau(P_{m\times k}) &= 2\sum_{r=2}^{m}\sum_{i=1}^{r-1}\sum_{t=2}^{k}\sum_{j=1}^{t-1}(p_{ij}p_{rt} - p_{it}p_{rj}) \\ &= 2\sum_{r=2}^{m}\sum_{i=1}^{r-1}\sum_{t=2}^{k}\sum_{j=1}^{t-1} det_{ijrt}. \end{aligned}$$

Then we set

$$\tau_{sgn}(P_{m\times k}) = \frac{4\sum_{r=2}^{m}\sum_{i=1}^{r-1}\sum_{t=2}^{k}\sum_{j=1}^{t-1} sgn(det_{ijrt})}{m(m-1)k(k-1)} \tag{8.52}$$

which has the following properties:

- $-1 \leq \tau_{sgn}(P) \leq 1$,
- $\tau_{sgn}(P) = 1$ iff $det_{ijrt} > 0$ for any $1 \leq i < j \leq m$, $1 \leq r < t \leq k$,
- $\tau_{sgn}(P) = 1 \Rightarrow \tau(P)/\tau_{abs}(P) = 1$,
- $\tau_{sgn}(P) = -1 \Rightarrow \tau(P)/\tau_{abs}(P) = -1$.

Thus, the value of $\tau_{sgn}(P_{m\times k})$ can be treated as the distance of $P_{m\times k}$ from the family of "strict TP_2 tables", i.e. tables with positivity of all determinants det_{ijrt}. It follows that $1-\tau_{sgn}$ evaluates departure of $P_{m\times k}$ from regularity, in a way similar to that of $1-\tau/\tau_{abs}$. The applications of τ_{sgn} will be considered in Chapter 9.

8.7 The hinged sequences of rows and columns

We are going to introduce two sequences, one consisting of rows and the other of columns appearing in $P_{\max}$ and $P^{\Leftarrow}_{\max}$. In the case of rows, for any $i = 1, \ldots, m$ we take into account either row i from $P_{\max}$ or its opposite appearing in $P^{\Leftarrow}_{\max}$. The basis of the selection is that the chosen sequence of rows has a possibly small global differentiation (so that these rows "behave similarly" w.r.t. the marginal column distribution in $P_{\max}$). An analogous procedure provides a suitable selection of columns from $P_{\max}$ and $P^{\Leftarrow}_{\max}$.

The procedure will be explained for rows in the case of $BRIT_{7\times 7}$ (this dataset is specifically dedicated to achieve this goal because it is very regular): $\rho^*(BRIT) = \rho^*_{\max}(BRIT)$, $\tau(BRIT) = \tau_{\max}(BRIT)$, the rows (as well as columns) are stochastically ordered, the table is almost TP_2. The stochastic ordering of rows is shown in Fig. 8.19 containing the concentration curves of row profiles w.r.t. *the column marginal.*

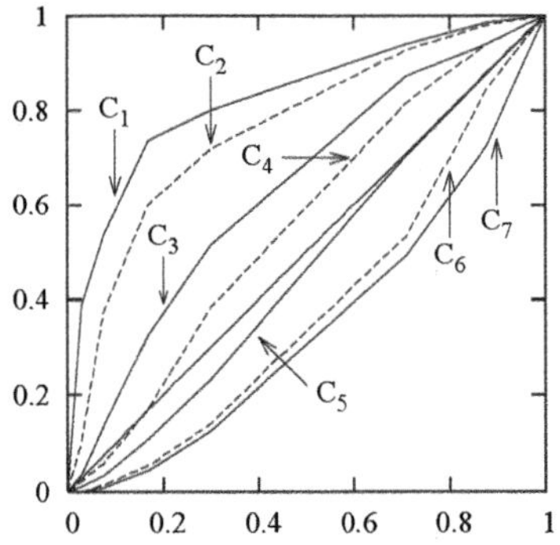

FIGURE 8.19. Concentration curves for row profiles w.r.t. the column marginal in $BRIT_{7\times 7}$ (curve C_i corresponds to row i)

Remark 8.9 *Please note that rows are stochastically ordered if and only if the concentration curves of particular rows w.r.t. any pattern distribution are stochastically ordered (i.e. lie one over another).*

The matrix $[ar(j : i; row)]$ is given in Table 8.8 (upper triangle). This matrix is matched with Fig. 8.19, namely in that for any cell (i, j) values of ar increase, so that $ar(j' : i) \geq ar(j : i)$ if $j' \geq j \geq i$, $ar(j : i') \geq ar(j : i)$ if $i' \leq i \leq j$.

The concentration curves in Fig. 8.19 are either above *Diag* (for rows 1,2,3) or under *Diag* (for rows 5,6,7), or intersect *Diag* (for row 4). Since *Diag* represents the column marginal in $BRIT_{7\times 7}$, the concentration indices

TABLE 8.8. The matrix $[ar(j : i, row)]$, $j = i + 1, \ldots, 7$ (upper triangle) and $[ar_{\max}(i : j; row)]$, $j = 1, \ldots, 7$, $i = j + 1, \ldots, 7$ (lower triangle) for $BRIT_{7\times7}$

	1	2	3	4	5	6	7
1		0.2608	0.5083	0.6198	0.7076	0.7854	0.8083
2	0.3453		0.3230	0.4769	0.5999	0.7078	0.7387
3	0.5415	0.3379		0.1788	0.3509	0.5036	0.5505
4	0.6357	0.4800	0.1980		0.1967	0.3755	0.4348
5	0.7076	0.5999	0.3553	0.2020		0.1852	0.2590
6	0.7854	0.7078	0.5210	0.3758	0.02103		0.0939
7	0.8083	0.7387	0.5505	0.4348	0.2590	0.1523	

$ar(row\ i : col.\ marg.)$ are expressed as negative double areas between the curve and $Diag$ (for rows 1,2,3), positive double areas between the curve and $Diag$ (for rows 5,6,7) and the suitable double difference of the areas under and above $Diag$ for row 4. So we obtain the respective values of ar : -0.6510, -0.5264, -0.2723, -0.1213, 0.0627, 0.2342, 0.3044. In view of formula (8.32), ar's can be linearly transformed into the values of the grade regression function $r^*_{col|row}$ which is shown in the form of vertical segments in the overrepresentation map of $BRIT_{7\times7}$ (Fig. 8.20 left), and also presented separately (Fig. 8.20 right).

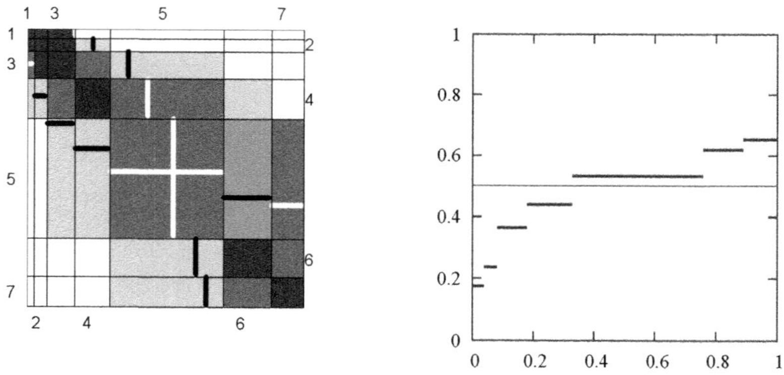

FIGURE 8.20. Overrepresentation map for $BRIT_{7\times7}$ showing the regression function (left box) and the grade regression function $r^*_{col|row}$ for $BRIT_{7\times7}$ (right box; it are vertical segments in left figure)

The negative values of ar coincide with the values of the regression function which are smaller than 1/2 (as we have here no value of ar equal to zero, there is no value of the regression function equal to 1/2).

So we obtain in $BRIT$ two subsets of rows: (1,2,3,4) and (5,6,7). The rows in the first subset will be called *left* to the column marginal in $BRIT_{7\times7}$, the rows in the second subset will be called *right* to the column marginal, and the whole subsets will be respectively called *left* and *right*. The rows in each subset behave similarly w.r.t. the column marginal. This means that rows in the left subset are more concentrated on the left categories than it happens in the column marginal, and in the right subset the tendency is opposite.

We will now exchange the rows in the right subset with their opposites from $BRIT^{\Leftarrow}$ and join them with the left subset. We get in this way rows $1, 2, 3, 4, 5^{\leftarrow}, 6^{\leftarrow}, 7^{\leftarrow}$ all of them with the same tendency w.r.t. the column marginal in $BRIT_{7\times7}$. The resulting set of concentration curves is shown in Fig. 8.21: they form the sequence C_1, C_2, $C_7^{\leftarrow}$, C_3, $C_6^{\leftarrow}$, C_4, $C_5^{\leftarrow}$, in which

- C_1 lies above all subsequent curves, and the same holds for C_2 and $C_7^{\leftarrow}$
- C_3 intersetcs $C_6^{\leftarrow}$ and $C_5^{\leftarrow}$ and lies above C_4
- $C_6^{\leftarrow}$ lies above C_4 and $C_5^{\leftarrow}$
- C_4 intersects *Diag* and lie above $C_5^{\leftarrow}$.

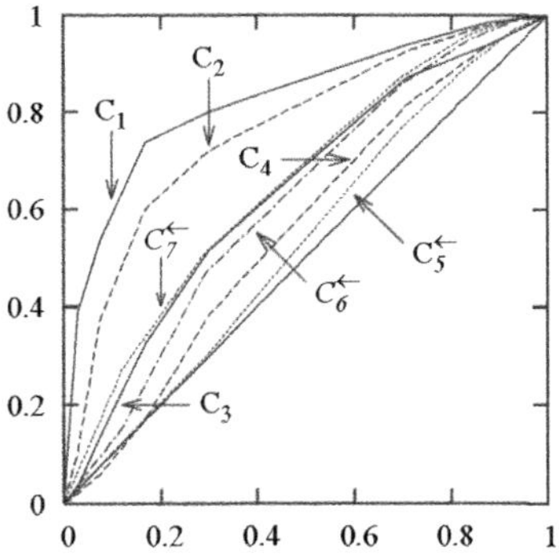

FIGURE 8.21. Set of concentration curves in the hinged row sequence for $BRIT_{7\times7}$

Thus, the curves in the hinged set are no longer stochastically ordered (although they are very close to it). A *linear ordering of the elements of hinged sequence* can be introduced by the values of *ar* corresponding to particular curves. This is equivalent to finding, for each category $i = 1, \ldots, 7$, the value of $min(r_i^*, 1 - r_i^*)$, where $r_i^* = r^*_{col|row}(i)$, and to ordering the categories according to these minima. According to the calculations presented in Table 8.10, the hinged row sequence is

$$1, 2, 7^{\leftarrow}, 3, 6^{\leftarrow}, 4, 5^{\leftarrow},$$

which agrees with the sequence as observed roughly from Fig. 8.21.

TABLE 8.9. The matrix of ar's (upper triangle) and $ar_{\max}$'s (lower triangle) in the row hinged sequence for $BRIT_{7\times7}$

ar	1	2	$7^{\leftarrow}$	3	$6^{\leftarrow}$	4	$5^{\leftarrow}$
1		0.2608	0.4664	0.5083	0.5512	0.6198	0.6304
2	0.3453		0.2780	0.3230	0.3838	0.4769	0.4986
$7^{\leftarrow}$	0.5069	0.3232		0.0352	0.0939	0.2084	0.2590
3	0.5415	0.3379	0.1339		0.0624	0.1788	0.2288
$6^{\leftarrow}$	0.5984	0.4176	0.1523	0.1594		0.1249	0.1852
4	0.6357	0.4800	0.2159	0.1980	0.1362		0.0655
$5^{\leftarrow}$	0.6350	0.5082	0.2590	0.2643	0.2103	0.1295	

TABLE 8.10. Construction of the hinged row sequence for $BRIT_{7\times7}$

i	$r_i^* = r^*_{col\|row}(i)$	$min(r_i^*, 1 - r_i^*)$	element appearing in the hinged row sequence	position in the hinged row sequence
1	0.175	0.175	1	1
2	0.237	0.237	2	2
3	0.364	0.364	3	4
4	0.439	0.439	4	6
5	0.531	0.469	$5^{\leftarrow}$	7
6	0.617	0.383	$6^{\leftarrow}$	5
7	0.652	0.348	$7^{\leftarrow}$	3

It is easy to see that C_3 is very close $C_7^{\leftarrow}$, which means that the occupational distribution of sons of higher supervisors is almost precisely opposite to that of sons of unskilled manual workers. Generally, presentation of the curves in the hinged set (Fig. 8.21) is useful to interpret the data and seems more useful than the original curves in Fig. 8.19. It provides a great deal of information about asymmetry.

It is interesting to compare the ar's among the elements in the initial set (upper triangle in Table 8.8) with those in the hinged set (upper triangle in Table 8.9). We see that the set of initial row profiles is more differentiated than the set of row hinged profiles.

The hinged column sequence for $BRIT_{7\times7}$ is equal to 1,2,3, $7^{\leftarrow}, 6^{\leftarrow}, 4, 5^{\leftarrow}$.

The hinged row sequence for $CARS_{16\times16}$ is equal to

$$15^{\leftarrow}, 3, 9, 6, 2, 5, 14^{\leftarrow}, 4, 1, 7, 11, 16^{\leftarrow}, 8, 13^{\leftarrow}, 12^{\leftarrow}, 10. \qquad (8.53)$$

The hinged column sequence for $CARS_{16\times 16}$ is equal to

$$15^{\leftarrow}, 6, 14^{\leftarrow}, 3, 9, 16^{\leftarrow}, 5, 4, 2, 11, 13^{\leftarrow}, 7, 1, 8, 10, 12.$$

8.8 Appendix: Bibliographical remarks

The general Bivariate Lilliputian Model introduced in Section 8.1 should generalize the main concepts appearing in the Univariate Lilliputian Model, in particular the concentration index ar. A direct attempt to introduce the bivariate $ar(Q:P)$ done in [63]. This problem is considered in [81] and in[30].

The density of the grade bivariate distribution is considered in the literature concerning copulas (see e.g. [50], [45] and in [40]) The overrepresentation maps which visualize the grade density might be considered to be introduced in [27] and in [29] (and then extensively used in many applicational papers by the authors of this book, such as [56], [55], [20], [24], [86].

A theoretical approach to the copulas is nicely presented in Nelsen's book ([80]) (see also [45]; information on Archimedean copulas can be found in [110].)

The links between Spearman's *rho* and Kendall's *tau* and different expressions for them are considered in many papers. In Section 8.4, these parameters were introduced geometrically in a strict analogy to the geometrical presentation of the concentration index ar. This approach seems to have been tackled earlier in [26]. This is an important aspect of the unified infrastructure of the grade methodology.

Important links between the grade regression function and the set of concentration indices between the particular conditional distributions and the marginal distribution (formula (8.32) in Section 8.5) are introduced in [54] and also in [1].

Section 8.6 (the most important for practical applications) was preceded by extensive research on part of the authors of the book (e.g. [59], [104], [86], [103],[99]).

The hinged sequences were introduced in Section 8.7 for the first time.

9

Grade Correspondence Analysis and outlier detection

Olaf Matyja, Wiesław Szczesny

9.1 Introduction

This chapter and the next present the main procedures used in grade methodology. They are based on grade correspondence analysis (GCA) which is introduced in Section 9.2. The basic procedure of GCA solves the problem of permuting the rows and columns of a probability table in order to maximize the value of the grade correlation ρ^*. This is done by alternately permuting the rows and columns according to the respective grade regression function until both these functions are nondecreasing and no further improvement of ρ^* is possible. Such a table provides a local maximum of ρ^*. GCA finds several local maxima starting from the original table after the random permutations of its rows and columns, and then selects the highest value of ρ^* as the global maximum (or at least a very good approximation of the global maximum).

Other local maxima of ρ^* are also interesting. Sometimes there appear local maxima of a special type, as we shell see interesting and demanding special consideration, called "folds." Presence of local maxima other than folds and distinctly different from the global maximum indicate that the global maximum table is not sufficiently regular and the respective trend is not unique. Local maxima can be analysed (as in [103]) to obtain further trends present in the data.

Sections 9.3-9.5 are devoted to different approaches to find existing trends. All attempts are exemplified by two tables discussed earlier in Chapter 8: the very regular table $\mathrm{BRIT}_{7\times7}$, and the very irregular table $\mathrm{CARS}_{16\times16}$.

The first approach (Section 9.3) is called Smooth Grade Correspondence Analysis (SGCA). It attempts to decompose the table into a number of more regular tables which sum to the global maximum (or to a table very close to the global maximum) after readjusting the orderings of rows and columns. Each of the resulting tables may have a different order of rows and columns.

The second approach consists in detecting and removing outliers, i.e. rows or columns which disturb the regularity of the global maximum. This approach is based on comparing the concentration indices ar and maximal concentration indices ar_{max} for ordered pairs of conditional distributions, separately among rows and among columns. The respective grade tools are introduced in Section 9.5. It is shown that prior rejection of most notorious outliers (which might have been created by a too detailed discretization) is beneficial for further analysis. The outliers can be handled separately and further decomposition can be attempted. This time, in contrast to the SGCA, the subsets of rows or of columns in the component tables are disjoint. This section also provides references to examples of the practical application of GCA and SGCA.

Most helpful for the authors of this chapter were the following two positions: [44] and [94].

9.2 Algorithms of GCA

9.2.1 *GCA algorithm based on Spearman's ρ^**

Figure 9.1 is an overrepresentation map of the discretized binormal distribution with $\rho^* = 0.6$ (This means that Spearman's ρ^* between the row and column variables equals 0.6). The row and column variables are positively dependent; the largest values of each row and each column are placed on the diagonal *Diag* of the unit square. Each white line on the chart marks the center of mass of a row or a column. Its coordinate is called the grade regression. As one might expect, larger values concentrated along the diagonal make the grade regression functions increasing.

A question: What will happen if we start from any table in which the values of the grade regressions for columns are not ordered, and we then sort the columns by the values of the regresssion? The regression for columns will be increasing, but the regression for rows will then change. Then, if we sort the regression for rows, the regression for columns will change.

In [27] there is a proof that each ordering according to the grade regression of rows or of columns increases the value of Spearman's ρ^*. The number of possible states (combinations of permutations of rows and columns) is finite (equal to $k!m!$), therefore the algorithm must stop.

EXAMPLE

Let us start from any permutations of rows and columns, as in Figure 9.2. After sorting of the grade regression $r^*_{row|col}$, we achieve Figure 9.3.

After sorting the $col|row$ grade regression (i.e. $r^*_{col|row}$), Figure 9.4 is achieved.

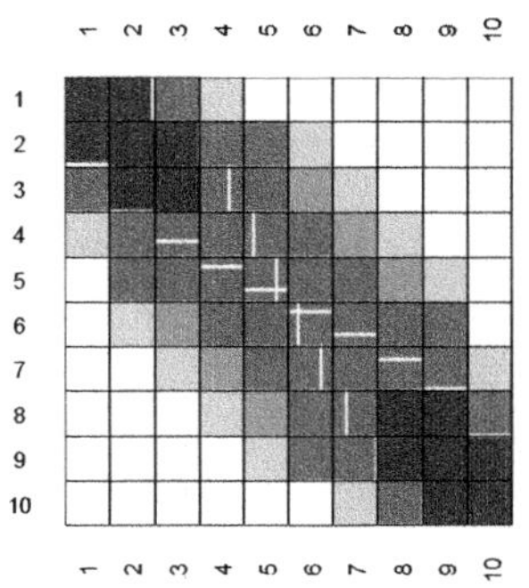

FIGURE 9.1. Grade regressions in a discretized binormal table

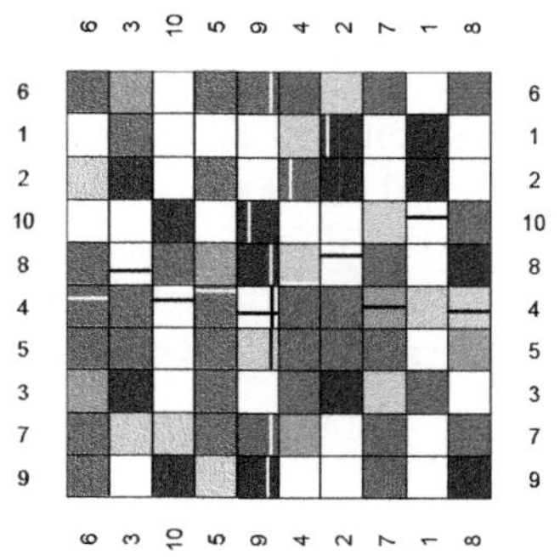

FIGURE 9.2. Random permutations of rows and columns of the table from Figure 9.1

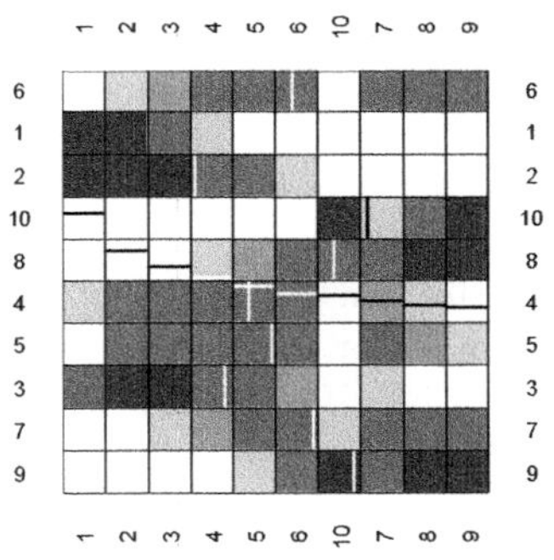

FIGURE 9.3. Table from Figure 9.2 after the first sorting

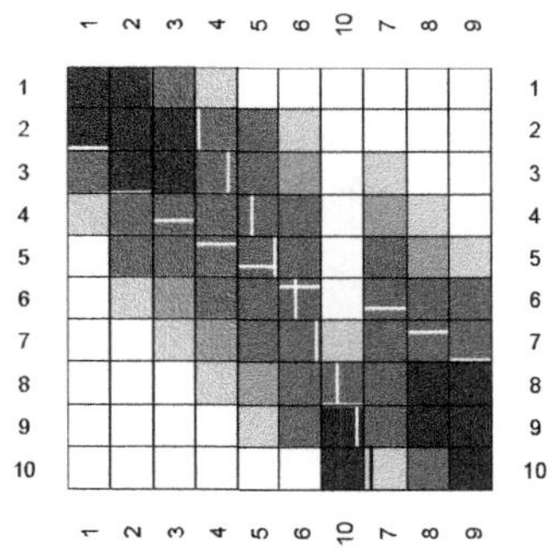

FIGURE 9.4. Table from Figure 9.2 after the second sorting

After another sorting the *col|row* grade regression we achieve Figure 9.1. This state is stable — both regression functions are increasing.

After each sorting the value of Spearman's *rho* increases. The last one is the largest, and after the last sorting no larger value of ρ^* may be achieved by changing the order of the rows alone or the columns alone.

The result of GCA depends on the starting permutations. With reversed starting permutations of rows and columns we will achieve a symmetrically reversed table. Such a pair of tables (the achieved table and the symmetrically reversed one) is called a local maximum of Spearman's *rho*.

In the example in Fig 9.1 there is only one local maximum of ρ^*. For more irregular examples and even in many TP_2 tables (see section 9.2.5) there is more than one maximum and it is necessary to try many random start permutations and choose the local maximum with the highest value of Spearman's *rho*. Such a maximum is called the largest local maximum of Spearman's ρ^*. If we check all start permutations, the result will be the global maximum of ρ^*, denoted ρ^*_{max}. In most cases however the number of all permutations of rows and columns (equal to $m!k!$) is too large to check it. We are able only to check a limited number of random start permutations (for example 100 or 1000) and treat the largest local maximum as the global maximum. In the unlikely event that the largest maximum found is not the global maximum (which is unlikely when a large number of permutations is checked), the result will be close to the global maximum.

* * *

As mentioned earlier, each local maximum is a pair of tables P and $P^{\Leftarrow}$. Each of the tables P and $P^{\Leftarrow}$ is equally likely to be reached by the GCA algorithm. It would be covenient, however, if the algorithm were able to choose one of the two tables as a representative of the maximum. To allow for easy comparing of the local maxima, the representatives should be chosen in such a way that the similarity of their order of columns or rows should be maximized.

The auto-revert algorithm can be called for any table P reached by the GCA algorithm. For each P it chooses either table P or $P^{\Leftarrow}$. The algorithm runs in three steps:

1. $Gini(p_X(X);\varphi)$ (the ordered Gini Index for the marginal density transform, introduced in Section 4.3.4) is calculated for P. If the result is smaller than zero, $P^{\Leftarrow}$ is taken as a representative. If the result is larger than zero, P is taken.

2. If the result is equal to zero, the same procedure is run for $Gini(p_Y(Y);\psi)$.

3. If this index also equals zero, $ar\left(\mathbf{p}_{\cdot+} : \frac{\mathbf{p}_{\cdot 1}}{\sum_{i=1}^{m} p_{i1}}; col\right)$ is calculated, where $\mathbf{p}_{\cdot 1} = (p_{11}, ..., p_{m1})$ and $\mathbf{p}_{\cdot+} = (p_{1+}, ..., p_{m+})$. This represents ar between the marginal distribution and the first (leftmost) column.
   Similarly $ar\left(\mathbf{p}_{\cdot+} : \frac{\mathbf{p}_{\cdot k}}{\sum_{i=1}^{k} p_{i1}}; col\right)$ between the marginal distribution and the last (rightmost) column is calculated.
   If $ar\left(\mathbf{p}_{\cdot+} : \frac{\mathbf{p}_{\cdot 1}}{\sum_{i=1}^{k} p_{i1}}; col\right) < ar\left(\mathbf{p}_{\cdot+} : \frac{\mathbf{p}_{\cdot k}}{\sum_{i=1}^{k} p_{i1}}; col\right)$, $P^{\Leftarrow}$ is taken, otherwise (if $\geq$) P.

The overrepresentation map drawn for the representative (P or $P^{\Leftarrow}$) chosen by the auto-revert algorithm for the maximum found by GCA is called the GCA overrepresentation map.

* * *

Calculation of the grade regression requires a non-zero sum of every row and column in a table; therefore, this requirement applies also to the GCA algorithm.

The algorithm of GCA runs as follows:

Input:

$P_{m\times k}$ — probability table

Parameters:

n — number of iterations

Output:

$List_i$ — List of permutations of table P being the local maxima of ρ^*. $\rho^*(List_1) \geqslant \rho^*(List_2) \geqslant ... \geqslant \rho^*(List_n)$

The algorithm:

Function $GCAForRho(P_{m\times k}$:table; n :integer) :list of *Permutations*
begin
    Let *List* be an empty list of *Permutations*
    Store the current permutations of the rows and columns of P
    Perform in a loop n times:
        Apply random permutations to the rows and columns
            of table P.
        repeat
            $Changed := false$
            Calculate the grade regression for each row of P.
            Sort rows of P according to the calculated regression.
            If order of rows changed then $Changed := true$
            Calculate the grade regression for each column of P.
            Sort columns of P according to the calculated regression.
            If order of columns changed then $Changed := true$
        until not *Changed*

        If $Gini(p_X(X);\varphi) < 0$ or $(Gini(p_X(X);\varphi) = 0$ and
          $(Gini(p_Y(Y);\psi) < 0$ or $(Gini(p_Y(Y);\psi) = 0$ and
            $ar\left(\mathbf{p}_{\cdot+} : \frac{\mathbf{p}_{\cdot 1}}{\sum_{i=1}^{k} p_{i1}}; col\right) < ar\left(\mathbf{p}_{\cdot+} : \frac{\mathbf{p}_{\cdot k}}{\sum_{i=1}^{k} p_{i1}}; col\right))$ then
                Reverse the order of the rows and columns of P.
        Add current permutations of P and $\rho^*(P)$ to *List*
    end
    Restore the initial permutations of the rows and columns of P.
    Sort descending local maxima stored on *List* according to
        their values of ρ^*.
    $GCAForRho := List$
End

9.2.2 GCA algorithm based on Kendall's τ

Searching for local maxima of ρ^* is easy because sorting by the grade regression cannot make ρ^* smaller, which guarantees reaching of the local maximum.

.

Searching for local maxima of Kendall's *tau* is not as straightforward as in the case of ρ^*; however, the properties of the τ presented in formulas 8.44 and 8.45 help in this task.

We can use an algorithm similar to that of GCA for ρ^*. However, instead of sorting by the grade regression, we sort rows and columns using any algorithm that compares only adjacent rows or columns (as in bubble sorting). Each time when we have to decide if adjacent rows or adjacent columns should be swapped, we calculate $ar((i+1) : i; row)$ or $ar((s+1) : s; col)$. If the result is negative, the rows (or columns) are swapped.

Remark 9.1 *GCA based on τ will be denoted GCA_τ. If GCA abbreviation is used in the text without index, assume that we mean GCA based on ρ^* (i.e., GCA_{ρ^*}).*

The formal description of the algorithm denoted GCA_τ (using the simplest version of the bubble sort algorithm):

Input:
    $P_{m\times k}$ — probability table
Parameters:
    n — number of iterations
Output:
    List — List of permutations of table P being the local maxima
        of τ. $\tau(List_1) \geqslant \tau(List_2) \geqslant ... \geqslant \tau(List_n)$

The algorithm:

Function $GCAForTau(P_{m\times k} : Table; n$:integer):list of *Permutations*
begin
    Let *List* be an empty list of *Permutations*
    Store current permutations of rows and columns of P
    Perform in a loop n times:
        Apply random permutations to the rows and columns
            of the table.
        repeat
            $Changed_1 := false$
            repeat
                $Changed_2 := false$
                for $i = 1...m-1$ do
                    if $ar((i+1) : i; row) < 0$ then begin

```
                Swap adjacent rows i and i+1
                Changed_1 := true
                Changed_2 := true
            end
        until not Changed_2
        repeat
            Changed_2 := false
            for s = 1...k-1 do
                if ar((s+1) : s; col) < 0 then begin
                    Swap adjacent columns s and s+1
                    Changed_1 := true
                    Changed_2 := true
                end
        until not Changed_2
    until not Changed_1
    If Gini(p_X(X); φ) < 0 or (Gini(p_X(X); φ) = 0 and
      (Gini(p_Y(Y); ψ) < 0 or (Gini(p_Y(Y); ψ) = 0 and
        ar(p.+ : p.1 / Σ_{i=1}^k p_{i1}; col) < ar(p.+ : p.k / Σ_{i=1}^k p_{i1}; col)) then
          Reverse the order of the rows and columns of P.
    Add current permutations of the rows and columns of P
        and τ(P) to List
  end
  Restore the initial permutation of the rows and columns of P.
  Sort descending local maxima stored on List according to
      their values of τ.
  GCAForTau := List
end
```

Notice that stored pairs of permutations are not exactly the local maxima of τ in the same sense as local maxima of ρ^*. The permutations found by GCA$_\tau$ often might be improved (in terms of value of τ) by changing order of rows alone or columns alone. The algorithm assumes only that the global maximum of τ, if achieved, will be recognized as a local maximum in this particular sense.

Remark 9.2 *The definition of the local maximum of τ may be narrowed (see Proposition 8.7). This definition would be, however, more complicated and has no obvious advantages over the one presented here.*

In practice, the results of the GCA$_\tau$ according to the algorithm above are quite similar to the results of GCA$_{\rho^*}$. The number of pairs of permutations found is usually larger than the number of local maxima of ρ^*, but the global maximum of τ is achieved equally easy.

Thus, the formal definition of a local maximum of τ will be a pair of probability tables P and $P^{\Leftarrow}$ (one being the reversed version of the other) in which $ar((i+1):i;row) \geqslant 0$ for each pair of adjacent rows, and $ar((s+1):s;col) \geqslant 0$ for each pair of adjacent columns in P (and thereby $P^{\Leftarrow}$). This definition also includes the global maximum of τ as a special case of a local maximum of τ, just as in the definition of the global maximum of ρ^*.

9.2.3 GCA algorithm based on τ_{sgn}

A version of the GCA algorithm maximizing τ_{sgn} (defined by formula 8.52, and presented there as a measure of regularity) is similar to GCA based on τ. However, instead of ar, another parameter denoted ar_{sgn} is used, which is defined as follows:

$$ar_{sgn}(t,s;col) = \frac{2\sum_{i=1}^{m}\sum_{j=i+1}^{m} sgn(p_{is}p_{jt} - p_{it}p_{js})}{m(m-1)} \tag{9.1}$$

$$ar_{sgn}(j,i;row) = \frac{2\sum_{t=2}^{k}\sum_{s=1}^{t-1} sgn(p_{is}p_{jt} - p_{it}p_{js})}{k(k-1)} \tag{9.2}$$

In [73] this algorithm is proved to achieve the local maxima of τ_{sgn}, defined in a way similar to that of GCA_τ, i.e. as a permutation of the table for which swapping a pair of adjacent rows or columns does not make the value of τ_{sgn} larger.

9.2.4 GCA and a mixture of permuted discretized binormal tables

In this section we recall the simulation study performed in [73]. The behaviour of the GCA_{ρ^*} algorithm confronting a mixture of two discretized binormal tables (denoted T^1 and T^2) has been examined. The order of the rows and columns were different in each table. These tables T^l have been added with different weights π_l.

Let $Q_{m\times k}$ be a probability table. Then for any table P, let $Perm(P)$ be a table obtained from P by a permutation of the rows and/or columns. The simulation study was performed on the basis of a decomposition model in which Q was represented as a linear combination of permuted binormal tables:

$$Q = \pi Perm_1(N_{m\times k}(\rho_1^*)) + (1-\pi)Perm_2(N_{m\times k}(\rho_2^*)) \tag{9.3}$$

where $0 < \pi < 0.6$, while $T^l = N_{m\times k}(\rho_l^*)$ were discretized binormal tables with ρ^* after the discretization equal to ρ_l^* and the same number of rows and columns as Q with either uniform marginal distributions in half of the cases, or random marginal distriution in another half of the cases.

Remark 9.3 *Notice, that in formula (9.3) $N_{m\times k}(\rho^*)$ means a discretized binormal table with ρ^* after the discretization, whereas throughout the book ρ^* in $N_{m\times k}(\rho^*)$ is in reference to the value before discretization.*

The GCA algorithm permutes rows and columns of the table to achieve the maximum dependence (maximum value of ρ^* or τ). The properties of GCA applied to a mixture of permuted binormal distribution tables will be examined below. This examination requires a measure to compare the table permuted by GCA $Perm_{GCA}(Q)$ with the components T^l of the sum (9.3) in order to determine which component (or at least its approximation) has been found.

We have to take into account, that in a half of the cases GCA reverses the order of the rows and columns.

Let the index in Q of row i of component T^l after its permuting, including into the sum (9.3) and applying GCA permutation $Perm_{GCA}(Q)$ will be denoted r_i^l.

Let a similar index in Q of column s will be denoted c_s^l.

In order to compare the permutation of the rows, we will apply Kruskal's γ to vectors of the row's indexes in T^l $(1,2,...,m)$ and in Q permuted by the GCA algorithm $(r_1^l, r_2^l, ..., r_m^l)$. Kruskal's γ has been chosed because of its connections (9.7) with a regularity index τ_{sgn}, which is better explained in [73]. Kruskal's γ for a pair of vectors $\begin{bmatrix} x_1 \\ \cdots \\ x_N \end{bmatrix}, \begin{bmatrix} y_1 \\ \cdots \\ y_N \end{bmatrix}$ is calculated according to the formula:

$$\gamma\left(\begin{bmatrix} x_1 \\ \cdots \\ x_N \end{bmatrix}, \begin{bmatrix} y_1 \\ \cdots \\ y_N \end{bmatrix}\right) = \frac{\sum_{i=2}^{N}\sum_{j=1}^{i-1} sgn(x_i - x_j)sgn(y_i - y_j)}{\binom{N}{2}} \tag{9.4}$$

Substituting i for x_i and r_i for y_i, we obtain

$$\gamma\left(\begin{bmatrix} 1 \\ \cdots \\ m \end{bmatrix}, \begin{bmatrix} r_1 \\ \cdots \\ r_m \end{bmatrix}\right) = \frac{\sum_{i=2}^{m}\sum_{j=1}^{i-1} sgn(r_i - r_j)}{\binom{m}{2}} \tag{9.5}$$

The same will be done for pairs of columns:

$$\gamma\left(\begin{bmatrix} 1 \\ \cdots \\ k \end{bmatrix}, \begin{bmatrix} c_1 \\ \cdots \\ c_k \end{bmatrix}\right) = \frac{\sum_{s=2}^{k}\sum_{t=1}^{s-1} sgn(c_s - c_t)}{\binom{k}{2}} \tag{9.6}$$

As a measure for comparing the permutation of Q after GCA with T^l their product will be used.

In [73] this product is proved to be the regularity index τ_{sgn} (see formula formula 8.52)applied to table T^l with rows and columns permuted first according to permutation $Perm_l$ and then according to $Perm_{GCA}$, which explains the notation. τ_{sgn} was introduced in the Chapter 8:

$$\tau_{sgn}(Perm_{GCA}(Perm_l(T_l))) = \\ = \gamma\left(\begin{bmatrix}1\\ \dots\\ m\end{bmatrix}, \begin{bmatrix}r_1\\ \dots\\ r_m\end{bmatrix}\right) \gamma\left(\begin{bmatrix}1\\ \dots\\ k\end{bmatrix}, \begin{bmatrix}c_1\\ \dots\\ c_k\end{bmatrix}\right) \tag{9.7}$$

If $Perm_{GCA}(Perm_l(T^l)) = T^l$, then this measure equals 1, which means that GCA reverted the permutation $Perm_l$ and found the original order of T^l. The value of τ_{sgn} does not change, when order of rows and columns is reversed simultaneously (Thus it covers also the reversed GCA results).

* * *

The simulation study leads to interesting conclusions regarding GCA properties.

Property (i) Let $\omega_l = \pi_l \rho_l^*$ be called the strength of the l-th mixture component of sum 9.3. The result of GCA with maximum ρ^* is almost always most similar to the component of the sum, especially when the discretization is uniform. Figure 9.5 illustrates this feature.

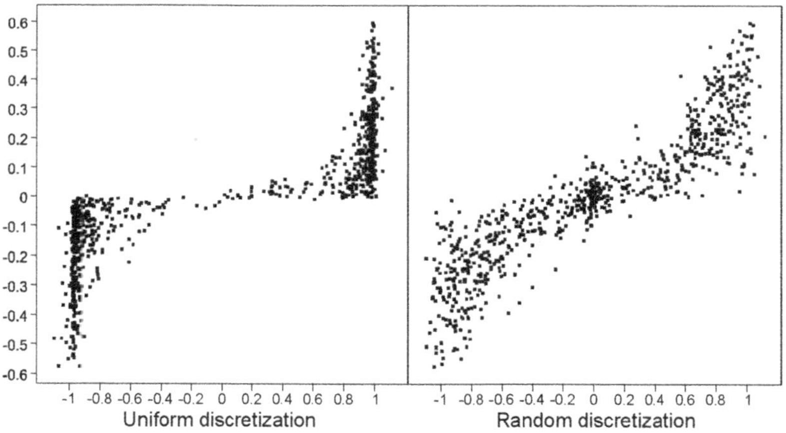

FIGURE 9.5. Results of GCA for ρ^* in the case of uniform and random discretization. Description of the axes is given in the text.

Values of

$$\Delta\omega = \pi\rho_1^* - (1-\pi)\rho_2^* \tag{9.8}$$

are placed on the vertical axis and values of

$$\Delta\tau_{sgn} = \tau_{sgn}(Perm_{GCA}(Perm_1(P_1))) - \\ -\tau_{sgn}(Perm_{GCA}(Perm_2(P_2))) \tag{9.9}$$

on the horizontal axis. Notice that the signs of $\Delta\tau_{sgn}$ and $\Delta\omega$ are almost always the same — the difference between the best GCA permutation and the permutation of a component is smaller when $\pi_l\rho_l^*$ for this component is larger. The GCA method usually selects the most important (with the largest $\pi\rho^*$) of the components rather than create a combination of them.

The same is true in the case of GCA$_\tau$:

Spearman's correlation between $\Delta\tau_{sgn}$ for GCA$_\tau$ and GCA$_{\rho^*}$ is perfect: $\rho^* = 0.999$ for this sample. There is almost no difference in the global maxima found by GCA$_{\rho^*}$ and GCA$_\tau$.

Property (ii) Let q be the index of the primary component, i.e.

$q\epsilon\{1,2,...n\}$, $\pi_q\rho_q^* \geqslant \pi_l\rho_l^*$ for $l = 1,...,n$. The following expressions are almost equal: $\rho_{max}^* \approx \pi_q\rho_q^*$, $\tau_{max} \approx \pi_q\tau_q$. Notice that they become exactly equal only if all permutations $Perm_l$ except $Perm_F$ produce no strength of dependence after the $Perm_{GCA}$ permuting.

The Spearman's correlations between the left and right parts of the approximations for this sample:

	uniform discretization	random discretization
$\rho^*(\rho_{max}^*, \pi_q\rho_q^*)$, GCA$_{\rho^*}$	0.997289	0.972351
$\rho^*(\tau_{max}, \pi_q\tau_q)$, GCA$_\tau$	0.997523	0.974039

In [73] these results are justified theoretically.

9.2.5 Folds

As shown in the previous section, the result of the GCA algorithm with the largest value of ρ^* almost always indicates the component with the largest value of $\pi\rho^*$. In this section other local maxima of ρ^* will be examined. One would expect that they indicate some other components of the sum (9.3) When the strengths of the components are similar, this expectation is sometimes fulfilled. However, even in this case, the number of local maxima found by GCA is much larger than the number of components. Clustering may be used to analyse them, as in paper [103]. Unfortunately, even if only one binormal component exists (i.e., if a discretized binormal table is processed), often there appear strange secondary maxima of ρ^*, which will be called "folds." The same effect appears for local maxima of τ. A similar effect can be observed also in

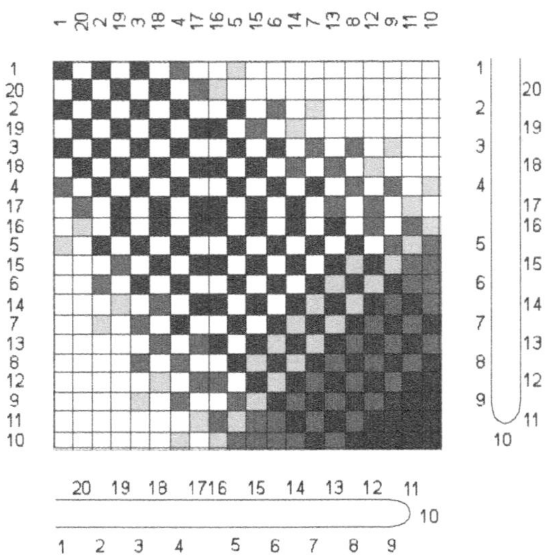

FIGURE 9.6. Mechanics of folds. Maximum number 10 in the GCA overview table, $\rho^* = 0.5242$, $\tau = 0.3634$

the case of τ_{sgn}, although some details are different. Details are presented in [73].

EXAMPLE: For a discretized binormal distribution table $N_{20\times 20}(0.8)$, GCA$_{\rho^*}$ has been performed. 1000 random start permutations were examined. As many as 971 times the algorithm reached the global maximum. A binormal table is very regular, so the existence of only one maximum would be expected. Unfortunately, other local maxima (folds) have also been reached as shown in the following table, called GCA overview table (Frequency refers to the number of start permutations that lead to a given local maximum):

Number	Frequency	ρ^*	τ	τ/τ_{abs}	τ_{sgn}
1	971	0.8	0.6029	1.0000	1.0000
2	6	0.5271	0.365	0.6053	0.0028
3	4	0.5271	0.365	0.6053	0.0028
4	4	0.5271	0.365	0.6053	0.0028
5	3	0.5271	0.365	0.6053	0.0028
6	3	0.5257	0.3637	0.6032	0.0018
7	3	0.5257	0.3637	0.6032	0.0018
8	2	0.5257	0.3637	0.6032	0.0018
9	1	0.5257	0.3637	0.6032	0.0018
10	1	0.5242	0.3634	0.6026	0.0001
11	1	0.5242	0.3634	0.6026	0.0001
12	1	0.3413	0.2385	0.3955	0.0018

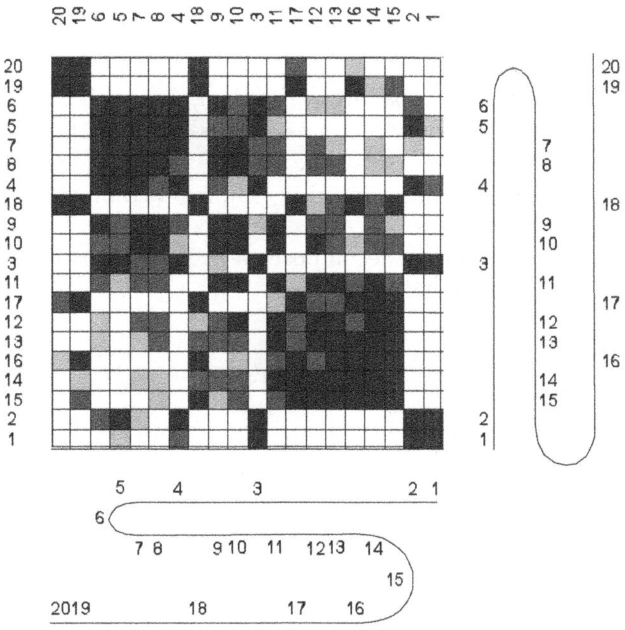

FIGURE 9.7. More complicated fold. Maximum number 12 in the GCA overview table, $\rho^* = 0.3413$, $\tau = 0.2385$

Folds have always a Spearman's *rho* and a Kendall's *tau* smaller than the true maxima. The largest difference, however, can be found in the values of τ_{sgn}. The easiest way to detect folds seems to be via the value of the τ_{sgn} coefficient. Its values are over a hundred times smaller in the case of folds than in the case of normal local maxima. They seem to be transformations of the global maximum, as in Figure 9.6. The array of rows and the array of columns are folded along some row and column and then interwoven with themselves as in the Figure 9.6, which clarifies the name "folds." Sometimes folds are more complicated. In Figure 9.7 the table has been folded two times.

In all examples of folds from this section a binormal distribution has been uniformly discretized, and the number of columns and rows is equal. Folds, however, exist regardless of the discretization and the table does not have to be square. Folds are likely to exist in non-binormal tables as well, but they are harder to distinguish from a normal local maxima. (In binormal tables we did not expect any maxima except the global maximum.)

The number of folds grows rapidly when the global maximum of ρ^* is large. Moreover, folds (especially in the case of ρ^*) seem to be more frequent in smaller tables. A more detailed examination of folds may be found in [73].

9.3 Algorithm for Smooth Grade Correspondence Analysis (SGCA)

The simple correspondence analysis family is an evolving set of statistical algorithms that consists of two groups — classic CA with its direct modifications based on Pearson's correlation, and the much more recent GCA based on ρ^*, τ, or τ_{sgn}. SGCA, introduced in [73], is based on the GCA framework, expanding it in many ways.

The algorithm of SGCA decomposes a table P into the mixture (convex combination) of n tables $Q^1 ... Q^n$:

$$P = \sum_{l=1}^{n} Q^l, \text{where } Q^l = Perm_l(T^l \pi_l) \text{ and} \sum_{l=1}^{n} \pi_l = 1 \tag{9.10}$$

Each table T^l, created by the algorithm, represents one trend, constitutes a local maximum of ρ^*, and to achieve Q^l should be permuted in a different way, described by a function $Perm_l$ and divided by π_l. Their cells will be denoted t^l_{ij}. SGCA tries to maximize the sum of $\pi_l \rho^*(T^l)$.

At the start, each cell t^l_{ij} in each working table is set to p_{ij}, and each table T^l is permuted randomly.

Four operations are then done in a loop: smoothing, arbitration, GCA, and comparing with the best result until now.

- Data smoothing algorithm is applied to T^l. The algorithm for each cell of the table:

  O calculates an average of the overrepresentation values (that is $t^l_{ij}/(t^l_{i*} t^l_{*j})$) in the neighbourhood of the cell; the definition of the neighbourhood is the first parameter of the algorithm; the average being used (like the arithmetic mean or median), is the second parameter.

  O multiplies the result by $t^l_{i*} t^l_{*j}$, and

  O puts this result into the corresponding cell of the smoothed table T^l (that is being created).

  Data smoothing is applied to each working table T^l. T^l permuted randomly in the first pass of the loop, and then according to the final GCA result. When the progress of the algorithms is too small, the tables T^l are randomly shuffled again to allow the procedure to try other random start permutations. In every moment function $Perm_l$ describes how to create table Q^l out of T^l.

- Arbitration is a process of normalization of the tables $Q^l = Perm_l(T^l \pi_l)$ to make sure, that $\sum_{l=1}^{n} Q^l = P$. This is done by multiplying each value q^l_{ij} by $p_{ij}/\sum_{l=1}^{n} q^l_{ij}$ Then new value of $\pi_l = \sum_{i,j} q^l_{ij}$ is calculated. Multiplying by π_l and the reverse permutation $Perm_l^{-1}$ are applied to Q^l to achieve back new values of T^l.

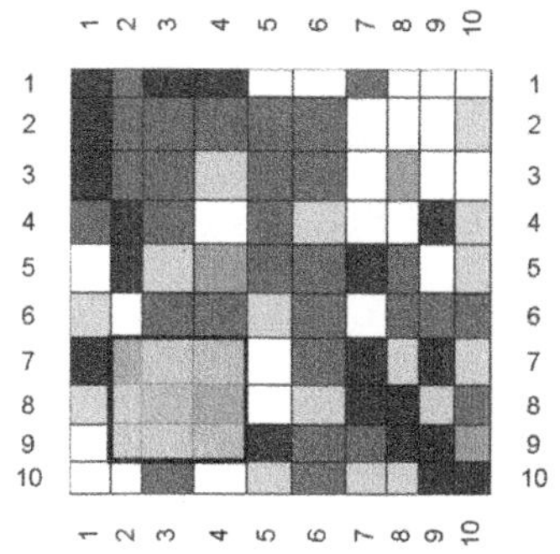

FIGURE 9.8. Example of neighbourhood rectangle 3 × 3 (rows 7-9, columns 2-4), mentioned in the SGCA algorithm

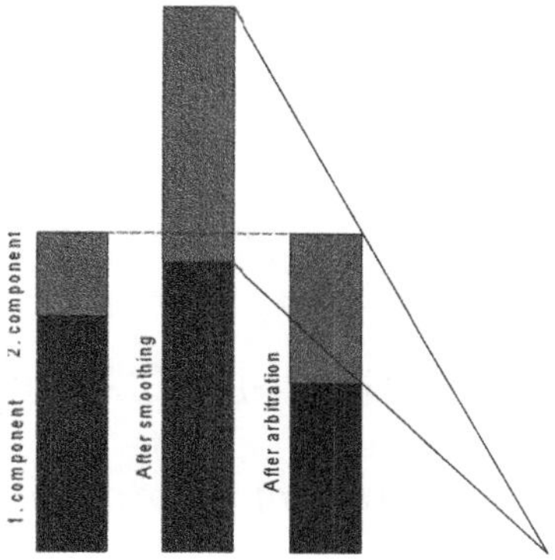

FIGURE 9.9. Visual explanation of the first two steps in the SGCA loop (smoothing changes the data in the tables, the arbitration process normalizes every cell of smoothed component tables to produce the same sum).

- On each table T^l GCA is performed to achieve the new permutation $Perm_l$.
- Algorithm maximizes the sum of $\pi_l \rho^*(T^l)$. The current sum is compared with the best result so far. The best result after a given time constitutes the result of the algorithm.

The algorithm in each pass of smoothing and arbitration shifts the mass of each cell that outlies from the neighbourhood cells to the other tables. Then GCA is performed, which corrects sequence of rows and columns after this shift. The algorithm of SGCA asymptotically approaches the result, thus the stop condition should be chosen. Unfortunately the number of possible results is much larger than in case of GCA, so that much more start permutations must be checked to obtain something near the global maximum.

The results of SGCA:

- Set of tables T^l representing trends. SGCA gives a unique opportunity to evaluate strength of influence of each trend l on each cell p_{ij} of the table P. It can be calculated as $q^l_{ij}/p_{ij} \times 100\%$. No other CA method produces a similar output.
- Set of values $\pi_l \rho^*(T^l)$ representing strengths of trends.
- Regression vectors $r^*_{col|row}$ and $r^*_{row|col}$ for each table T^l ordered after calculation according to $Perm^l$. The resulting values constitute coordinates of each row or column ordered by a given l-th trend. These

coordinates for T^1 and T^2 may be compared on a scatterplot to show the latent structures in the data.

For the formal definition and a detailed analysis of SGCA see [73].

9.4 Examples of GCA and SGCA results

9.4.1 A mixture of binormals

Suppose that we start with the two discretized binormal distribution tables shown in Figures 9.10 and 9.11.

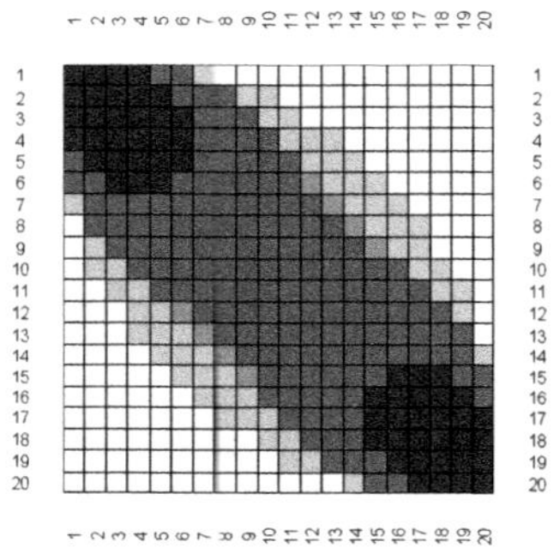

FIGURE 9.10. Component A, $\rho^* = 0.6$ (discretized binormal)

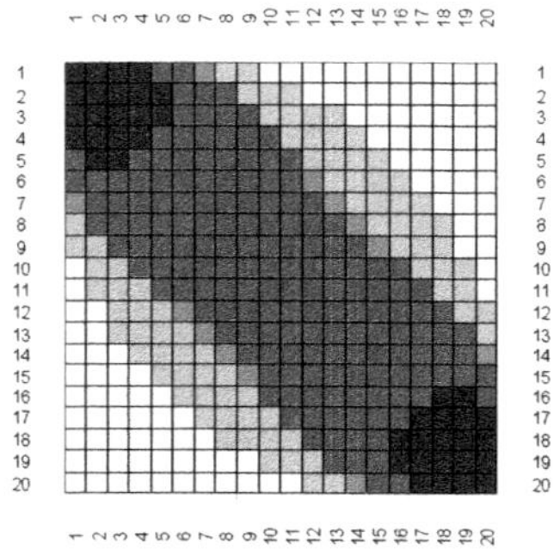

FIGURE 9.11. Component B, $\rho^* = 0.5$ (discretized binormal)

Then, let the rows and columns of the component tables be permuted in a random way which leads to Figures 9.12 and 9.13.

Next the linear combination $0.6 \cdot A_{perm} + 0.4 \cdot B_{perm}$ is created and presented in Figure 9.14.

In Figures 9.14, 9.15, 9.16, 9.17, and 9.18.the first number in the label of the rows and columns indicates the coresponding row or column in component A, the second number refers to component B.

GCA and SGCA results

Thousand iterations of GCA produced 8 local maxima.

Those most similar to components A and B are shown on Figures 9.15 and 9.16.

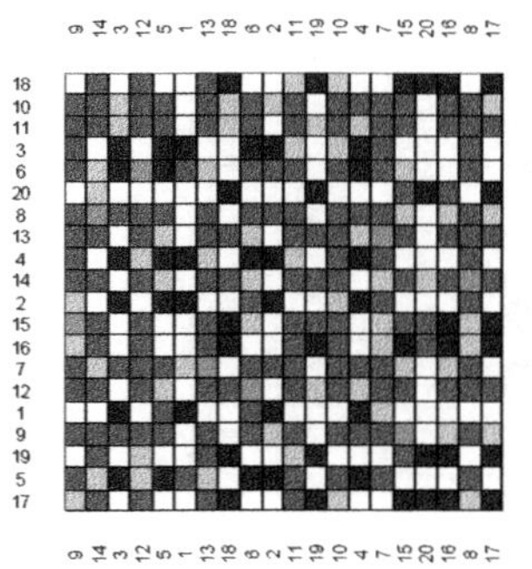

FIGURE 9.12. A_{perm} (Permuted component A)

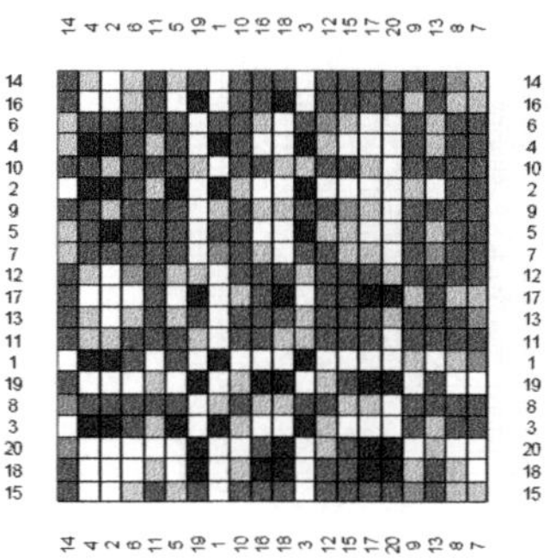

FIGURE 9.13. B_{perm} (Permuted component B)

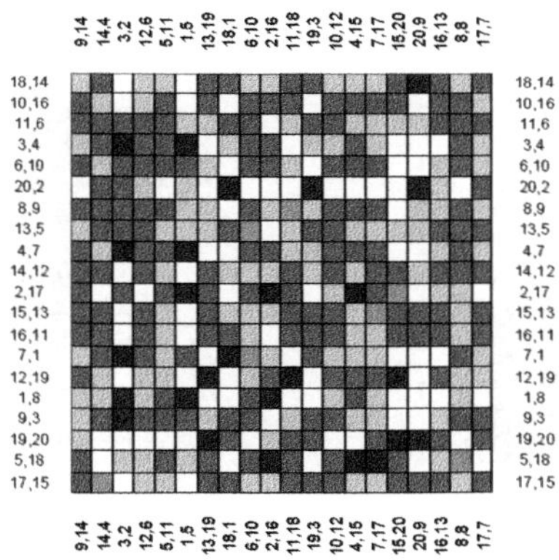

FIGURE 9.14. Mixture of permuted components

After processing the mixture by the SGCA algorithm the components shown on Figures 9.17 and 9.18 have been obtained.

Notice that the reconstruction of the components is almost perfect.

9.4.2 $BRIT_{7\times7}$ and $CARS_{16\times16}$

These datasets have been previously considered in Chapter 8.

GCA results

For the $BRIT_{7\times7}$ dataset there are three local maxima of ρ^*. In each the order of the rows is the same as the order of the columns. (This is rather unusual and confirms the high regularity of this dataset.) The local maxima are shown in the form of a GCA overview table:

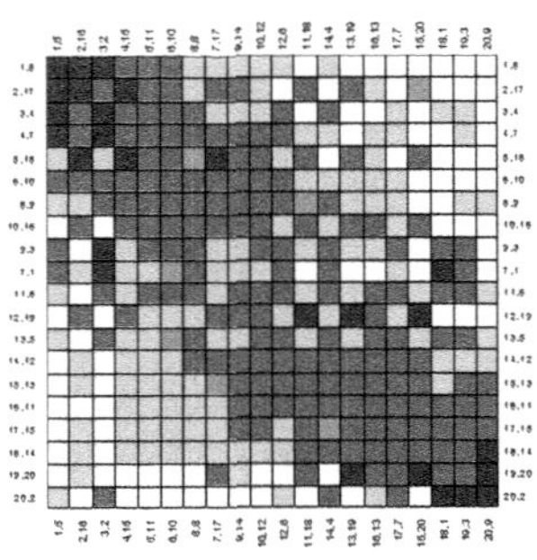

FIGURE 9.15. Mixture: the local maximum most similar to the component A, $\rho^* = 0.3041$

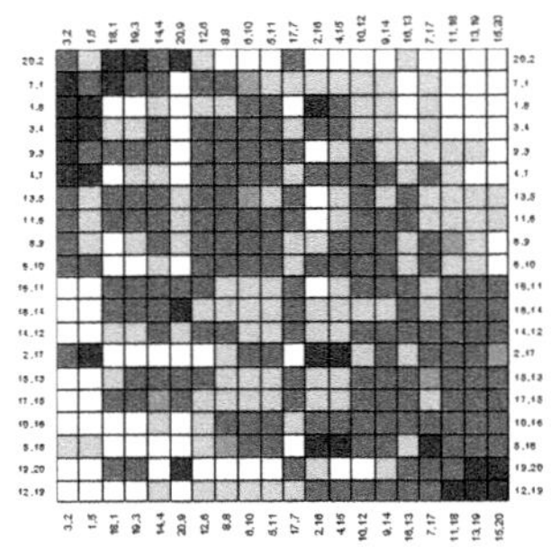

FIGURE 9.16. Mixture: the local maximum most similar to the component B, $\rho^* = 0.2546$

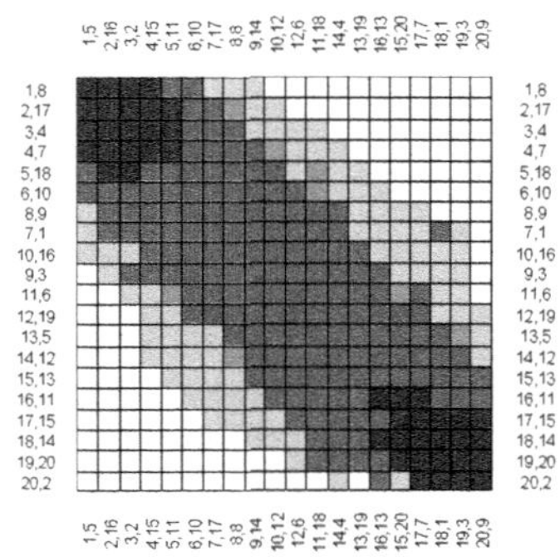

FIGURE 9.17. First component of SGCA resulting from the mixture of binormal tables

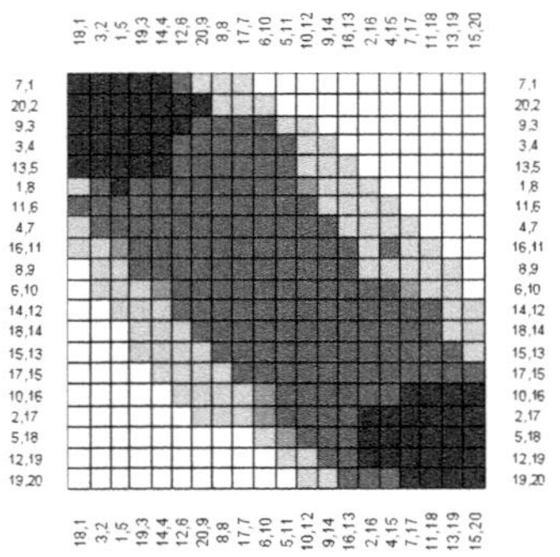

FIGURE 9.18. Second component of SGCA resulting from the mixture of binormal tables

No.	Order							Frequency	ρ^*	τ	τ/τ_{abs}
1	1	2	3	4	5	6	7	990	0.3720	0.2566	0.97
2	1	4	6	5	3	7	2	6	0.1080	0.0742	0.28
3	2	6	3	5	7	4	1	4	0.0955	0.0654	0.25

The main thing to be pointed out is the frequency of the first maximum. In 99 percent of the cases the GCA algorithm starting from the random permutation ends with this order of rows and columns. The parameters ρ^*, τ and τ/τ_{abs} are much larger than in the case of the other local maxima. This means that the first maximum is the global maximum. All three maxima are shown in Figures 8.8 (the global maximum), 9.19 and 9.20.

In the case of $CARS_{16\times16}$, thousand iterations of GCA produced 63 local maxima, which form a few subsets, each containing very similar ordering of

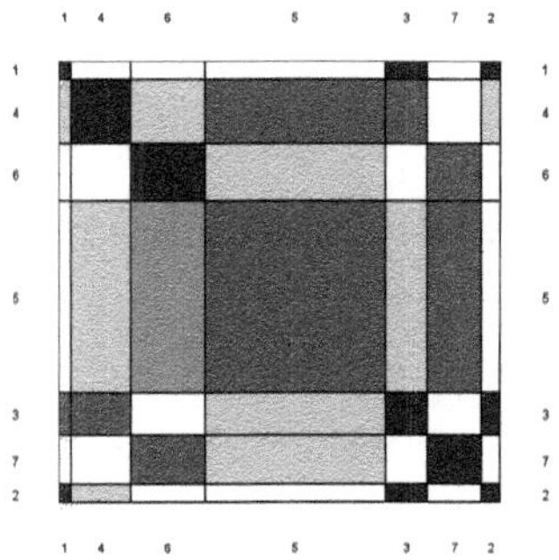

FIGURE 9.19. $\text{BRIT}_{7\times 7}$: second local maximum of ρ^* in the $\text{BRIT}_{7\times 7}$ example, $\rho^* = 0.1080$

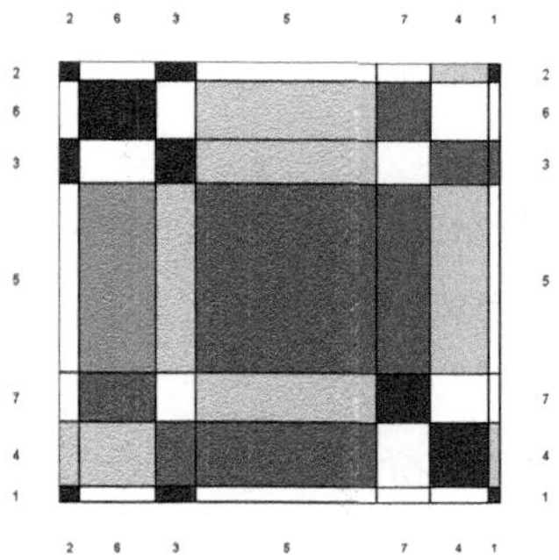

FIGURE 9.20. $\text{BRIT}_{7\times 7}$: third local maximum of ρ^* in the $\text{BRIT}_{7\times 7}$ example, $\rho^* = 0.0955$

rows and columns. The global maximum with $\rho^* = 0.424$ is shown on Figure 8.13.

The table $\text{CARS}_{16\times 16}$ with such a large number of maxima may be considered an irregular, while $\text{BRIT}_{7\times 7}$ is a regular table with only 3 local maxima and the global maximum achieved in 99 percent of cases.

SGCA results

Due to regularity, SGCA decomposes the table $\text{BRIT}_{7\times 7}$ into two almost identical tables with the same order of rows and columns as in the global maximum found by GCA. (The global maximum is presented in Figure 8.8. Results of SGCA are not presented here, because of their similarity to this Figure).

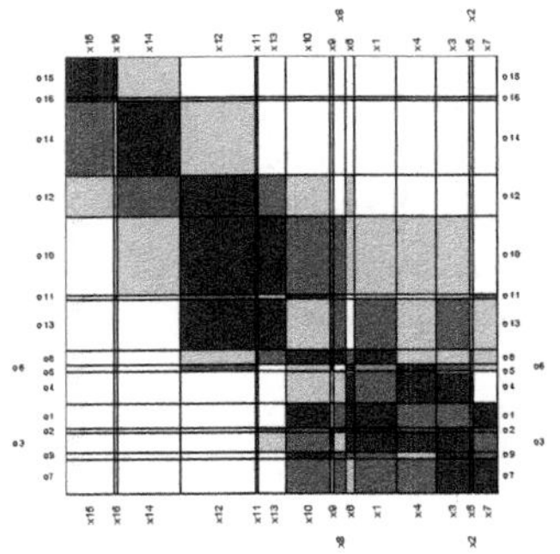

FIGURE 9.21. First SGCA decomposition table in the $\text{CARS}_{16\times 16}$ example

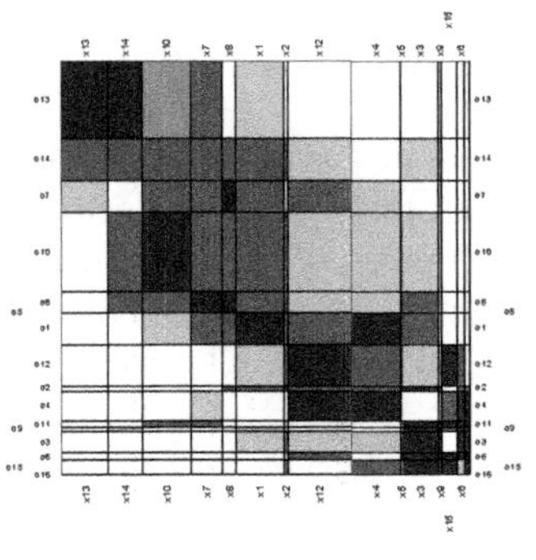

FIGURE 9.22. Second SGCA decomposition table in the $\text{CARS}_{16\times 16}$ example

SGCA decomposition of $\mathrm{CARS}_{16\times 16}$ resulted in the tables shown on Figures 9.21 and 9.22.

9.5 Detection of rows and columns outlying the main trend

9.5.1 Scatterplots for rows and for columns

The role of the TP_2 family was stressed in Chapters 7 and 8. The most important features were presented in Proposition 7.2 It might be said that in a TP_2 table there is just one monotone trend (called "*main trend*") between the row variable and column variable.

The closer a table is to a TP_2 table, the more similar is the set of $ar(j:i;P)$ to the set of $ar_{\max}(j:i;P)$, where the indices ar and $ar_{\max}$ are calculated either for all ordered pairs of row profiles (row j appearing after row i), or for all ordered pairs of column profiles (column j appearing after column i). Therefore, the distance of a table from TP_2 can be evaluated from two scatterplots of points:

- **scatterplot for rows**

$$\{(ar(j:i;row(P)),\ ar_{\max}(j:i;row(P)))\,,\ \ i{=}1,...,m,\ j{=}i{+}1,...,m\} \tag{9.11}$$

- **scatterplot for columns**

$$\{(ar(j:i;col(P)),\ ar_{\max}(j:i;col(P)))\,,\ \ i{=}1,...,k,\ j{=}i{+}1,...,k\} \tag{9.12}$$

Both scatterplots for $\mathrm{BRIT}_{7\times 7}$ are shown in Figure 9.23.

It is seen in Figure 9.23 that all points concentrate either on the line $ar = ar_{\max}$ or just above it. Recall that in a TP_2 table the equality $ar = ar_{\max}$ holds for *all* pairs of ordered row profiles, as well as for *all* pairs of ordered column profiles. The scatterplots for $\mathrm{BRIT}_{7\times 7}$ show geometrically how close $\mathrm{BRIT}_{7\times 7}$ is to being a TP_2 table.

The analogous scatterplots for table $\mathrm{CARS}_{\max}$ are shown in Figure 9.24. They look very different than in the case of $\mathrm{BRIT}_{7\times 7}$ since,apart from a few exceptions, the points lie above the line $ar = ar_{\max}$ and many of the points are very distant from that line. This confirms that the table $\mathrm{CARS}_{\max}$ is distinctly different from a TP_2 table.

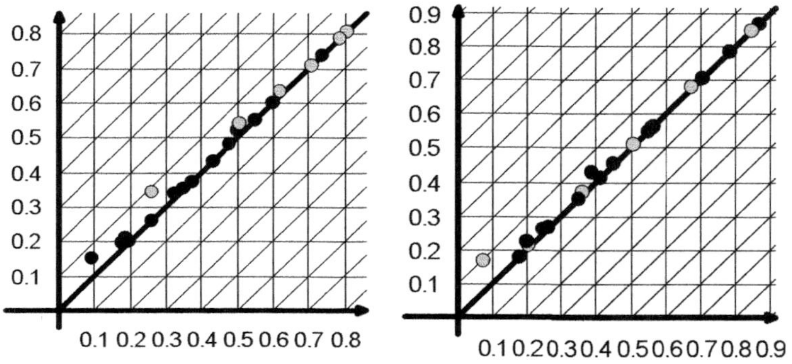

FIGURE 9.23. Scatterplots concerning outliers from TP_2 for $BRIT_{7\times 7}$, in the case of rows (left) and columns (right). The grey points correspond to row number 6 ("semiskilled manual fathers") and to column number 6 ("semiskilled manual sons").

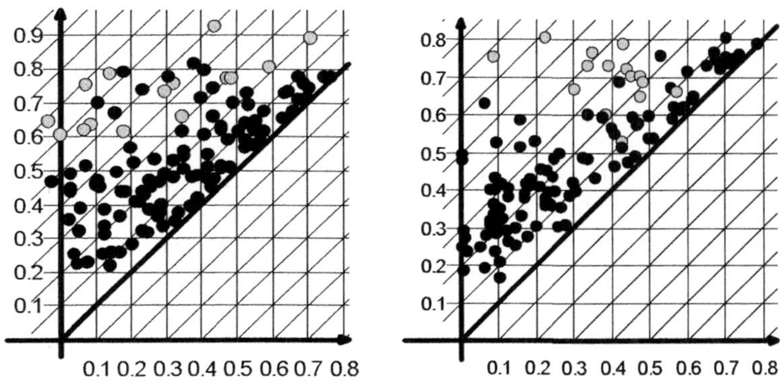

FIGURE 9.24. Scatterplots concerning outliers from TP_2 for $CARS_{max}$, in the case of rows (left) and columns (right). The grey points correspond to row number 16 and to column number 16 (car category "luxury import").

Remark 9.4 *The scatterplots defined by the formulas (9.11) and (9.12) can be developed for any table not necessarily satisfying the condition* $\rho^* = \rho^*_{\max}$. *However, we are interested here in how distant is a considered table* P *from* TP_2, *and therefore we develop the scatterplots not for* P *but for* $P_{\max}$, *since it is* $P_{\max}$ *which exhibits the monotone version of the main trend in* P. *We recall that* $\rho^*(BRIT_{7\times 7}) = \rho^*_{\max}(BRIT_{7\times 7})$ *and therefore no prior permuting of rows and columns was needed in this case.*

Remark 9.5 *Please note that the rows in the scatterplot defined by the formula (9.11) are labeled* $1, ..., m$ *according to the order of rows appearing in table* P *under consideration, and the same applies to the columns in the scatterplot (9.12). When the considered table* P *is* $BRIT_{7\times 7}$, *labels 1,...,7 are identical with the names of categories in* $BRIT_{7\times 7}$ *(Number 1 refers to professionals, ..., Number 7 to unskilled manuals). When the considered ta-*

ble is $CARS_{\max}$, labels 1,...,16 are different from the names of categories in $CARS_{16\times16}$. So in the scatterplot for rows of $CARS_{\max}$, label i refers to the ith category appearing in the sequence of rows in $CARS_{\max}$; in the scatterplot for columns of $CARS_{\max}$, label i refers to the ith category appearing in the sequence of columns in $CARS_{\max}$. We see from Fig (9.26) that row car categories in positions 1 through 16 of $CARS_{\max}$ are 3, 9, 6, 2, 5, 4, 1, 7, 11, 8, 10, 12, 13, 16, 14, 15, and the sequence for columns can also be read from Figure (9.26).

9.5.2 Measures of departure from TP_2

Information contained in the scatterplots defined by the formulas (9.11) for rows and (9.12) for columns will now be reduced in order to evaluate numerically, for each row and column separately, how distant it is from the TP_2 pattern.

Two different measures of departure will be presented, each of them in two analogous forms dedicated to rows or to columns. All four measures are based on subsets of points in the scatterplot which correspond to a particular row or column of table P. A subset corresponding to a row contains $m-1$ points, a subset corresponding to a column contains $k-1$ points. The Euclidean distance from the line $ar = ar_{\max}$ is calculated for every point.

The expectation of all distances in a particular subset is denoted $AvgDist_{row}$ or $AvgDist_{col}$ (it depends on whether the subset refers to a row or to a column); the maximal value of all distances in a subset is denoted $MaxDist_{row}$ or $MaxDist_{col}$.

So we have:

$$AvgDist_{row}(i;P) = \sum_{s=1}^{i-1} \frac{ar_{\max}(i:s;row(P)) - ar(i:s;row(P))}{(m-1)\sqrt{2}} + $$
$$+\sum_{s=i+1}^{m} \frac{ar_{\max}(s:i;row(P)) - ar(s:i;row(P))}{(m-1)\sqrt{2}}, \quad i=1,...,m \qquad (9.13)$$

$$AvgDist_{col}(j;P) = \sum_{t=1}^{j} \frac{ar_{\max}(j:t;col(P)) - ar(j:t;col(P))}{(k-1)\sqrt{2}} + $$
$$+\sum_{t=j+1}^{j} \frac{ar_{\max}(t:j;col(P)) - ar(t:j;col(P))}{(k-1)\sqrt{2}}, \quad j=1,...,k \qquad (9.14)$$

$$
\begin{aligned}
MaxDist_{row}(i;P) = & \frac{1}{\sqrt{2}} \max([ar_{\max}(i:1;row) - ar(i:1;row)], ..., \\
& [ar_{\max}(i:i-1;row) - ar(i:i-1;row)], \\
& [ar_{\max}(i+1:i;row) - ar(i+1:i;row)], ..., \\
& [ar_{\max}(m:i;row) - ar(m:i;row)])
\end{aligned} \tag{9.15}
$$

$$
\begin{aligned}
MaxDist_{col}(j;P) = & \frac{1}{\sqrt{2}} \max([ar_{\max}(j:1;col) - ar(j:1;col)], ..., \\
& [ar_{\max}(j:j-1;col) - ar(j:j-1;col)], \\
& [ar_{\max}(j+1:j;col) - ar(j+1:j;col)], ..., \\
& [ar_{\max}(k:j;row) - ar(k:j;col)])
\end{aligned} \tag{9.16}
$$

Remark 9.6 *Rows and columns appearing in formulas (9.13) and (9.14) defining AvgDist and MaxDist are labeled according to the same rule which was mentioned in Remark 9.5. Please note that, contrarily, in the plots of these functions, the names of categories given in the original table appear. This convention was applied to the plot in Figure 9.25.*

Figure 9.25 (upper part) contains the plots of *AvgDist* and *MaxDist* for the rows of the dataset $\text{CARS}_{\max}$ when the rows are permuted according to non-decreasing values of *AvgDist*. It follows that the row category with the smallest value of *AvgDist* (and therefore the least distant from the TP_2 pattern and therefore from the main trend) is car category number 15, the next is car category 14, and so on.

The departures evaluated according to *MaxDist* are much larger than the departures according to *AvgDist* and also differently ordered (although the trend is similar). The values of *MaxDist* are considerably enlarged by the exceptionally big differentiation of the sizes of the rows in $\text{CARS}_{\max}$. The distribution of the categories of cars in a relatively small group of new car buyers can be drastically different from such distributions for other rows in the $\text{CARS}_{\max}$ table.

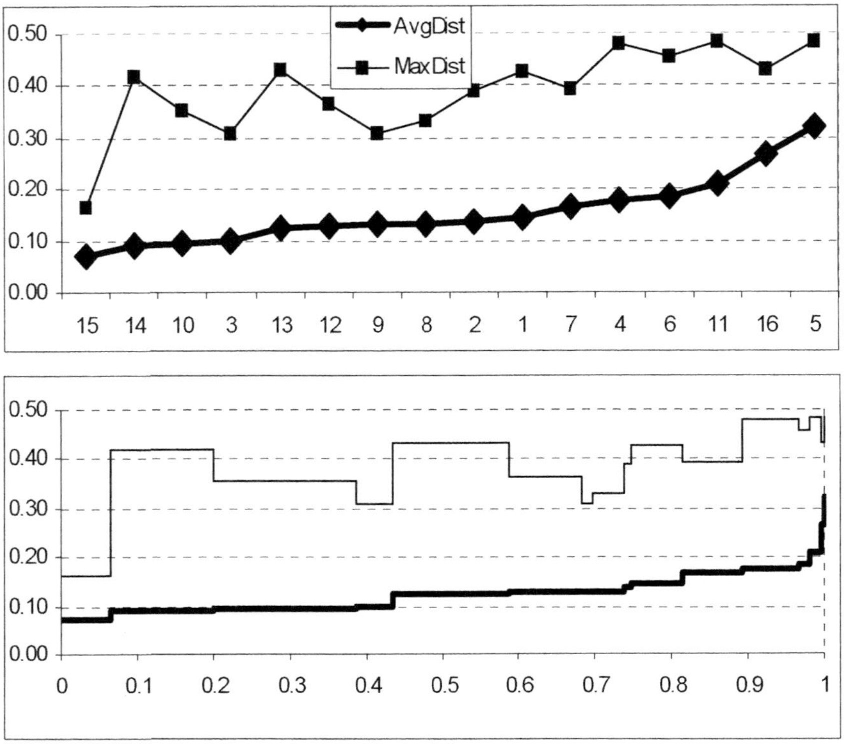

FIGURE 9.25. **Upper part:** $AvgDist_{row}(\bullet, \mathrm{CARS}_{\max})$ and $MaxDist_{row}(\bullet, \mathrm{CARS}_{\max})$, plotted according to non-decreasing values of *AvgDist*. Numbers on the horizontal axis denote the names of car categories which refer to particular rows. **Lower part:** the same plots transferred to the interval (0,1) so that each row is represented by an interval of length equal to its size.

Remark 9.7 *The influence of big differentiation of the marginals on the values of $MaxDist$ has been observed before by the authors of this Chapter in a simulation study concerning finite random sampling from a very regular, but wildly discretized bivariate distribution.*

The consequences of removing the rows with very small marginal totals and high values of $AvgDist$ and $MaxDist$ will be discussed in Section 9.5.3. Here we will only transform the plots of $AvgDist$ and $MaxDist$ in the way presented in the lower part of Figure 9.25. Each row in this plot is represented as an interval of length equal to its size. The lower part of Figure 9.25 shows that there are six rows (car categories 5, 16, 11, 6, 2, 9) with very small sizes, and four of them (5, 16, 11, 6) have the largest values of $AvgDist$ and also have maximal or very close to maximal values of $MaxDist$.

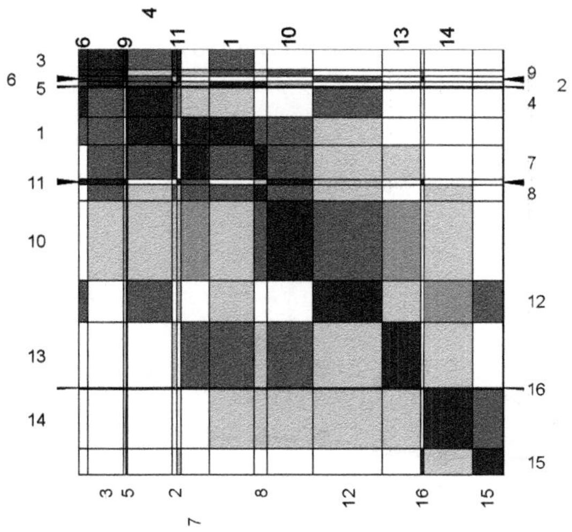

FIGURE 9.26. GCA overrepresentation map for $CARS_{16\times16}$ (with arrows indicating row outliers 6, 5, 11, also 16). The respective columns are also very narrow and strongly outlie from TP_2.

Remark 9.8 *Overall departures of rows and columns from TP_2 can be introduced as a pair of weighted means of the values of $AvgDist_{row}$ and $AvgDist_{col}$, with weights equal to the sizes of rows or respectively of columns. We obtain in this way the pair (0.0070, 0.0063) for $BRIT_{7\times7}$ and the pair (0.1211, 0.1117) for $CARS_{\max}$ (which confirms once more how close to TP_2 is the dataset $BRIT_{7\times7}$ and how distant from TP_2 is the dataset $CARS_{\max}$). Please note that the result (0, 0) is obtained if and only if a table belongs to TP_2.*

It is also possible to find weighted sums of the components of ($MaxDist_{row}$, $MaxDist_{col}$), equal to (0.0268, 0.0284) for $BRIT_{7\times7}$ and (0.3802, 0.3148) for $CARS_{\max}$.

The values of $MaxDist$ (and hence of $AvgDist$) cannot exceed $\sqrt{2}/2$.

9.5.3 Rejecting outlying rows and columns

The most outlying rows for $CARS_{\max}$ are marked with arrows in the overrepresentation map in Figure 9.26. Their counterparts among the columns also have very small sizes and are also detected as substantial outliers due to their values of *AvgDist* and *MaxDist*. So it is reasonable to remove these car categories from the rows and columns and consider the resulting table $(CARS_{12\times12})_{\max}$. It is still a very irregular table, but the respective plots of

$AvgDist_{row}$ and $MaxDist_{row}$ (Figure 9.27) are placed under such plots in Figure 9.25 (the plots of $MaxDist_{row}$ lies below 0.3 instead of 0.5, the plot of $AvgDist_{row}$ lies below 0.15 instead of 0.33).

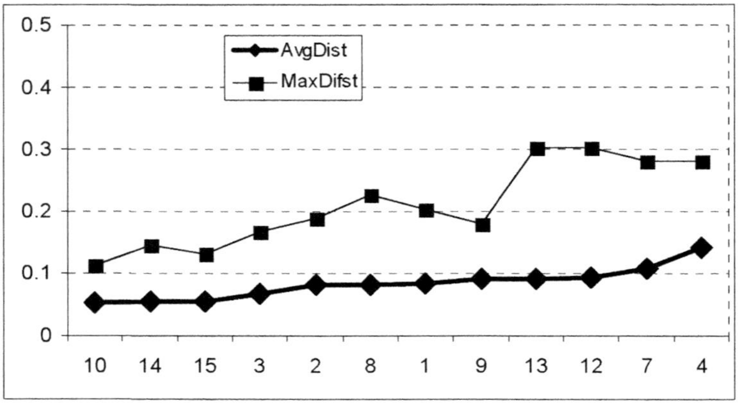

FIGURE 9.27. $AvgDist_{row}$ and $MaxDist_{row}$ for $(\mathrm{CARS}_{12\times 12})_{\max}$, plotted according to increasing values of $AvgDist$.

When the rows and columns which are most outlying - due to their exceedingly small sizes - are eliminated, it is possible to detect the rows and columns which are the *closest* to the main trend. According to Figure A9.27, the most close rows are: 10, 14, 15, 3, 2; while at the other end there are the most outlying rows: 4, 7, 12, 13. Rows 8, 1, 9 are in the middle.

The same procedures were then applied to the columns. The plots for columns were similar to those for rows, but the orderings of categories are slightly different. This is mostly due to category 12: among rows it is found to be very distant from the main trend, while among columns not so very distant. Special difficulties with category 12 are also confirmed visually when comparing Figs 9.26 and Figure 8.17 (right) which presents the binormal analogue of $\mathrm{CARS}_{16\times 16}$. Please notice in Figure 9.26 that there is also a considerable difference in the sizes of the row and column corresponding to category 12.

But one can say roughly that categories 15, 14, 10, 3, 2 fit the main trend both for rows and for columns. Therefore we expect that they will form a *regular subtable* which fits the main trend. Such a subtable denoted $\mathrm{CARS}_{5\times 5}$, has been formed by deleting the remaining rows and columns in the initial table $\mathrm{CARS}_{16\times 16}$. The GCA overrepresentation map of $\mathrm{CARS}_{5\times 5}$ presented in Figure 9.28 (left) is certainly very regular; in fact it is almost TP_2. The overall departures from TP_2 based on $AvgDist_{row}$ and $AvgDist_{col}$ are very smal; 0.0245 and 0.0349; the overall departures from TP2 based on $MaxDist_{row}$ and

$MaxDist_{col}$ are also small: 0.0472 and 0.0662. The orderings of the categories in $(\text{CARS}_{5\times5})_{\text{max}}$ are the same for rows and for columns and they agree with the order of these categories in the main trend related to CARS_{max}.

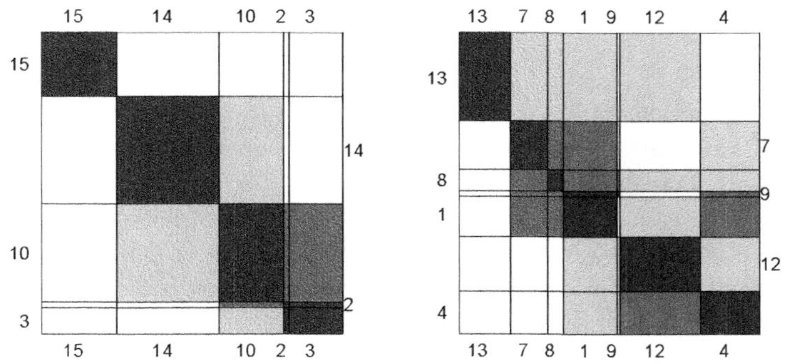

FIGURE 9.28. GCA overrepresentation maps for $\text{CARS}_{5\times5}$ (left) and $\text{CARS}_{7\times7}$ (right).

Figure 9.28 (right) provides the GCA overrepresentation map for the table extracted from the initial table in the same way as $\text{CARS}_{5\times5}$ but for the 7 remaining categories. This new table will be called $\text{CARS}_{7\times7}$. The orderings of the categories in rows and columns are practically the same (up to an insignificant exchange of adjacent categories 9 and 1), but they are not concordant with the ordering provided by the main trend (columns in $\text{CARS}_{7\times7\,\text{max}}$ are ordered 13, 7, 8, 1, 9, 12, 4, while in the main trend they were ordered 13, 12, 8, 1, 7, 4, 9). Clearly, the essential disturbance is caused by category 12, which is mainly responsible for the decrease of regularity and departure from the main trend.

Table $\text{CARS}_{7\times7}$ is less close to TP_2 than $\text{CARS}_{5\times5}$: the respective overall departures are $(0.0787, 0.0853)$ for *AvgDist* and $(0.2090, 0.203)$ for *MaxDist*. Categories 12 and 9 were detected to be the most outlying rows in $\text{CARS}_{7\times7}$. If they are both rejected, the remaining 5×5 table with categories 13, 7, 8, 1, 4 also becomes quite regular.

In summary, CARS_{max} was decomposed into *three* subtables: $\text{CARS}_{7\times7}$ (for categories 15, 14, 10, 2, 3), a subtable for categories 13, 7, 8, 1, 4 extracted from $\text{CARS}_{7\times7}$, and a subtable for categories 5, 6, 11, 16, 12, 9. The first subtable is very regular and fits the main trend; the second is less regular (category 8 is still a small outlier) and does not fit the main trend, although it is rather close to it. The third table is very irregular: it contains "complete outliers." On the other hand, $\text{BRIT}_{7\times 7}$ is not decomposable in this way because it is regular enough and no second or third subtable can be extracted from it.

Remark 9.9 *Every subtable derived from a TP_2 table by removing any number of rows and columns is still TP_2.*

In a regular subtable each category is "the most associated" with its nearest neighbors, and its associacion with the other categories diminishes with the distance between the positions of these categories in the main trend of the table. It is seen from Figure 9.28 (left) that an owner of a car in category 14, when choosing to buy among cars with categories 15, 10, 2, 3, is most interested in buying a car from the same category as before (14) or in category 10; the owner of a car in category 10 (under the same circumstances) is the most interested in cars in categories 10, 2, 3 from on the right and then - slightly - in a car of category 14 on the left. If the trend from category 15 through category 3 of the new car corresponds to price of the car (as it seems to be), it follows that there is a trend for buyers to look for a car which is *less* expensive but *close* to the price standard of the formerly owned car. This type of analysis is easy based on the overrepresentation map in a regular table.

* * *

Both examples presented in this chapter (i.e., $\text{BRIT}_{7\times 7}$ and $\text{CARS}_{16\times 16}$) deal with a table $\text{T}_{m\times m}$, in which each row refers to one and only one column. That is, the same occupation in the case of fathers and sons, or the same car category in the case of previous and present purchase. The decomposition of such a table into regular subsets is specially difficult, in that, it is preferable to get component tables $\text{T}_{s\times s}$ related to disjoint subsets of s categories. We usually take a less specific approach when decomposing a table $\text{T}_{m\times k}$, in which the row and column categories are *not* related. In this case, the tools and methods of outlier detection described in this chapter can be used. In addition, there is more freedom in selecting the proper subsets of rows and columns for the first and most important subtable, i.e., the one "fitting the main trend."

There is no way there to escape some subjectivity of decisions corresponding to the special needs of the researcher. The plots of $AvgDist$ and $MaxDist$ and the overrepresentation maps of the selected component tables are no more than useful prerequisites of decisions. Most often two subtables are extracted: one called FIT-MT (a regular one which FITs the Main Trend) and the residual subtable called OUT-MT (not necessarily regular, with OUTliers from the Main Trend). Subtables FIT-MT and OUT-MT will be constructed and thoroughly analysed in two case studies in Chapter 13 concerning multivariate datasets.

9.6 Appendix - Bibliographical remarks

The detailed examination of classic CA can be found in [94].

The exact algorithm of GCA has been for the first time published in [27]. The GCA algorithm for τ has been for the first time published in [64]. A version of GCA algorithm maximizing τ_{sgn} has been proposed in [73]. Versions of GCA and SGCA algorithms, that handle missing data have been also proposed in [73].

For practical applications of GCA see some papers written by the authors of the present book ([73], [25], [26], [76], [98], [103], [101]) or papers from the Department of Applied Economics, University of Cambridge ([107], [108], [97]). For more detailed description of folds, comparison of GCA based on ρ^* and τ, and more detailed explanation of mentioned simulation studies see [73].

Dataset $\mathrm{BRIT}_{7\times7}$ has been analysed in [55]. Dataset $\mathrm{CARS}_{16\times16}$ has been analysed in [21]. SGCA has been proposed in [73]. The algorithms of detecting outlying rows and/or columns have been proposed in [99].

10

Cluster analysis based on GCA

Alicja Ciok

10.1 Introduction

This chapter presents three variants of a clustering method for two-way data tables. The method is a generalization of the discretization procedure, described in Chapter 6 for pairs of random variables. Although the theoretical background assumes that the data must have the form of two-way probability tables, the method is applicable to almost any two-dimensional table which contains non-negative values. For example, input data tables can contain similarity measurements for pairs of categories or values of attributes which characterize particular objects. (Let us note that in the latter case multivariate data tables are considered.) Moreover, infinite data tables can be taken into account - for example, bivariate densities or cdfs for continuous random variables. The formalizm used can be applied to any of these cases, however the interpretation of the results must be adapted to the contents of the table.

The only restriction imposed on the input data is that the data values must reflect some order (for example, an order determinded by the intensity of an attribute). In consequence, if the data tables include values of variables, then these variables must be at least ordinal. Each nominal variable should be replaced by a set of dummy variables corresponding to the particular categories.

All clustering methods presented in this chapter are based on the results of grade correspondence analysis (GCA) introduced in Chapter 9 and assume that the number of clusters is given. Determination of the proper number of clusters is a very difficult problem which will be discussed in the next chapter.

Sec. 10.2 introduces the theoretical background for the first clustering procedure - single grade clustering (SGCL). In this procedure, the adjacent (according to the optimal permutations provided by the GCA) rows (or columns) of the data table are aggregated optimally. The optimality criterion in this case is the maximal value of the correlation coefficient ρ^* calculated for the aggregated data table. If two single clusterings are applied separately for rows and for columns of the same table, such a procedure will be called double grade clustering (DGCL). It will be shown on the data example of time allocation, how both these clustering methods (SGCL and DGCL) work.

When aggregations of rows and columns are considered simultaneously, the maximization of the correlation ρ^* is provided for by the so called optimal grade clustering (OGCL). This method is discussed in Sec. 10.3, and the respective algorithm is given in the appendix to this chapter. The data example (the same as analysed in Sec. 10.2) will be used to illustrate that this method can provide different results than the DGCL method, although similar with respect to the criterion value of ρ^*. It will be shown that the OGCL procedure, although more complicated and time consuming, is worthy of consideration because it can provide more precise and in-depth knowledge about the data than can DGCL. On the other hand, if great precision is not necessary and the data table is large, the much simpler and faster DGCL is absolutely sufficient.

The name GCCA (grade correspondence-cluster analysis) is used jointly for all variants of the grade clustering methods.

Sections 10.4 and 10.5 are devoted to the analysis of two real data tables by means of grade clusterings. The results show that although the data aggregations can lead to a loss of information, they still provide for the possibility of the expression of complicated data structures and their clear interpretation. Very often any loss of information is fully compensated for by the easy identification of trends for which formal models are difficult to construct.

Analysis of the first data (Sec. 10.4) turns out to be particularly interesting. Its results shows that the data tables having the low values of the grade correlations are worthy of further consideration. Usually such tables are interpreted as not interesting due to the lack of structure. This example shows that a similar effect can be caused by the existence of several different structures in one data table. The greater the differences, the stronger the impression that there are no strong correlations in the data table. More formally, the example shows that the GCCA can be very useful in the separation of mixtures (of subpopulations or of distributions - in our grade framework these tasks are equivalent), which are characterized by different structures of dependence.

The second analysis (Sec. 10.5) shows how the clusterings can help in detection of imprecisely defined trends in repeated measurement data. These clusterings also allow one to determine data variability or stability in time. Such possibilities are particularly useful when the analyzed data are also very "rough" (i.e., variables defined on a nominal or ordinal scale with very few categories and values express imprecisely, for example, subjective opinions, etc.).

The applicability of the clustering methods to various data will be discussed in Sec. 10.6. The analysis of the examples from the previous sections confirms that the GCCA is particularly useful in data mining in that:

- GCCA is applicable to various data types;
- GCCA provides significant insight into the nature of complex structures present in the data;
- GCCA can provide solutions to a broad range of practical problems.

10.2 Single and double grade clustering

In this section our considerations will be restricted to the clusterings of rows only or columns only. Such clustering will be called single grade clustering (SGCL). Moreover, as the single cluster procedure for rows is identical with that for columns (rows and columns just exchange roles) further explanations will concern only the clustering of rows.

Let us assume that the input data table $P = (p_{ij})_{m \times k}$ has the form of a bivariate probability table where the variable X describes its rows, and variable Y corresponds to the columns. Instead of X and Y the new variables X^* and Y^* are considered (cf Chapter 8). According to the definition (cf Sec. 3.3.3 and Sec. 3.5.1) X^* and Y^* are the results of the grade transformations applied to X and Y respectively. Both new variables are always continuous and are uniformly distributed on $[0, 1]$.

The cdf for the pair (X, Y) is also replaced by the cdf for (X^*, Y^*) (it is called the copula of (X, Y)). The grade correlation coefficient $\rho^*(X, Y)$ is equal to $corr(X^*, Y^*)$ (cf Proposition 8.1). It measures the monotone dependence between X and Y.

In this chapter we assume that the initial data table is optimally arranged by the grade correspondence analysis (GCA) introduced in Chapter 9. The GCA maximizes the positive dependence between X and Y as measured by the grade correlation coefficient $\rho^*(X, Y)$ in the set of all pairs of permutations of categories of X (rows) and Y (columns).

Our goal is to find such clusters of rows and/or columns which yield the minimal loss of information about the dependence between X and Y. In other words, it is a problem of maximization of the correlation $\rho^*(X, Y)$ after aggregation according to the clustering results.

> In single clustering the maximization for the set of all possible aggregations is not performed. In this clustering procedure only adjacent rows (columns) in the optimal permutations provided by the GCA are aggregated. An important property of the GCA used here is that similar rows (columns) are always placed close to each other.

The generated clusters form categories of the new row variable. The rows of the new probability table are the aggregates (sums) of the initial rows corresponding to the respective clusters. In this sense, we will say that the rows are aggregated.

The clustering method is based on the optimal discretization procedure described in Chapter 6. This procedure generates partitions of the variable values in such a way that the corresponding concentration index is maximal. Therefore if one wishes to apply this procedure to two-way tables, the counterparts of the convex concentration curves and the concentration index are needed. Note that these roles can be played by the grade correlation curves and the grade correlation index ρ^* (cf Chapter 8). They can be expressed by the following formulas:

$$\begin{aligned}\rho^*(X,Y) &= 3ar(2C^*_{cor(Y:X)}) = 6\int_0^1 (u - 2C^*_{cor(Y:X)}(u))du = \\ &= 3ar(2C^*_{cor(X:Y)}) = 6\int_0^1 (u - 2C^*_{cor(X:Y)}(u))du\end{aligned}$$

where $C^*_{cor(Y:X)}(t) = \int_0^t r^*_{Y|X}(u)du$ is the grade correlation curve and $r^*_{Y|X}(t) = E(Y^* \mid X^* = t)$ is the grade regression function of Y on X defined (cf Chapter 8) as the regression function of Y^* on X^*. $2C^*_{cor(Y:X)}$ belongs to $\mathbb{ULM}$, it will be called the normalized grade correlation curve.

Note that $\rho^*(X,Y)$ is 3 times the concentration index calculated for curve $2C^*_{cor(Y:X)}$, therefore the maximization of ρ^* is equivalent to the maximization of the *ar* for the normalized correlation curve.

The role of the derivative for the concentration curve is now played by $2r^*_{Y|X}$. An important property of the GCA is that both regression functions are nondecreasing (cf Chapter 9), hence the normalized correlation curve calculated for the optimally arranged data table is always convex. Let us also note that the maximization of ρ^* is equivalent to the maximal differentiation of both grade regressions: $r^*_{Y|X}$ and $r^*_{X|Y}$.

Assuming that the number of row clusters is given, the optimal discretization procedure described in Chapter 6 can be applied to the correlation curve $C^*_{cor(Y:X)}$. The categories obtained for this curve form the clusters of rows for the input table. This procedure will be called the single grade clustering (SGCL). Of course the identical procedure applied to $C^*_{cor(X:Y)}$ provides the clusters for the columns.

Generally, in cluster analysis the basic concept is the dissimilarity of analyzed objects (whatever they are). In our case, the dissimilarity between rows i and j can be measured by the difference between the values of the regression function: $r^*_{Y|X}(t_i) - r^*_{Y|X}(t_j)$, where values t_i, $t_i \in [0,1]$ correspond to the respective rows. Therefore we can say that the GCA arranges the rows according to their similarity. For this reason we may aggregate only those rows which are adjacent in the optimal permutation provided by the GCA.

Of course the same clustering procedure can be applied to the columns of the data table.

A pair of two single clusterings which are independently applied to the initial table (for rows and for columns) will be called double clustering (DGCL).

Example 1

The data come from the paper [78] and characterize how people in the Netherlands spend their time, cross-classified by gender, age and year. In each cell the average number of minutes per day spent on a certain activity by groups of people is given. The groups of people differ by gender and age and year of survey (i.e. 1975, 1980 or 1985). There are 30 groups corresponding to all combinations of gender, age and year (rows of the data table) and 18 activities (columns). Tables 10.2 and 10.1 show the characteristics of the groups of people as well as the activities. These tables show the optimal permutations of rows and columns.

Figure 10.1 presents the overrepresentation map for the optimally arranged data table. The optimal permutation of activities forms a scale with paid work and education on one end and domestic work on the other. It is typical for the GCA results that the extreme positions are the dominant characteristics of the generated scales.

For groups of people (rows) the opposite ends are formed by: men and very young women on one end, and women (with the exception of the very young) on the other. While the dependency pattern between age and gender on one hand, and the prefered activity on the other is visible in Fig. 10.1, a simple interpretation is not so easy. Concentrating on just a few groups of people characterized by different types of behaviour will make the task much easier.

In an effort to make things more clear, double clustering was applied to the data table; the results, when four clusters were chosen for rows as well as for columns, are presented in Tables 10.2 and 10.1. There was only a small decrease in the value of ρ^* caused by the aggregations. The ρ^* value after the GCA is equal to 0.2582; after the DGCL it decreases to 0.2369 (it is 8.2 percent decrease from the initial value). A much simpler and clearer interpretation of the results is undoubtedly the gain.

TABLE 10.1. The GCA permutations and the clusters for columns

Number in opt. perm.	Activity code	Activity	Cluster number
1	8	education	1
2	1	paid work	1
3	11	going out	1
4	14	recreation outside	2
5	15	tv, radio, audio	2
6	7	sleeping, resting	2
7	18	other	2
8	17	relaxing	2
9	6	eating, drinking	3
10	5	personal needs	3
11	13	gardening, pets	3
12	12	sports, hobbies	3
13	9	volunteer work	3
14	16	reading	3
15	10	social contacts	3
16	4	shopping	4
17	3	caring for members household	4
18	2	domestic work	4

TABLE 10.2. The optimal GCA permutations and the optimal clusters for rows

Number in optimal permutation	Ident. code	Gender	Age	Year	Cluster number
1	111	M	12-24	75	1
2	113	M	12-24	85	1
3	112	M	12-24	80	1
4	213	F	12-24	85	1
5	121	M	25-34	75	1
6	133	M	35-49	85	1
7	123	M	25-34	85	1
8	122	M	25-34	80	2
9	212	F	12-24	80	2
10	131	M	35-49	75	2
11	132	M	35-49	80	2
12	211	F	12-24	75	2
13	141	M	50-64	75	2
14	142	M	50-64	80	2
15	143	M	50-64	85	2
16	153	M	>64	85	3
17	152	M	>64	80	3
18	151	M	>64	75	3
19	223	F	25-34	85	3
20	253	F	>64	85	4
21	221	F	25-34	75	4
22	243	F	50-64	85	4
23	222	F	25-34	80	4
24	251	F	>64	75	4
25	233	F	35-49	85	4
26	232	F	35-49	80	4
27	231	F	35-49	75	4
28	252	F	>64	80	4
29	241	F	50-64	75	4
30	242	F	50-64	80	4

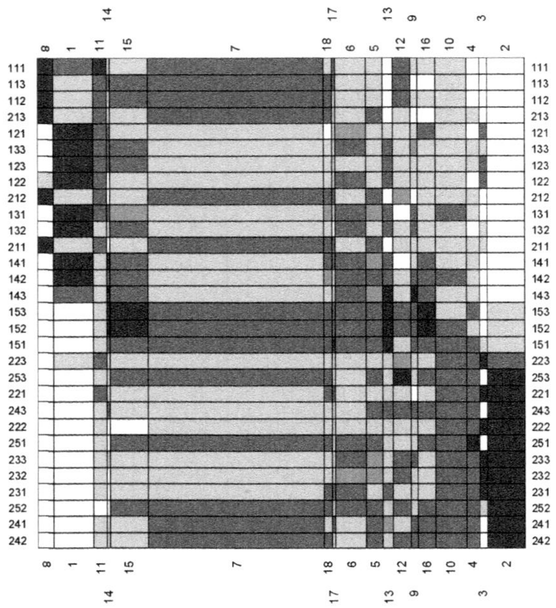

FIGURE 10.1. The overrepresentation map for the time allocation data arranged according to the GCA (gender-age-year combinations on the verical axis, activity codes on the horizontal axis)

The overrepresentation map of the aggregated data according to the DGCL is shown in Fig. 10.2. The clusters (rows and columns visible in this figure) can be characterized in the following way:

Rows:

1. Very young men and women
2. Working men and very young women
3. Men of pensionable age and young women
4. Women (very young excluded)

Columns:

1. Education, paid work, going out
2. Rest, relaxation
3. Personal interests and hobbies, volunteer work, social contacts
4. Domestic duties

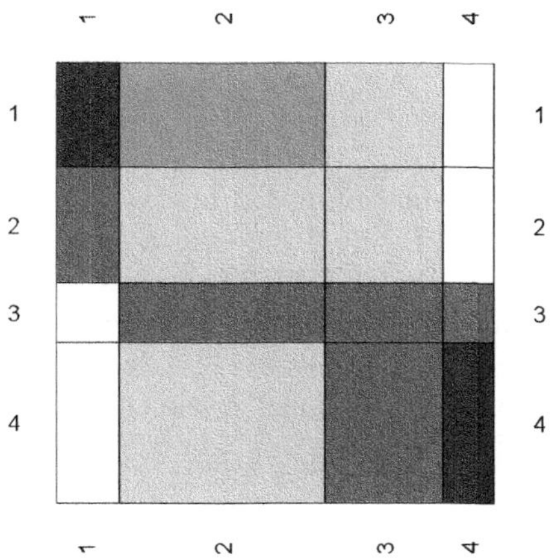

FIGURE 10.2. The overrepresentation map of the time allocation data after DGCL (demographic cluster codes on the vertical axis, activity cluster codes on the horizontal axis)

The dependence pattern is now very clear (cf Fig. 10.2): young women and men (with the exception of the old) are concentrated (i.e., have high overrepresentation values) on education or paid work. The occupations of women and men of pensionable age are linked (i.e. concentrated) with domestic duties and personal interests. Additionally, the men spend much time on various forms of relaxation.

The exception to the rule that groups of people are ordered (clustered) according to gender and age, is that the group of very young women is scattered among the other groups. One may jump to the conclusion that this group has no specific profile, but this is not true as will be shown in the next section.

10.3 Optimal grade clustering

As implied by its name, the optimal grade clustering (OGCL) maximizes the correlation coefficient $\rho^*(X,Y)$ in the set of pairs of aggregations for rows and for columns (aggregations of rows and columns are considered simultaneously). Also in this case the numbers of clusters for rows and columns must be given, and only adjacent rows (columns) in the considered permutations are aggregated. However, in the OGCL procedure permutations other than those provided by the GCA are considered. Therefore the clusters may not preserve the optimal GCA arrangements. (In this case, these arrangements just serve as a good starting point for the optimization procedure. This maximization procedure was specially designed and is presented in the appendix of this chapter.

Example 2

This example is a continuation of the analysis of the time allocation data from the previous section. Tables 10.3, 10.4 and Fig. 10.3 present the clustering results obtained by the OGCL. The corresponding value of ρ^* for the aggregated table is equal to 0.2374, so the difference in the criterion function between the two clustering methods is almost negligible (0.0005). It is interesting that the optimal clusterings in the OGCL method are not optimal according to the DGCL procedure.

TABLE 10.3. The optimal OGCL clusters for rows

Cluster	Gender	Age	Year
1	M	12-49	75, 80, 85
2	F	12-24	75, 80, 85
	M	50-64	75, 80, 85
3	M	>64	75, 80, 85
	F	25-34	85
4	F	25-34	75, 80
	F	>34	75, 80, 85

TABLE 10.4. The optimal OGCL clusters for columns

Cluster	Acitivity numbers	Activities
1	8, 1, 11	education; paid work; going out
2	14, 15, 13, 7, 6	recreation outside; tv, radio, audio; gardening, pets; sleeping, resting; eating, drinking
3	9, 18, 17, 5, 16, 12, 10	volunteer work; other; relaxing; personal needs; reading; sports, hobbies; social contacts
4	4, 3, 2	shopping; caring for members of household; domestic work

While the results appear to be almost equivalent, this is not the case. The difference between the obtained clusters (rows or columns) is definitely not negligible, although the general trend imposed by the GCA exists in both cases. The OGCL results have some significant advantages. One is that the generated clusters differ more among themselves. This is particularly visible in

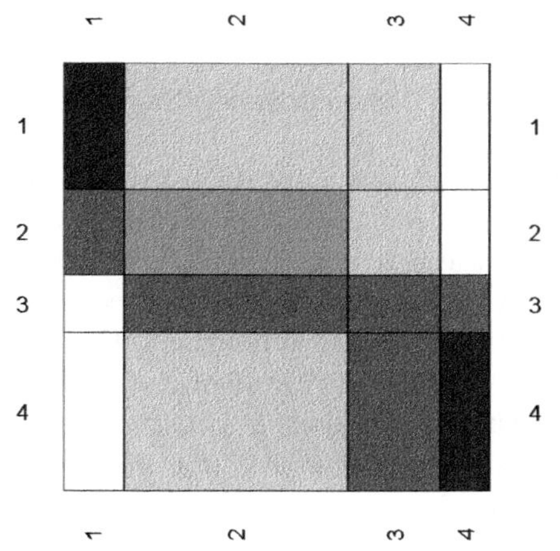

FIGURE 10.3. The overrepresentation map of the time allocation data after OGCL (demographic cluster codes on the vertical axis, activity cluster codes on the horizontal axis)

the case of row clusters: 1 and 2 (compare Fig. 10.2 and Fig. 10.3). Moreover, the row OGCL clusters are more homogeneous with respect to gender and age, which consequently provides a better characterization of the subpopulations. The group of very young women is particularly interesting here. These women are less scattered among the clusters produced by the OGCL - they are included in only one cluster jointly with the men of age 50-64. Moreover, they occupy an intermediate position between two groups: younger men, and men older than 64 and women. This means that this group is a specific unit, although it combines characteristics from both ends of the activity ordering.

10.4 Cluster analysis in the detection of mixtures

10.4.1 Straight and reverse regular structures

The typical approach of data analysts to bivariate dependence is to focus on data sets which are characterized by high correlations measured by various indices (in our case it is the maximal grade correlation $\rho^*_{\max}$). The other datasets are recognized as those characterized by lack of dependence and therefore they are not considered in further analyses. However, sometimes in statistical literature low correlations may be explained by the specific form of the dependence. In this section we will analyze an example where a mixture of several data tables with various characteristics explains the low correlations in the combined data.

Let us compare two data tables with identical marginal sums. Both are arranged according to the GCA, that is, their respective correlations measured by values of ρ^*, are maximal. Moreover, these ρ^*_{max} values are identical and very small (equal to 0.085). The corresponding overrepresentation maps are shown in Figs 10.4 and.10.5

Let us note how greatly these maps differ even though the correlations are equal. The map in Fig. 10.4 represents a discretized 4×23 binormal probability table. The overrepresentations (grade density values) form a saddle surface with a non-decreasing hummock from the upper left corner to the bottom right corner of the unit square. Overrepresentation tends to decrease when one proceeds from any point of the hummock towards the sides of the square.

The second map (Fig. 10.5) looks also quite regular, but definitely there is no monotone regular trend here. At first sight, it is even difficult to be sure that it is a GCA table, since it seems that some other row permutation (in particular 2,1,4,3) might improve the correlation ρ^*. This, however, is not true, because the permutation 1,2,3,4 was checked to be optimal for this table and any change can only reduce ρ^* (eg., ρ^* for row permutation 2,1,4,3 and column permutation 1,2,3,4,5 is about half of the maximal value of ρ^*). A close examination of both figures shows that they are - in a way - the reverse of each other. Instead of the hummock formed by the darker rectangles in Fig. 10.4, there is a depression formed by the lighter rectangles in Fig. 10.5; this goes from the bottom left to the upper right corner of the unit square; the overrepresentation tends to increase when going from any point of the depression towards the sides of the square.

Data structures of the type shown in Fig. 10.4 will be called structures with high straight regularity, and structures of the type visible in Fig. 10.5 - structures with high reverse regularity. The first type represents a "one-way" structure where natural latent variables (one for rows and one for columns) exist and determine the orderings of the rows and the columns. Moreover, there is no doubt that the only potential aggregation is to join adjacent rows (and adjacent columns as well). For structures of the second type the situation is different. For these, one should separate them into two or more components each of which possesses strong straight regularity, each component representing another latent variable which orders the rows (columns).

A structure with a high reverse regularity and a very small maximal ρ^* was unexpectedly found in the survey data concerning the condition of small business servicing firms in Poland described below. Figure 10.5 shows the overrepresentation maps for these data after the GCA rearrangement and the following aggregation according to the single clustering. This example illustrates that data structures with low correlations - but having high regularity - should be carefully investigated.

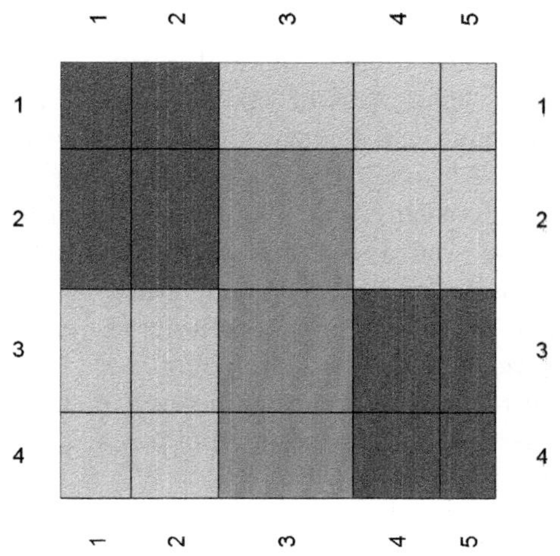

FIGURE 10.4. The overrepresentation map for a straight regular structure

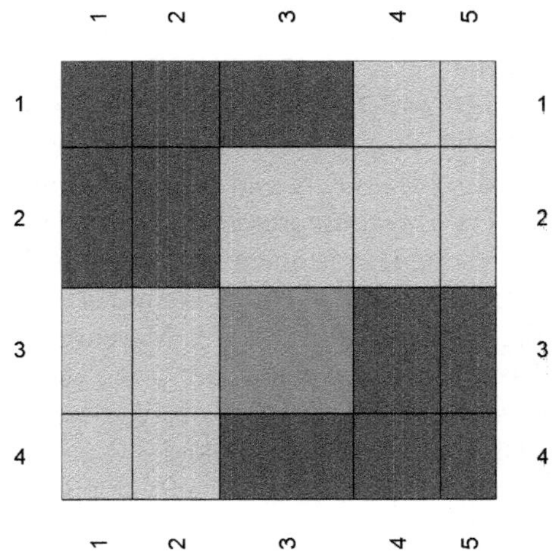

FIGURE 10.5. The overrepresentation map for a reverse regular data table (the small business data after DGCL aggregations)

10.4.2 Survey of small business servicing firms

The main goal of the research carried on by Dr L. Kuczewska from the Institute of Home Market and Consumption in 1999, was to detect factors which were conducive to the development of small business in Poland. The sample included 89 Polish servicing firms of small and medium sizes and from four branches (1 - hairdressing and beauty salons, 2 - travel offices, 3 - language courses, 4 - dentists). Although the questionnaire included many questions, only 23 of them (presented in Table 10.5) were chosen here. The set of questions can be partitioned into two subsets. The first includes questions (3-13) which characterize the actual development status of the firm. High values mean that conditions of the firm are worsening; the exception is question 13, where the scale is reversed. The remaining questions (16.1-18.5) serve to assess various factors with respect to their significance to the firm's business activity. All factors are assessed on the same 5 point scale, where 5 corresponds to "very important".

The table of questionnaire answers was normalized before application of the grade procedures. In this case, division by the total sum of elements in the data table was used, which enables us to treat the data as a probability table.

10.4.3 SGCL results for the whole sample

The GCA applied to the 89×23 table provided the following optimal permutation of columns (variables): 17.1 17.4 13 16.5 17.3 17.2 18.4 18.5 18.1 16.1 16.4 21 18.3 18.2 6 12 5 16.2 16.3 7 17.6 3 17.5, with the corresponding value of maximal ρ^* equal to 0.096. Fig. 10.6 shows the overrepresentation map for the data table transformed by the GCA and aggregated according to the single clustering into 4 clusters of rows representing firms. Due to this aggregation the general structure of the data is much more visible. This structure cannot be improved by any permutation of columns and/or of aggregated rows (the GCA determined the maximal value of ρ^*). Moreover, it is highly reversely regular. Its discretized binormal counterpart (with the same value of ρ^* and the same marginals) shown in Fig. 10.7, is straight-regular.

Notice in Fig. 10.6, that row clusters 1 and 3 form a natural pair in which the characteristics of cluster 1 are almost exactly opposite to those of cluster 3. In cluster 1 there are high overrepresentations for the first thirteen questions in the optimal permutation listed at the beginning of this section (from 17.1 to 18.3). In cluster 3 high overrepresentations occur almost exclusively in the remaining questions. Consequently, in cluster 1 high values prevail in the first group of questions, in cluster 3 they prevail in the remaining questions. So the overrepresentation map for this 2×23 subtable is close to a straight-regular structure.

TABLE 10.5. Questions from the small business questionnaire

Number	Question
3	Has the territorial scope of activities declined?
5	Has the number of employees declined?
6	Is the activity unprofitable?
7	Has the level of profitability of services declined?
12	Is there a strong competition in the market?
13	What is the firm's position in the market?
How important are factors deciding the level of services?	
16.1	Personnel qualifications
16.2	Up-to-dateness of equipment
16.3	Raw materials and other materials
16.4	Technologies for service provision
16.5	Location of the firm
How important are elements of the business environment?	
17.1	Bank services
17.2	Tax system
17.3	Possibilities of choice of suppliers
17.4	Possibilities of gaining employees
17.5	Governmental policy
17.6	Local government's policy
How important are elements affecting the firm's development?	
18.1	Ever increasing requirements of customers
18.2	Growing competition
18.3	Greater financial possibilities
18.4	Better conditions in terms of premises
18.5	Personnel qualifications development
21	Is the firm affiliated?

A similar situation occurs for the 2×23 subtable formed by clusters 2 and 4. Both clusters have opposite characteristics, the only difference is in the partition of the set of questions which determines the structure. In this case, the first question subset includes the first eight questions (from 17.1 to 18.5).

Such a property of data tables (always treated as bivariate probability tables) can be expressed in the language of mixtures. In our case the data table is the sum of two probability tables corresponding to straight-regular structures:

$$p_{i,j} = a_1 p^1_{i,j} + a_2 p^2_{i,j},$$

where $p^k_{i,j} = 0$ if row i does not belong to the k-th structure, $k = 1, 2$.

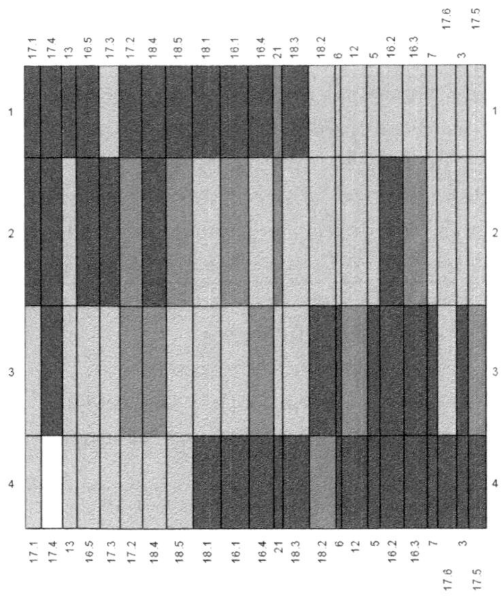

FIGURE 10.6. The overrepresentation map for the small business data, after SGCL and the aggregation of rows (question number on the horizontal axis, cluster number of firms on the vertical axis)

When the table contains values of variables instead of probabilities this property means that the analyzed population can be partitioned into two subpopulations characterized by different data structures.

Now the question arises: why does a firm assume a particular structure. To answer this question a new variable is generated which characterizes the affiliation of a given firm to one of the four row clusters provided by the SGCL. This variable is found to be correlated with several variables from the same questionnaire which characterize the firms. The respective two-way contingency tables are calculated and the GCA applied to them. The most interesting result is obtained for the variable which represents the branches. Therefore, the next hypothesis is that the detected pair of structures in the data is just the effect of mixing the different branches in the data.

10.4.4 SGCL results for the particular branches

The GCA procedure was applied to the data subtables restricted to particular branches (cf Sec. 10.4.2). Table 10.6 shows the optimal permutations of variables for these subtables as well as for the whole data (that is for the table which includes the pooled three years data). As the particular branches

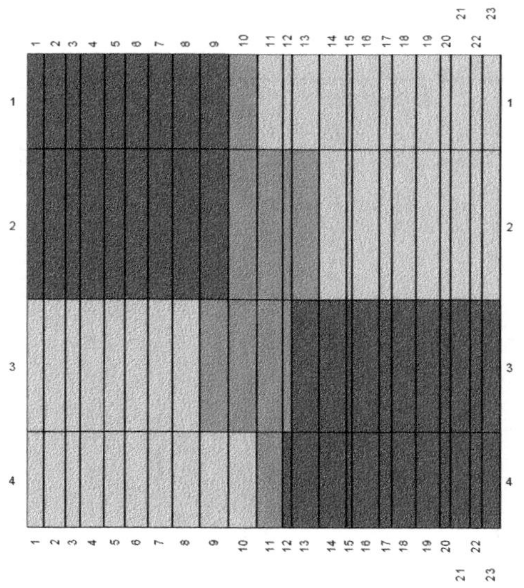

FIGURE 10.7. The overrepresentation map for discretized binormal probability table with the same marginals and correlation coefficent as in the table from Fig. 10.6

have specific requirements, it is not surprising that the positions of particular variables in the optimal permutations differ significantly. The extreme positions are particularly important, because they practically define the detected structures.

Particularly great differences are visible at one end of these permutations. It is interesting that the positions are changed not only by factors influencing the development of these firms, but also by variables which characterize the firms' status. For example, the very important variable 13 (firm's position in the market) may be meaningless in the detected structure when all answers are similar - this occurs in branch 4. The conclusion is therefore that important development factors as well as the description of the development status are specific for every branch. On the other hand, at the opposite end of the permutations, the three variables 17.5, 17.6 and 3 usually appear (for each branch). Consequently, this effect is also present in the whole data table. Moreover, the structures detected for particular branches appear also reversely regular with characteristics similar to those described in the previous section. Therefore the branch structures are also mixtures of structures. Why firms choose one structure or another is still an open question. The analysis of the questionnaire does not provide the answer. The small number of firms in the

TABLE 10.6. Positions of the variables in the optimal permutations obtained for the branch subtables.

Var. no	Whole data	Branches 1	2	3	4	Var. no	Whole data	Branches 1	2	3	4
17.1	1	19	2	3	19	18.3	13	12	12	12	15
17.4	2	10	21	1	1	18.2	14	14	14	16	4
13	3	2	3	4	13	6	15	17	7	7	12
16.5	4	5	13	2	3	12	16	18	19	6	18
17.3	5	1	1	17	5	5	17	13	18	18	21
17.2	6	20	6	14	2	16.2	18	16	8	21	6
18.4	7	6	5	9	8	16.3	19	8	16	20	11
18.5	8	7	4	8	7	7	20	15	22	15	20
18.1	9	9	11	10	14	17.6	21	23	23	22	16
16.1	10	11	10	11	9	3	22	21	17	19	22
16.4	11	4	15	5	10	17.5	23	22	20	23	23
21	12	3	9	13	17						

sample did not allow further partitions. Therefore further investigations with a larger sample of firms and an expanded questionnaire would be interesting.

10.4.5 Some final remarks

It should be emphasized that the identification of mixtures is a complex and difficult task. The example presented above shows that the GCCA can help to separate monotone trends in data. In particular, this is possible when the objects linked with different trends form separate clusters (this follows from the nature of cluster analysis). Generally, a monotone trend present in data can be distorted by the influence of various factors:

- correlated groups of variables forming different structures,
- gross outliers,
- small, random but numerous deviations from the main trend,
- different monotone trends.

The identification of these factors is not easy and needs further investigation and development of special methods, but these problems are beyond the scope of this chapter, which concentrates on the grade cluster analysis. A method for detection of outliers is described in Chapter 9.

10.5 Cluster analysis and the detection of an imprecisely defined trend

Let us note that the GCCA results, however formally expressed , can also provide a very "rough" description of trends or relations (i.e., dependence structures) which are not precisely formalized. Very often researchers must work with concepts which cannot be precisely measured (i.e., the social and psychological sciences), and strict formulation of complex, multivariate relations is impossible. Moreover, a strict formulation is not always necessary. Very often practitioners are satisfied given only a general trend, which can be very helpful in explanation of a complex phenomenon.

Just how the GCCA methods can be helpful in such cases will be illustrated using the results of the application of the grade clustering methods for the analysis of data concerning Polish retail trade firms described in the next section.

10.5.1 The use of sources of capital by retail trade firms in Poland

The data table analyzed here comes from the research of Dr A. Mielczarek (1999) of the Institute of Home Market and Consumption. The aim of this research was to detect:

- which capital sources are preferable, and by what kind of firms,
- what are the effects of the choices on the firm's economic condition,
- whether the detected structures (relations) are stable in time.

As one can see, these goals are not expressed formally in mathematical language. The same non formal language is also expected in the formulation of results.

The questionnaire included questions of two kinds. The first characterized various forms of capital sources obtainable in Poland. The second provided information about inner capital stuctures of the analyzed firms and their profitability. The list of all questions used in this analysis is included in Table 10.7. For all questions, the firms provided answers for three consecutive years: 1996-1998. The initial data table was restricted to those firms which provided a complete set of answers, in consequence, the sample used is composed of 66 firms (rows in the data table).

10.5.2 Typology of firms for the pooled, three-year data

The GCA applied to this data table revealed that it has two specific properties. First, variables which correspond to the same question, but for different years

TABLE 10.7. Analyzed questions

Question no	Description
1	Generation of financial surplus
3	Profitability level
4	Share of firm's own capital
5	Share of long-term obligations
6	Share of retail obligations
7	Bank credit
11	Credit guaranty fund
12	Expansion of partner base
15	Leasing
19	Factoring

are usually ordered by GCA one after the other. The only exceptions were questions 12 and 19. This means that the similarity within years is much stronger than the similarity among questions.

The second observation is that two questions: 19 and 1 provide numerous exceptions (i.e., there are many dark rectangles far away from the diagonal) from the straight regularity rule. It seems that these questions form outliers in this structure. The deletion of either 1 or 19 from the data table increases the value of the maximal correlation coefficient ρ^*. After the removal of both, the maximal ρ^* increases from the initial value 0.2233 to 0.2502. As the removal of significant variables diminishes the dependence in the data table, this increase in ρ^* confirms that both of these questions are not strongly correlated with the revealed structure. Therefore, these two questions are not taken into account in further analyses.

It is interesting that the optimal permutation of the remaining columns (i.e., questions by year) did not change, and was as follows: 15(97), 15(98), 15(96), 7(98), 7(96), 7(97), 5(96), 5(97), 5(98), 6(97), 6(98), 6(96), 4(96), 4(97), 4(98), 12(96), 11(96), 11(97), 11(98), 12(97), 12(98), 3(98), 3(97), 3(96). The corresponding overrepresentation map is shown in Fig. 10.8. It shows that similar firms are placed close to one another.

To an optimally arranged data table, the single cluster analysis was applied. The horizontal lines in Fig. 10.8 mark the boundaries of particular clusters. In this case, three was arbitrarily chosen as the number of clusters.

The distribution of dark and light rectangles provides for an interpretation of the generated classes. Comparative analysis of the average values of the variables in the particular clusters (shown in Table 10.8) confirms the following interpretation:

- CLUSTER 1: In this cluster firms are characterized by the greatest number of leasing contracts (15) and bank credits (7), and a high share

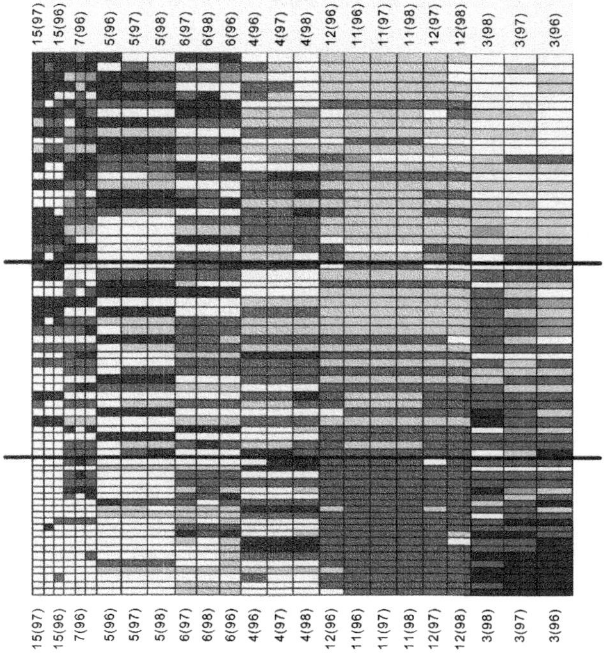

FIGURE 10.8. The overrepresentation map for the pooled three-year capital sources data after GCA (question by year codes on the horizontal axis)

of obligations of both types (5 and 6). All these characteristics are linked with the lowest average profit (3).

- CLUSTER 3: A very low number of leasing contracts and bank credits; also a very low share of both obligations, and the highest average profits.

- CLUSTER 2: This cluster is an intermediary level between cluster 1 and 3. The numbers of leasing contracts and bank credits are lower than those in class 1 and higher than those in class 3. Also the share level of retail obligations and profitability are lower than those in class 1 and higher than those in class 3. However, the long-term obligations are similar to those in class 1.

Note that the cluster interpretation does not include three questions: 4, 11 and 12. The reason is that their averages and distributions are very similar in all classes. Moreover, the averages for questions 11 and 12 are very near to their maximal values, which means that they are almost constant. Maximal values denote in this case, that these capital sources were not used. In other words, credit guaranty funds and the expansion of the partner base are used very rarely by Polish firms. In consequence, all three questions were acknowledged

TABLE 10.8. Average values of the variables in the capital sources data analysis by clusters, and across clusters

Quest. code	Clusters 1	2	3	All clust.	Quest. code	Clusters 1	2	3	All clust.
15(97)	1.65	0.78	0.00	0.85	4(96)	2.26	2.04	1.65	2.00
15(98)	1.35	0.65	0.05	0.71	4(97)	2.17	2.04	1.70	1.98
15(96)	1.39	0.74	0.10	0.77	4(98)	1.96	2.00	1.70	1.89
7(98)	1.48	0.70	0.30	0.85	12(96)	1.87	1.91	1.85	1.88
7(96)	1.39	0.61	0.40	0.82	11(96)	1.96	2.00	2.00	1.98
7(97)	1.30	0.74	0.35	0.82	11(97)	1.91	2.00	2.00	1.97
5(96)	2.43	2.09	1.15	1.92	11(98)	1.91	2.00	2.00	1.97
5(97)	2.35	2.35	1.10	1.97	12(97)	1.87	1.96	1.90	1.91
5(98)	2.39	2.39	1.10	2.00	12(98)	1.83	1.91	1.90	1.88
6(97)	2.09	1.61	1.30	1.68	3(98)	1.65	2.74	2.90	2.41
6(98)	2.13	1.70	1.30	1.73	3(97)	1.87	2.65	3.20	2.55
6(96)	2.00	1.52	1.30	1.62	3(96)	2.00	2.74	3.40	2.68

as insignificant in the generated trend. However, their behaviour should be interpreted with care and verified in future analyses using a larger sample.

Nevertheless, it should be noted that the quality of the presented typology is good, because the aggregation of the data table according to the cluster analysis diminishes the value of the correlation ρ^* only slightly - from 0.2502 to 0.2215. This means, that the aggregation resulted in only a small loss of information.

10.5.3 Firm typologies for annual data

The next step in the analysis was the application of the same method as before, but now for each of the three data subtables separately. Each data subtable contained data restricted to one year only. The optimal permutations of questions as well as the maximal values of ρ^* are shown in Table 10.9. The values

TABLE 10.9. Results of the GCCA for annual data tables

Year	Optimal permutations								max ρ^*
1996	15	7	5	6	4	11	12	3	0.2540
1997	15	7	5	6	4	11	12	3	0.2587
1998	7	15	5	6	11	12	4	3	0.2620

of ρ^* are very similar. The permutations for 1996 and 1997 are identical; the permutation for 1998 differs a little. However, the interpretation of the particular firm typologies, which follows, remains identical to the interpretation generated for the pooled, three-year data. (As in the previous analysis, three was chosen as the number of clusters.) Based on the interpretation, the firm typology provided by the SGCL is stable in time. In other words, there is only one typology of firms, invariant across time.

10.5.4 Relationship between the generated firm typology and the firm profitability

Profitability level appears to be one of the most important firm characteristics in the typology presented above. It occupies the extreme position in the optimal permutation of the variables, which confirms its great discriminating power among the firms. Although this extreme position is stable in the analyzed years, this does not necessarily mean that firm profits are also stable.

The SGCL method was applied once more, this time to the data table which includes only profit values (3(96) to 3(98)). An interesting result is that the optimal permutation of the variables is identical with the ordering of years. This suggests that some trend across time exists. This trend is described well by the result of the cluster analysis. Fig. 10.9 shows the overrepresentation map of the data table after the optimal permutations; the horizontal lines mark the cluster boundaries. In this case three clusters was chosen because a greater number does not change the interpretation of the data trend. The problem of the determination of the proper number of clusters will be discussed in the next chapter.

The analysis of this map and the average profit in particular clusters leads to the following typology.

- CLUSTER 1: Formed by firms with decreasing profits. Moreover, this decrease is rather strong and the starting levels (in 1996) were comparatively high.
- CLUSTER 2: Firms in this class are characterised by various profitability levels, which for individual firms remain stable across time.
- CLUSTER 3: This cluster has characteristics opposite to cluster 1. It includes successful firms with continually increasing profits. Moreover, these firms started from rather low profit levels. Unfortunately, this cluster contains only seven firms.

Let us note that this results illustrate very well that the inferences based on overrepresentations and the inferences based on average values of variables in particular clusters are not identical. Looking at cluster 2 in the overrepresentation map it seems that in this clusters the profitability increases. But comparing profitability for particular firms we see that they are constant in

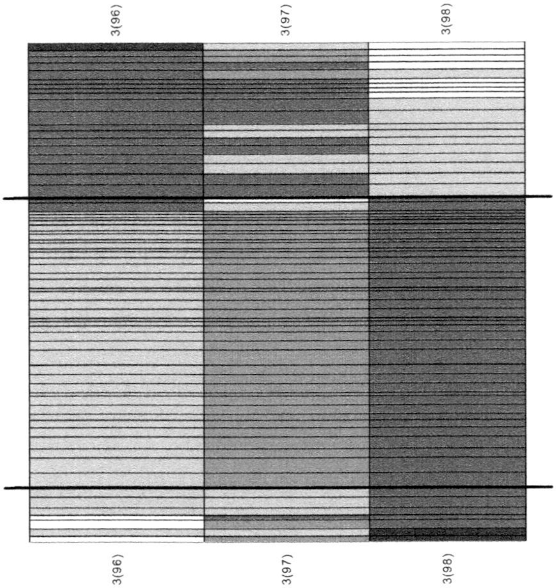

FIGURE 10.9. The overrepresentation map for the profitability data table after GCA (question code on the horizontal axis)

TABLE 10.10. Average values for the variables in the typology of profitability data

Cluster	3(96)	3(97)	3(98)
1	3.421	2.8947	1.9473
2	2.45	2.4	2.45
3	2	2.4285	3.4285

time for almost all firms from this cluster. Previous experience shows that both inferences usually agree. This example is rather the exception. It is caused by the discretization used for the overrepresentation map. Thanks to it the overrepresentations are classified into 3 categories although they are very similar (and all near to 1). Therefore the analysis of object-variable tables should always take into account the initial values of variables.

Now, let us create a new variable which will be equal to the cluster number for this typology. A similar variable is generated for the typology obtained for all questions (cf Sec. 10.5.2). The maximal correlation between these new variables is equal to 0.1867. However, the overrepresentation map (Fig.10.10), generated for the respective contingency table, represents a structure with straight regularity (cf Sec. 10.4). This means that these typologies are correlated, but that the dependence is not very strong.

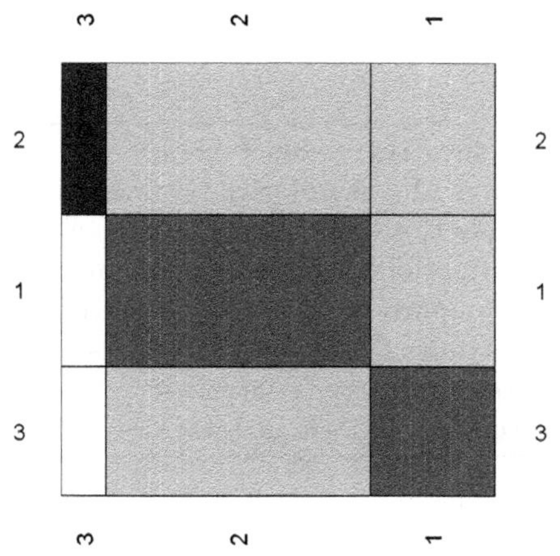

FIGURE 10.10. The overrepresentation map for the contingency table of variables: cluster numbers for the pooled three years data (on the vertical axis), cluster numbers for the profitability data (on the horizontal axis)

10.6 On GCCA application to various data sets

For any finite data table with finite and nonnegative values, the GCCA procedures are always calculable regardless of the meaning of the data. Formalists can ask whether application of the GCCA to data sets different from bivariate probability tables is justified; perhaps the GCCA procedure applied to the data containing values of several variables are just calculable.

The following citation from ([35]) helps to answer this question - "Any classification (clustering) is a division of the objects into groups based on a set of rules - it is neither true nor false and should be judged on the usefulness of results". As it is hard to imagine useful results without practical interpretation, interpretability becomes a main criterion for the evaluation of clustering procedures.

The GCCA, like all clustering methods, divides the set of objects which correspond to the rows or columns of the input data table into clusters whose members are similar. As described previously (see Sec. 10.2), the difference in regression functions for pairs of objects (or their formal counterparts for non probability data tables) .serves as the measure of dissimilarity between them. If the data table contains values of variables which characterize the set of objects, then the "regression function" (the notation "regression function" will be used for nonprobability tables) for objects is just the weighted sums of these values. Hence the regression differences work similarly to any distance function commonly used in cluster analysis. That is, for two objects characterized by similar values of particular variables, the regression difference (disimilarity) is near zero.

> The differences in the "regression function" for pairs of objects serve as the measures of dissimilarity between them . This is the main interpretation of this function in the GCCA context. This excludes only the straightforward use of nominal variables, because these comparisons require ordered values. In other words, these functions and consequently the GCCA results are always meaningful (i.e., interpretable) in the context of correspondence-cluster analysis if the variables used are at least ordinal.

The interpretation imply also another. When a data table (objects-variables) possesses a strong monotone trend, then consecutive groups of objects in the optimal ordering (and consequently the GCCA clusters) are characterized by particularly great (albeit relatively) values of corresponding, also consecutive groups of variables. Let us note that if the values of variables are treated as "measurements of similarity to some standard" (i.e., not in the sense of mathematical theory of measurement), this interpretation becomes almost identical with the interpretation for probability data. This provides additional support for the argument that the GCCA results are always interpretable.

The problem of variable normalization is linked with the regression based dissimilarity measure. The total sums of values for particular variables play the role of weights in the GCA. Therefore, one may suppose that by the appropriate normalization all these sums should become 1. Based on this author experience, three possibilities exist:

- the original data and the normalized data provide very similar GCCA results (at least relative to the results interpretation),
- the role of significant variables may be diminished when the variables are not normalized,
- the significance of variable may be overestimated by the GCCA for normalized data. This frequently occurs when a variable has only the values of 0 and 1 (where 1 indicates a very rare attribute). Generally, it depends on the data whether an inference about the monotone trend can be based on such an attribute. In cluster analysis this situation corresponds to the problem of the identification of very small clusters, i.e., are they really a cluster or only outliers?

The obvious conclusion is that there are no apriori reasons as to whether the normalization should be applied or not. Only the comparison of the results for normalized and nonnormalized data can provide the answer. The only practical hint is that a great disparity in value ranges for particular variables is not advisable. If a significant monotone trend exists in the data, small differences in these ranges have no significant influence on the final results. Of course there are many methods of variable normalization, but any of them introduces a subjective factor whose influence on the results should be very carefully investigated.

10.7 Appendix

10.7.1 An algorithm for optimal clustering

Let X and Y be the respective row and column variables corresponding to the data table $T(X, Y)$. The consecutive steps of the procedure are the following:

1. Let $X_0 = X$ and $Y_0 = Y$.

2. The rows of $T(X, Y_i)$ are aggregated according to the single clustering for rows (let us recall that table T must be arranged according to the GCA before aggregation). As a result, we obtain the new row variable X_{i+1} and the new table $T(X_{i+1}, Y)$. Note that the columns of this new table are not aggregated (we return to the columns of the initial table).

3. $T(X_{i+1}, Y)$ are permuted by the GCA, and its columns are aggregated according to the SGCL for columns. We get the new column variable Y_{i+1} and table $T(X, Y_{i+1})$.

4. Steps 2 and 3 are repeated for $i = 1, ...$ until the consecutive solutions (pairs of clusters for rows and for columns) stabilize.

Proposition 10.1 *The procedure for optimal clustering always converges.*

Proof:

Let us note that the values of ρ^* for the consecutive (aggregated) tables $T(X_i, Y_i)$ form a nondecreasing sequence. If the data table is finite then the number of possible clustering results is also finite. Consequently the procedure must end. For infinite tables the proof is identical to that for the proposition 6.5.

Our previous experience shows that the procedure converges quickly; however, much more testing on very large data tables will be necessary.

10.7.2 Bibliographical remarks

The literature concerning cluster analysis is enormous. Even a short review of the subject would require a separate monograph. A good example of such a book is [3]. A survey on recent advances can be found in [11].

The clustering methods discussed in this chapter were developed by the author of this chapter and are described in [22] and [19]. There are many applications of these clustering methods in various real data analyses. Those applications shown in this chapter are derived from papers [18], [16] and [19].

11

Regularity and the number of clusters

Alicja Ciok

11.1 Introduction

In this chapter the concept of ideally regular two-way data tables will be introduced and discussed. It is a generalization of the ideal regularity of curves introduced in Chapter 6. It will be shown that members of a parametrized family of two-dimensional distributions from the Bivariate Lilliputian Model ($\mathbb{BLM}$) which are based on parabolas are ideally regular data tables (Sections 11.2-11.3).

In the $\mathbb{BLM}$ this family plays a similar role to that played by the parabola family in the $\mathbb{ULM}$. Its members form the reference points in the characterization of the cluster structures present in data tables. The properties of this family are particularly useful in the detection of the proper number of clusters for rows or columns of data tables (see Sections 11.4-11.5).

The definition of natural clusters is introduced via the intervals characterized by constant grade regression functions. As in the univariate case, ideal regularity is associated with infinite divisibility of data tables lacking natural clusters. Finite data tables, even transformed into continuous copulas, always have natural cluster structures. Therefore, they can be considered as the approximations of ideally regular tables at most. A procedure which helps to determine whether a finite data table is aproximatelly regular is presented in Sec. 11.5. This procedure helps also to recognize a clustering structure present in the data.

11.2 Generalization of the parabola family from the $\mathbb{ULM}$

Let X and Y be continuous random variables from the $\mathbb{ULM}$. Let their joint distribution be described by cdf F and density f. We would like to construct a distribution which has the following properties:

- all conditional distributions belong to the parabola family (cf. Chapters 5 and 6),

- the marginal distributions are uniform on $[0, 1]$ (i.e., the distribution belongs to the Dependence Lilliputian Model ($\mathbb{DLM}$)),
- it is symmetric: $f(x, y) = f(y, x)$ for $x, y \in [0, 1]$.

Let us assume that the density f has the following form:

$$f(x, y) = a_1 xy + a_2 x + a_3 y + a_4$$

Since the density corresponding to the cdf in the form of a parabola:

$$\alpha x^2 + (1 - \alpha)x$$

is equal to a linear function:

$$2\alpha x + 1 - \alpha$$

our conditional densities must take this form. The assumption of the uniformity of the marginals leads to the condition:

$$\int_0^1 f(x, y)dx = \int_0^1 f(x, y)dy = 1 \text{ for every } x, y \in [0, 1]$$

Consequently the bivariate density and both conditional densities are equal. These conditions and the specific form of the density leads to the following family of pairs of variables (X_β, Y_β). Their joint densities f_β are given by the formula :

$$f_\beta(x, y) = \beta + 2(1 - \beta)(x + y - 2xy) \quad x, y \in [0, 1]. \tag{11.1}$$

Of course $f_\beta(x, y) = f_\beta(x|Y_\beta = y) = f_\beta(y|X_\beta = x)$.

The respective cdf F_β is equal to:

$$F_\beta(x, y) = xy(\beta + (1 - \beta)(x + y - xy)) \quad x, y \in [0, 1].$$

A pair of variables from the $\mathbb{BLM}$ which has the joint density given by formula (11.1) will be called the paraboloidal.

The paraboloidal pair of variables (X_β, Y_β) fulfills all requirements formulated above. In particular, the conditional densities are given by the linear formulas (cf. (11.1)), consequently the conditional cdfs are parabolas.

Since a density must be nonnegative, parameter β must fulfill the following requirements: $f_\beta(0, 0) = f_\beta(1, 1) = \beta \geq 0$ and $f_\beta(1, 0) = f_\beta(0, 1) = 2 - \beta \geq 0$. This means that the parameter β must belong to the interval $[0, 2]$.

- If $\beta = 1$ then the density f_β is constant and consequently the distribution is uniform on the unit square ($F_1(x,y) = xy$).
- If the order of variable X_β (or Y_β) is changed into its opposite, the density preserves its form, but the parameter β changes into $2-\beta$:

$$f_\beta(1-x,y) = f_\beta(x,1-y) = f_{2-\beta}(x,y)$$

The conditional densities $f_\beta(y|X_\beta = x)$ for $\beta = 1.5$ and several values of x are shown in Fig. 11.1.

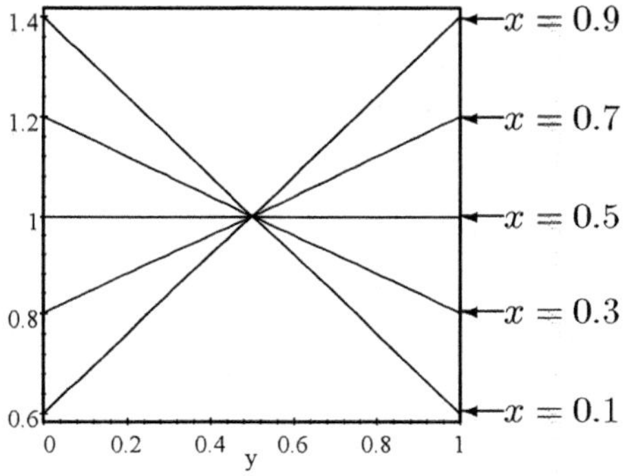

FIGURE 11.1. The conditional densities $f_\beta(y|X_\beta = x)$ for $\beta = 1.5$ and for $x = 0.1, 0.3, 0.5, 0.7, 0.9$

The conditional cdfs for the same parameter $\beta = 1.5$ and for the same values of X_β are shown in Fig. 11.2.
Figure 11.2 illustrates a general property of the paraboloidal family of bivariate distributions. The conditional distributions are ordered according to the respective values of the concentration index *ar*. The cdfs corresponding to small values of X_β ($x < 0.5$) are concave and lie over the diagonal (the respective values of *ar* are negative). The distribution in the middle ($x = 0.5$) is uniform and the cdf lies on the diagonal. The cdf corresponding to the larger values ($x > 0.5$) are convex, lie below the diagonal, and the values of *ar* are positive.

Another property of this paraboloidal family is that their members (X_β, Y_β) are TP_2 (totally positive dependent of order two - cf. Chapter 7) for $\beta \in (1,2)$ and they are TN_2 (totally negative dependent) for $\beta \in (0,1)$.

The density f_β for $\beta = 1.5$ is shown in Fig. 11.3. The next figure (11.4) shows the corresponding cdf F_β.

Since the grade regression functions (cf. Definition 8.5), in the case of continuous variables from the ULM, are equal to the common regression: $E(Y_\beta|X_\beta = x)$ then

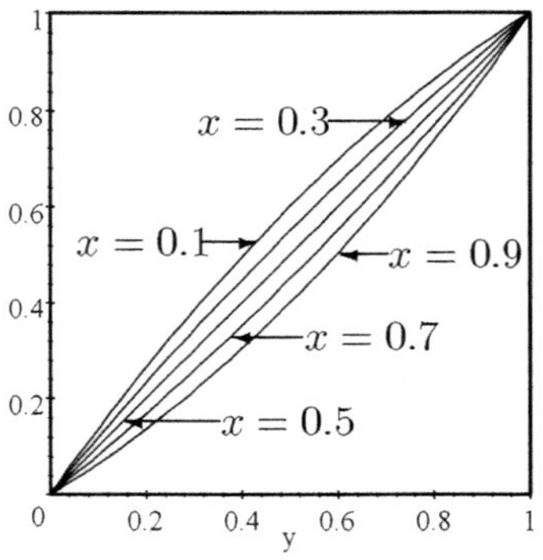

FIGURE 11.2. The conditional cdfs $F_\beta(y|X_\beta = x)$ for $\beta = 1.5$ and for $x = 0.1, 0.3, 0.5, 0.7, 0.9$

$$r^*_{Y_\beta|X_\beta}(x) = \frac{1}{6}(4 - \beta + 2x(\beta - 1)), \quad x \in [0,1]$$

These regressions are increasing (decreasing) functions of x iff $\beta > 1$ (iff $\beta < 1$). If $\beta = 1$ then $r^*_{Y_\beta|X_\beta} \equiv 0.5$. The relationship between parameter β and $2 - \beta$ can be expressed in the form:

$$r^*_{Y_\beta|X_\beta}(x) = 1 - r^*_{Y_{2-\beta}|X_{2-\beta}}(x), \quad x \in [0,1].$$

Due to the symmetry, the analogous formula can be calculated for $r^*_{X_\beta|Y_\beta}$.

The normalized grade correlation curve (cf. Definition 8.6 and Chapter 10) is given by the formula:

$$2C^*_{cor(Y_\beta:X_\beta)}(x) = \frac{x}{3}\left(4 - \beta + (\beta - 1)x\right), \quad x \in [0,1]. \tag{11.2}$$

This normalized correlation curve is a parabola; if $\beta > 1$ it is convex (if $\beta < 1$ it is concave). Of course the identical calculation can be made for $C^*_{cor(X_\beta:Y_\beta)}$. Due to the symmetry, the repective formula is analogous to formula (11.2).

When the correlation curves are known, it is easy to calculate the grade correlation coefficient $\rho^*(X_\beta, Y_\beta)$ (cf. Proposition 8.6), which reduces to the simple formula:

$$\rho^*(X_\beta, Y_\beta) = \frac{\beta - 1}{3}$$

Note that in this case $\rho^*(X_\beta, Y_\beta) = \rho^*_{\max}(X_\beta, Y_\beta)$. Since $\beta \in [0,2]$, the values of $\rho^*(X_\beta, Y_\beta)$ must belong to the interval $[-\frac{1}{3}, \frac{1}{3}]$. The value of $\rho^*(X_\beta, Y_\beta) > 0$

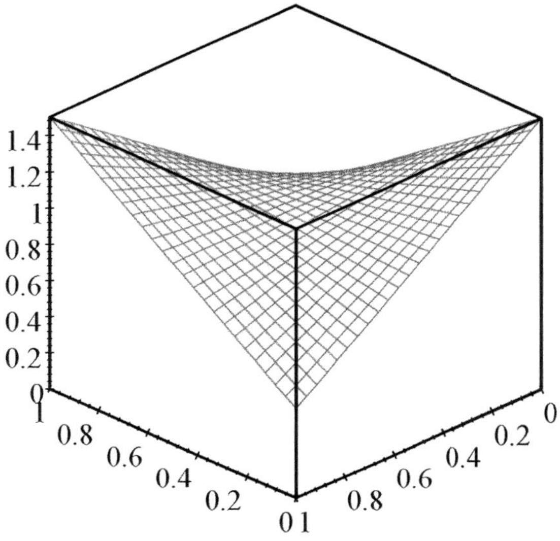

FIGURE 11.3. Density $f_{1.5}(x, y)$

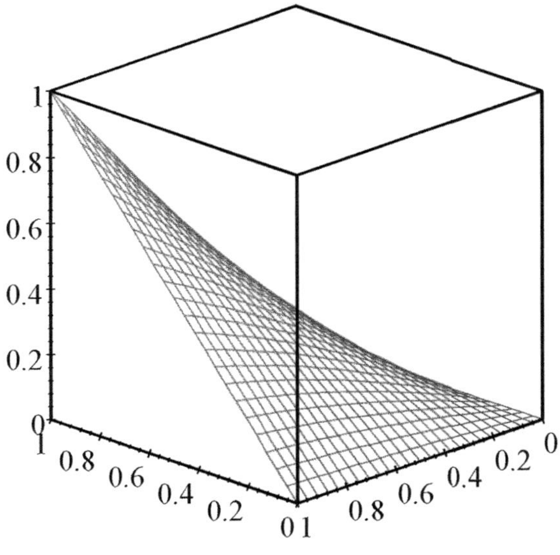

FIGURE 11.4. Cdf corresponding to density $f_{1.5}$

iff $\beta > 1$. Once more, the duality between β and $2-\beta$ can be observed now in the values of the correlation coefficient ρ^*:

$$\rho^*(X_{2-\beta}, Y_{2-\beta}) = -\rho^*(X_\beta, Y_\beta)$$

This equality is in agreement with the fact described above that changing from β to $2-\beta$ is equivalent to the reversal of value order for X_β or Y_β.

It would be interesting to observe the overrepresentation maps for the paraboloidal pairs of variables (X_β, Y_β). For this purpose both variables must be discretized. This discretization problem will be considered in the next section.

11.3 The ideal regularity of two-way data tables

Let us consider the single clustering procedure (SGCL) described in Chapter 10. According to the SGCL the new categories for the row (column) variables are determined by the optimal discretization of the respective grade correlation curves. The first step in this procedure is to arrange the values of these row and column variables in nondecreasing order of the respective grade regressions r^*. Paraboloidal variables have their regression functions properly arranged if $\beta > 1$. In the following we will consider only the paraboloidal variables with such parameters.

Generally, in the family of paraboloidal variables, both normalized correlation curves are parabolas which are convex for $\beta > 1$. It is shown in Chapter 6 that for parabolas, the uniform discretization is optimal. Recall that a uniform discretization means the partitioning of $[0,1]$ into subintervals of identical lenghts. Figure 11.5 shows the overrepresentation map for the paraboloidal pair of variables (X_β, Y_β), where $\beta = 1.5$ and each variable is uniformly (hence optimally) discretized into 20 categories.

Figure 11.6 shows the overrepresentation map for the same pair of paraboloidal variables also uniformly discretized, but now into 10 categories. The loss of information corresponding to both discretizations is very small. The value of ρ^* for the initial continuous distribution was equal to 0.166(6); for the discretized variables the value of ρ^* decreased only to 0.1658 and to 0.1633 respectively for 20 and 10 categories.

Comparing both overrepresentation maps it is difficult to find significant differences other than the granularity. This property also holds when other numbers of clusters are used.

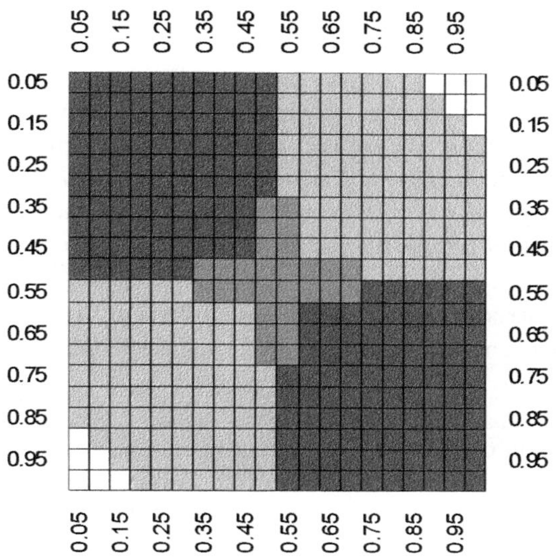

FIGURE 11.5. The overrepresentation map for the variables $X_{1.5}$ and $Y_{1.5}$ discretized into 20 categories

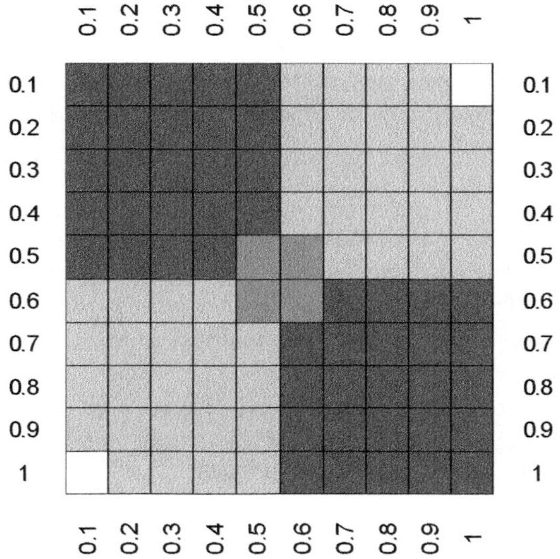

FIGURE 11.6. The overrepresentation map for the variables $X_{1.5}$ and $Y_{1.5}$ discretized into 10 categories

The comparison of the clusters provided by the SGCL procedure, with m and mk as the numbers of clusters for rows (columns), reveals a behaviour identical to that presented in Chapter 6. The clustering for mk does not change the cluster boundaries, but only further divides the clusters obtained by the clustering for m. Since this property does not depend on the values of m and k, such a property can be interpreted as *infinite divisibility* (analogous to the univariate case). Consequently, in the bivariate case the concept of ideal regularity can be analogously introduced.

Before the introduction of the formal definition, let us note that the uniform grade transformation reduces any bivariate distribition to an element of $\mathbb{BLM}$.

The definition of a paraboloidal family can be generalized to the case of any bivariate distribution. It is defined then as the set of bivariate distributions whose grade distributions are defined by formula (11.1)

The grade correlation curves are given by (11.2) and all conclusions drawn in the $\mathbb{BLM}$ case are still valid. Let us also recall that two-way data tables can be treated (after appropriate normalization) as bivariate probability tables, and they may be finite as well as infinite (as discussed in Chapter 10). Therefore, the definition of regularity can be introduced in the general case.

Definition 11.1 *Two-way data tables will be called ideally regular if both corresponding normalized grade correlation curves are ideally regular.*
If only one correlation curve is ideally regular, the data table will be called ideally semi regular.

According to definition probability (density) tables corresponding to pairs of paraboloidal variables (they will be called paraboloidal tables) are ideally regular. It stems from the fact that in this family all correlation curves are parabolas and that all parabolas are ideally regular.

In the univariate case the ideal regularity concept and deviations from it were exploited in the determination of the proper number of clusters. We will now try to repeat this for the bivariate case.

11.4 Regularity and cluster detection

Let us start with the formalization of the idea of natural clusters. It is a generalization of the natural categories of variables introduced in Chapter 6.

These categories are determined by the intervals with constant derivatives of the concentration curves.

> Normalized grade correlation curves are concentration curves. Therefore it is natural to regard the intervals of values for row (column) variables as clusters if they are characterized by constant values of the grade regression functions.

Identical vectors of conditional densities imply that the corresponding values of the row (column) variable belong to the same cluster. However, equality of regressions can coexist along with the diversity of their components.

It should be emphasized that differences between the values of the grade regression functions (precisely their modules) play the role of the dissimilarity measures between the values (i.e., categories) of the variables. Since the GCA always arranges these values (rows or columns of the bivariate probability table) in nondecreasing order of the grade regression functions, we are able to say that they are ordered according to their similarities.

The main expectation in any cluster analysis is to be able to divide the data into blocks of very similar elements, with blocks differing as greatly as possible. In Fig. 11.7 we see an extreme example of such a data structure. This figure shows the overrepresentation map with visible regressions for both discrete variables. Each variable has 9 categories. There are three natural clusters of rows (and columns as well). The vectors of conditional probabilities and consequently the regressions for rows and for columns are identical inside each particular cluster.

This example was created in such a way that each cluster differs strongly from the others. The values of the regressions (0.1667, 0.5, 0.8333) corresponding to particular clusters of rows (columns too, since the table is symmetric) reflect well these differences. Moreover, due to these differences the grade correlation ρ^* attains the high value of 0.8889.

> It is a very important property that when the probabilities inside particular clusters of rows and columns are aggregated (summed), the regressions and the value of ρ^* do not change. The respective overrepresentation map after aggregation looks like the map before aggregation, except the granulation. In other words, in this case the optimal aggregation does not cause any loss of information.

An immediate conclusion is that the data table after aggregation according to the structure of the natural clusters retains the monotone trend present in the data, but loses the cluster structure. Therefore, cluster structures and dependence structures are different characteristics of data tables.

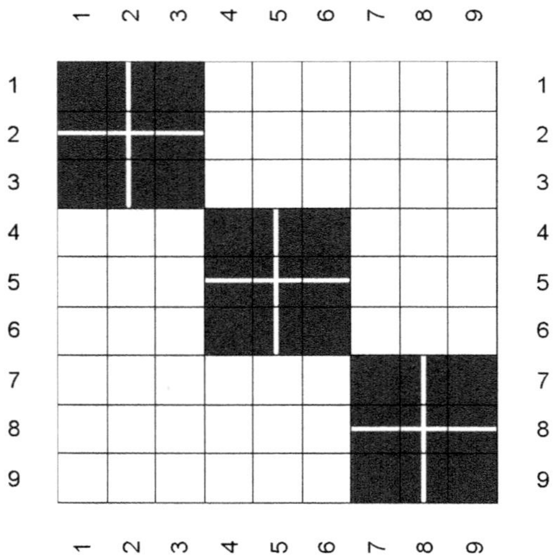

FIGURE 11.7. An example of an overrepresentation map for the three cluster data structure

> The complete description of two-way data tables (bivariate distributions or otherwise) must include the following components:
>
> - the measures of monotone dependence between row and column variables (e.g., $\rho^\star$);
> - the type of monotonicity (straight or reverse);
> - a granularity, meaning the existence (or absence) of natural clusters.

The paraboloidal tables as well as all ideally regular data tables have no natural clusters. Therefore, for ideally regular tables, any number of clusters is fine if the deviation of the aggregated data (according to the clustering results) from the original table is acceptable. If tables are only semi regular then the same conclusion is valid only for rows or only for columns (it depends on which correlation curve is regular)

Like the parametrized family of ideally regular curves in the univariate case, the parametrized family of paraboloidal tables can be used to confirm that there is no natural granularity in the data. For infinite data tables (i.e., bivariate continuous distributions) one may exploit the procedure similar to that used for concentration curves. Instead of checking the formula from the definition of ideal regularity, it is much easier to find the paraboloidal table which is most similar to the given table. If the similarity is great, the data can be regarded as approximately ideally regular.

When the distribution is not ideally regular, but the grade regressions (the grade correlations as well) and the value of ρ^* do not change after aggregation according to the clustering, then the real structure (the proper number of clusters and the proper clusters) are revealed.

In the case of finite data tables (the most commonly used), a straightforward comparison with the paraboloidal continuous family is not possible. To this purpose an additional procedure is needed which will be presented in the following section.

11.5 Cluster detection in finite data tables

The paraboloidal data table is a perfect example of ideal regularity. However, the data tables one usually works with are finite and nonsymmetric. These tables are transformed into copulas which are continuous distributions; obviously these copulas have the natural cluster structure corresponding to the initial data. Therefore, they can not be ideally regular but they can approximate ideally regular (or semi regular) tables.

Let f be the density of a bivariate distribution corresponding to the pair of continuous variables (X, Y) from the $\mathbb{BLM}$. Let us assume that these variables are uniform (after transformation into the copula they are always uniform) and that the density has a more general form than the densities of the paraboloidal tables (in particular it can be nonsymmetric).

$$f(x,y) = a_1xy + a_2x + a_3y + a_4, \ x, y \in [0,1].$$

For a fixed x (y) this density as well as the regression function $r^*_{Y|X}$ ($r^*_{X|Y}$) is a linear function of y (x); the corresponding correlation curve is a parabola and consequently this curve is ideally regular.

Let $(\widehat{X}, \widehat{Y})$ be the same variable pair, but after discretization which has been determined by respective vectors $(x_1, ..., x_n)$ and $(y_1, ..., y_k)$. Let us note that both variables remain uniform after discretization. The regression function $r^*_{\widehat{Y}|\widehat{X}}$ (cf. formula (8.31)) is given by the following formula:

$$2r^*_{\widehat{Y}|\widehat{X}}(x_{i+1}) = \sum_{j=1}^{k} \frac{(y_{j+1} + y_j)p_{i+1,j+1}}{p_{i+1,\bullet}},$$

where p_{ij} $(i = 1, ..., n;\ j = 1, ..., k)$ are the probabilities of this new bivariate distributions and $p_{i+1,\bullet} = \sum_j p_{i+1,j} = x_{i+1} - x_i$. According to definition, the correlation curve consists of linear segments determined by the points $(\sum_i x_i, 2\sum_i r^*_{Y|X}(x_i)p_{i\bullet})$. Then the slopes of particular segments are equal to $2r^*_{Y|X}(x_{i+1})$. For our distribution

$$p_{i+1,j+1} = \frac{1}{2}(x_{i+1} - x_i)(y_{j+1} - y_j)[a_1(x_{i+1} + x_i)(y_{j+1} + y_j) + \\ + a_2(x_{i+1} + x_i) + a_3(y_{j+1} + y_j) + 2a_4]$$

and

$$2r^*_{\widehat{Y}|\widehat{X}}(x_{i+1}) = \frac{1}{2}\sum_{j=1}^{k}(y^2_{j+1} - y^2_j)[a_1(x_{i+1} + x_i)(y_{j+1} + y_j) + \\ + a_2(x_{i+1} + x_i) + a_3(y_{j+1} + y_j) + 2a_4].$$

Then $2r^*_{\widehat{Y}|\widehat{X}}(x_{i+1})$ for fixed y_j, $j = 1, ..., k$ is a linear function of $\frac{1}{2}(x_{i+1} + x_i)$. This provides a simple rule for determining whether a given correlation curve is derived from a parabola via discretization (in other words, whether it can be regarded as an approximation of a parabola).

> If points $(\frac{1}{2}(x_{i+1} + x_i), r^*_{\widehat{Y}|\widehat{X}}(x_{i+1}))$ for $i = 1, ..., n$ lie on a line then we can say that the corresponding correlation curve is approximately ideally regular and the rows do not form clusters.

Of course the same determination rule can be created for columns of data tables.

To show how this rule works, the profitability data table analysed in Sec. 10.5.4 will be considered once again. Its overrepresentation map with the boundaries of the chosen 3 clusters is shown in Fig. 10.9. Figure 11.8 shows the points $(\frac{1}{2}(x_{i+1} + x_i), r^*_{\widehat{Y}|\widehat{X}}(x_{i+1}))$ calculated for rows of this data table. The vertical lines mark the boundaries of the same clusters visible in Fig. 10.9.

Cluster 2 is almost ideal. The regression values in this cluster are almost constant and the only two exceptions differs very slightly from the common value. In cluster 3 the approximation of regression values by a line is also very good. Therefore we may say that this cluster rather does not contain natural subclusters. Moreover, further partitions provide the same interpretation of the obtained subclusters.

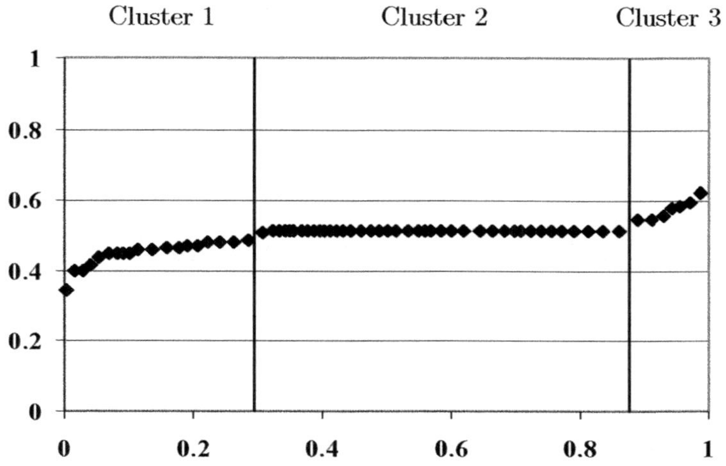

FIGURE 11.8. Points $(\frac{1}{2}(x_{i+1}+x_i), r^*_{\hat{Y}|\hat{X}}(x_{i+1}))$ calculated for rows of the profitability data table from Sec. 10.5.4

The analogous approximation by a line for cluster 1 is not very good. The first four points (on the left) differ considerably from this line, the deviation for the first point is particularly great. It is interesting that the clustering procedure for four clusters provides a partition of cluster 1 into two subclusters, the other two clusters (2 and 3) remain the same. Now the question arises: which number of clusters is better: three or four. The generation of subclusters in cluster 1 does not lead to a broader interpretation (at least if only the profitability data is considered). On the other hand, four clusters provide a better approximation of the initial data table than three clusters. The quality of this approximation is measured by the values of ρ^* after data aggregation. In this case the gain is not great (ρ^* is equal to 0.1115 and 0.1189 after aggregation of rows into 3 and 4 clusters respectively). The interpretation of particular clusters is more clear in the case of 3 clusters. Overall, it seems that three clusters is a better choice.

This data example illustrates well difficulties in determining the proper number of clusters. A clustering procedure can reveal various types of clusters in one data table.

- (almost) natural clusters,
- clusters which have an inner subcluster structures,
- clusters which are not natural and have no natural subclusters (they correspond to inner ideally regular substructures).

The proper choice needs consideration of formal aspects (quality of approximation) as well as the interpretation aspects, which are particularly important in any cluster analysis. Therefore, a procedure which determines automatically the proper number would be difficult to develop. The material presented in this chapter is a first step in this direction.

11.6 Appendix - Bibliographical remarks

In the opinion of eminent specialists "the estimation of the true number of classes is one of the most difficult problems in cluster analysis" (Bock [12]). The enormous volume of literature devoted to this problem confirms this opinion. Even a short review of the subject is beyond the scope of this book. Such a review can be found in [3].

The ideal regularity for curves and data tables; the families of ideally regular curves and data tables, the methods which help to determine the proper number of categories or clusters (introduced in Chapters 6 and 11) are ideas of the author of the present chapter.

12

Grade approach to the analysis of finite data matrices

Wiesław Szczesny

12.1 Introduction

Chapters 12, 13 and 14 explain the grade approach to the analysis and understanding of multivariate data matrices. Chapters 12 and 14 are dedicated to readers oriented on practice of data analysis, Chapter 13 is dedicated to readers looking for the common frame of models and methods in three domains: data analysis, probability and statistics.

Data matrices are objects/variables tables. Examples were given before in Chapter 2 (examples concerning budget expenses and test scores), and also considered in Chapter 10 (examples concerning time allocation data, survey of small business servicing firms, and sources of capital by retail trade firms).

Example A in Chapter 2 has provided the first dataset used in this book to illustrate grade methods and models. This dataset, contained in Table 2.1, dealt with budget expenses in two countries: Country A and Country B. The expenses were divided into five categories: *Food, Housing, Utilities, Transportation, Health Care.* Thus the data formed a 2×5 table, with two rows representing countries A and B, and five columns representing variables such as expenses on food (variable *Food*), ..., expenses on health care (variable *Health*). Each variable had just two values, referring to country A and country B.

Generally, a $2 \times k$ data matrix provides two *records,* each being a vector of values of k variables ($k \geq 2$). So, there are two rows (called objects, subjects, individuals, etc) and k columns (called variables, items, features, etc).

The differences between the records of a $2 \times k$ table can be described by the concentration curve $C(1:2; row)$ (or equivalently, by its mirror reflection $C(2:1; row)$) and by the respective concentration indices $ar(1:2; row)$ or $ar(2:1; row)$. If the curve under consideration is not convex, it might be transformed into a convex one by a suitable reordering of columns; the reordered curve is denoted $C_{\max}$ and is accompanied by the maximal concentration index ar_{max}.

In Example A, curves $C(1:2; row)$ (which is not convex) and $C_{\max}(1:2; row)$ (which is convex) are presented in Chapter 2 in Figures 2.1 and 2.2; the concentration indices are $ar(1:2; row) = -0.1175$ and $ar_{\max}(1:2; row) = 0.3025$.

The differences between any two variables (e.g., *Food* and *Health*) are described by the triangular concentration curves and respective concentration indices referring to columns.

Note: The selected curves and the 5×5 matrix of concentration indices for columns in Example A were presented in Chapter 7. However, the calculations and figures in Chapter 7 were addressed not to example A but to Example B concerning Letter Grades in Geography in East City School and West City School. This was possible since the numbers in examples A and B were deliberately chosen in Chapter 2 to be exactly the same (the budget expenses were rounded to form data which could be taken as frequencies in a 2×5 contingency table).

The 2×5 table (common for Examples A and B) was then analyzed once more in Chapter 8, Section 8.2. This time, the aim was to calculate the grade distribution of the table and visualize it by means of the 2×5 overrepresentation map. The map was also produced for the data table with columns reordered in the way shown in Figure 2.2. This configuration indicates the main trend which associates budget categories with particular countries. The reordering of columns in a $2 \times k$ table which is optimally permuted (i.e., ensures that the concentration curve for row 1 against row 2 is convex), is the same as that obtained as a result of Grade Correspondence Analysis (GCA) described in Chapter 9. The maximal strength of the main trend is evaluated in Example A by $\rho^*_{\max} = 0.2178$.

Data matrices with m records and k variables, considered in Chapter 10, concerned Grade Correspondence Cluster Analysis (GCCA): as the variables were non-negative, it was possible to perform the calculations and obtain the overrepresentation maps in exactly the same way as we did in Chapter 8 in the case of $m \times k$ contingency tables.

The examples of data matrices provided up to now in Chapter 2 and 10 will be supplemented in the present chapter (in Section 12.2) by two other insight examples.

All data matrices corresponding to the examples appearing in the present chapter and in Chapter 2 and 10 *have a finite number m of records.* Such data matrices will be called *finite* data matrices or *finite datasets.* Non-finite data matrices (i.e. matrices with k variables and a non-finite number of records) will be considered in Chapter 13. They refer to k-variate probability distributions in which at least some variables have a non-finite number of values. If, however, one deals with samples of size m drawn from any k-variate probability distribution, the resulting sample data matrices are finite, with m rows corresponding to m sampled records.

Applicability of procedures based on GCA for finite data matrices is considered in Section 12.3, which reflects the point of view oriented on the junction between measurement scales and data analysis. The considerations in Section 12.3 are analogous to those in Chapter 4 which concerned sets $\mathbb{RATIO}$ and $\overline{\mathbb{RATIO}}$. Set $\mathbb{RATIO}$ contains variables measured on a ratio scale, set $\overline{\mathbb{RATIO}}$ contains non-negative random variables with finite expectations. We remind that the inequality measures (Lorenz Curve, Gini Index, etc) *are meaningfully defined* for variables from the first set but are still *calculable* for those from the second set.

Variables from $\mathbb{RATIO}$ represent a direct extension of counts (exemplified by variables in Example B (Chapter 2) which count how many pupils in each City School got a particular Letter Grade in Geography). A data matrix with additive count variables form *a contingency table.*

A counterpart of the $\mathbb{RATIO}$ set is the set of finite data matrices with additive ratio variables. This set is a direct extension of the set of contingency tables, and this is where GCA can be *meaningfully applied.* The set of finite data matrices with non-negative variables is a counterpart of set $\overline{\mathbb{RATIO}}$: this is where GCA is only *calculable.*

Note: Variables in one of the insight examples in Section 12.2 (the A/B data) form a set of additive ratio variables, while the other insight example (the C/J data) holds this requirement only approximately.

Transformations provided to make GCA calculable and its results "approximately meaningful" results are mentioned in Section 12.6.

The models for finite data matrices are introduced in Section 12.5. First, Section 12.4 revisits the Lorenz Curves introduced in Chapter 4, to prepare a smoth passage from the curves to Lorenz Surfaces. The Lorenz Surfaces are *specific concentration surfaces* – just like Lorenz Curves are specific concentration curves. Analogously as the Lorenz curves were meaningfully introduced for ratio variables, i.e., for the set $\mathbb{RATIO}$, and then extended onto set $\overline{\mathbb{RATIO}}$, the Lorenz Surfaces will be meaningfully introduced for the finite data matrices with additive ratio variables and then extended onto finite data matrices with non-negative variables.

We remind that the Lorenz Curve measures inequality between the distribution of a ratio variable and the corresponding first moment distribution. Analogously, the Lorenz Surface measures inequality between a two-way contingency table related to a set of additive ratio variables and the corresponding (bivariate) first moment distribution.

Generally, the Lorenz Surface is *not* a copula, i.e., it does not belong to the Dependence Lilliputian Model ($\mathbb{DLM}$), although it belongs to the Bivariate Lilliputian Model ($\mathbb{BLM}$). The grade transform of the Lorenz Surface, called the Grade Lorenz Surface, belongs to $\mathbb{DLM}$ and hence can be visualized by overrepresentation maps as described in Chapter 8 for the elements of $\mathbb{DLM}$. These maps were used before in the examples provided in Chapter 10 and will be extensively used in the present chapter and two next.

Two main aspects of the analysis of finite multivariate data matrices are: level of *strength of the correspondence between records and variables*, and level of *regularity of this correspondence*. Regularity will be intuitively discussed on the basis of the insight examples from Section 12.2, one of which is extremely regular, the other quite irregular. Advantages of regularity are visible on this background. Therefore, it is desirable to decompose data sets into more regular subsets and to extract outliers. Grade methods of this type will be presented in Chapter 14 on two experimental datasets of size 601×10 and 633×18. An introductory decomposition is presented in Section 12.7, in which outliers are detected and removed from the irregular dataset introduced in Section 12.2.

12.2 Insight Examples

Warning: The insight examples are completely hypothetical "staged" examples. Any resemblance to the "real thing" is purely intentional. We will provide more details as we proceed.

12.2.1 The Competitors-Judges Data (C/J Example)

There is a new sporting event for which 100 competitors pre-register. (They are numbered from 1 to 100.) There is a pool of 50 potential judges. (They are numbered from 1 to 50.)

Twenty competitors are selected at random. Five judges are selected at random. The judges independently assign points for "skill" in the event. The data are shown in Table 12.1. The competitors are listed in the random order in which they participated. The judges were originally listed alphabetically, but we show only their ID numbers. (We will assume that the alphabetical order is random.)

Note: Table 12.1 as presented, has a "two-way layout." The scores are natural numbers. This table layout will be analysed in the same way as the two-way contingency table layouts in Chapter 7 and Chapter 8 (see, e.g., datasets $BRIT_{7\times7}$ or $CARS_{16\times16}$). It is important to keep in mind, that when GCA is performed, N (i.e., the total for the table) will be considered equal to 21,625, (i.e., the sum of all the scores).

TABLE 12.1. C/J New Sporting Event (points awarded by judges) with 20 competitors listed in order of random participation, and 5 judges randomly ordered.

Comp.	Judge ID					
ID	10	33	17	41	49	Total
43	169	60	355	71	287	**942**
82	208	134	207	116	240	**905**
60	229	379	71	201	144	**1024**
6	264	216	197	165	262	**1105**
35	210	528	39	224	101	**1102**
32	216	260	100	162	167	**905**
77	264	292	135	190	215	**1096**
10	222	195	151	144	210	**922**
17	270	131	385	131	375	**1292**
11	164	70	276	75	248	**832**
9	288	295	162	201	247	**1193**
67	241	144	267	130	294	**1076**
87	283	214	231	171	294	**1193**
88	129	36	372	48	259	**844**
28	316	416	130	248	231	**1341**
76	174	335	44	164	99	**816**
70	320	224	288	186	350	**1368**
45	249	135	310	128	323	**1145**
7	264	385	95	218	180	**1142**
18	334	316	207	224	301	**1382**
Total	**4814**	**4765**	**4022**	**3197**	**4827**	**21625**

Note: For the specific random orderings of rows and columns shown, there is no trend (ρ^* = -0.0030). A bold horizontal line separates the first 10 competitors from the second 10, and is referred to later.

TABLE 12.2. C/J New Sporting Event (points awarded by judges). Rows and columns permuted by GCA to obtain maximal value of $\rho^* = 0.36$

Comp.	Judge ID					
ID	33	41	10	49	17	Total
35	528	224	210	101	39	**1102**
76	335	164	174	99	44	**816**
60	379	201	229	144	71	**1021**
7	385	218	264	180	95	**1142**
28	416	248	316	231	130	**1341**
32	260	162	216	167	100	**905**
77	292	190	264	215	135	**1096**
9	295	201	288	247	162	**1193**
18	316	224	334	301	207	**1382**
10	195	144	222	210	151	**922**
6	216	165	264	262	197	**1104**
87	214	171	283	294	231	**1193**
70	224	186	320	350	288	**1368**
82	134	116	208	240	207	**905**
67	144	130	241	294	267	**1076**
45	135	128	249	323	310	**1145**
17	131	131	270	375	385	**1292**
11	70	75	164	248	276	**833**
43	60	71	169	287	355	**942**
88	36	48	129	259	372	**844**
Total	**4765**	**3197**	**4814**	**4827**	**4022**	**21625**

TABLE 12.3. C/J New Sporting Event. Overrepresentation values for points awarded by the judges, with rows and columns permuted by GCA to obtain $\rho^*_{\max}$=0.36. **Note:** the overrepresentation values (i.e., observed/expected) are calculated from Table 12.1 or 12.2; for example, using competitor 35 under judge 33, the observed 528 divided by (1102/21625)*4765 (i.e., divided by the expected value under fair representation) is equal to 2.17 .

Comp.	Judge ID					
ID	33	41	10	49	17	marginal
35	2.17	1.37	0.86	0.41	0.19	**0.051**
76	1.86	1.36	0.96	0.54	0.29	**0.038**
60	1.68	1.33	1.00	0.63	0.37	**0.047**
7	1.53	1.29	1.04	0.71	0.45	**0.053**
28	1.41	1.25	1.06	0.77	0.52	**0.062**
32	1.30	1.21	1.07	0.83	0.59	**0.042**
77	1.21	1.17	1.08	0.88	0.66	**0.051**
9	1.12	1.14	1.08	0.93	0.73	**0.055**
18	1.04	1.10	1.09	0.98	0.81	**0.064**
10	0.96	1.06	1.08	1.02	0.88	**0.043**
6	0.89	1.01	1.07	1.06	0.96	**0.051**
87	0.81	0.97	1.07	1.10	1.04	**0.055**
70	0.74	0.92	1.05	1.15	1.13	**0.063**
82	0.67	0.87	1.03	1.19	1.23	**0.042**
67	0.61	0.82	1.01	1.22	1.33	**0.050**
45	0.54	0.76	0.98	1.26	1.46	**0.053**
17	0.46	0.69	0.94	1.30	1.60	**0.060**
11	0.38	0.61	0.88	1.33	1.78	**0.039**
43	0.29	0.51	0.81	1.36	2.03	**0.044**
88	0.19	0.38	0.69	1.37	2.37	**0.039**
marginal	**0.22**	**0.15**	**0.22**	**0.22**	**0.19**	1

The ρ^* value (grade correlation value) for Table 12.1 (with its "random" ordering of rows and columns) is -0.003. This low value for ρ^* suggests a lack of dependence between the row and column configuration. That is, there is no noticeable trend relationship between rows and columns.

Next, GCA will permute the rows and columns of Table 12.1 to obtain the maximum possible value of ρ^*. Table 12.2 (with new row and column permutations) has ρ^* equal to 0.36 (now referred to as ρ^*_{max}). The row and column configuration in Table 12.2 is strongly positively related.

Table 12.3 shows the overrepresentation values (i.e. the densities of the grade distribution) for the row and column permutations which yield the maximal value of ρ^*.

Doing a little examination of Table 12.3: based on the size of the over-representation values, it would appear that judge number 33 (left side of the table) tends to greatly over-scoring competitors 35, 76, and 60, while greatly under-scores competitors 88, 43, and 11. On the other hand, for the same six competitors, judge number 17 (right side of the table) tends to under-score the first three, while over-scoring the last three.

We also find decreasing values in column 1 (judge 33) and increasing values in column 5 (judge 17). The second column also decreases, and the fourth column increases. (Remember, these data were "faked" to demonstrate these nice smooth increases and decreases.)

While perhaps not so obvious, there is a tendency for high overrepresentation values (i.e., over-scoring) to run from the upper-left corner to the bottom-right corner along a diagonal path. That is, there is a trend of *over-scoring on the diagonal, in particular on both corners.*

Assuming that the competitors' scores (at least the sums of the five judges) actually have something to do with their competitive "skill" performance, then what does this structure of over/under scoring represent?

Let us go back and look at the total score column in Table 12.2. The total scores are not increasing or decreasing with the GCA trend. Remember that the ordering of the table (i.e., columns and rows) is based on the grade trend. IF the grade trend represented performance "skill," then the totals would be nicely ordered from top to bottom, let us say, in increasing order. That is, the competitors would be listed in a different order than shown, with their total scores increasing. This is not the case, however. Therefore, the GCA trend structure in the data seems not related to skill of performance.

Note: The judges' total scores (shown at the bottom of Table 12.2) also do not seem to be ordered in any obvious way that would account for this trend structure.

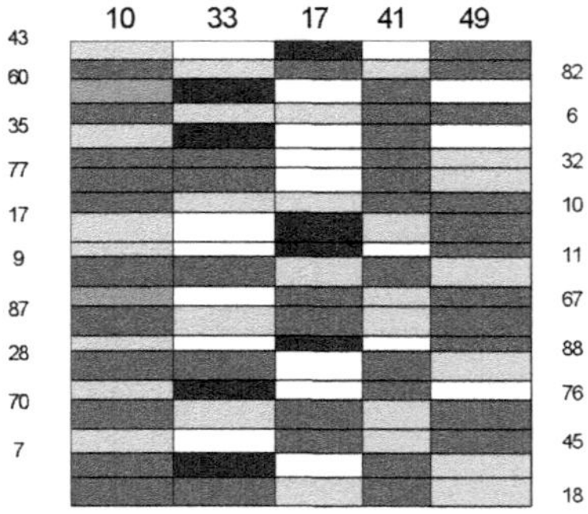

FIGURE 12.1. The overrepresentation map for the C/J New Sporting Event data from Table 12.1, in which judges (columns) and competitors (rows) are "randomly ordered"; $\rho^* = -0.0030$ (according to the participation order)

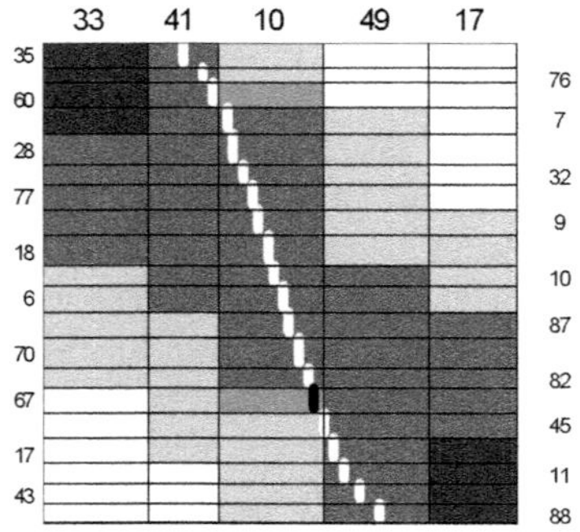

FIGURE 12.2. The GCA overrepresentation map for the C/J New Sporting Event data from Table 12.2 (20 competitors, $\rho^* = \rho^*_{\max} = 0.3600$). Vertical black segment for competitor 67 and vertical white segments for other competitors indicate the values of the grade *col|row* regression.

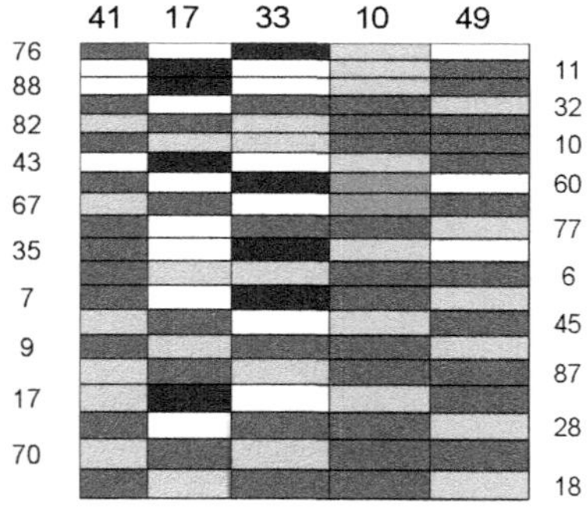

FIGURE 12.3. The overrepresentation map for the C/J New Sporting Event when both competitors (rows) and judges (columns) are ordered according to increasing totals as read from Table 12.1 (then $\rho^* = -0.0272$)

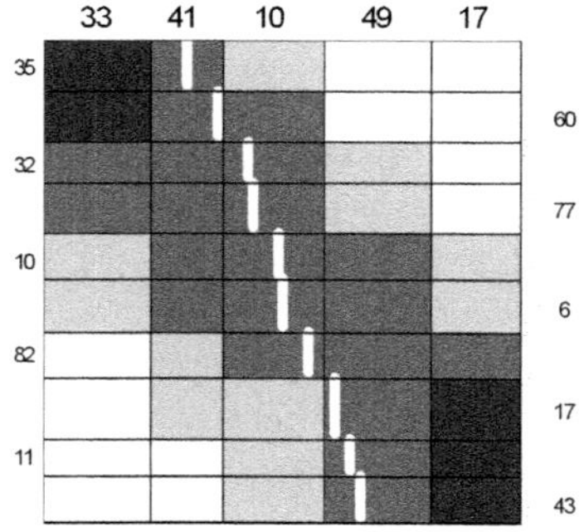

FIGURE 12.4. The GCA overrepresentation map for the C/J New Sporting Event data from the upper half in Table 12.1 (10 competitors, $\rho^* = \rho^*_{\max} = 0.3914$). Vertical white segments indicate the values of the grade *col|row* regression.

The overrepresentation map in Figure 12.1 reflects the initial ordering of competitors and judges as presented in Table 12.1. Judge ID is indicated above the columns. Competitor ID is listed down the sides. The overrepresentation values contained in Table 12.3 are displayed in the overrepresentation map in Figure 12.2. The ordering of rows and columns is the same as in Table 12.3.

Recall that five shades are used: the larger the overrepresentation value, the darker the shading. The darker shading along the diagonal in Figure 12.2 represents the "over-scoring" trend in Table 12.3. (Note: This is an artificially created very "regular" trend.)

The different widths of the columns and rows represent the different row and column totals. This provides a "view" of the differentiation of the row and column marginal distributions. The different row widths reflect the different competitor totals. The different column widths reflect the different totals for judges. That is, the tendency for a given judge to give higher or lower scores.

In Figure 12.3, the competitors are ordered according to their increasing skill scores (represented by row totals), the judges are ordered according to increasing column totals. No trend is visible ($\rho^* = 0.0128$). The widths of rows as well as of columns are rather weakly differentiated; the respective Gini Indices are equal to 0.0901 (for competitors) and 0.0642 (for judges).

Comment: We purposely called this a "New" sporting event to introduce the "possibility" that the judges - although considered to be experts in this new sport - are awarding points, perhaps, somewhat subjectively. Of course, with this in mind, the organizers of the event took the precaution of using at least five judges, in hope that the totals of five scores would have a more credible meaning.

If the grade trend does not represent performance skill, then what does it represent? The trend certainly exist, and it refers to with something purly associated with performance skill.

Since the example is hypothetical, we offer a hypothetical - but plausible - interpretation of the grade trend: it may have to do with the "style" of performance, as opposed to the competitor's "skill." That is, the competitors have a style which a given judge likes, or does not like. Assuming that higher scores are better, judge 33 likes the style of competitors 35, 76, 60, etc. (i.e., those near the top of the table). Judge 17 likes the style of competitors 88, 43, 11, etc. (i.e., those near the bottom). It can also be said that some judges (33, 17) take style seriously into account while others (in particular judge 10) are not sensitive to performance style.

Comment: If this were a skating event, we might claim that judges were awarding points based on their preference for the competitor's music, costumes, personality, charisma, etc., rather than skating skill.

We now have insight into the structure of the data. There is apparently a competitor/judge style preference at work. How would one use this information?

A competitor might want to consider their style characteristics, and make changes where possible (i.e., in order to "play" to particular judging styles). In other words, make changes, or adapt, based on this new knowledge.

A judge might try to eliminate (or isolate) this "style" influence on their awarding of points for skill (review the technical "skill" points of the sport related to difficulty, etc.)

A data analyst has learned something about the data. This something is tangible; the trend actually exists; it is not a theoretical construct. The data analyst might recommend the collection of data actually dealing with style. Competitors might be classified (or matched with trainers) by style type.

Please note that we are dealing with two important features (variables) describing the set of competitors:

- skill, i.e., the sum of points assigned by the five judges, and
- style, evaluated by the value of the *grade col|row regression function* (i.e. $r^*_{col|row}$). The regression values for style are shown in Figure 12.2 in the form of the vertical white segments (more precisely, the segments are white in black or dark grey rectangles, and black otherwise).

Skill will be denoted $Z^{(20)}$, *style* $R^{(20)}$. The overscript (20) refers to the fact that we are dealing with 20 competitors drawn at random from the set of 100 competitors. Generally, we can consider $Z^{(n)}$ and $R^{(n)}$ for n competitors randomly drawn from the whole set (without replacement). It is important to know how the values of $R^{(n)}$ and the distribution of $Z^{(n)}$ depend on n.

Suppose that we take into account only the ten competitors who participated first (they are placed in Table 12.1 above the horizontal line which divides the table into halves). Proceeding as before, the ten competitors and five judges are permuted to obtain the strongest style trend. The result is shown in Figure 12.4 as the related GCA overrepresentation map. The ordering of the competitors by the five judges is *exactly the same* as in the case of 20 competitors. That is, if you look at the ordering of 20 competitors, and cross out the identification of any competitor competing in positions 11 through 20, the ordering of the remaining ten competitors is exactly the same as if only those ten competitors had been ordered by GCA (see Table 12.2 and Figure 12.2). The *values* of the grade *col|row* regression function $R^{(10)}$ (white vertical segments in Figure 12.4) are not *quite* the same as of $R^{(20)}$, but comparing the vertical segments in Figures 12.2 and 12.4 one can see that the differences are negligible.

The distributions of $R^{(10)}$ and $R^{(20)}$ are very close one to another, and the same holds for the distributions of $Z^{(10)}$ and $Z^{(20)}$. As to the joint distributions of $(R^{(n)}, Z^{(n)})$ for $n = 10$ and $n = 20$, we limit ourselves to finding the values of their correlation coefficients:

$$\operatorname{corr}(R^{(10)}, Z^{(10)}) = -0.174, \qquad \operatorname{corr}(R^{(20)}, Z^{(20)}) = -0.147,$$
$$\rho^*(R^{(10)}, Z^{(10)}) = -0.214, \qquad \rho^*(R^{(20)}, Z^{(20)}) = -0.092.$$

Conclusion: suitable permutations of the rows and columns of Table 12.1, supplemented by further calculations and visualizations, provide insight into the C/J data (as demonstrated for various contingency tables in Chapter 8 and in chapters to follow).

In the case of contingency tables, the GCA and related procedures were shown before (in Chapter 8 and 9) useful and meaningful. What about the *meaningfulness* of GCA in the case of multivariate data matrices? We will try to answer this question in Section 12.3.

12.2.2 The Annual Bonus Data (A/B Example)

Table 12.4 contains the data for the annual bonuses (in thousands of dollars) of 27 employees, earned in five successive years. Each row refers to a particular employee, each column to a particular year. There are no negative entries in the table. The total sum is equal to 349.5 (the sum of bonuses obtained by all employees during the five years). Dividing each entry by this sum we get a table which *technically* can be processed according to GCA.

The employees in Table 12.4 are ordered according to the constrained GCA, under the constraint that the years are *ordered chronologically*. This means that the rows (employees) are permuted to get the maximal value of ρ^* provided that the columns (years) are kept in the chronological order. The related chronological trend is the strongest possible for chronologically ordered columns (the value of ρ^* is 0.2362).

Thus the employees in Table 12.4 are ordered according to the chronological trend. It is seen that the first employee (with ID 19) had a comparatively *large* bonus in the first year, but a comparatively *small* bonus in the last year, and generally there was a trend according to which he was *decreasingly* rewarded during the five year period. (Note: Employee 19 is positioned first in Table 12.4 by the constrained GCA based on the overrepresentations. See Figure 12.5). Contrarily, the employee in the last position (with ID 3) had a comparatively *small* bonus in the first year, but a comparatively *large* bonus in the last year, and generally there was a trend according to which he was *increasingly* rewarded during the five year period. (Note: Employee 3 is positioned last in Table 12.4 by the constrained GCA based on the overrepresentations. See Figure 12.5).

TABLE 12.4. The Annual Bonus Data of 27 employees (with ID's from 1 through 27) ordered by constrained GCA from the least to the most highly rewarded during five successive years (in thousands of dollars).

Empl. ID	YEAR					Total	Empl. ID	YEAR					Total
	1	2	3	4	5			1	2	3	4	5	
19	3	1.5	2	0.5	0.5	7.5	4	1	2.5	2	0.5	3	9
22	3	2.5	2	1	1	9.5	11	2	1.5	3	4	2.5	13
24	3	2.5	4	2	0.5	12	15	2	2.5	4	2	4.5	15
25	3	3.5	2	1.5	1.5	11.5	17	2	3.5	3	1.5	5.5	15.5
16	2	3.5	2	2	2	11.5	9	1	3.5	4	4	3.5	16
7	1	3.5	2	2.5	1	10	6	1	2.5	4	1	4.5	13
21	3	1.5	4	3	1.5	13	8	1	3.5	3	2.5	5	15
26	3	3.5	3	3	3	15.5	10	2	1.5	2	4	4	13.5
18	2	3.5	4	1	3.5	14	2	1	1.5	3	0.5	4	10
14	2	2.5	3	3.5	2	13	12	2	1.5	4	3	5.5	16
13	2	2.5	2	2.5	2.5	11.5	1	1	1.5	2	1.5	4	10
27	3	3.5	4	1	5	16.5	5	1	2.5	3	3.5	6	16
23	3	2.5	3	3.5	3.5	15.5	3	1	1.5	4	1.5	6	14
20	3	1.5	3	0.5	4.5	12.5	Total	54	67.5	81	57	90	349.5

The overrepresentation map in Figure 12.5 visualises the chronological trend. The trend is certainly very weak and irregular, although years $1, 2, 3, 5$ preserve some regularity and it is only year 4 which distinctly outlies from a regular structure. This observation is confirmed by the values of the *AvgDist* procedure for columns.

Remark 12.1 *Recall that the functions AvgDist (for rows and for columns separately) were introduced in Section 9.6.2 in order to distinguish outliers among rows or columns of a non-negative two-way table. The functions are usually applied to tables transformed by GCA, but it is also possible to use them with respect to any fixed order of rows or of columns. Note that we used the $AvgDist_{col}$ to the A/B table with chronologically ordered columns (Table 12.4). Based on a visual inspection of Figure 12.5 we would expect column 4 to be the main outlier, and in fact the value of AvgDist for this column was found to be* 0.1689, *while the remaining values of $AvgDist_{col}$ were found to be considerably smaller (*0.0823, 0.1063, 0.1017, *and* 0.0604 *for years* $1, 2, 3, 5$*).*

The GCA *without constraints* changed the chronological succesion of years to $1, 4, 3, 2, 5$ as shown in Figure 12.6. The value of ρ^* increased from 0.2362 for the chronological trend to $\rho^*_{\max} = 0.2474$ for the GCA trend, which caused a rather dramatic change in the appearance of the overrepresentation maps, i.e., Figure 12.5 versus Figure 12.6. Year 4 in Figure 12.6 still outlies from the GCA trend, but not so strongly as in Figure 12.5.

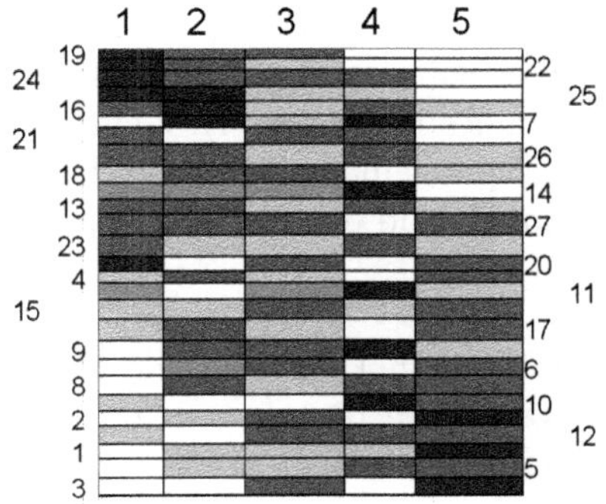

FIGURE 12.5. The overrepresentation map of the A/B data transformed by GCA under the constraint that years are ordered chronologically ($\rho^* = 0.2362$).

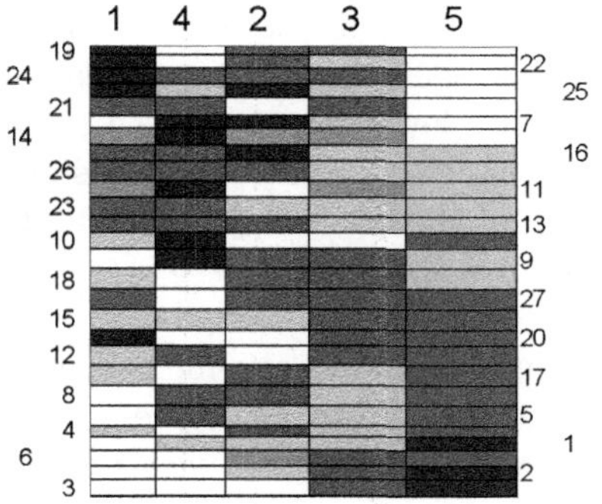

FIGURE 12.6. The GCA overrepresentation map of the A/B data when transformed by GCA with no constraint ($\rho^* = \rho^*_{\max} = 0.2474$).

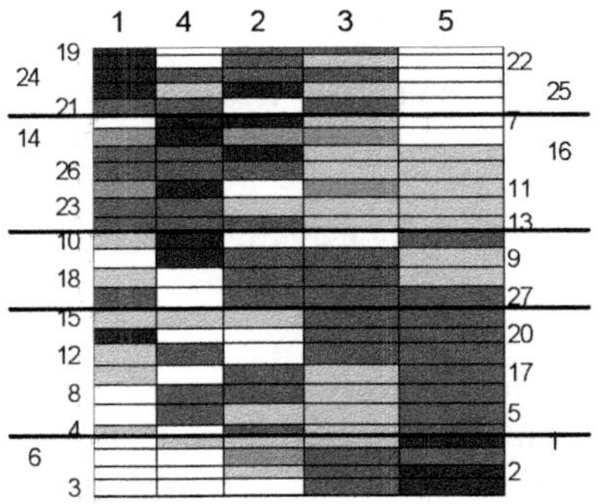

FIGURE 12.7. Clusters (separated by bold lines) of employees ordered by progress category: unsatisfactory, rather unsatisfactory, average, rather satisfactory, and satisfactory, based on the GCA overrepresentation map (Figure 12.6).

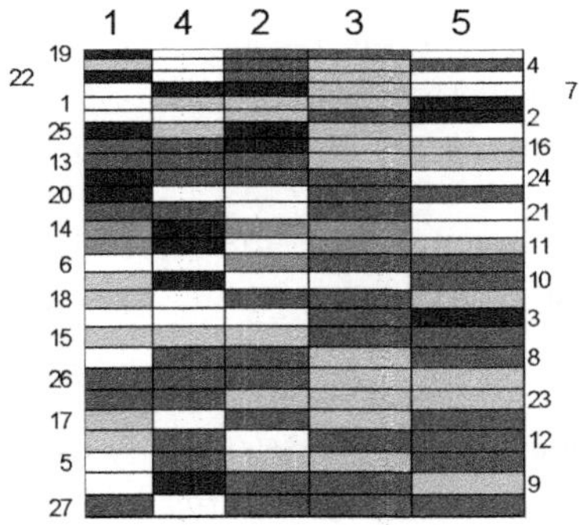

FIGURE 12.8. The A/B overrepresentation map of the data table with rows and columns ordered non-decreasingly according to their widths (i.e., columns are ordered by the sums of the annual bonuses); $\rho^* = 0.0930$, Gini Indices: 0.0967 for employees, 0.1069 for years.

Let us compare the GCA permutation (of employees) without the chronological restriction with the GCA permutation of employees with years in chronological order. Four employees distinctly changed their positions, jumping "up" or "down' from the position due to chronological trend to the position due to the GCA trend. The most spectacular "ups" are the employees with IDs 10 and 11, both with a very large overrepresentation of the bonus in year 4 (black rectangles in the column corresponding to year 4 in Figures 12.5 and 12.6); the most spectacular "downs" are employees with IDs 4 and 18 with exceptionally low bonus in year 4 (white rectangles in column 4). Generally, however, changes of positions are insignificant; in particular, there is no change at the top of the list (employees 19, 22, 24, and 25) and on the bottom of the list (employee 3).

The underlying measurement by which employees in Table 12.4 and Figure 12.5 are ordered from employee 19 to employee 3 will be referred to as the *chronological advancement measurement.* This measurement is the value of the grade regression function $r^*_{col|row}$, denoted R_{chron}.

However, there is no obligation to preserve chronology in a strict way. It might be desirable to use also the unconstrained GCA table (Figure 12.6). It is necessary in this case to choose this one of the two twin GCA tables $P_{\max}$ and $P^{\Leftarrow}_{\max}$ (notation introduced in Section 8.6 and used throughout Chapter 9), in which the order of the columns is *most close* to the chronological order. As the GCA column permutation $(1,4,2,3,5)$ is obviously more close to the chronological than its reverse $(5,3,2,4,1)$, the former was chosen as seen in Figure 12.6.

Let R denote the grade regression function for unconstrained GCA. It is clear that R is different than R_{chron}.

Figure 12.7 is a copy of Figure 12.6 with the rows divided into five non-overlapping clusters which classify the set of employees according to R into five groups consisting of those *progressing unsatisfactorily, rather unsatisfactorily, averagely, rather satisfactorily,* and *satisfactorily.* Those employees who are average have values of R close to $1/2$.

Another important feature is the *total bonus* which is the sum of the employee annual bonuses for the five years. This feature will be denoted Z. Figure 12.8 shows the overrepresentation map of the A/B data with columns ordered according to nondecreasing total sums, and employees ordered according to the total bonus Z. The differentiation of Z is relatively small (Gini Index equal to 0.0967); the years are also mildly differentiated (similar to that of the judges in the C/J data); the Gini Index is equal to 0.1069.

The correlation coefficients for the pair R_{chron} and Z and the pair R and Z are:

$$corr(R_{chron}, Z) = 0.4958, \qquad \rho^*(R, Z) = 0.4375.$$

This means that employees with large total bonus (Z) tend to progress satisfactorily, while those with small total bonus tend to progress unsatisfactorily. Of course, it is just a specific feature of the A/B data which shows the relationship between the main descriptive statistics: R (or R_{chron}) and Z representing level of progression and level of total bonification.

12.3 Applicability of GCA

The analyses performed for the A/B data and the C/J data indicate that the grade methods introduced in Chapter 9 to analyze $m \times k$ contingency tables can also be usefully applied to various non-negative datasets. Latent variables, outliers, strength of trends (constrained and unconstrained) can be determined. But on what basis?

Let us repeat that a finite data matrix (a dataset) is a table with m rows corresponding to objects (individuals), and k columns corresponding to variables which describe these objects. Rows (called *records*) are vectors of variables values.

Notice that a $m \times k$ *contingency table* can be considered as a specific data matrix in which variables are *counts* of individuals (and so the values of each variable are natural numbers or zero). Let us take as an example a table with numbers of votes cast by voters in different election districts (rows) on behalf of particular political parties (columns). Let districts form a set Ω_0 of objects. Then there are k functions from Ω_0 to $\mathbb{N}$ (where $\mathbb{N}$ is the set of natural numbers supplemented by zero), each corresponding to one party and assigning to districts the numbers of votes obtained by this party. A record describing a voting district consists of k numbers of votes.

The text below concerns important links existing between finite data sets and measurement scales.

Remark 12.2 *Information concerning measurement scales was given in Section 4.5.1 and was used extensively in Sections 4.2 amd 4.3.*

Votes are counted *on the absolute scale* (hence there is the same scale for each political party). Thus in this example there is *a unique* function defined on *all* pairs $\langle$*region,political party*$\rangle$, which assigns to any such pair the related number of votes.

A similar situation holds in the second insight example (i.e., A/B data) presented in Section 12.2.2. Suppose first that there is established the monetary unit ($, thousand of dollars, euro etc) related to the annual bonus. Then, as in the case of votes, there is also a *unique* function defined on the set of pairs (employee, year). However, this measurement is *on the ratio scale* (and *not* on the absolute scale), since - contrary to counting votes - there exists a possibility to change the monetary unit and so change bonus quantity - but the empirical and numerical ratio relations like "this bonus is t times larger than that bonus" will remain unchanged.

This two-argument function, defined on the product of the set Ω_0 of employees and the set $\{1, ...5\}$ with elements representing years, generates five functions, each defined on Ω_0. Each of them "measures" the bonus on a common ratio scale (e.g., all provide the bonus in thousands of dollars). Therefore every employee is described by five ratio measurements performed repetitively in successive years. The function values are *additive*; the *sums of values* for a given employee (in one row), as well as *the sums of values* for a given year (in one column) are empirically and numerically *meaningful* and are both measured on the same ratio scale as the values of the column variables. Therefore it is fully justified to apply the GCA procedure (with or without a constraint) to the A/B dataset.

> Generally, the application of GCA will be fully justified for *any finite dataset* obtained by *repetitive, additive, non-negative measurements, performed on a specified ratio scale for a given set of objects.* The GCA procedures (possibly supplemented by grade clustering, table decomposition, outlier detection, latent traits analysis, etc.) will be exactly the same as those procedures used for contingency table analysis.

However, neither additivity nor repetitiveness of ratio measurements is fulfilled in the first insight example (the C/J data). Moreover, the measurement scale is *not the same for each judge,* while the scales seem not to be *ratio,* but rather *ordinal.* On the other hand, we learned in Section 12.2.1 that the GCA procedures had been beneficially applied to that dataset. Was this legitimate?

The results of GCA for the C/J data suggest that the differences among judges are not due to inadequate training aimed at standardizing their evaluations; instead, they suggest that judges measure the quality of performance from their own perspective, reflecting their individual *sensitivity to style.* It is *style* of performance *which optimally orders competitors,* as well as *sensitivity to style which optimally orders judges.* It follows that the measurement performed on the competitors refers to *overall quality involving style and skill.* The decisions of particular judges are influenced by these two aspects in a subjective individual way. Their evaluations are *expert opinions;* so we deal with five different functions applied simultaneously to each competitor. Each function can be treated as a random variable defined on the set of competitors (set Ω_0).

We assume that each judge will deliver his opinion according to his own inner random mechanism, for each of the one hundred competitors independently. The opinions are non-negative; they might be more or less *close* to ratio measurements, but *are not exactly ratio.*

But let us recall what was written about *ratio variables* in Section 4.2.1. Variables measured on a ratio scale (in the sense introduced in measurement theory), such as earnings, were called *ratio* variables in the strict sense. Further, the set $\overline{\mathbb{RATIO}}$ of nonnegative (random) variables with finite positive expectation was introduced as an extension of the set of (random) variables which are ratio in the strict sense, and the definitions of the first moment distribution and the Lorenz Curve were extended onto the set $\overline{\mathbb{RATIO}}$.

When variables belong to $\overline{\mathbb{RATIO}}$ but are not ratio in the strict sense, their parameters and transformations might not fulfill specific "ratio" requirements (such as specific invariance), and are only *calculable.* Thus, the GCA results are *calculable* whenever all variables forming the columns of a finite dataset belong to the set $\overline{\mathbb{RATIO}}$.

Thus, technically, the whole body of data analysis based on GCA can be applied to any finite dataset obtained from a number of non-negative measurements with finite positive expectations. We have no doubt that the value of the interpretational and the descriptive role of such analyses - although, strictly speaking, not legitimate - should be appriociated highly.

The final requirement of a fully meaningful GCA is additivity of variables. This holds for the A/B data, but it is rather doubtful for the C/J data. Generally, additivity is an assumption which is desirable, but does not affect the calculations involved in GCA. In the case of judges one can argue that the sums of their opinions are somehow meaningful (and, moreover, these sums differ only slightly for particular competitors). But it could occur in other data matrices that the variables are quite different and the adding of the values is completely inapproppriate. In this case, it might be possible to transform the variables so that their expectations are equal, or to standardize them according to various other techniques. These problems will be shortly considered in Section 12.6 and illustrated in the case studies in Chapter 14.

12.4 A revisit of the univariate data

The inequality measures for a single variable $X \in \overline{\mathbb{RATIO}}$ (i.e., the Lorenz Curve and Gini Index as well as their φ-ordered versions) were introduced and discussed in Sections 4.2 and 4.3. An important goal of Chapter 4 was to indicate the links between measures of concentration and measures of inequality (in particular, to indicate that the Lorenz Curve of the overrepresentation $OV_{Y:X}$ is equal to the maximal concentration curve $C_{\max}(Y : X)$).

In the present text we revisit the inequality measures of a single variable X, restricting the considerations to *finite univariate* datasets. Selected single columns in the datasets C/J and A/B will be used as illustrations. The material in this section regarding finite univariate datasets, although seemingly too detailed, is a necessary introduction to the finite multivariate case.

Let $X : \Omega_0 \rightarrow \mathbb{X}$ belong to set $\overline{\mathbb{RATIO}}$, let Ω_0 contain n objects, and let $\mathbb{X} = (x_1, ..., x_m)$ where $x_1 \leq ... \leq x_m$. We assume that the objects in Ω_0 are "uniformly distributed", i.e., each of them can be treated as randomly drawn with the same probability $1/n$. Note that set Ω_0 is usually considered a subset of a larger population $\mathbb{UNIV}$ and X may be a function from $\mathbb{UNIV}$ into $\mathbb{X}$.

Illustration (C/J data):

$\mathbb{UNIV}$ is the set of 100 competitors pre-registered for a new sporting event and Ω_0 is a subset of $\mathbb{UNIV}$ consisting of $n = 20$ competitors who participated first in the competition. The competitors in $\mathbb{UNIV}$ were labeled according to a 1-1 function $id : \mathbb{UNIV} \rightarrow \{1, ..., 100\}$; those who participated first were labeled 43, 82, 60, 6, ..., 19 and they were listed in this order in Table 12.1. Thus order of participation is based on a function $\varphi : \mathbb{UNIV} \rightarrow \{1, ..., 100\}$, where $\varphi(\omega; id(\omega) = 43) = \varphi(id^{-1}(43)) = 1, ..., \varphi(\omega; id(\omega) = 19) = \varphi(id^{-1}(19)) = 20$.

Generally, for any function $\varphi : \mathbb{UNIV} \rightarrow \mathbb{R}^+$, the members of Ω_0 can be listed according to increasing values of φ. This listing can be transferred onto the set of the related values of X, and then the set $\{X(\omega), \omega \in \Omega_0\}$ can be presented as a sequence

$$D_{X;\varphi} = (x_{i;\varphi},\ i = 1, ..., n) \tag{12.1}$$

where $x_{i;\varphi}$ is the value of X assigned to this $\omega \in \Omega_0$ which occupies the i-th position when ordered according to ascending values of φ.

The sequence $D_{X;\varphi}$ will be called *the dataset on Ω_0 related to $X : \mathbb{UNIV} \rightarrow \mathbb{X}$ and to $\varphi : \mathbb{UNIV} \rightarrow \mathbb{R}^+$*. When φ is equal to X, we write D_X instead of $D_{X;X}$.

Remark 12.3 *When it is important to indicate which subset* Ω_0 *of* $\mathbb{UNIV}$ *is considered, we will use the notation* $D_{X;\varphi,\Omega_0}$ *and* $D_{X;\Omega_0}$ *instead of* $D_{X;\varphi}$ *and* D_X. *When it is more convenient to indicate only the number of elements of* Ω_0, *we will write* $D^{(n)}_{X;\varphi}$ *and* $D^{(n)}_X$.

Illustration (C/J data):

Suppose that the set of five judges is reduced to one judge with ID = 10, and let the values of $X : \mathbb{UNIV} \to \mathbb{X}$ be the scores delivered by this judge. When φ enumerates competitors according to their order of participation, the dataset on Ω_0 related to X and φ forms the first column in Table 12.1, headed *judge 10*. If one wanted to indicate the competitors in Ω_0 according to the increasing value of number of their identifiers, then $\varphi = id$, and one can see in Table 12.1 that $D^{(20)}_{X;id} =$ (264, 288, ..., 129) since the identifiers of the successive competitors increase from 7 to 88. If we put $\varphi =$ X, $D^{(20)}_X =$ (129, 164, ..., 334), i.e., the rows are ordered from the smallest score to the greatest.

Illustration (A/B data):

In the A/B data, Ω_0 is the set consisting of $n = 27$ employees (set $\mathbb{UNIV}$ is identified with Ω_0). Each member of Ω_0 has an ID attached by a function id (which can be given any suitable interpretation). Let X be the bonus obtained in the fourth year. Set $\mathbb{X}$ contains eight numbers:

$$\mathbb{X} = \{x_1, ..., x_8\} = \{0.5, 1, \ 1.5, 2, ..., 4\}$$

The function φ which orders Ω_0 so that the related dataset forms column 4 in Table 12.4 will be specified further by formula (12.16). Please note in Table 12.4 that the dataset related to $\varphi = id$ forms the sequence $D_{X;id} =$ (1.5, 0.5, 1.5, 0.5, 3.5,..., 1.5, 3, 1) and that the dataset D_X related to $\varphi = X$ consists of 27 numbers $\mathrm{x}_{i;X}$ such that

$\mathrm{x}_{1;X}, ..., \mathrm{x}_{4;X}$	are equal to	$\mathrm{x}_1 = 0.5$,
$\mathrm{x}_{5;X}, ..., \mathrm{x}_{8;X}$	are equal to	$\mathrm{x}_2 = 1$,
....	....	... ,
$\mathrm{x}_{25;X}, ..., \mathrm{x}_{27;X}$	are equal to	$\mathrm{x}_8 = 4$.

Any dataset D_X can be rewritten in an aggregated form as

$$\overline{D}_X = [(x_i, n_i), i = 1, ..., m] \tag{12.2}$$

where n_i is the frequency of x_i, and m is the number of different values of X which appear in D_X (generally $m \le n = \sum_{i=1}^{n} n_i$, with $m = n$ when the values $X(\omega)$, $\omega \in \Omega_0$ are all different).

Let $p_i = n_i/n$; the vector $\mathbf{p}_X = (p_1, ..., p_m)$ defines the distribution of X. The expectation $E(X) = \sum_{i=1}^{m} x_i p_i$ is equal to $\frac{1}{n}\sum_{i=1}^{n} x_{i;\varphi}$ for any order function φ. The products $x_i p_i$ form the partition of $E(X)$ and therefore the ratios

$$p_X^{(1)}(i) = \frac{x_i p_i}{E(X)} \quad , i = 1, ..., m \tag{12.3}$$

are called the *first moment probabilities of* X (see formula (4.2) in Chapter 4). For an arbitrary order function φ, partition of $E(X)$ is formed by the vector

$$\mathbf{p}_{X;\varphi}^{(1)} = \left(p_{X;\varphi}^{(1)}(i) = \frac{x_{i;\varphi}}{nE(X)} \quad , i = 1, ..., n \right), \tag{12.4}$$

called the first moment distribution of X related to order φ.

Measures and visual presentations of inequality between the first moment distribution of X (*"redistribution of goods"*) and the distribution of X, are based on the comparison of vectors $\mathbf{p}_{X;\varphi}^{(1)}$ and $\mathbf{p}_{X;\varphi} = \left(\frac{1}{n}, ..., \frac{1}{n}\right)$; if $\varphi = X$, we compare $\mathbf{p}_X^{(1)}$ and $\mathbf{p}_X = (p_1, ..., p_m)$. Each of the vectors $\mathbf{p}_{X;\varphi}^{(1)}$ and $\mathbf{p}_{X;\varphi}$ determines a discrete Lilliputian variable; and so we compare the distributions of these variables. The related concentration curve

$$L_{X;\varphi} = C\left(\mathbf{p}_{X;\varphi}^{(1)} : \mathbf{p}_{X;\varphi}\right)$$

is an ascending curve in the unit square, consisting of n segments joining points

$$\left(\frac{i-1}{n}, \sum_{s=1}^{i-1} p_{X;\varphi}^{(1)}(s)\right) \text{ and } \left(\frac{i}{n}, \sum_{s=1}^{i} p_{X;\varphi}^{(1)}(s)\right), \tag{12.5}$$

where $i = 1, ..., n$ and $p_{X;\varphi}^{(1)}(0) = 0$.

The curve is called the φ-ordered Lorenz Curve. The slope of the *i-th* segment of the Lorenz Curve is the ratio

$$p_{X;\varphi}^{(1)}(i) \Big/ \frac{1}{n} = \frac{x_{i;\varphi}}{E(X)} \,. \tag{12.6}$$

The Lorenz Curve $L_{X;\varphi}$ is the Lilliputian Curve of a continuous element of $\mathbb{ULM}$, called t*he Lorenz variable corresponding to* X and φ. The values $\frac{x_{i;\varphi}}{EX}$

are the successive values of the Lorenz density $\lambda_{X;\varphi}$ (i.e., the density of the Lorenz variable), defined as

$$\lambda_{X;\varphi}(u) = \frac{x_{i;\varphi}}{E(X)} \quad \text{for} \quad u \in \left(\frac{i-1}{n}, \frac{i}{n}\right), i = 1, ..., n. \tag{12.7}$$

If $\varphi = X$,

$$\lambda_X(u) = \frac{x_i}{E(X)} \quad \text{for } u \in \left(\sum_{s=1}^{i-1} p_s, \sum_{s=1}^{i} p_s\right), \ i = 1, ..., m.$$

We have

$$L_{X;\varphi}(u) = \int_0^u \lambda_{X;\varphi}(t)dt, \ u \in (0,1) \tag{12.8}$$

$$\lambda_{X;\varphi}(u) = dL_{X;\varphi}(u)/du, \quad u \in (0,1) \backslash \{\frac{i}{n}, i = 1, ..., n\} \tag{12.9}$$

The numerical measure of inequality is the Gini Index:

$$Gini(X;\varphi) = ar(L_{X;\varphi}), \tag{12.10}$$

which attains its maximal value in the set of all order functions φ when $\varphi = X$; we use the notation $Gini(X) = Gini(X;X) = ar(L_X)$.

Illustration (C/J data):
Figure 12.9 presents the Lorenz Curves $L_{X_{10}}$ and $L_{X_{33}}$ for variables X_{10} and X_{33}, corresponding to scores delivered by judges 10 and 33; moreover, Figure 12.9 contains $L_{X_{33};\varphi}$ where φ corresponds to the order of participation. The Gini Indices are: $Gini(X_{10}) = 0.1273$, $Gini(X_{33}) = 0.3030$, $Gini(X_{33};\varphi) = 0.0359$.

Illustration (A/B data):
Figure 12.10 presents the Lorenz Curves L_{X_1} and L_{X_4} for variables corresponding to the bonus obtained by employees in the first, fourth and fifth year. Moreover, Figure 12.10 contains $L_{X_4;id}$ corresponding to variable X_4 ordered by employee identification.

The slopes of the segments show the related values of the Lorenz densities.

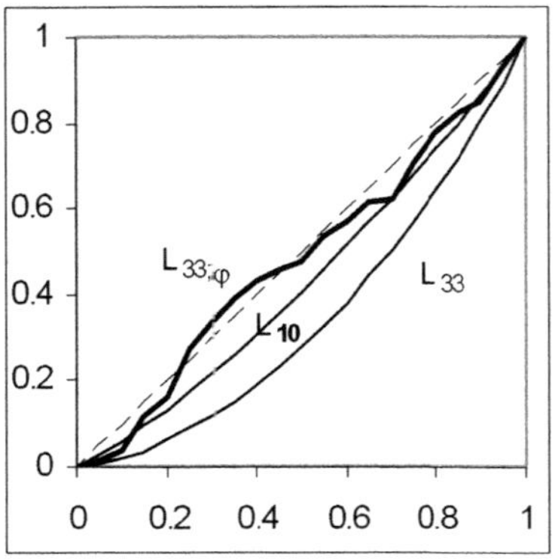

FIGURE 12.9. The C/J data: the Lorenz Curves $L_{X_{10}}$, $L_{X_{33}}$ and $L_{X_{33;\varphi}}$.

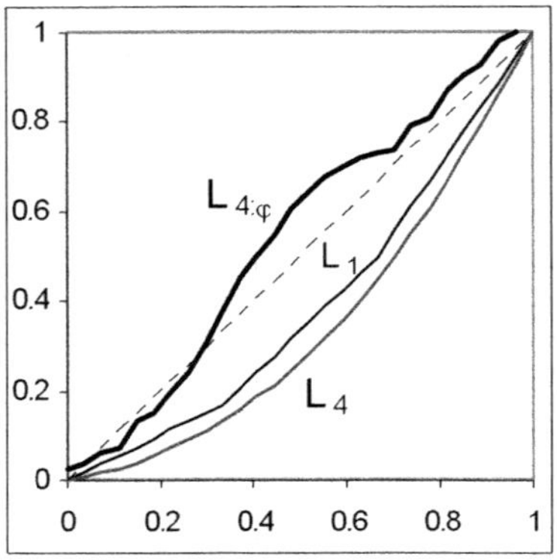

FIGURE 12.10. The A/B data: the Lorenz Curves L_{X_1}, L_{X_4} and $L_{X_{4;id}}$.

Please note that the Lorenz Curve L_{X_4} consists of *eight* segments since there are just $m = 8$ values of X_4 (many adjacent segments have the same slope). We have $\lambda_{X_4}(u) = 0.2368$ for $u \in (0, \frac{4}{27})$; $\lambda_{X_4}(u) = 0.4737$ for $u \in (\frac{4}{27}, \frac{8}{27})$; ...; $\lambda_{X_4}(u) = 1.8947$ for $u \in (\frac{24}{27}, 1)$; the corresponding probabilities are $\frac{4}{27}, \frac{4}{27}, \ldots, \frac{3}{27}$. The Gini Indices are: $Gini(X_1) = 0.2222$, $Gini(X_4) = 0.3132$, $Gini(X_4; id) = -0.0123$.

* * *

The definitions of the φ-ordered Lorenz Curve and φ-ordered Gini Index of $X \in \overline{\mathbb{RATIO}}$ were introduced in Chapter 4, Section 4.3.3 in the general case (i.e., without the restrictions introduced in the present section that Ω_0 and $\mathbb{X}$ are finite and that $P(\omega) = \frac{1}{n}$ for any $\omega \in \Omega_0$). However, a review of this simple case was needed to introduce in the next section the finite multivariate datasets and related inequality measures for vectors of discrete ratio variables.

Note that random samples of size n provide finite subsets $\Omega_0 \subset \mathbb{UNIV}$ with a uniform sample distribution $P^{(n)}_{\Omega_0}(\omega) = \frac{1}{n}$; sample distributions are described by the *sample inequality measures* which converge to the related population measures when n tends to infinity. This shows that if set $\mathbb{X}$ is finite, there is a perfect agreement between the univariate inequality measures used in the data analysis approach, the probabilistic inference approach, and the statistical approach.

12.5 Finite multivariate datasets and related inequality measures

12.5.1 Finite data tables and their grade regression functions

The contents of Section 12.4 will be now extended to deal with *finite multivariate datasets* (i.e., with datasets with a finite number of records (rows, objects) and k variables ($k \geq 2$).

Instead of a single variable $X \in \overline{\mathbb{RATIO}}$, let $\mathbf{X} = (X_1, ..., X_k)$ be a vector with k components such that

$$\mathbf{X} = (X_1, ..., X_k) \in \overline{\mathbb{RATIO}}^k, \quad X_j : \mathbb{UNIV} \to \mathbb{X}^{[j]}.$$

The assumption $\mathbf{X} \in \overline{\mathbb{RATIO}}^k$ will hold in Section 12.5-12.7 without further mentioning.

Let Ω_0 be a set containing n elements of $\mathbb{UNIV}$; let all variables X_j be discrete; let the set $\mathbb{X}^{[j]}$ ($j = 1,...,k$) contain m_j elements denoted $x_1^{[j]}, ..., x_{m_j}^{[j]}$. Then a record corresponding to $\omega \in \Omega_0$ is equal to $(X_1(\omega), X_2(\omega), ..., X_k(\omega))$, and the total number of different records in the dataset cannot exceed the minimum of two numbers: n (the total number of records in the dataset) and the product $m_1 \cdot m_2 \cdot ... \cdot m_k$) (maximal possible number of different records formed by $X_1,...,X_k$).

Analogously as in the case of a single variable X, we introduce a $n \times k$ table

$$\mathbf{D}_{\mathbf{X};\varphi} = [x_{ij;\varphi} : i = 1, ..., n \ , \ j = 1, ..., k] \tag{12.11}$$

(where $x_{ij;\varphi}$ is the value of variable X_j assigned to this $\omega \in \Omega_0$ which is on the *i-th* position when ordered according to ascending values of φ).

Table $\mathbf{D}_{\mathbf{X};\varphi}$ will be called *the dataset on* Ω_0 *related to vector* $\mathbf{X}$ *and order function* $\varphi : \mathbb{UNIV} \to \mathbb{R}^+$. Rows of $\mathbf{D}_{\mathbf{X};\varphi}$ are called *records:* the i-th record according to φ is $x_{i;\varphi} = (x_{i1;\varphi}, ..., x_{ik;\varphi})$.

The records (rows) of dataset $\mathbf{D}_{\mathbf{X};\varphi}$ are ordered according to φ, while the columns correspond to the order of the components of vector $\mathbf{X}$. When the component variables $X_1, ..., X_k$ are permuted according to a permutation $\psi = (\psi_{(1)}, ..., \psi_{(k)})$, we get the permuted vector

$$\mathbf{X}_\psi = \left(X_{\psi(1)}, ..., X_{\psi(k)}\right). \tag{12.12}$$

The permuted vector also belongs to $\overline{\mathbb{RATIO}}^k$.

Symbol $\mathbf{D}_{\mathbf{X}_\psi;\varphi}$ will denote a dataset on Ω_0 related to vector $\mathbf{X}_\psi$ and order function φ which orders records.

Any dataset $\mathbf{D}$ can be normed to form a probability table, i.e., a table with terms from [0,1] which sum to 1. This is achieved by dividing each term by the sum of all terms $S = \sum_i \sum_j x_{ij;\varphi}$.

Definition 12.1 *(probability data table). Table*

$$\mathbf{T}_{\mathbf{X};\varphi} = \frac{1}{S}\mathbf{D}_{\mathbf{X};\varphi}, \quad S = \sum_i \sum_j x_{ij,\varphi} \tag{12.13}$$

(where $\mathbf{D}_{\mathbf{X};\varphi}$ is a dataset on Ω_0 related to X and φ) will be called the probability data table related to $\mathbf{X}$ and φ. (The word "probability" will be usually omitted whenever this will cause no confusion).

The words dataset and data table are used rather interchangeably. Moreover, we will use the term *grade regression functions of a dataset* meaning such functions of the corresponding probability data table.

We remind that for a given probability table $[p_{ij}; i = 1, ..., m, j = 1, ..., k]$ with row sums $p_{i+}, i = 1, ..., m$, column sums $p_{+j}, j = 1, ..., k$, and $S_i = p_{1+} + \cdots + p_{i+}$, $T_j = p_{+1} + \cdots + p_{+j}$, the grade *col|row* regression in point $u \in (0,1)$ according to (8.31) is equal

$$\sum_{j=1}^{k}(T_{j-1} + \frac{1}{2}p_{+j})\frac{p_{ij}}{p_{i+}}, \quad u \in (S_{i-1}, S_i),\ i = 1, ..., m$$

and *row|col* regression in point $v \in (0,1)$ is equal to

$$\sum_{i=1}^{m}(S_{i-1} + \frac{1}{2}p_{i+})\frac{p_{ij}}{p_{+j}}, \quad v \in (T_{j-1}, T_j),\ j = 1, ..., k.$$

Each grade *col|row* regression function of $\mathbf{T}_{\mathbf{X};\varphi}$ is functionally related to φ in the following way: the vertical interval [0,1] in the overrepresentation map for $\mathbf{T}_{\mathbf{X};\varphi}$ is divided into n intervals corresponding to records; the grade *col*|*row* regression of $\mathbf{T}_{\mathbf{X};\varphi}$ is constant on these intervals. If $r^*_{col|row}(u;\mathbf{T}_{\mathbf{X};\varphi})$ is nondecreasing, it is an order function equivalent to φ.

Similarly, the grade *row*|*col* regression function of $\mathbf{T}_{\mathbf{X};\varphi}$ establishes reordering of the components of vector $\mathbf{X}$.

The GCA procedure can be applied to data tables in the same way it was applied to the probability tables in Chapters 8 and 9.

Illustration (C/J data):

The components of vector $\mathbf{X} = (X_{10}, X_{33}, X_{17}, X_{41}, X_{49})$ are scores delivered by the respective judges. Sets $\mathbb{UNIV}$ and Ω_0 are as defined earlier in Section 12.2.1 (100 registered competitors and first 20 participants). Sets $\mathbb{X}_j$ are finite since $\mathbb{UNIV}$ is finite; their sizes m_j are close to 100; in the subset Ω_0 the number of different values is 20 for each variable. Table 12.1 is the dataset $\mathbf{D}_{\mathbf{X};\varphi}$ for Ω_0 related to $\mathbf{X}$ and to the participation order φ. Table 12.2 is the dataset of the permuted vector $\mathbf{X}_\psi = (X_{33}, X_{41}, X_{10}, X_{49}, X_{17})$ (so that $\psi = (\psi(1), ..., \psi(5)) = (2, 4, 1, 5, 3)$). We recall (Section 12.2.1) that Table 12.2 is the GCA transform of Table 12.1 and therefore both rows and columns of this table are ordered according to the increasing values of the respective grade regression functions:

$$r^*_{col|row}(u; Table\ 12.2) = \begin{cases} 0.3005 & for \quad u \in [0.0000; 0.0510] \\ 0.3407 & for \quad u \in (0.0510; 0.0887] \\ .... & .... \\ 0.6710 & for \quad u \in (0.9174; 0.9610] \\ 0.7100 & for \quad u \in (0.9610; 1.0000] \end{cases} \tag{12.14}$$

and

$$r^*_{row|col}(u; Table\ 12.2) = \begin{cases} 0.3553 & for \quad u \in [0.0000; 0.2203] \\ 0.4231 & for \quad u \in (0.2203; 0.3681] \\ 0.4881 & for \quad u \in (0.3681; 0.5908] \\ 0.5766 & for \quad u \in (0.5908; 0.8140] \\ 0.6548 & for \quad u \in (0.8140; 1.0000] \end{cases} \tag{12.15}$$

Thus Table 12.2 forms $\mathbf{D}_{\mathbf{X}_\psi;\varphi}$ for φ given by (12.14) and $\psi = (2, 4, 1, 5, 3)$.

Illustration (A/B data):

The components of vector $\mathbf{X} = (X_1, ..., X_5\ )$ denote the bonus obtained by the employees in chronologically ordered years. Set Ω_0 consists of 27 employees; $\mathbb{UNIV}$ is here identified with Ω_0; further, $\mathbb{X}_1 = \{1, 2, 3\}$, $\mathbb{X}_2 = \{1{,}5, 2.5, 3.5\}$, $\mathbb{X}_3 = \{2, 3, 4\}$, $\mathbb{X}_4 = \{\frac{1}{2}\text{i}, \text{i}=1,..., 8\}$, $\mathbb{X}_5 = \{\frac{1}{2}\text{i}, \text{i}=1,..., 12\}$.

The dataset $\mathbf{D}_{\mathbf{X};id}$ in which rows are ordered according to successive identificators 1,...,27 was not presented in Section 12.2.2 as a separate table but can be easily reconstructed from Table 12.4.

As stated in Section 12.2.2, Table 12.4 is a constrained GCA transformation of $\mathbf{D}_{\mathbf{X};id}$ in which the years (columns) are kept chronologically ordered, but rows are ordered according to the related *col*|*row* grade regression function, which is equal to

$$r^*_{col|row}(u; Table\ 12.4) = \begin{cases} 0.3069 & for \quad u \in [0.0000; 0.0215] \\ 0.3493 & for \quad u \in (0.0215; 0.0486] \\ .... & .... \\ 0.6023 & for \quad u \in (0.9141; 0.9599] \\ 0.6091 & for \quad u \in (0.9599; 1.0000] \end{cases} \tag{12.16}$$

The overrepresentation map of dataset $\mathbf{D}_{\mathbf{X};\varphi}$ related to φ given by (12.16) is presented in Figure 12.5.

Dataset $\mathbf{D}_{\mathbf{X};id}$ transformed by *full* (*unconstrained*) GCA can be visualized by the map in Figure 12.6. The column permutation $\psi = (1, 4, 2, 3, 5)$ corresponds to the following *row*|*col* grade regression of the GCA table (visualized in Figure 12.5):

$$r^*_{row|col}(u; Figure\ 12.5) = \begin{cases} 0.3989 & for \quad u \in [0.0000; 0.1545] \\ 0.4451 & for \quad u \in (0.1545; 0.3176] \\ 0.4668 & for \quad u \in (0.3176; 0.5107] \\ 0.5070 & for \quad u \in (0.5107; 0.7425] \\ 0.6140 & for \quad u \in (0.7425; 1.000] \end{cases} \tag{12.17}$$

The row permutation in this GCA table (visualized in Figure 12.5) corresponds to the *col*|*row* grade regression of the GCA table:

$$r^*_{col|row}(u; Figure\ 12.5) = \begin{cases} 0.3546 & for \quad u \in [0.0000; 0.0215] \\ 0.3819 & for \quad u \in (0.215; 0.0486] \\ .... & .... \\ 0.6181 & for \quad u \in (0.9313; 0.9599] \\ 0.6276 & for \quad u \in (0.9599; 1.0000] \end{cases} \tag{12.18}$$

* * *

Let $z_{i;\varphi}$ be the values of the sum

$$Z = \sum\nolimits_{j=1}^{k} X_j \tag{12.19}$$

in the i-th row of data probability table $\mathbf{T}_{\mathbf{X};\varphi}$. The row sums in $\mathbf{T}_{\mathbf{X};\varphi}$ are equal to $z_{i;\varphi}/nE(Z)$, $i = 1, ..., n$, the column sums are equal to $E(X_j)/E(Z)$, $j = 1, ..., k$. Denoting

$$v_j = \sum\nolimits_{s=1}^{j} E(X_s)/E(Z), \quad j = 1, ..., k, \tag{12.20}$$

$$b_{i;\varphi} = \sum_{s=1}^{i} z_{s;\varphi}/(nE(Z)), \quad i = 1, ..., n \tag{12.21}$$

we find that, for $u \in (b_{i-1,\varphi}, b_{i;\varphi})$, the *col|row* grade regression function for $\mathbf{T}_{\mathbf{X};\varphi}$ is equal to

$$\sum_{j=1}^{k} \left(v_{j-1} + \frac{1}{2} E(X_j)/E(Z) \right) \left(\frac{x_{ij;\varphi}/nE(Z)}{z_{i;\varphi}/nE(Z)} \right).$$

This shows that the *col|row* regression function for $\mathbf{T}_{\mathbf{X};\varphi}$ may be treated as defined on the set of records. A detailed justification goes as follows:

For $\omega \in \Omega_0$, let

$$R_{\mathbf{X}}(\omega) = \sum\nolimits_{j=1}^{k} \frac{\left(v_{j-1} + \frac{1}{2} E(X_j)/E(Z)\right) X_j(\omega)}{Z(\omega)} \tag{12.22}$$

define the order function $\mathbf{R}_{\mathbf{X}}$ which will play an important role in describing any multivariate dataset $\mathbf{D}_{\mathbf{X};\varphi}$. *Please note that this is a function of vector* $\mathbf{X}$ *and of its expectation* $E(\mathbf{X}) = (E(X_1), ..., E(X_k))$; *therefore* $R_{\mathbf{X}}$ *is in fact defined on the set of records* (regardless of how these records are ordered) and so we may use the notation $R_{\mathbf{X}}(\mathbf{x})$ for any record $\mathbf{x}$. Suppose now, that the elements $\omega_1, ..., \omega_n$ in Ω_0 are ordered according to φ. Then, due to formula (8.31) in Chapter 8, the *col|row* grade regression function for dataset $\mathbf{D}_{\mathbf{X};\varphi}$ is functionally connected with $R_{\mathbf{X}}$:

$$R_{\mathbf{X}}(\omega_i) = r^*_{col|row}(u\,; \mathbf{D}_{\mathbf{X};\varphi}) \text{ for } u \in (b_{i-1;\varphi},\ b_{i;\varphi}],\ i = 1, ..., n. \tag{12.23}$$

If $\varphi = \mathbf{R}_{\mathbf{X}}$ then $\mathbf{R}_{\mathbf{X}}(\omega_i) = \mathbf{R}_{\mathbf{X}}(x_{i1}, \dots, x_{ik}) = \mathbf{R}_{\mathbf{X}}(\mathbf{x}_i)$ where

$$\mathbf{R}_{\mathbf{X}}(\mathbf{x}_i) = \sum_{j=1}^{k} \frac{\left(v_{j-1} + \frac{1}{2} E(X_j)/E(Z)\right) x_{ij}}{z_i}$$

Illustration (C/J data):

Let $\varphi = \mathbf{R}_{\mathbf{X}}$ so that we consider Table 12.2. The object ω_1 is a competitor with ID=35, record $\mathbf{x}_1 = (528, 224, 210, 101, 39)$ and, according to formula 12.22,

$$\mathbf{R}_{\mathbf{X}}(\mathbf{x}_1) = 0.3305.$$

Remark 12.4 *The expectations $E(X_j)$ appearing in 12.22 are related to the given set of objects denoted Ω_0. (For instance $E(X_{33})$ (scores of judge 33) is related to the chosen subset of competitors. Thus the values $E(X_{33})$ will be different in the set of ten first competitors in upper part of Table 12.1 and in the set of twenty first competitors (whole Table 12.1)). It follows that* $\mathbf{R_X}$ *is also related to a given dataset. Whenever needed, it is indicated to use the notation like* $\mathbf{R_X}(\mathbf{x}; \mathbf{D}_{\mathbf{X};\varphi})$, $E_{\Omega_0}(X_{33})$, $\rho^*_{col|row}(u;$ *Table* 12.4) *as in formula (12.16), etc.*

12.5.2 Lorenz Surfaces

The values of $Z = \sum_i X_i$ are usually presented in an additional column of dataset $\mathbf{D}_{\mathbf{X};\varphi}$ which follows the column with the values of the last component of vector $\mathbf{X}$. This *marginal column* forms the univariate dataset $D_{Z;\varphi}$. The partition of the total mean $E(Z)$, given by

$$E(Z) = \frac{1}{n} \sum_{i=1}^{n} z_{i;\varphi}, \tag{12.24}$$

gives rise to the marginal first moment distribution $\mathbf{p}^{(1)}_{Z;\varphi}$ (according to formula 12.3), and then also to the Lorenz density $\lambda_{Z;\varphi}$ and the Lorenz Curve $L_{Z;\varphi}$ (according to formulas (12.7)-(12.8)).

A *more detailed partition of the total mean*,

$$E(Z) = \frac{1}{n} \sum_{i=1}^{n} \sum_{j=1}^{k} x_{ij;\varphi} \ , \tag{12.25}$$

leads to the *bivariate first moment distribution* with probabilities

$$p^{(1)}_{ij;\varphi} = \frac{x_{ij;\varphi} p_{i;\varphi}}{E(Z)} = \frac{x_{ij;\varphi}}{nE(Z)} \quad i = 1, ..., n, \ j = 1, ..., k \ . \tag{12.26}$$

Since

$$\frac{x_{ij;\varphi}}{nE(Z)} = \frac{x_{ij;\varphi}}{\sum_s \sum_t x_{st;\varphi}},$$

the table of the first moment probabilities is equal to the probability data table $\mathbf{T}_{\mathbf{X};\varphi}$ given by the formula (12.13), i.e.,

$$\left[p^{(1)}_{ij;\varphi}, \ i = 1, ..., n, \ j = 1, ..., k\right] = \mathbf{T}_{\mathbf{X};\varphi}$$

Similarly as in the univariate case, let us replace $p^{(1)}_{ij;\varphi}$ by the ratios

$$p^{(1)}_{ij;\varphi}/p_{ij;\varphi} = \frac{x_{ij;\varphi}}{E(X_j)}\ ,\ i = 1, ..., n\ ,\ j = 1, ..., k. \tag{12.27}$$

where $p_{ij;\varphi} = p_{i;\varphi}E(X_j)/E(Z)$ and $p_{i;\varphi} = \frac{1}{n}$.

Definition 12.2 *(Lorenz Density, Lorenz Surface and Lorenz Pair). The ratios given by (12.27) define on the unit square the Lorenz Density related to vector* **X** *and order function* φ*, denoted* $\boldsymbol{\lambda}_{\mathbf{X};\varphi}$ *:*

$$\boldsymbol{\lambda}_{\mathbf{X};\varphi}(u, v) = x_{ij;\varphi}/E_{\Omega_0}(X_j) \tag{12.28}$$

for $(u, v) \in (a_{i-1}, a_i] \times (v_{j-1}, v_j]$, $i = 1, ..., n$, $j = 1, ..., k$, *where*

$$a_0 = v_0 = 0, \quad a_i = \sum\nolimits_{s=1}^{i} p_{s;\varphi}, \tag{12.29}$$

and v_j *are given by formula (12.20).*

The bivariate cdf formed from density $\boldsymbol{\lambda}_{\mathbf{X};\varphi}$ *will be called the Lorenz Surface related to* **X** *and* φ*, and will be denoted* $\mathbf{L}_{\mathbf{X};\varphi}$.

The element of set $\mathbb{BLM}$ *(Bivariate Lilliputian Model introduced in Chapter 8) with the density* $\boldsymbol{\lambda}_{\mathbf{X};\varphi}$ *will be called the Lorenz Pair related to vector* **X** *and function* φ*. So the Lorenz Pair is defined by the Lorenz Density or, equivalently, by the Lorenz Surface.*

Note that

$$\int_0^1 \boldsymbol{\lambda}_{\mathbf{X};\varphi}(u, v)du = 1, \quad \int_0^1 \boldsymbol{\lambda}_{\mathbf{X};\varphi}(u, v)dv = \frac{z_{i;\varphi}}{E(Z)}$$

for $u \in (\frac{i-1}{n}, \frac{i}{n})$, $i = 1, ..., n$. Hence the Lorenz Surface $\mathbf{L}_{\mathbf{X};\varphi}$ can be expressed by the set of φ-ordered Lorenz Curves of the components of **X**: for any $(u, v) \in (0, 1] \times (v_{j-1}, v_j]$,

$$\begin{aligned}\mathbf{L}_{\mathbf{X};\varphi}(u, v) &= \sum\nolimits_{s=1}^{j-1}(v_s - v_{s-1})L_{X_s;\varphi}(u) + (v - v_{j-1})L_{X_j;\varphi}(u),\\ &= \mathbf{L}_{\mathbf{X};\varphi}(u, v_{j-1}) + (v - v_{j-1})L_{X_j;\varphi}(u),\end{aligned} \tag{12.30}$$

where v_j are defined by (12.20).

The marginal horizontal distribution of the Lorenz Pair (i.e. the marginal distribution which corresponds to columns) is uniform:

$$\mathbf{L}_{\mathbf{X};\varphi}(1,v) = v = Diag(v), \qquad v \in [0,1]. \tag{12.31}$$

The vertical marginal distribution of the Lorenz Pair (the one concerning records) has a *cdf* equal to the φ-ordered Lorenz Curve of $Z = \sum_j X_j$:

$$\mathbf{L}_{\mathbf{X};\varphi}(u,1) \ = L_{Z;\varphi}(u), \ u \in (0,1). \tag{12.32}$$

Definition 12.3 *The element of* $\mathbb{BLM}$ *with the Surface* $\mathbf{L}^{\perp}_{\mathbf{X};\varphi}$ *defined by*

$$\mathbf{L}^{\perp}_{\mathbf{X};\varphi}(u,v) = L_{Z;\varphi}(u) \times Diag(v) \tag{12.33}$$

will be called the Independent Lorenz Pair related to $\mathbf{X}$ *and* φ.

The cdf of the Independent Lorenz Pair is the product of the marginal cdf's of the Lorenz Surface $\mathbf{L}_{\mathbf{X};\varphi}$, which means that it is defined by $L_{Z;\varphi}$ only. If $\varphi = Z$ and $\mathbf{L}_{\mathbf{X};\varphi}$ is close to $\mathbf{L}^{\perp}_{\mathbf{X};Z}$ (see Figure 12.12), it can be roughly said that Z orders records so that ordering of vector components is "almost independent" of the ordering of records.

> The vertical (row) marginal distribution of the Lorenz Pair related to $\mathbf{X}$ and φ is uniform *if and only if the distribution of Z is concentrated at one point.* This happens if and only if the sums of the record components are the same for all records in Ω_0. This is *the neccessary and sufficient condition* that the Lorenz Pair related to $\mathbf{X}$ and φ belongs to the Dependence Lilliputian Model ($\mathbb{DLM}$) which is contained in $\mathbb{BLM}$ (see Chapter 8).
>
> The comparison of $\mathbf{L}^{\perp}_{\mathbf{X};\varphi}$ and $Diag_2$ provides information about the degree of inequality of Z ordered according to φ.
>
> If the Lorenz Surface $\mathbf{L}_{\mathbf{X};\varphi}$ is not a member of $\mathbb{DLM}$ (i.e., it is not a copula) it can be transformed to such by a *grade transformation of the marginal row distribution into a uniform distribution.* This leads to *the grade distribution of the Lorenz Pair,* which is directly derivable from the probability data table $\mathbf{T}_{\mathbf{X};\varphi}$ (due to formula (12.34)). It follows that the cdf of the grade distribution of $\mathbf{L}^{\perp}_{\mathbf{X};\varphi}$ is $Diag_2$.

Definition 12.4 *(Grade Lorenz Density, Grade Lorenz Surface, and Grade Lorenz Pair). The function* $\overset{*}{\boldsymbol{\lambda}}_{\mathbf{X};\varphi}$ *defined by*

$$\overset{*}{\boldsymbol{\lambda}}_{\mathbf{X};\varphi}(u,v)=\left(\frac{x_{ij;\varphi}}{n\,E(Z)}\right)\Big/\left(\frac{z_{i;\varphi}}{n\,E(Z)}\,\frac{E(X_j)}{E(Z)}\right)=\frac{x_{ij;\varphi}}{E(X_j)}\Big/\frac{z_{i;\varphi}}{E(Z)} \qquad (12.34)$$

$$\text{for } u\in\left(\textstyle\sum_{s=1}^{i-1}\frac{z_{s;\varphi}}{n\,E(Z)},\ \sum_{s=1}^{i}\frac{z_{s;\varphi}}{n\,E(Z)}\right),\ v\in(v_{j-1},v_j)$$

will be called the Grade Lorenz Density related to vector $\mathbf{X}$ *and order function* φ.

The bivariate cdf obtained from grade density $\overset{*}{\boldsymbol{\lambda}}_{\mathbf{X};\varphi}$ *will be called Grade Lorenz Surface related to* $\mathbf{X}$ *and* φ, *and will be denoted* $\overset{*}{\mathbf{L}}_{\mathbf{X};\varphi}$.

The element of set $\mathbb{DLM}$ *(Dependence Lilliputian Model introduced in Chapter 8) with density* $\overset{*}{\boldsymbol{\lambda}}_{\mathbf{X};\varphi}$ *will be called the Grade Lorenz Pair related to* $\mathbf{X}$ *and* φ. *So the Grade Lorenz Pair is defined by the Grade Lorenz Density or, equivalently, by the Grade Lorenz Surface.*

> The Grade Lorenz Density $\overset{*}{\boldsymbol{\lambda}}_{\mathbf{X};\varphi}$ can be visualized by means of maps on the unit square which discretize the values of $\overset{*}{\boldsymbol{\lambda}}$ (see Figures 12.1-12.4 for the C/J data and Figures 12.5-12.8 for the A/B data). Recall that the visualization of a grade distribution in the form of an overrepresentation map was introduced in Chapter 8, Section 8.5.

Proposition 12.1 *(i)*

$$\mathbf{L}_{\mathbf{X};\varphi}(u,v)=\overset{*}{\mathbf{L}}_{\mathbf{X};\varphi}(L_{Z;\varphi}(u),v)$$

and

$$\mathbf{L}_{\mathbf{X};\varphi}(L_{Z;\varphi}^{-1}(u),v)=\overset{*}{\mathbf{L}}_{\mathbf{X};\varphi}(u,v)$$

for any $(u,v)\in(0,1)\times(0,1)$.

(ii) For $u\in(0,1)$, $v\in(v_{j-1},v_j)$, $j=1,\ldots,k$

$$\overset{*}{\boldsymbol{\lambda}}_{\mathbf{X};\varphi}(u,v)=\frac{\boldsymbol{\lambda}_{\mathbf{X};\varphi}(u',v)}{\lambda_{Z;\varphi}(u',v)},$$

where $u'=L_{Z;\varphi}^{-1}(u)$. *Here, the Grade Lorenz Density* $\overset{*}{\boldsymbol{\lambda}}_{\mathbf{X};\varphi}$ *is the ratio of the Lorenz Density* $\boldsymbol{\lambda}_{\mathbf{X};\varphi}$ *and the marginal Lorenz Density* $\lambda_{Z;\varphi}$, *provided that the* *first argument of both densities is transformed by the inverse of* $L_{Z;\varphi}$. *This transformation ensures that the vertical marginal is uniform.*

(iii) As the Grade Lorenz Surface $\mathbf{L}^*_{\mathbf{X};\varphi}$ *arises from the Lorenz Surface* $\mathbf{L}_{\mathbf{X};\varphi}$ *when the marginal cdf* $L_{Z;\varphi}$ *is transformed to Diag, then after this transformation the marginal cdf* $L_{Z;\varphi}$ *in the surface* $\mathbf{L}_{\mathbf{X};\varphi}$ *is replaced by the concentration curve*

$$C(L_{Z;\varphi} : L_{Z;\varphi}) = L_{Z;\varphi}(L^{-1}_{Z;\varphi}) = Diag.$$

Moreover, the concentration curves

$$C(L_{X_j;\varphi} : L_{Z;\varphi}) = L_{X_j;\varphi}(L^{-1}_{Z;\varphi}), \; j = 1, ..., k \tag{12.35}$$

are the conditional cdfs $L^*_{X_j;\varphi}$ *of* $\mathbf{L}^*_{\mathbf{X};\varphi}$.

We see from formula (12.34) that the conditional cdfs mentioned in proposition 12.1(iii) have densities equal to the ratios of the densities of cdfs $L_{X_j;\varphi}$ and $L_{Z;\varphi}$.

Illustration (C/J data):

The five components of vector $\mathbf{X}$ in C/J data are scores delivered by judges 10, 33, 17, 41, 49. Let us choose to illustrate a conditional grade Lorenz distribution (see Figure 12.11).

$$L^*_{X_{33};Z}(u) = \begin{cases} 1.8632u & for \quad u \in [0.0000; 0.0377] \\ 0.0703{+}0.3814(u - 0.0377) & for \quad u \in (0.0377; 0.0763] \\ \ldots\ldots & \ldots \quad \ldots\ldots \\ 0.8867{+}0.7431(u - 0.8728) & for \quad u \in (0.8728; 0.9361] \\ 0.9337{+}1.0377(u - 0.9361) & for \quad u \in (0.9361; 1.0000] \end{cases} \tag{12.36}$$

$$\lambda^*_{X_{33};Z}(u) = \begin{cases} 1.8632 & for \quad u \in [0.0000; 0.0377] \\ 0.3814 & for \quad u \in (0.0377; 0.0763] \\ \ldots & \ldots \\ 0.7431 & for \quad u \in (0.8728; 0.9361] \\ 1.0377 & for \quad u \in (0.9361; 1.0000] \end{cases} \tag{12.37}$$

The intersections of the Grade Lorenz Surface $\mathbf{L}^*_{(X_{33},x_{41},X_{10},X_{49},X_{17});\varphi}$ by the planes $v = v_j$ with v_j given by the formula (12.20) are presented in Figure 12.11 when $\varphi = Z$, and in Figure 12.13 when $\varphi = R_{\mathbf{X}}$.

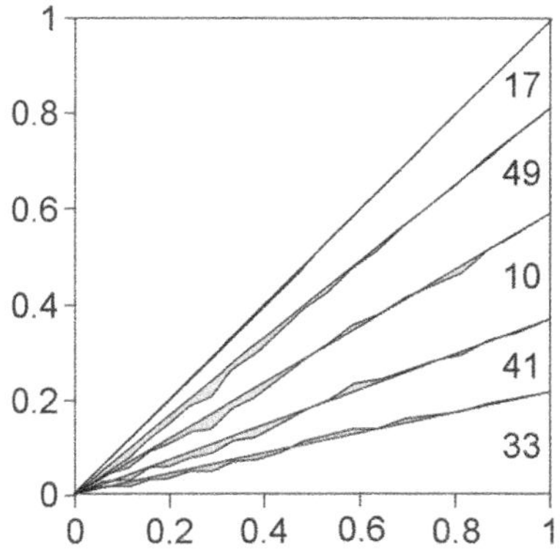

FIGURE 12.11. The C/J data: the intersection of the Grade Lorenz Surface $\mathbf{L}^*_{(X_{33},X_{41},X_{10},X_{49},X_{17});Z}$ by the planes $v = v_j$, $j = 1, ..., 5$ and the relative intersections of $Diag_2$. The lengths of the segments of the right vertical side are equal to $(E(X_{33}/E(Z), \dots, E(X_{17})/E(Z))$.

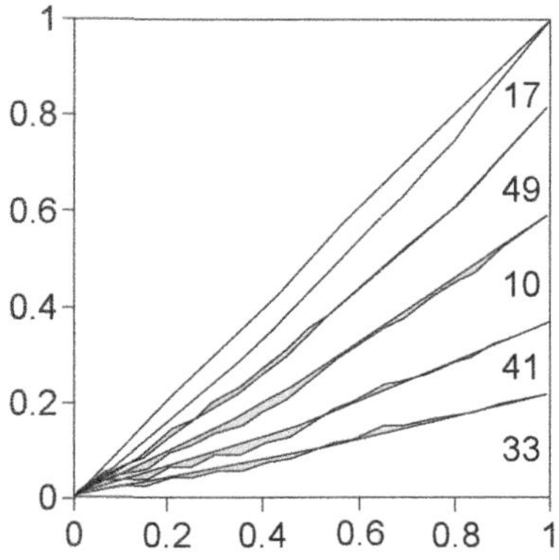

FIGURE 12.12. The C/J data: there are five pairs of curves; one curve in each pair refers to the intersection of the Lorenz Surface $\mathbf{L}_{(X_{33},X_{41},X_{10},X_{49},X_{17});Z}$ by a plane $v = v_j$, $j = 1, ..., 5$ (for $j = 5$ the intersection (curve at the top) is equal to L_Z); second curve in each pair refers to the analogous intersection of $\mathbf{L}^{\perp}_{X_{33},...,X_{17};Z}$.

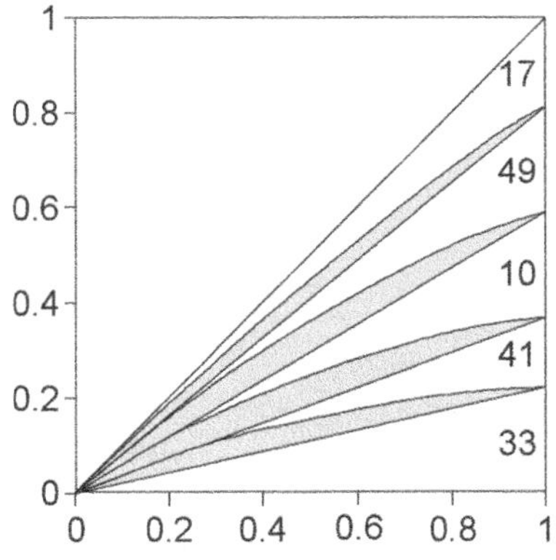

FIGURE 12.13. The C/J data: the intersections of the Grade Lorenz Surface $\mathbf{L}^*_{(X_{33},\dots,X_{17});R_{\mathbf{X}}}$ (upper curve) and of the respective intersections of $Diag_2$ (lines under the curve).

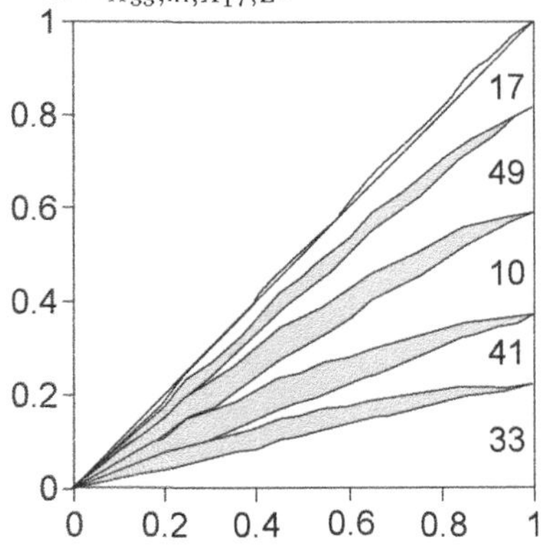

FIGURE 12.14. The C/J data: the intersections of Lorenz Surface $\mathbf{L}_{(X_{33},\dots,X_{17});R_{\mathbf{X}}}$ by the planes $v = v_j$, $j = 1, ..., 5$ and the related intersections of $\mathbf{L}^{\perp}_{(X_{33},\dots,X_{17});R_{\mathbf{X}}}$.

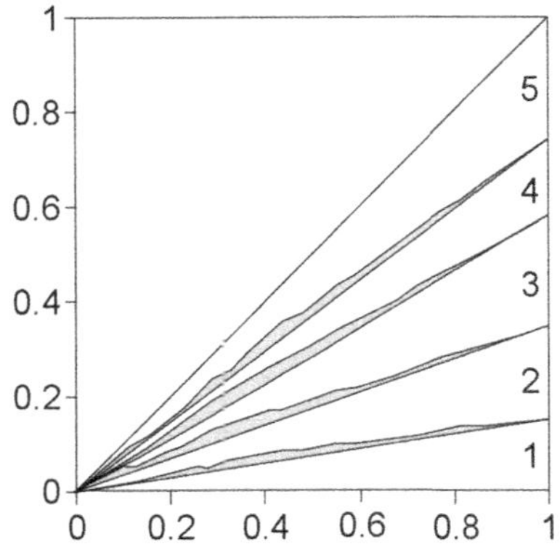

FIGURE 12.15. The A/B data: the intersection of the Grade Lorenz Surface $\mathbf{L}^*_{\mathbf{X};Z}$ and of $Diag_2$ by the planes $v = v_j$, $j = 1, ..., 5$.

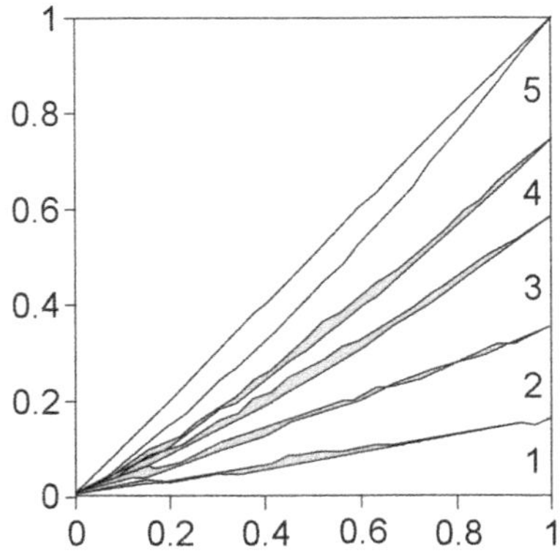

FIGURE 12.16. The A/B data: the intersections of the Lorenz Surface $\mathbf{L}_{\mathbf{X};Z}$ and of the $\mathbf{L}^{\perp}_{\mathbf{X};Z}$ by the planes $v = v_j$, $j = 1, ..., 5$ (for $j = 5$ the intersection (curve at the top) is equal to L_Z).

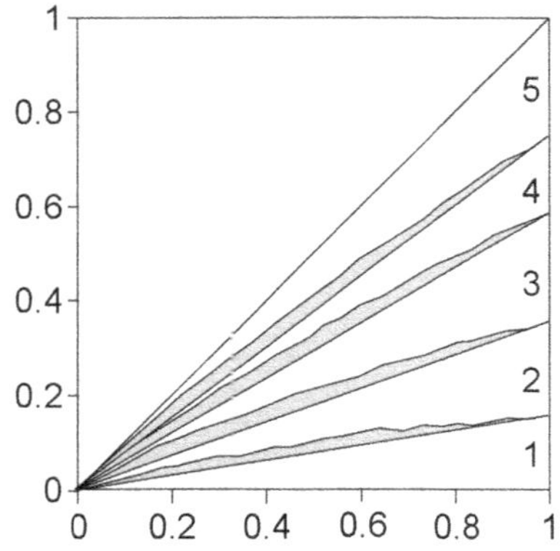

FIGURE 12.17. The A/B data: the intersections of the Grade Lorenz Surface $\mathbf{L}^*_{\mathbf{X};R_{\mathbf{X}}}$ and of $Diag_2$ by the planes $v = v_j; j = 1, \ldots, 5$.

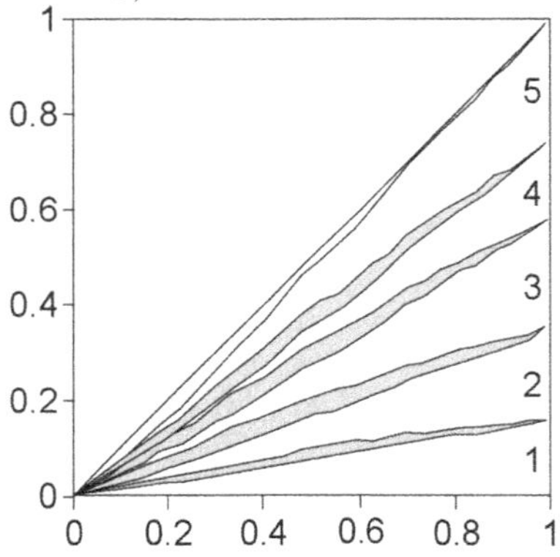

FIGURE 12.18. The A/B data: the intersections of Lorenz Surface $\mathbf{L}_{\mathbf{X};R_{\mathbf{X}}}$ and of $\mathbf{L}^{\perp}_{\mathbf{X};R_{\mathbf{X}}}$ by the planes $v = v_j$, $j = 1, ..., 5$. The curve at the top (intersecting $Diag$) is $L_{Z;R_{\mathbf{X}}}$.

Figures 12.12 and 12.14 present the intersections of the Lorenz Surface $\mathbf{L}_{(X_{33},x_{41},X_{10},X_{49},X_{17});\varphi}$ for C/J data when $\varphi = Z$ (Figure 12.12) and $\varphi = R_{\mathbf{X}}$ (Figure 12.14). The curves distinctly differ for $\varphi = Z$ and for $\varphi = \mathbf{R}_{\mathbf{X}}$. In the first case (Figure 12.12), the intersections related to $\mathbf{L}_{(X_{33},...,X_{17});Z}$ almost coincide with the intersections related to independence between records and variables, i.e., to $\overset{\perp}{\mathbf{L}}_{(X_{33},...,X_{17});Z}$. In the second case (Figure 12.14), the intersections related to $\overset{\perp}{\mathbf{L}}_{(X_{33},...,X_{17});R_{\mathbf{X}}}$ are visibly different from the curves underneath related to $\overset{\perp}{\mathbf{L}}_{(X_{33},...,X_{17});R_{\mathbf{X}}}$.

In other words, the Grade Lorenz Surface (and the Lorenz Surface) for $\varphi = Z$ demonstrates almost complete lack of dependence between variables and records ordered by φ. But for $\varphi = \mathbf{R}_{\mathbf{X}}$ there exists a distinct positive dependence (positive since the curves are *above* the dotted curves, and in the bivariate case the concentration surfaces in $\mathbb{QD}^+$ or $\mathbb{TP}_2$ lie above $Diag_2$).

The shadowed areas in the Figures (12.11 and 12.12) are approximately equal; similar approximate equality holds for the shadowed areas in the Figures (12.13 and 12.14). Generally, the situation is analogous (mutatis mutandis) to that concerning the approximate equality of area $\int_0^1 (C_1(u) - C_2(u))du$ and $ar(C_2 : C_1)$ when C_1 and C_2 are two Lilliputian Curves (see expressions (3.48) and (3.49) in Chapter 3). The distance between two Lilliputian surfaces in $\mathbb{BLM}$ might be measured by the volume between them (e.g., between $\mathbf{L}_{\mathbf{X}}$ and $\overset{\perp}{\mathbf{L}}_{\mathbf{X}}$) as well as by the volume between $\overset{*}{\mathbf{L}}_{\mathbf{X}}$ and $Diag_2$. Generally, one volume is almost the same as the other; they tend to be identical with growing regularity of $\mathbf{L}_{\mathbf{X}}$.

We remind that in the univariate case, the conditions sufficient for equality of the two areas are given in Proposition 3.3(ii) and (iii).

Illustration (A/B data):

The notation $\mathbf{L}_{\mathbf{X}}$ refers to vector $\mathbf{X}$ consisting of the bonuses in years 1, 2,...,5, with records ordered by function (12.15). Figure 12.16 shows the intersections of the respective Lorenz Surface $\mathbf{L}_{\mathbf{X}}$ by the planes $v = v_j$, j = 1,..., 5. Generally, Figures 12.15-12.18 are analogous to Figures 12.11-12.14 for the C/J data. It is seen that for the A/B data the shaded areas in Figures 12.17 and 12.18 are smaller than in Figures 12.13 and 12.14 for the C/J data.

12.5.3 Global differentiation and its decomposition

Strength of dependence between records and variables which form the probability data table $\mathbf{T}_{\mathbf{X};\varphi}$ is evaluated by $\rho^*(\mathbf{T}_{\mathbf{X};\varphi})$. The properties of ρ^* were thoroughly considered in Chapter 8, Section 8.4 and it was stressed there that ρ^* evaluates how differentiated are column profiles and simultaneously how differentiated are row profiles. A visualization of ρ^* in a copula was provided in Figure 8.11 down. It was said there that ρ^* is the integral of ar's for all intersections. When the copula belongs to $\mathbb{QD}^+$ the intersection of the copula and the plane always lied *above* uv_0; and the afore said this integral referres to the areas between the copula and $Diag_2$. When the copula belongs to $\mathbb{QD}^-$

the intersection of the copula and the plane is situated below uv_0 and the afore said this integral referres to the areas between the copula and $Diag_2$, taken with negative sign.

Illustration (C/J data when $\mathbf{X} = (X_{33}, X_{41}, X_{10}, X_{49}, X_{17})$):
An attempt to visualize $\rho^*(\mathbf{T}_{\mathbf{X};\varphi})$ is given in Figures 12.11 and 12.13. When $\varphi = \mathbf{R_X}$ the copula belongs to $\mathbb{QD}^+$ (Figure 12.13) and so $\rho^*(\mathbf{T}_{\mathbf{X};\mathbf{R_X}})$ is positive (equal to 0.36). For $\varphi = Z$ (Figure 12.11), the copula does not belong to $\mathbb{QD}^+$ and is close to $Diag_2$, and $\rho^*(\mathbf{T}_{\mathbf{X};Z})$ is equal to -0.027.

Let us now proceed from $\rho^*(\mathbf{T}_{\mathbf{X};\varphi})$ to $\max_\varphi \rho^*(\mathbf{T}_{\mathbf{X};\varphi})$. When φ is any function defined on the set of records due to Remark 8.3 in Chapter 8 and the fact that $\mathbf{R_X}$ corresponds to the grade regression function when records are ordered by $\mathbf{R_X}$), this maximum is attained for $\varphi = R_{\mathbf{X}}$. Hence

$$\max_\varphi(\rho^*(\mathbf{T}_{\mathbf{X};\varphi})) = \rho^*(\mathbf{T}_{\mathbf{X};\mathbf{R_X}}).$$

Let us introduce the notation

$$\mathbf{D_X} \stackrel{df}{=} \mathbf{D}_{\mathbf{X};R_\mathbf{X}}, \quad \mathbf{T_X} \stackrel{df}{=} \mathbf{T}_{\mathbf{X};R_\mathbf{X}}, \tag{12.38}$$

$$\mathbf{L_X} \stackrel{df}{=} \mathbf{L}_{\mathbf{X};R_\mathbf{X}}, \quad \boldsymbol{\lambda}_\mathbf{X} \stackrel{df}{=} \boldsymbol{\lambda}_{\mathbf{X};R_\mathbf{X}} \tag{12.39}$$

$$\overset{*}{\mathbf{L}}_\mathbf{X} \stackrel{df}{=} \overset{*}{\mathbf{L}}_{\mathbf{X};R_\mathbf{X}}, \quad \overset{*}{\boldsymbol{\lambda}}_\mathbf{X} \stackrel{df}{=} \overset{*}{\boldsymbol{\lambda}}_{\mathbf{X};R_\mathbf{X}} \tag{12.40}$$

Due to (12.38),

$$\max_\varphi(\rho^*(\mathbf{T}_{\mathbf{X};\varphi})) = \rho^*(\mathbf{T_X}). \tag{12.41}$$

We will call $\rho^*(\mathbf{T_X})$ the *global differentiation of record profiles* in the dataset related to vector $\mathbf{X}$ and function $R_\mathbf{X}$; it is also simultaneously the maximal differentiation of the distributions of variables when the variables follow the order of the components of $\mathbf{X}$.

The next step is to allow column permutations and find the global maximum of ρ^* (full GCA). In this case records and variables are ordered according to the respective $col|row$ and $row|col$ grade regression functions corresponding to the GCA table.

The function φ which orders records in dataset $\mathbf{D}_{\mathbf{X};\varphi}$ according to full GCA applied to $\mathbf{D}_{\mathbf{X}}$ will be denoted $R_{\mathbf{X}}^{GCA}$. (12.42)

So we will write

$$\mathbf{D}_{\mathbf{X}}^{GCA} \stackrel{df}{=} \mathbf{D}_{\mathbf{X}_{\psi};R_{\mathbf{X}}^{GCA}} \tag{12.43}$$

for ψ referring to the optimal order of columns due to GCA.

It follows that the grade correlation of the probability data table $\mathbf{D}_{\mathbf{X}}^{GCA}$ is the *global differentiation of the set of profiles of optimally permuted components of vector* $\mathbf{X}$:

$$\rho^*(\mathbf{D}_{\mathbf{X}}^{GCA}) = \max_{\psi} \rho^*(\mathbf{D}_{\mathbf{X}_{\psi}}). \tag{12.44}$$

Illustration (C/J data):
The function $R_{\mathbf{X}}^{GCA}$ is given by formula (12.14). The resulting value of $\mathbf{Dif}_{\mathbf{X}}^{GCA}$ is 0.3600. The intersections of the Lorenz Surface $\mathbf{L}_{\mathbf{X}}$ by the respective planes are presented in Figure 12.14.

Slightly simplifying the notation and using **Dif** for *differentiation,* we put

$$\mathbf{Dif}_{\mathbf{X};\varphi} \stackrel{df}{=} \rho^*(\mathbf{D}_{\mathbf{X};\varphi}), \tag{12.45}$$

$$\mathbf{Dif}_{\mathbf{X}} \stackrel{df}{=} \rho^*(\mathbf{D}_{\mathbf{X}}), \tag{12.46}$$

$$\mathbf{Dif}_{\mathbf{X}}^{GCA} \stackrel{df}{=} \rho^*(\mathbf{D}_{\mathbf{X}}^{GCA}), \tag{12.47}$$

and let $\mathbf{L}_{\mathbf{X}}^{GCA}$ (resp. $\boldsymbol{\lambda}_{\mathbf{X}}^{GCA}$) denote the Lorenz Surface (resp. Lorenz Density) for $\mathbf{D}_{\mathbf{X}}^{GCA}$.

Notation (12.45)-(12.5.3) will be also used when datasets

$$\mathbf{D}_{\mathbf{X}_{\psi}},\ \mathbf{D}_{\mathbf{X};R_{\mathbf{X}}^{GCA}},\ \mathbf{D}_{\mathbf{X}}^{GCA},\ \mathbf{D}_{\mathbf{X}},\ \mathbf{D}_{\mathbf{X};\varphi},$$

are respectively replaced by the probability data tables

$$\mathbf{T}_{\mathbf{X}_{\psi}},\ \mathbf{T}_{\mathbf{X};R_{\mathbf{X}}^{GCA}},\ \mathbf{T}_{\mathbf{X}}^{GCA},\ \mathbf{T}_{\mathbf{X}},\ \mathbf{T}_{\mathbf{X};\varphi}.$$

Remark 12.5 *Property (8.37) in Chapter 8 concerning* ρ^* *(which presents one third of* ρ^* *as parameter ar between the row grade correlation function and the row marginal cdf), can be applied to* $\mathbf{Dif}_X$ *and* $\mathbf{Dif}_X^{GCA}$; *it is only needed to select for* $\mathbf{Dif}_X$ *and* $\mathbf{Dif}_X^{GCA}$ *a proper row grade correlation function and a proper row marginal distribution.*

The links between $\mathbf{Dif}_{\mathbf{X}}$ *and ar and between* $\mathbf{Dif}_{\mathbf{X}}^{GCA}$ *and* $ar_{\max}$ *contribute to the integrity of grade infrastructure.*

* * *

At the end of this subsection, we will introduce aggregated data tables $\overline{\mathbf{T}}_{\mathbf{X};\varphi}$.

Definition 12.5 *For any order function* φ

$$\overline{\mathbf{T}}_{X;\varphi} = \left[\left(\left(\frac{E_{\Omega_0}(X_j; \varphi = t)}{E(Z)}, j = 1, ..., k \right), p_t \right), t = 1, ..., n_{\varphi}, \right], \qquad (12.48)$$

where $\Omega_0 = \{\omega_1, ..., \omega_n\}$, p_t *is the fraction of records with* $\varphi(\omega)$ *equal to* t, $n_{(\varphi)} = \#\{\varphi(\omega), \omega \in \Omega_0\}$, *and* $E_{\Omega_0}(X_j; \varphi = t\}$ *for* $t = 1, ..., n_{(\varphi)}$ *is the mean of* $x_{i,j;\varphi}$ *in dataset* $\mathbf{D}_{\mathbf{X};\varphi}$ *for all* i *from the set* $\{1, ..., n\}$ *such that* $\varphi(\omega_i) = t$.

If $n_{(\varphi)} = 1$ (i.e., all records have the same value of φ), the aggregated dataset $\overline{\mathbf{T}}_{\mathbf{X};\varphi}$ is reduced to only one row (namely to $(EX_1/EZ, EX_2/EZ, ..., EX_k/EZ)$).

Illustration (C/J and A/B data):

For all datasets in the C/J and A/B data considered up to now (i.e., datasets with rows and columns ordered in many ways), no value of φ appeared for more than one record; so, for all considered functions φ, $n_{(\varphi)}$ was equal to n (20 for C/J or 27 for A/B data).

Data tables ordered and aggregated w.r.t. $R_{\mathbf{X}}$ are specially important. They will be denoted $\overline{\mathbf{T}}_{\mathbf{X}}$:

$$\overline{T}_{\mathbf{X}} \stackrel{df}{=} \overline{\mathbf{T}}_{\mathbf{X};R_{\mathbf{X}}}.$$

Records which were adjacent in $\mathbf{T}_{\mathbf{X}}$ and have the same value of $R_{\mathbf{X}}$ are replaced by the mean record (consisting of expectations over this set which correspond to particular variables).

The notation $\overline{\mathbf{T}}_{\mathbf{X}}^{GCA}$ will be used for $\mathbf{T}_{\mathbf{X}}^{GCA}$ aggregated w.r.t. $R_{\mathbf{X}}^{GCA}$.

We will exploit property of ρ^* introduced in Proposition 8.6 in Chapter 8, for stating an important property of $\overline{\mathbf{T}}_{\mathbf{X}}$ and $\mathbf{T}_{\mathbf{X}}^{GCA}$. It was formulated in Proposition 8.6(i). According to it, aggregation due to the regression function of a contingency table has no impact on the value of the grade correlation. Then, by Proposition 8.4(i) and 8.6(i)

$$\rho^*(\overline{\mathbf{T}}_{\mathbf{X}}) = \rho^*(\mathbf{T}_{\mathbf{X}}) = \mathbf{Dif}_{\mathbf{X}}, \tag{12.49}$$

and

$$\rho^*(\overline{\mathbf{T}}_{\mathbf{X}}^{GCA}) = \rho^*(\mathbf{T}_{\mathbf{X}}^{GCA}) = \mathbf{Dif}_{\mathbf{X}}^{GCA}. \tag{12.50}$$

Remark 12.6 *The aggregated data tables $\overline{\mathbf{T}}_{X}$ and $\overline{\mathbf{T}}_{X}^{GCA}$ will be introduced in Section 13.3 of Chapter 13, relative to non-finite sets of records produced by vectors of non-negative random variables with finite expectations. Then, the grade correlations for aggregated data tables become grade parameters of the related multivariate distributions.*

The Lorenz Surfaces for aggregated probability data tables will be denoted $\overline{\mathbf{L}}_{\mathbf{X};\varphi}$, $\overline{\mathbf{L}}_{\mathbf{X}}$, $\overline{\mathbf{L}}_{\mathbf{X}}^{GCA}$; they satisfy $\overline{\mathbf{L}}_{\mathbf{X};\varphi} \leq \mathbf{L}_{\mathbf{X};\varphi}$ for any φ. Adjacent columns and adjacent rows with identical profiles can be aggregated with no loss in value of the parameter **Dif** and no effect on $\overline{\mathbf{L}}_{\mathbf{X}}$, $\overline{\mathbf{L}}_{\mathbf{X}}^{GCA}$ or $\mathbf{L}_{\mathbf{X}}$, $\mathbf{L}_{\mathbf{X}}^{GCA}$.

12.5.4 Decomposition of $\mathbf{Dif}_{\mathbf{X}}$

Remark 12.7 *A decomposition of $\mathbf{Dif}_{\mathbf{X}}$ implied by a decomposition of ρ^* given in Proposition 8.6 (formula 8.37) in Chapter 8 can be rewritten as*

$$\mathbf{Dif}_{\mathbf{X}} = 3\sum_{j=1}^{k} \frac{(v_{j-1}+v_j)E(X_j)}{E(Z)}\, ar(L_{X_j;R_{\mathbf{X}}} : L_{Z;R_{\mathbf{X}}}). \tag{12.51}$$

Even more important is another decomposition of this $\mathbf{Dif}_{\mathbf{X}}$ based on the decomposition formula for ρ^* introduced in Chapter 8 (Proposition 8.2, formula (8.14). When rewritten here in reference to a data table $\mathbf{T}_{\mathbf{X}}$ and its subtables of size $m \times 2$ corresponding to pairs of components of $\mathbf{X}$, we get:

$$\mathbf{Dif}_{\mathbf{X}} = 2\sum_{j=1}^{k}\sum_{i=1}^{j-1} \frac{v_{j-1}+v_j-v_{i-1}-v_i}{2}\,\frac{(E(X_i+X_j))^2}{(E(Z))^2}\mathbf{Dif}_{(X_i,X_j);R_{\mathbf{X}}}\,. \tag{12.52}$$

Please note that for any (i,j), $i<j$,

$$\mathbf{Dif}_{(X_i,X_j);R_{\mathbf{X}}} \leq \mathbf{Dif}_{(X_i,X_j)},$$

and equalities for all pairs (i,j) hold if and only if the data table $\mathbf{T_X}$ is TP_2. Therefore the two differentiation matrices

$$\mathbf{DIFF}(\mathbf{T_X}) \stackrel{df}{=} \left[\mathbf{Dif}_{(X_i,X_j);R_{\mathbf{X}}}, i=1,...,m, j=1,...,k\right] \tag{12.53}$$

and

$$\mathbf{supDIFF}(\mathbf{T_X}) \stackrel{df}{=} \left[\mathbf{Dif}_{(X_i,X_j)}, i=1,...,m, j=1,...,k\right] \tag{12.54}$$

can be used to describe and measure how close the data table $\mathbf{T_X}$ is to being a TP_2 table.

Let $sup\mathbf{Dif_X}$ be calculated from the matrix $\mathbf{supDIFF}(\mathbf{T_X})$ analogously as $\mathbf{Dif_X}$ is calculated from the matrix $\mathbf{DIFF}(\mathbf{T_X})$ by formula 12.52. Then the measure of closeness of $\mathbf{T_X}$ to TP_2 will be defined as the ratio

$$\frac{\mathbf{Dif_X}}{sup\mathbf{Dif_X}} \cdot 100\%. \tag{12.55}$$

Remark 12.8 *For any two component vector* $\mathbf{X}=(X_1,X_2)$, *the data table* $\mathbf{T_X}$ *is* TP_2 *(*TN_2*) since the order function* $R_{(X_1,X_2)}$ *is an increasing (decreasing) function with respect to the ratio* X_2/X_1 *and therefore this order is equivalent to order induced by the likelihood ratio for two columns in table* $\mathbf{T_X}$.

Remark 12.9 *If* $\mathbf{T_X}$ *is* TP_2 *and* $k>2$, *the subtables* $\mathbf{T}_{(X_i,X_j);R_{\mathbf{X}}}$ *are also* TP_2 *(see Proposition 8.2(ii)) and hence for any* $i<j$ *the order function* $R_{(X_i,X_j)}$ *is an increasing function of* $R_{\mathbf{X}}$. *Therefore,*

$$\mathbf{Dif}_{(X_i,X_j);R_{\mathbf{X}}} = \mathbf{Dif}_{(X_i,X_j);R_{(X_i,X_j)}} = \mathbf{Dif}_{(X_i,X_j)}$$

and, for $i'<i<j<j'$

$$\mathbf{Dif}_{(X_i,X_j)} \leq \mathbf{Dif}_{(X_i,X_{j'})},\ \mathbf{Dif}_{(X_{i'},X_j)} \leq \mathbf{Dif}_{(X_{i'},X_{j'})}.$$

The matrix $\mathbf{DIFF}(\mathbf{T_X})$ can be treated as a specific dissimilarity matrix for the set of variables $X_1,...,X_k$. The elements of $\mathbf{Dif}(\mathbf{T_X})$ are less than corresponding elements of $\mathbf{supDIFF}(\mathbf{T_X})$. Remark 12.9 implies the following proposition:

Proposition 12.2 *If table* $\mathbf{T_X}$ *is* TP_2 *then* $\mathbf{DIFF}(\mathbf{T_X})$ *is anti-Robinson (i.e., its elements increase monotonically when moving away from the matrix diagonal which consists of the nulls).*

To be anti-Robinson is a desirable property for a *dissimilarity matrix.*

Illustration (C/J and A/B data):

According to Table 12.7, the C/J data are TP_2, while the A/B data are rather far from being TP_2.

TABLE 12.5. The C/J data (Table 12.2, Fig. 12.2): the differentiation matrices **DIFF** (lower triangle) and **supDIFF** (upper triangle)

	33	**41**	**10**	**49**	**17**
33	0	0.1051	0.2077	0.3321	0.4298
41	0.1051	0	0.0974	0.2249	0.3422
10	0.2077	0.0974	0	0.1375	0.2556
49	0.3321	0.2249	0.1375	0	0.1252
17	0.4298	0.3422	0.2556	0.1252	0

TABLE 12.6. The A/B data (Fig. 12.6): the diferentiation matrices **DIFF** (lower triangle) and **supDIFF** (upper triangle)

	1	4	2	3	5
1	0	0.2960	0.2213	0.2015	0.3128
4	0.0832	0	0.2699	0.2424	0.3040
2	0.1010	0.0287	0	0.1800	0.2643
3	0.1508	0.0878	0.0585	0	0.1926
5	0.2982	0.2529	0.2157	0.1545	0

TABLE 12.7. The ratios of $\mathbf{DIF}_X$ and *sup*$\mathbf{Dif_X}$: measures of the closeness of the data tables C/J and A/B to a TP_2 table

	$\mathbf{Dif_X}$	sup $\mathbf{Dif_X}$	%
C/J	0.3600	0.3600	100.0%
A/B	0.2474	0.3403	72.7%

12.6 Transformations of variables

It has been assumed through this chapter that the expectations $E(X_j)$ for $j = 1, ..., k$, can be different. In practice it is often desirable (primarily for reasons mentioned in Section 12.3) to replace vector $\mathbf{X}$ by

$$\mathbf{Y} = (Y_1, ..., Y_k) = \left(\frac{X_1}{E(X_1)}, \cdots, \frac{X_k}{E(X_k)}\right), \tag{12.56}$$

so that $E(Y_j) = 1$. This implies that the expressions for $R_{\mathbf{Y}}$, $\mathbf{T}_{\mathbf{Y}}$, $\boldsymbol{\lambda}_{\mathbf{Y}}$, $\mathbf{L}_{\mathbf{Y}}$, etc., are simpler than the respective expressions for vector $\mathbf{X}$.

Proposition 12.3 *If* $\mathbf{Y}$ *is given by (12.56), then data table* $\overline{\mathbf{T}}_{\mathbf{X}}^{GCA}$ *is* TP_2 *if and only if data table* $\overline{\mathbf{T}}_{\mathbf{Y}}^{GCA}$ *is* TP_2.

The proof is immediate.

For data tables less regular than TP_2 tables, there may appear discrepancies between the orderings of records and variables derived for the vectors $\mathbf{X}$ and $\mathbf{Y}$.

Illustration (C/J data and A/B data):
In the case of the C/J data, $Y_j = X_j/E(X_j)$ is the overrepresentation divided by the mean score delivered by judge j. The data table $\mathbf{T}_{\mathbf{X}}$ is TP_2 and therefore the orderings of rows and of columns in $\mathbf{T}_{\mathbf{X}}$ and $\mathbf{T}_{\mathbf{Y}}$ remain unchanged; the value $\mathbf{Dif}_{\mathbf{Y}}$ is 0.3611 which is almost the same as $\mathbf{Dif}_{\mathbf{X}} = 0.3600$.

In the case of the A/B data, $Y_j = X_j/E(X_j)$ is the bonus overrepresentation over the respective yearly mean bonus. As the data table $\mathbf{T}_{\mathbf{X}}^{\mathbf{GCA}}$ is not TP_2, the ordering of rows in data table $\mathbf{T}_{\mathbf{Y}}^{\mathbf{GCA}}$ (see Figure 12.19) is different than in $\mathbf{T}_{\mathbf{X}}^{\mathbf{GCA}}$ (see Figure 12.6), but the differences are rather small. The value 0.238 of $\mathbf{Dif}_{\mathbf{Y}}^{\mathbf{GCA}}$ is slightly smaller than $\mathbf{Dif}_{\mathbf{X}}^{\mathbf{GCA}} = 0.247$.

Another important data transformation is *ranking*: the values of each variable separately are replaced by the respective ranks. This transformation can be used for ordinal variables as well as for variables which are not nonnegative; in this way the use of GCA will become technically possible. The reader

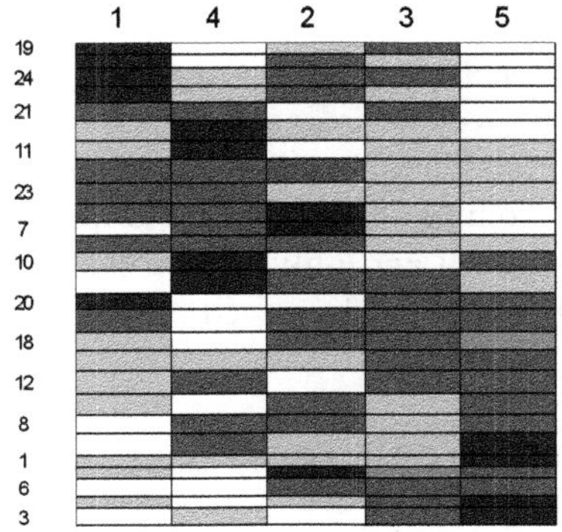

FIGURE 12.19. The A/B example: the GCA map for bonuses divided by their yearly mean ($\rho^* = 0.2381$)

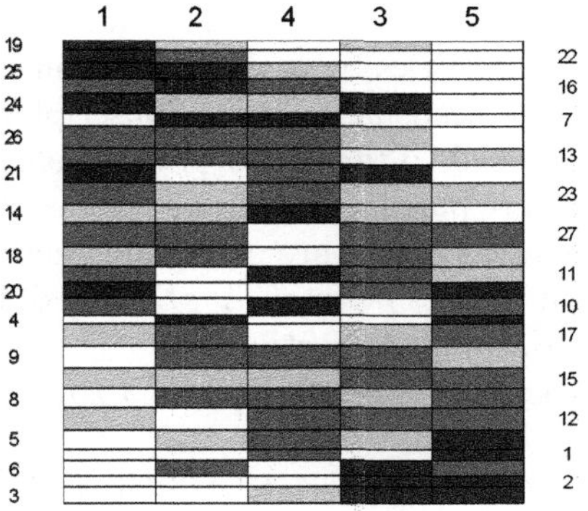

FIGURE 12.20. The A/B example: the GCA map when bonuses are replaced by their ranks calculated within each year separately ($\rho^* = 0.3000$)

should realize (or even check on some tables for the C/J and A/B data) that the sums and also the mean values of ranks are the same for each variable. Therefore ranking provides another way of transforming variables to ones with equal means.

Illustration (C/J and A/B data):

Would ranking be applied to the C/J data, the columns of the GCA data tables will remain to be ordered in the same way as before ranking, while the ordering of rows will be slightly changed with respect to just one pair of adjacent competitors with ID's 67 and 82 (before ranking, competitor 82 precedes competitor 67; after ranking, this order is reversed). Moreover, the TP_2 property possessed under GCA before ranking, does not hold after ranking.

In the case of the A/B data, the ordering after ranking (see Figure 12.20) is changed more than in the case of C/J data: not only rows, but also columns are reordered ((1,2,4,3,5) instead of (1,4,2,3,5)).

12.7 Detection of outliers and decomposition of a dataset

The insight examples (C/J and A/B) are of similar size (20×5) and (27×5), but with different degrees of regularity: the C/J data are very regular, the A/B data very irregular. But it is possible to detect outliers in the A/B data and reject them to improve regularity. The detection tool was described in Chapter 9, where it was used to detect outlying rows and/or columns in contingency tables. The same tool is applicable to detect outlying objects (rows) and/or outlying variables (columns) in a data probability table $\mathbf{T_X}$ or $\mathbf{T_X}^{GCA}$.

The statistics to detect outliers in Chapter 9 were called: $AvgDist$ and $MaxDist$, and had a lower subscript row or col according to whether they were applied to the sets of rows or of columns. These statistics measure how distant are particular rows (or, respectively, columns) from a TP_2 table. Both functions measures are equal identically to zero if and only if a table is TP_2. The differences between $AvgDist$ and $MaxDist$ were described and exemplified in Chapter 9, Section 9.5.

Since the C/J data table is TP_2,

$$AvgDist_{row} = MaxDist_{row} = AvgDist_{col} = MaxDist_{col} \equiv 0;$$

the graphs of $AvgDist_{row}$ and $MaxDist_{row}$ for the A/B data are presented in Figure 12.21.

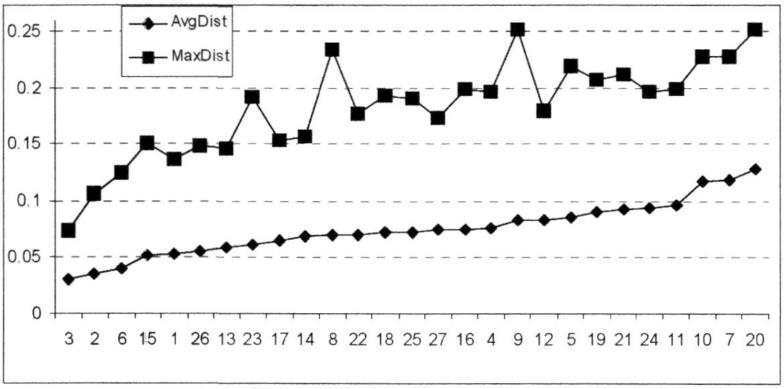

FIGURE 12.21. The A/B data: the plots of $AvgDist_{row}$ and $MaxDist_{row}$ by employee id ordered according to $AvgDist_{row}$.

The employees are ordered by $AvgDist$ from those best fitted to the main trend (3, 2, 6, 15, 1) to definite outliers (20, 7, 10, 11, 24). The ordering corresponding to $MaxDist$ is similar to that corresponding to $AvgDist$ except for employees 23, 8 and 9.

Figure 12.21 helps the data analyst to spot employees who should be looked at with special care. Suppose that the analyst's aim is to select a few ordered clusters that include employees who range from those whose bonus tends to diminish across years, to those whose tends to increase across years. However, such a clustering is possible only in a sufficiently regular table. It is therefore desirable first to detect the employees who spoil regularity and then to form clusters due to an optimal discretization of the grade col|row regression function (see Grade Correspondence Cluster Analysis in Chapter 10). The subset of outliers should be inspected separately.

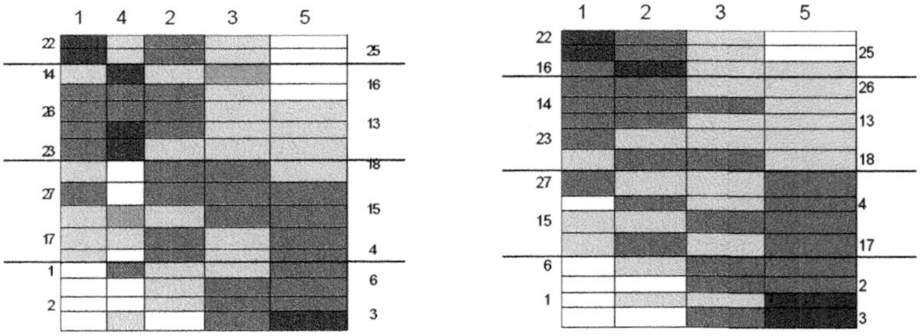

FIGURE 12.22. **Left:** the GCA map for the 16 × 5 data table of the A/B data after rejecting 11 rows. **Right:** the GCA map for the 16 × 4 data table of the A/B data after rejecting 11 rows, for years 1, 2, 3, 5.

It is up to the data analyst as to which objects are classified as outliers. The graphs of $AvgDist_{row}$ and $MaxDist_{row}$ just indicate various possibilities.

Figure 12.22 left shows the effect of rejecting eleven employees with values of $AvgDist_{row}$ larger than employee four (i.e., employees 9, 12, 5, ..., 10, 7, 20) and, additionally, employee 8. This group of outliers contained most employees with large values of $AvgDist_{row}$ and/or $MaxDist_{row}$. The GCA procedure is then applied to the remaining set of 16 employees. The resulting map for the 16 × 5 table is presented in Figure 12.22 (left). The map (see Figure 12.7 for the 27 × 5 table) is visually much more regular than before , the ordering of years is the same: 1, 4, 2, 3, 5, and the column corresponding to the 4th year is still distinctly outlying from a TP_2 regularity (perhaps even more distinctly now than before); the definite departure of the 4th year from regularity is also confirmed by a high value of $AvgDist_{col}$ and $MaxDist_{col}$.

The map in Figure 12.22 (right) shows how the map changes when the bonuses in the 4th year are not taken into account. The corresponding 16 × 4 table is not far from being a TP_2 table.

The resulting 16 × 5 and 16 × 4 tables are clustered into four subsets of employees denoted by the bold horizontal lines. Please note how similar are the rows in particular clusters and how relatively small ordering changes arise when the 4th year bonuses are omitted.

This mini study of outlier detection is an introduction to the decomposition of a set of objects into at least two more regular subsets, with a possible addition of a subset of outliers with no trend at all. Such decomposition is a basic procedure in grade multivariate data analysis. The grade $row|col$ regression functions constitute the key variables describing particular subsets; the grade $col|row$ regression functions indicate objects with extremal values of a latent variable which is ordered like the grade $col|row$ regression.

Decomposition performed in two real-data case studies will be described in Chapter 14.

13

Inequality measures for multivariate distributions

Wiesław Szczesny

13.1 Introduction

The previous chapter introduced multivariate inequality measures (Lorenz Surfaces, Grade Lorenz Surfaces and the related numerical measures) for multivariate datasets with *finite* sets of records. These measures are easy to percept for those who are familiar with the concepts of the Lorenz Curve and Gini Index which are so important in the univariate case.

Non-finite sets of records appear mostly in probabilistic models known as *multivariate distributions*. The aim of this chapter is to extend the definitions of various multivariate inequality measures (shortly referred to as MIM) onto a suitably defined family of multivariate distributions.

The natural questions are: *how* such extension can be done, and *why* (*what for*) it is worth to do it. Let us say a bit about *how*, and then about *what for*.

The question *how* can be immediately answered for multivariate distributions with finite set of records, related to non-negative discrete variables. In this case the definitions of MIMs are almost the same as those introduced in Section 12.5. There are just two changes.

The first change is the use of the term "probability of a record in a multivariate distribution" instead of the term "fraction of identical records in a multivariate dataset". The second change is that the set of functions φ admitted to order records has to be *limited to functions defined on the set of records* (instead more generally on the set Ω_0 of considered objects). In other words, a random variable φ defined on a set of objects is not admitted to order records if the same record can co-appear with two different values of φ. Among these functions which can be admitted, three were introduced before in Chapter 12, and denoted there by Z (the sum of the components of vector $\mathbf{X}$), $R_{\mathbf{X}}$ (defined by (12.22)), and $R_{\mathbf{X}}^{GCA}$ (defined by (12.42)).

Functions Z, $R_{\mathbf{X}}$ and $R_{\mathbf{X}}^{GCA}$ play crucial role in constructing the inequality measures for multivariate distributions. In the case of multivariate distributions *with finite sets of records* this role is analogous to that which they play for finite multivariate datasets, as illustrated in Section 13.2 for a random vector consisting of three non-negative discrete variables: the values of $R_{\mathbf{X}}$, $R_{\mathbf{X}}^{GCA}$, Z are derived, the probability data tables $\overline{\mathbf{T}}_{\mathbf{X}}$ and $\overline{\mathbf{T}}_{\mathbf{X}}^{GCA}$ (aggregated respectively according to $R_{\mathbf{X}}$ and $R_{\mathbf{X}}^{GCA}$) are constructed, the Lorenz measures $\overline{\lambda}_{\mathbf{X}}$, $\overline{\lambda}_{\mathbf{X}}^{GCA}$, $\overline{\mathbf{L}}_{\mathbf{X}}$, $\overline{\mathbf{L}}_{\mathbf{X}}^{GCA}$, $\overline{\mathbf{Dif}}_{\mathbf{X}}$, $\overline{\mathbf{Dif}}_{\mathbf{X}}^{GCA}$ are calculated.

Section 13.3 extends the definitions of *aggregated* MIMs as given in Section 13.2 onto the set of multivariate distributions with non-finite sets of records. These measures refer to the aggregated data probability tables $\overline{\mathbf{T}}_{\mathbf{X};\varphi}$.

The aggregated data tables were introduced before in Chapter 12 in Section 12.5.3 (see formula (12.48)), but they played there only a secondary role because, for finite sets of records, data tables which are *non-aggregated with respect to an order function* φ are always available. However, for non-finite sets of records, *only aggregated data tables can be constructed* (with aggregation applied to records with a common value of the order function). This provides in particular the aggregated data tables for the order functions $R_{\mathbf{X}_\psi}$, related to selected permutations ψ of the components of vector $\mathbf{X}$. Then it is possible to introduce the corresponding MIMs (including global differentiation $\mathbf{Dif}_{\mathbf{X}}$ and GCA global differentiation $\mathbf{Dif}_{\mathbf{X}}^{GCA}$).

Section 13.4 contains several examples of MIMs for chosen continuous bivariate distributions. In some simpler cases calculations are performed to the end and supplemented by visual presentation. In more complicated cases MIMs are obtained for large random samples drawn from the considered distributions. In particular, sequences of samples with increasing sizes are drawn, and the overrepresentation maps related to particular samples visualize *convergence of sample* MIMs *to the related theoretical* ("population") MIM.

Three examples in Section 13.4 refer to family of pairs of continuous variables *initially transformed by their cdf's,* so that the starred pairs

$$(X_1^*, X_2^*) = (F_{X_1}(X_1), F_{X_2}(X_2))$$

from the Dependence Lilliputian Model ($\mathbb{DLM}$) are considered. Moreover, components X_1 and X_2 in this families are symmetrically and regularly dependent. For each family, the plot of global differentiation $\mathbf{Dif}_{(X_1^*, X_2^*)}$ with respect to the grade correlation $corr(X_1^*, X_2^*)$ is derived. Please note that

$corr(X_1^*, X_2^*)$ measures *similarity* of X_1^* and X_2^*, while $\mathbf{Dif}_{(X_1^*,X_2^*)}$ measures their *dissimilarity.*

The result of the comparison of plots is remarkable: the three plots are practically identical and, moreover, the shape of the average plot (given special considerations in the Appendix in Section 13.7): is very close to a decreasing *ideally regular curve.* This result leads to a hypothesis that, in families of sufficiently regularly dependent pairs (X_1^*, X_2^*), that ideally regular decreasing function directly transforms the similarity measure $corr(X_1^*, X_2^*)$ onto the dissimilarity measure $\mathbf{Dif}_{(X_1^*,X_2^*)}$.

The important next step is to find out how grade similarity and dissimilarity measures behave when there is more than two variables in vector $\mathbf{X}$. This problem is tackled in Section 13.5 for grade multinormal distributions and in Section 13.6 for grade Moran distributions.

Two types of measures are taken into account: $k \times k$ matrices and single numerical indices. In the first case grade correlation matrices $[corr(X_i^*, X_j^*)]$ (similarity measures) are compared with the grade differentiation matrices $[\mathbf{Dif}_{(X_i^*,X_j^*)}]$ (dissimilarity measures). In the second case, the global differentiation $\mathbf{Dif}_{\mathbf{X}^*}$ is primarily used as a numerical dissimilarity measure. For sufficiently regular distribution of vector $\mathbf{X}$ this index becames a function of the differentiation matrix and hence is close to be a function of the grade correlation matrix.

A suitable standardization of $\mathbf{Dif}_{\mathbf{X}^*}$ is hoped to help in neutralizing the influence of the number of components of vector $\mathbf{X}$.

* * *

The material in this chapter shows *how* the inequality measures for *finite multivariate datasets* are extended to construct inequality measures for *multivariate distributions with finite and also for non-finite sets of records.* The answer to the question "*what for*" stems from the following properties of MIMs:

(i) their meaningfulness and calculability is clearly formulated,

(ii) they are universally applicable in data analysis, probability theory and statistical inference (with the common restriction concerning meaningfulness and calculability),

(iii) their derivation is based on a general concept of concentration representing "variability" of a data matrix (between records as well as between variables),

(iv) they possess simple geometric representations and can be accordingly visualized,

(v) they are well suited for inter-population comparisons and for comparisons with a given probability model of data population,

(vi) their values can be effectively calculated,

(vii) the methods based on GCA applied to multivariate data tables (obtained from samples drawn from populations as well as from various non-sample data sources) solve satisfactorily diverse problems of multivariate analysis.

* * *

A unification of multivariate data analysis and probabilistic or statistical inference for multivariate distributions would be fruitfully performed on the basis of MIMs. Let us repeat that Chapter 12 concerns MIMs in the framework of data analysis, while the present chapter will concern MIMs in the probabilistic framework. The readers will find that the links between the two chapters are evident as far as the construction of MIMs is concerned.

The next step should be to formulate and solve analogously various problems predicting, detecting trends and outliers, decomposing data probability matrices into more regular ones, introducing suitably standardized numerical measures of variability, etc. For finite real data tables, some of these problems were tackled before in examples in Chapter 12 while further will be demonstrated in the case studies in Chapter 14. Similar demonstration of problems solving for non-finite data tables is theoretically possible but computationally would be extremely difficult. In particular, no probabilistic analog of Chapter 14 is included in the present book.

Let just mention however that various probabilistic optimization problems might be solved at the level of large random samples whenever such samples are available, while artifical reference sample models might be investigated in suitably organized simulation studies.

This research area (for data analysis and for probabilistic and statistical inference) has been only started. It could be called Lorenz Multivariate Data Analysis (LMDA). The importance of LMDA stems from properties (i)-(vii) concerning MIMs.

Most material presented in Chapters 12 and 13 is new. Helpful inspirations were found in [38], [4], [14], [36], [37], [40], [42], [45], [46], [47], [51], [50], [52], [53], [54], [57], [73], [74],[75], [77], [79], [82], [92], [97], [99], [100], [98], [105], [106], [107], [108], [109], [112], [113], [5], [8], [15], [24], [27], [28], [59], [63], [76], [80], [90], [102].

13.2 Inequality measures for multivariate distributions with finite sets of records

Let $\mathbf{X} = (X_j,\ j = 1, ..., k)$ be a vector of non-negative discrete random variables $X_j : \Omega \to \mathbb{X}_j$, and let each set $\mathbb{X}_j$ be finite with m_j elements. Then there is a finite number m of different records produced by vector $\mathbf{X}$ $(m \leq m_1 \cdot ... \cdot m_k)$. The distribution of $\mathbf{X}$ is specified by the probabilities attached to each record.

Let $\mathbb{X}$ denote the set of records $\{\mathbf{x}_1, ..., \mathbf{x}_m\}$ in which subscripts $1, ..., m$ are arbitrarily chosen, and let x_{sj} for $j = 1, \ldots, k$, $s = 1, \ldots, m$, be the value of variable X_j in record $\mathbf{x}_s$. Record $\mathbf{x}_s = (x_{s1}, ..., x_{sk})$ is accompanied by probability $p_s = \Pr(\mathbf{X} = \mathbf{x}_s)$.

Records can be ordered according to any measurable function $\varphi : \mathbb{X} \to \mathbb{R}$. The random variable $\varphi(\mathbf{X})$ will be shortly denoted $\varphi_{\mathbf{X}}$. Note that this convention was introduced in Chapter 12 for function $R_{\mathbf{X}}$ (Section 12.5, formula (12.22)), although it was not used for $Z = \sum X_j$. Both, Z and $R_{\mathbf{X}}$ are specially important examples of an order function $\varphi_{\mathbf{X}}$.

Remark 13.1 *The order functions φ in Section 12.5 were defined on the set of objects denoted Ω_0. Function $R_{\mathbf{X}}$ was also introduced initially as a function on Ω_0 but then shown to be related to Ω_0 only through the values of the components of vector $\mathbf{X}$. Therefore, it was redefined as a function on the set of records.*

Definition 13.1 *Let $\mathbb{X}$ consist of m records $\mathbf{x}_1, ..., \mathbf{x}_m$. The data table $\mathbf{T}_{\mathbf{X};\varphi_{\mathbf{X}}}$ related to random vector $\mathbf{X} : \mathbb{UNIV} \to \mathbb{X}$ and function $\varphi_{\mathbf{X}} : \mathbb{X} \to \mathbb{R}$ is the table*

$$\mathbf{T}_{\mathbf{X};\varphi_{\mathbf{X}}} = \left[\left(\left(\frac{x_{i(t),1} p_{i(t)}}{E(Z)}, ..., \frac{x_{i(t),k} p_{i(t)}}{E(Z)} \right);\ p_{i(t)} \right),\ t = 1, ..., m \right] \qquad (13.1)$$

where $(i(1), ..., i(m))$ is a permutation of $(1, ..., m)$ such that $i(t) < i(t')$ iff $\varphi_{\mathbf{X}}(\mathbf{x}_{i(t)}) \leq \varphi_{\mathbf{X}}(\mathbf{x}_{i(t')})$.

Definition 13.2 *The aggregated data table $\overline{\mathbf{T}}_{\mathbf{X};\varphi}$ related to random vector $\mathbf{X} : \mathbb{UNIV} \to \mathbb{X}$ and function $\varphi_{\mathbf{X}} : \mathbb{X} \to \mathbb{R}$ is the table*

$$\overline{\mathbf{T}}_{\mathbf{X};\varphi_{\mathbf{X}}} = \left[\left(\left(\frac{E(X_1; \varphi_{\mathbf{X}} = t)}{E(Z)}, ..., \frac{E(X_k; \varphi_{\mathbf{X}} = t)}{E(Z)} \right);\ p_t \right),\ t = 1, ..., n_\varphi \right] \qquad (13.2)$$

where $p_t = \Pr(\varphi_{\mathbf{X}} = t)$ and $n_\varphi = \#\{\varphi(\mathbf{x}_1), ..., \varphi(\mathbf{x}_m)\}$.

The only difference between the Definition 13.2 (related to the probabilistic framework for a random vector with a *finite set of records*), and the definition of $\overline{\mathbf{T}}_{\mathbf{X};\varphi}$ in Chapter 12 by formula (12.48) (related to the data analysis framework), follows from the fact that in Section 12.5 the order functions are defined directly on Ω_0 and are not necessarily able to be redefined as functions on the set of records $\mathbb{X}$ which is required in the present section.

The inequality measures corresponding to the data tables $\mathbf{T}_{\mathbf{X};\varphi}$ and $\overline{\mathbf{T}}_{\mathbf{X};\varphi}$ are defined for vectors of non-negative discrete random variables with finite sets of values according to the same formulas as in Section 12.5. It is only necessary to have in mind that p_t in Definition 13.2 is not "the fraction of records with the value of φ equal to t"as in Section 12.5 but "the probability that $\varphi_{\mathbf{X}} = t$, induced by the distribution of vector $\mathbf{X}$."

Illustration (trivariate discrete distribution):
Let $\mathbf{X} = (X_1, X_2, X_3)$, where $\mathbb{X}_1 = \mathbb{X}_2 = \{0.4, 0.8, 1.25, 2.5\}$, $\mathbb{X}_3 = \{0.5, 2\}$, $m_1 = m_2 = 4$, $m_3 = 2$, $m = 32$. The distribution of vector $\mathbf{X}$ is given in Table 13.1. The records with ID's denoted (111) up to (442) are listed in column (2) of this table according to the order function which will be denoted ID; their probabilities are shown in column (3). The expectations of X_1, X_2, X_3 are:

$$E(X_1) = 1.1789,\ E(X_2) = 1.0877,\ E(X_3) = 1.4000$$

so that $E(Z) = 3.6666$.

The probability data table $\mathbf{T}_{\mathbf{X};ID}$ is then

0.0048	0.0048	0.0060;	0.044
0.0072	0.0072	0.0360;	0.066
...	...	...	...
0.0450	0.0450	0.0360;	0.066

Obviously, the agregated dataset $\overline{\mathbf{T}}_{\mathbf{X};ID}$ is in this exemplary distribution equal to the non-agregated dataset $\mathbf{T}_{\mathbf{X};ID}$.

The records in $\mathbf{T}_{\mathbf{X};ID}$ can be reordered in different ways; the values of the important order functions Z , $R_{\mathbf{X}}$, and $R_{\mathbf{X}}^{GCA}$ are given in columns (4), (5), and (6) of Table 13.1. The corresponding data tables are denoted $\mathbf{T}_{\mathbf{X};Z}$, $\mathbf{T}_{\mathbf{X}}$ and $\mathbf{T}_{\mathbf{X}}^{GCA}$ according to the notation proposed in Section 12.5.

The grade distributions of the data tables $\mathbf{T}_{\mathbf{X};Z}$ and $\mathbf{T}_{\mathbf{X}}$ are presented in Figures 13.1 and 13.2 (in the form of the overrepresentation maps). The related values $\rho^*(\mathbf{T}_{\mathbf{X},Z})$ and $\rho^*(\mathbf{T}_{\mathbf{X}})$, which according to the notation proposed in Section 12.5 are denoted $\mathbf{Dif}_{\mathbf{X};Z}$ and $\mathbf{Dif}_{\mathbf{X}}$, are 0.0771 and 0.3036. This shows that the grade distribution of $\mathbf{T}_{\mathbf{X};Z}$ is practically uncorrelated, while the grade distribution with records reordered due to R_X provides a positively correlated copula.

TABLE 13.1. Probability data table of $\mathbf{X} = (X_1, X_2, X_3)$.

ID (1)	record **x** (2)	$p(\mathbf{x})$ (3)	$Z(\mathbf{x})$ (4)	$R_{\mathbf{X}}(\mathbf{x})$ (5)	$R_{\mathbf{X}}^{GCA}(\mathbf{x})$ (6)
111	(0.40, 0.40, 0.50)	0.044	1.30	0.505	0.498
112	(0.40, 0.40, 2.00)	0.066	2.80	0.668	0.664
121	(0.40, 0.80, 0.50)	0.020	1.70	0.497	0.415
122	(0.40, 0.80, 2.00)	0.030	3.20	0.643	0.600
131	(0.40, 1.25, 0.50)	0.016	2.15	0.491	0.359
132	(0.40, 1.25, 2.00)	0.024	3.65	0.622	0.544
141	(0.40, 2.50, 0.50)	0.044	3.40	0.483	0.282
142	(0.40, 2.50, 2.00)	0.006	4.90	0.583	0.443
211	(0.80, 0.40, 0.50)	0.020	1.70	0.424	0.488
212	(0.80, 0.40, 2.00)	0.048	3.20	0.605	0.639
221	(0.80, 0.80, 0.50)	0.020	2.10	0.433	0.423
222	(0.80, 0.80, 2.00)	0.078	3.60	0.590	0.584
231	(0.80, 1.25, 0.50)	0.052	2.55	0.439	0.375
232	(0.80, 1.25, 2.00)	0.018	4.05	0.576	0.536
241	(0.80, 2.50, 0.50)	0.016	3.80	0.449	0.300
242	(0.80, 2.50, 2.00)	0.006	5.30	0.551	0.444
311	(1.25, 0.40, 0.50)	0.024	2.15	0.369	0.482
312	(1.25, 0.40, 2.00)	0.042	3.65	0.550	0.616
321	(1.25, 0.80, 0.50)	0.052	2.55	0.385	0.429
322	(1.25, 0.80, 2.00)	0.030	4.05	0.542	0.570
331	(1.25, 1.25, 0.50)	0.020	3.00	0.398	0.387
332	(1.25, 1.25, 2.00)	0.090	4.50	0.535	0.528
341	(1.25, 2.50, 0.50)	0.004	4.25	0.419	0.317
342	(1.25, 2.50, 2.00)	0.024	5.75	0.521	0.445
411	(2.50, 0.40, 0.50)	0.036	3.40	0.292	0.473
412	(2.50, 0.40, 2.00)	0.018	4.90	0.451	0.576
421	(2.50, 0.80, 0.50)	0.016	3.80	0.311	0.439
422	(2.50, 0.80, 2.00)	0.024	5.30	0.452	0.543
431	(2.50, 1.25, 0.50)	0.012	4.25	0.328	0.408
432	(2.50, 1.25, 2.00)	0.030	5.75	0.453	0.513
441	(2.50, 2.50, 0.50)	0.004	5.50	0.360	0.349
442	(2.50, 2.50, 2.00)	0.066	7.00	0.456	0.448

The GCA procedure when applied to $\mathbf{T}_{\mathbf{X};ID}$ produced the GCA overview table (as introduced in Section 9.2.5) equal to

No.	column permutation	ρ^*	frequency	τ/τ_{abs}
1	2, 1, 3	0.3197	2	0.7607
2	1, 2, 3	0.3036	2	0.7217
3	1, 3, 2	0.2753	2	0.6535

According to this overview, the optimal variable permutation defined by GCA is $\psi = (2, 1, 3)$. Function $R_{\mathbf{X}}^{GCA}$ appearing in Table 13.1 refers to the optimally permuted vector, i.e. $R_{\mathbf{X}}^{GCA} = R_{(X_2, X_1, X_3)}$ (the map of data table $\mathbf{T}_{\mathbf{X}}^{GCA}$ is shown in Figure 13.3). All permutations (three pairs of permutations with ρ^*=0.3197, 0.3036, 0.2753) form local maxima of ρ^*. Moreover, frequencies of particular maxima are identical, and the values of the related grade correlations ρ^* are only slightly differentiated. This suggests that Table 13.1 is not regular and contains three different (possibly meaningful) trends. A decomposition of Table 13.1 into more regular subtables could be tried using the methods of outliers detection presented in Chapter 9, but it would be rather risky in a so small data table.

The calculations of the first moment distribution, Lorenz Density, Lorenz Surface, Grade Lorenz Density and Grade Lorenz Surface for functions $\varphi_{\mathbf{X}}$ equal to Z, $R_{\mathbf{X}}$ and $R_{\mathbf{X}}^{GCA}$ were performed according to the formulas introduced in Section 12.5.

The map corresponding to the Grade Lorenz Surface for $\mathbf{X}_{GCA} = (X_2, X_1, X_3)$ is visualized in Figure 13.4. There are shown the intersections of $\mathbf{L}_{(X_2,X_1,X_3)}$ by three planes $v = v_i$, $i = 1, 2, 3$, where $v_1 = 1.0877/3.667$, $v_2 = (1.0877 + 1.1789)/3.667$, $v_3 = 1$. Each curve $\mathbf{L}_{(X_2,X_1,X_3)}(u, v_j)$ is accompanied by the line $v = uv_j$, which is the intersection of $Diag_2$ by the plane $v = v_j$. The curves in Figure 13.4 for $j = 1, 2$ (for $j = 3$ both intersections are equal to $Diag$). This means that the Grade Lorenz Surface $\mathbf{L}^*_{(X_2,X_1,X_3)}$ belongs to $\mathbb{QD}^+$.

It remains to inspect the correlation matrix $\mathbf{CORR}(X_2, X_1, X_3)$ and also the differentiation matrices $\mathbf{DIFF}(X_2, X_1, X_3)$ and $\mathbf{supDIFF}(X_2, X_1, X_3)$.

TABLE 13.2. Correlation matrix of vector $\mathbf{X}_{GCA} = (X_2, X_1, X_3)$

CORR	2	1	3
2	1	0.18317	0.01167
1		1	0.11602
3			1

The correlation matrix of vector $\mathbf{X}_{GCA} = (X_2, X_1, X_3)$ (shown in Table 13.2) is regular in the sense that it is a Robinson matrix (we recall that a matrix is called a *Robinson matrix* if the magnitude of the elements decrease monotonically when moving away from the matrix diagonal). This property (desirable for any similarity matrix and hence for any grade correlation matrix since it contains indices of similarity for pairs of variables) is fulfilled whenever the probability table is total positive of order two (TP_2). However, the reverse implication does not hold in general. It is certainly not true for the vector under consideration, since Figure 13.3 shows that data table $\mathbf{T}_{X_2,X_1,X_3}$ is very irregular. This table is far from TP_2.

The degree of departure of $\mathbf{T}_{(X_2,X_1,X_3)}$ from TP_2 will be evaluated on the basis of Table 13.3 which provides the values of $\mathbf{Dif}_{(X_i,X_j)}$ and $\mathbf{Dif}_{(X_i,X_j);R_{\mathbf{X}}}$.

For $\mathbf{X}_{GCA} = (X_2, X_1, X_3)$ we have $v_1 = E(X_2)/E(Z) = 0.2967$, $v_2 = (E(X_2 + X_1))/E(Z) = 0.6182$, $v_3 = E(X_2 + X_1 + X_3)/E(Z) = 1$. The values of $\mathbf{Dif}_{\mathbf{X}}^{GCA}$ and $sup\mathbf{Dif}_{\mathbf{X}}^{GCA}$ are calculated on the basis of Table 13.3:

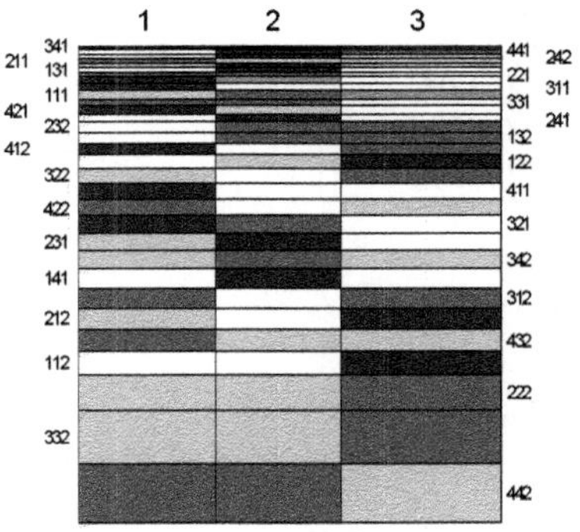

FIGURE 13.1. Trivariate discrete distribution: the overrepresentation map for grade distribution derived from $\mathbf{T}_{\mathbf{X};Z}$ ($\mathbf{Dif}_{\mathbf{X};Z} = 0.0771$).

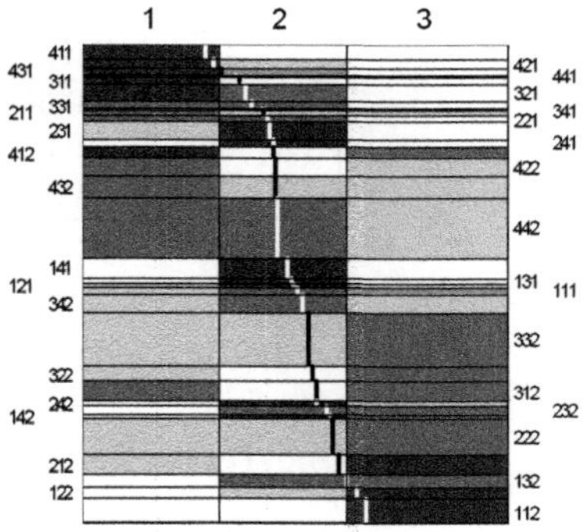

FIGURE 13.2. Trivariate discrete distribution: the overrepresentation map for the grade distribution derived from $\mathbf{T}_{\mathbf{X}}$ ($\mathbf{Dif}_{\mathbf{X}} = 0.3036$). The sequence of vertical segments visualises the grade *col*|*row* regression which corresponds to $R_{\mathbf{X}}$.

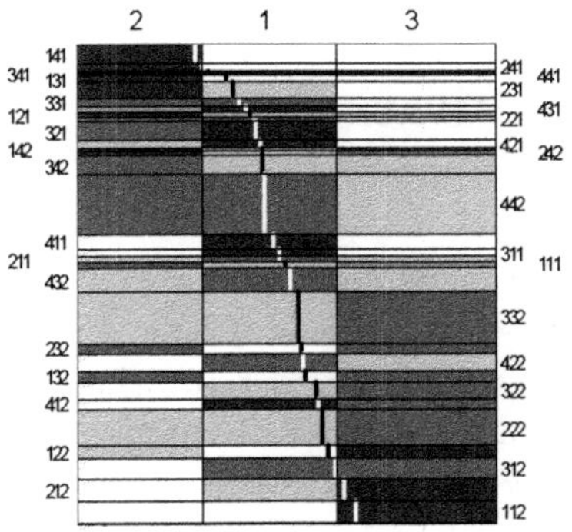

FIGURE 13.3. Trivariate discrete distribution: the overrepresentation map for grade distribution derived from $\mathbf{T}_{\mathbf{X}}^{GCA} = \mathbf{T}_{(X_2,X_1,X_3)}$ ($\mathbf{Dif}_{(X_2,X_1,X_3)} = 0.3197$). Function $R_{\mathbf{X}}^{GCA}$ is visualised in the same way as $R_{\mathbf{X}}$ in Figure 13.2.

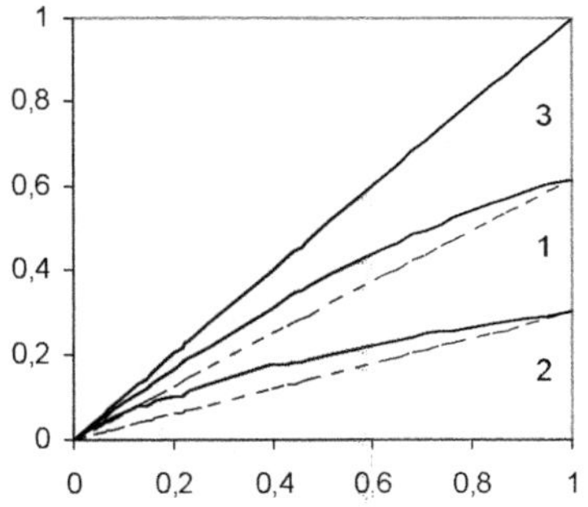

FIGURE 13.4. Trivariate discrete distribution: the intersections of the Grade Lorenz Surface $\mathbf{L}^{*}_{(X_2,X_1,X_3)}$ (solid curves) and of $Diag_2$ (dotted straight lines) by the three planes defined by $v = \frac{EX_2}{EZ}$, $v = \frac{E(X_2+X_1)}{EZ}$, and $v = 1$. The widths of segments numbered 2,1,3, which lie on the right side of the unit square, are equal to $E(X_2)/E(Z)$, $E(X_1)/E(Z)$, $E(X_3)/E(Z)$.

TABLE 13.3. Trivariate discrete distribution: values of the differentiations $\mathbf{Dif}_{(X_i,X_j)}$ and $\mathbf{Dif}_{(X_i,X_j;R_\mathbf{X})}$

	$\mathbf{Dif}_{(X_i,X_j)}$	$\mathbf{Dif}_{(X_i,X_j);R_\mathbf{X}}$
(X_2,X_1)	0.3088	0.1497
(X_1,X_3)	0.3159	0.2333
(X_2,X_3)	0.3360	0.3340

$$\begin{aligned}
\mathbf{Dif}_\mathbf{X}^{GCA} &= (v_2+v_1-v_1)\frac{(E(X_2+X_1))^2}{(E(Z))^2}\mathbf{Dif}_{(X_2,X_1);R_\mathbf{X}} \\
&+(v_3+v_2-v_1)\frac{(E(X_2+X_3))^2}{(E(Z))^2}\mathbf{Dif}_{(X_2,X_3);R_\mathbf{X}} \\
&+(v_3+v_2-v_2-v_1)\frac{(E(X_1+X_3))^2}{(E(Z))^2}\mathbf{Dif}_{(X_1,X_3);R_\mathbf{X}} \\
&= 0.3197, \\
sup\mathbf{Dif}_\mathbf{X}^{GCA} &= 0.3873
\end{aligned}$$

(where $sup\mathbf{Dif}_\mathbf{X}^{GCA}$ is calculated like $\mathbf{Dif}_\mathbf{X}^{GCA}$ but with $\mathbf{Dif}_{(X_i,X_j);R_\mathbf{X}}$ replaced by $\mathbf{Dif}_{(X_i,X_j)}$). Hence $\mathbf{Dif}_\mathbf{X}^{GCA}/sup\mathbf{Dif}_\mathbf{X}^{GCA} = 82.6\%$. As such ratio is 100% if and only if the respective probability data table is TP_2, vector (X_2,X_1,X_3) is not TP_2, and its closeness to TP_2 measured as above slightly surpasses 80%.

13.3 Inequality measures for multivariate distributions with non-finite sets of records

The present section extends the definition of the aggregated data table $\overline{\mathbf{T}}_{\mathbf{X};\varphi_\mathbf{X}}$, introduced previously for vectors $\mathbf{X}$ with finite set of records. This is a necessary introduction to deal with selected families of continuous bivariate and multivariate distributions, considered in the next sections. However, some readers could find this section too technical at the start so that they might prefer to read example contained in Section 13.4.1 first and then come back to the general case introduced here.

In the case of a multivariate distribution with a finite set of records, the finite *aggregated* data table $\overline{\mathbf{T}}_{\mathbf{X};\varphi_\mathbf{X}}$ introduced in Definition 13.2 can be rewritten as

$$\overline{\mathbf{T}}_{\mathbf{X};\varphi_\mathbf{X}} = \left[\left(\left(\frac{E(X_j;\varphi_\mathbf{X}=t_i)}{E(Z)},\ j=1,...,k\right),\ p_i\right),\ i=1,...,m,\right] \tag{13.3}$$

where $t_1 < ... < t_m$ are the values of $\varphi_\mathbf{X}$, $(E(X_j;\varphi_\mathbf{X}=t_i), j=1,...,k)$ is the mean value of records for which $\varphi_\mathbf{X}=t_i$, $p_i=\Pr(\varphi_\mathbf{X}=t_i)$.

An immediate extension of such finite aggregated data table to cover the distributions with *non-finite sets of records and with continuous random variables* $\varphi_{\mathbf{X}}$ is a configuration with k columns and non-finite set of rows denoted by numbers t belonging to $\mathbb{R}$:

$$\overline{\mathbf{T}}_{\mathbf{X};\varphi_{\mathbf{X}}} = \left[\left(\left(\frac{E(X_j; \varphi_{\mathbf{X}} = t)}{E(Z)}, j = 1, ..., k \right), f_{\varphi_{\mathbf{X}}}(t) \right), \ t \in \mathbb{R} \right] \tag{13.4}$$

Such specific data tables correspond to bivariate distributions which are discrete-continuous in the sense that the "row" (vertical) variable is continuous and the "column" (row) variable is categorical. This is why each row is accompanied by the value of the density of variable $\varphi_{\mathbf{X}}$ which is obtained from the distribution of vector $\mathbf{X}$.

Remark 13.2 *The definition of* $\overline{\mathbf{T}}_{\mathbf{X};\varphi_{\mathbf{X}}}$ *can be also extended to cover cases when random variable* $\varphi_{\mathbf{X}}$ *is not continuous but discrete-continuous. However, this extension will be not considered here.*

Please note that the *finite* aggregated data tables have an important property: they are equal for functions $\varphi_{\mathbf{X}}$ and $\varphi'_{\mathbf{X}}$ if the latter is an increasing function of the former. It follows from (13.4) that the non-finite aggregated data tables do not have this property. To obtain equality, random variables $\varphi_{\mathbf{X}}$ and $\varphi'_{\mathbf{X}}$ might be replaced by their uniform grade transforms $(\varphi_{\mathbf{X}})^*$ and $(\varphi'_{\mathbf{X}})^*$ (see Sections 3.3.3 and 3.5.1). This will be explicitly calculated in Section 13.4.1 for the order function $R_{\mathbf{X}}$ and a specially chosen function denoted $g_{\mathbf{X}}$ which is an increasing function of $R_{\mathbf{X}}$.

The aggregated Lorenz Density $\overline{\lambda}_{\mathbf{X};\varphi_{\mathbf{X}}}$ corresponding to the random vector $\mathbf{X}$ with *non-finite set of records* arises from the aggregated data table $\overline{\mathbf{T}}_{\mathbf{X};\varphi^*_{\mathbf{X}}}$ when this table is "stretched" horizontally over the unit square: for $u \in (0,1)$, and $v \in (v_{j-1}, v_j)$

$$\overline{\lambda}_{\mathbf{X};\varphi_{\mathbf{X}}}(u,v) = \frac{E(X_j; \varphi^*_{\mathbf{X}} = u)}{E(X_j)}. \tag{13.5}$$

In other words: given random variable $\varphi_{\mathbf{X}}$ (with the distribution induced by that of $\mathbf{X}$), it is possible to calculate, for any u, the expectation of component X_j over the set of records for which the grade transform of $\varphi_{\mathbf{X}}$ is equal to u. This expectation divided by $E(Z)$ is the value in point (u,v) of the aggregated Lorenz Density corresponding to $\mathbf{X}$ and to $\varphi_{\mathbf{X}}$, where v satisfies $v \in (v_{j-1}, v_j)$ and

$$v_j = E(X_1 + \cdots + X_j)/E(Z), \ \ j = 1, ..., k, \ v_0 = 0. \tag{13.6}$$

Further integrated measures of inequality are derived from $\overline{\lambda}_{\mathbf{X};\varphi_{\mathbf{X}}}$ according to formulas analogous to those in the finite case:

- $\overline{\mathbf{L}}_{\mathbf{X};\varphi_{\mathbf{X}}}(u,v) = \int_0^u \int_0^v \overline{\lambda}_{\mathbf{X};\varphi_{\mathbf{X}}}(u',v')du'dv'$,
- $\overline{\mathbf{L}}^*_{\mathbf{X};\varphi_{\mathbf{X}}}$ and $\overline{\lambda}^*_{\mathbf{X};\varphi_{\mathbf{X}}}$ are the *cdf* and density of the Grade Lorenz Pair (i.e. the cdf and density of the *grade distribution* of the Lorenz Pair) corresponding to $\mathbf{X}$ and $\varphi_{\mathbf{X}}$,
- $\mathbf{Dif}_{\mathbf{X};\varphi_{\mathbf{X}}}$ is the grade correlation ρ^* of the Lorenz Pair (or, equivalently, the correlation coefficient of the Grade Lorenz Pair) corresponding to $\mathbf{X}$ and $\varphi_{\mathbf{X}}$.

Analogously as in the case of finite sets of records, the abbreviate notation $\overline{\lambda}_{\mathbf{X}}, \overline{\lambda}^*_{\mathbf{X}}, \overline{\mathbf{L}}_{\mathbf{X}}, \overline{\mathbf{L}}^*_{\mathbf{X}}, \mathbf{Dif}_{\mathbf{X}}$ will be used instead of $\overline{\lambda}_{\mathbf{X};R_{\mathbf{X}}}, \overline{\lambda}^*_{\mathbf{X};R_{\mathbf{X}}}, \overline{\mathbf{L}}_{\mathbf{X};R_{\mathbf{X}}}, \overline{\mathbf{L}}^*_{\mathbf{X};R_{\mathbf{X}}}$, $\mathbf{Dif}_{\mathbf{X};R_{\mathbf{X}}}$ etc. Further, the abbreviation $\overline{\mathbf{T}}_{\mathbf{X}}$ will be used instead of $\overline{\mathbf{T}}_{\mathbf{X};R_{\mathbf{X}}}$ where

$$R_{\mathbf{X}} = \sum_{j=1}^{k} \frac{(v_{j-1} + \frac{1}{2}E(X_j)/E(Z))X_j}{Z} \tag{13.7}$$

and v_j is defined by (13.6.

The table $\overline{\mathbf{T}}_{\mathbf{X}}$ will be shortly called *the aggregated data table of vector* $\mathbf{X}$ (the word "probability" being omitted for short).

Proposition 13.1 *If* $\mathbf{X} = (X_1, ..., X_k)$ *and the distribution of* $\mathbf{X}$ *is symmetric (i.e.* $(X_1, ..., X_k) \sim (X_{i_1}, ..., X_{i_k})$ *for any permutation* $(i_1, ..., i_k)$ *of* $(1, ..., k)$*), then* $Z = \sum_{j=1}^k X_j$ *and the order function* $R_{\mathbf{X}}$ *defined by (13.7) are uncorrelated.*

Proof: due to symmetry of the distribution of $\mathbf{X}$, $(E(X_i)/E(Z)) = 1/k$ and hence $R_{\mathbf{X}}$ is equal to

$$\sum_{j=1}^{k} \left(\frac{2j-1}{2k}\right) \frac{X_j}{Z}$$

and the constants $a_t \stackrel{df}{=} E(X_j|Z=t)$ are equal to t/k. This implies that $E(R_{\mathbf{X}}|Z = t) = \frac{k\,a_t}{2\,t} \equiv \frac{1}{2}$ which completes the proof.

Due to Proposition 13.1, the order functions $R_{\mathbf{X}}$ and Z are specially important summary descriptors of vector $\mathbf{X}$: under symmetry of the distribution of $\mathbf{X}$, they are uncorrelated descriptive statistics with interesting interpretation.

If $k = 2$, the aggregated data table $\overline{\mathbf{T}}_{(X_1,X_2)}$ is totally positive dependent of order two (TP_2). The proof follows from the fact that $R_{(X_1,X_2)}$ is an increasing function of the ratio X_2/X_1.

Examples of the aggregated data tables for families of pairs (X_1, X_2) with continuous distributions will be considered in the next section. All families considered there have regular symmetric distributions, belonging to the set $\mathbb{TP}_2$. Pairs investigated in Sections 13.4.1-13.4.2 belong to $\mathbb{DLM}$, while pairs considered in Section 13.4.3 belong to $\mathbb{BLM}$ but not to $\mathbb{DLM}$. Investigation of each pair (X_1, X_2) from outside $\mathbb{DLM}$ is accompanied by investigation of the pair grade transforms (X_1^*, X_2^*).

The relationship between the parameters $corr(X_1^*, X_2^*)$ and $\mathbf{Dif}_{(X_1,X_2)}$ is established for each family. Formula (12.51) for $\mathbf{Dif}_{(X_1,X_2)}$ introduced for the pair (X_1, X_2) with a finite set of records can be extended onto set of pairs with non-finite sets of records:

$$\mathbf{Dif}_{(X_1,X_2)} = 3\frac{E(X_1)E(X_2)}{(E(X_1+X_2))^2}\, ar(L_{X_2;R^*_{\mathbf{X}}} : L_{X_1;R^*_{\mathbf{X}}}). \qquad (13.8)$$

Index ar in (13.8) can be replaced by $ar_{\max}$, since the data table $\overline{\mathbf{T}}_{(X_1,X_2)}$ is TP_2.

For the three families which are contained in $\mathbb{DLM}$, the plots of $\mathbf{Dif}_{(X_1^*,X_2^*)}$ versus $corr(X_1^*, X_2^*)$ are shown in Figure 13.12 (curves (2), (3),(4)). It is seen there that they lie very close to certain symmetric decreasing function mapping interval [-1,1] to [0,1] (i.e., mapping the domain of $corr$ onto the domain of $\mathbf{Dif}$). The plots for the remaining families (which are not contained in $\mathbb{DLM}$) are shown in Figure 13.12 as curves (1) and (5).

The symmetric continuous multivariate distributions with $k > 2$ will be investigated in Sections 13.5 and 13.6. They are: grade multinormal distributions, Moran distributions, and grade Moran distributions. Special attention will be given to the links existing between the grade correlation matrices $\mathbf{CORR}(\mathbf{X}^*)$ where $\mathbf{X}^* = (X_1^*, ..., X_k^*)$, and the differentiation matrices describing the aggregated data tables:

$$\overline{\mathbf{T}}_{\mathbf{X}},\ \overline{\mathbf{T}}_{\mathbf{X}}^{GCA},\ \overline{\mathbf{T}}_{\mathbf{X}^*},\ \overline{\mathbf{T}}_{\mathbf{X}^*}^{GCA}.$$

This is an extension of the similarity/dissimilarity investigation in the case of $k = 2$. The key problem will be how regularity of the aggregated data table is reflected by regularity of the grade correlation matrix. Is it possible to establish the properties of multivariate distributions which are so regular that the aggregated data table $\overline{\mathbf{T}}_{\mathbf{X}}$ is TP_2? What set of correlation matrices is admitted for such distributions? When would it be possible to express the global differentiation $\mathbf{Dif}_{\mathbf{X}^*}$ by the elements of $\mathbf{CORR}(\mathbf{X}^*)$?

Sections 13.4.4-13.6 provide partial answers to these questions. According to them, exact or almost exact links for suitable similarity/dissimilarity measures exist if vector $\mathbf{X}$ is sufficiently regularly distributed.

13.4 Inequality measures for continuous bivariate distributions

13.4.1 A pair of independent uniform Lilliputian variables

Let $\mathbf{X} = (U, V)$ be a pair of independent uniform Lilliputian variables ($F_{(U,V)} = Diag_2$). The set of records is now equal to the set of points in the unit square. We will consider two order functions: the sum $Z = U + V$ and the $col|row$ regression $R_{\mathbf{X}}$; according to (13.7),

$$R_{\mathbf{X}} = (0.25U + 0.75V)/Z. \tag{13.9}$$

Our aim is to construct the aggregated data tables $\overline{\mathbf{T}}_{\mathbf{X};Z}$ and $\overline{\mathbf{T}}_{\mathbf{X}}$, and to calculate the related inequality measures. Functions $R_{\mathbf{X}}$ and Z are descriptors of (U, V) which provide weakly related and well interpretable information about the records.

The Lorenz Curve L_Z.

The Lorenz Curve L_Z (i.e., the concentration curve $C(F_Z^{(1)} : F_Z)$ which is shown in Figure 13.5) is given by

$$L_Z(u) = \begin{cases} \frac{1}{3}\sqrt{8u^3} & \text{if } 0 \le u \le \frac{1}{2}, \\ -\frac{1}{3}\left(2 - \sqrt{2-2u}\right)^3 + \left(2 - \sqrt{2-2u}\right)^2 - \frac{1}{3} & \text{if } \frac{1}{2} < u < 1 \end{cases} \tag{13.10}$$

and the respective Gini Index is equal to 0.233.

Details of the computation of L_Z:
The expectation $E(Z)$ equals 1. The density and cdf of Z and the cdf of $Z^{(1)}$ are given by the following formulas:

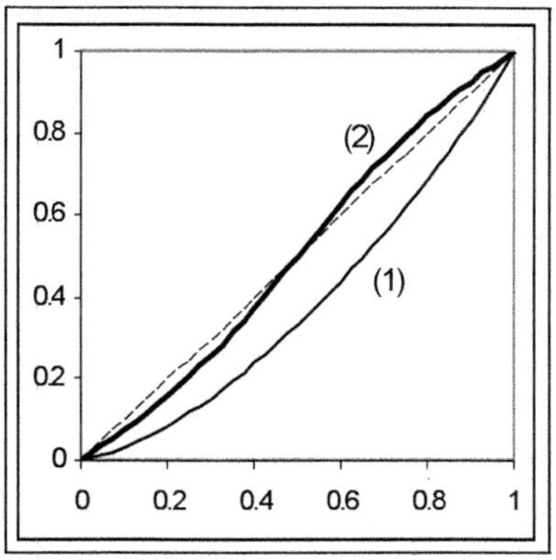

FIGURE 13.5. The Lorenz Curves: (1) L_{U+V} and (2) $L_{U+V;R_{\mathbf{X}}}$ for independent uniform Lilliputian variables U and V. Gini Indices are Gini(Z) = 0.233 and Gini(Z;R$_{\mathbf{X}}$) = 0.

$$f_Z(z) = \begin{cases} z & \text{if } 0 \leq z \leq 1 \\ 2 - z & \text{if } 1 < z \leq 2 \end{cases}, \tag{13.11}$$

$$F_Z(z) = \begin{cases} \frac{1}{2}z^2 & \text{if } 0 \leq z \leq 1 \\ -\frac{1}{2}z^2 + 2z - 1 & \text{if } 1 < z \leq 2 \end{cases}, \tag{13.12}$$

$$F_Z^{(1)}(z) = \begin{cases} \frac{1}{3}z^3 & \text{if } 0 \leq z \leq 1 \\ -\frac{1}{3}z^3 + z^2 - \frac{1}{3} & \text{if } 1 < z \leq 2 \end{cases}. \tag{13.13}$$

Curve L_Z consists of points $\{(F_Z(z), F_Z^{(1)}(z));\ z \in \mathbb{R}\}$ (curve (1) in Figure 13.5). The Gini Index $Gini(Z) = 0.233$. The Lorenz Curve $L_{Z;R_{\mathbf{X}}}$ is presented in Fig. 13.5 as a curve (2), with $Gini(Z; R_{\mathbf{X}}) = 0$.

Sample data tables $\overline{\mathbf{T}}_{\mathbf{X}}^{(n)}$ and $\mathbf{Dif}_{\mathbf{X}}^{(n)}$

Before constructing the aggregated data table $\overline{\mathbf{T}}_{\mathbf{X}}$ and the global differentiation $\mathbf{Dif}_{\mathbf{X}}$ random samples were drawn from the distribution of (U, V), of size $n = 20$, 100 and 1000. Each sample was then ordered according to GCA, which resulted the sample data tables $\overline{\mathbf{T}}_{\mathbf{X}}^{(n)}$. The respective maps are shown in Figures 13.6 - 13.8.

The sets of points $\{R_{\mathbf{X}}^{(n)}(u), u \in (0,1)\}$, which correspond to the graphs of the *col*|*row* regression of the sample data tables ordered according to GCA, are visible in the maps for $n = 100$ and $n = 1000$. They form curves which

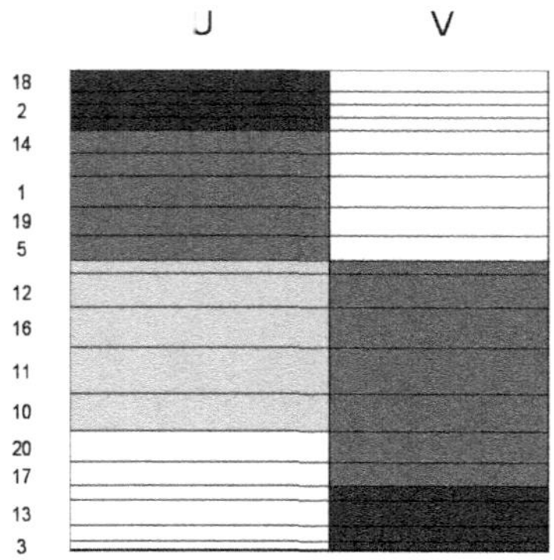

FIGURE 13.6. The map of sample GCA data table $\overline{\mathbf{T}}^{(20)}_{(U,V)}$ when U, V are independent and uniformly distributed on (0,1); $\mathbf{Dif}^{(20)}_{(U,V)} = 0.3966$.

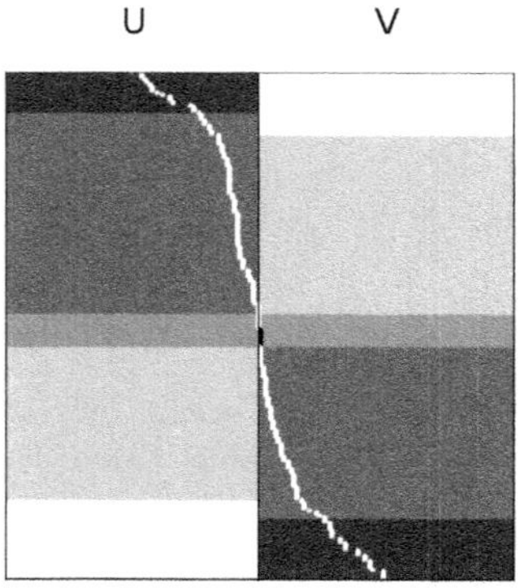

FIGURE 13.7. The map of sample GCA data table $\overline{\mathbf{T}}^{(100)}_{(U,V)}$ when U, V are independent and uniformly distributed on (0,1); $\mathbf{Dif}^{(100)}_{(U,V)} = 0.3166$

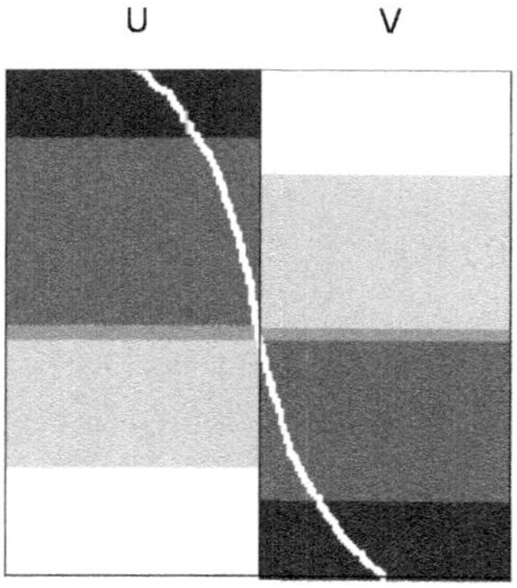

FIGURE 13.8. The map of sample GCA data table $\overline{\mathbf{T}}^{(1000)}_{(U,V)}$ when U, V are independent and uniformly distributed on (0,1); $\mathbf{Dif}^{(1000)}_{(U,V)} = 0.3663$

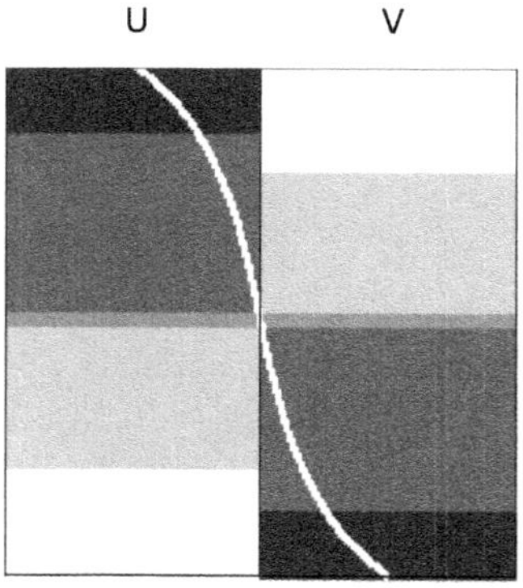

FIGURE 13.9. The map of the data table $\overline{\mathbf{T}}_{(U,V)}$ for two independent uniform Lilliputian variables; $\mathbf{Dif}_{(U,V)} = 0.3611$

decrease from the left upper corner of the unit square to the right lower corner, so that – in view of the convention introduced in Chapter 8 that point $(0, 0)$ is situated in the left upper corner and the vertical axis is directed down – the functions $R_{\mathbf{X}}^{(n)}$ are non-decreasing. Their graphs become more regular when n increases. The same can be roughly said about increasing regularity of the shading of the maps.

The limit to which tend the maps of the sample data tables, i.e., the "population" map of $\overline{\mathbf{T}}_{\mathbf{X}}$, is presented in Figure 13.9. The exact formula for the data table $\overline{\mathbf{T}}_{\mathbf{X}}$ is derived further in this section, together with the value 0.3611 of $\mathbf{Dif}_{\mathbf{X}}$. The values of $\mathbf{Dif}_{\mathbf{X}}^{(n)}$, equal to 0.3966, 0.3166, 0.3663, oscillate around 0.3611; for $n = 1000$ the departure of $\mathbf{Dif}_{\mathbf{X}}{}^{(n)}$ from $\mathbf{Dif}_{\mathbf{X}}$ is only 0.0018.

The values of $R_{\mathbf{X}}$ differ from 0.25 to 0.75.

The aggregated data table $\overline{\mathbf{T}}_{(U,V)}$

According to the general scenario described in the previous section, the construction of $\overline{\mathbf{T}}_{\mathbf{X}}$ should start with finding the joint distribution of vector $\mathbf{X}$ and $R_{\mathbf{X}}$ or of any increasing function of $R_{\mathbf{X}}$. In the present section, for $\mathbf{X} = (U, V)$, it will be convenient to use firstly the order function $g_{\mathbf{X}}$, defined for any record (u, v) by

$$g_{\mathbf{X}}(u, v) = \frac{v}{u}; \tag{13.14}$$

due to (13.9), $g_{\mathbf{X}}$ is an increasing function of $R_{\mathbf{X}}$:

$$g_{\mathbf{X}}(u, v) = \frac{R_{\mathbf{X}}(u, v) - 0.25}{0.75 - R_{\mathbf{X}}(u, v)}. \tag{13.15}$$

As the joint density of $\mathbf{X} = (U, V)$ is identically equal to 1, then (due to the Fubini Theorem) the joint density of $(U, g_{\mathbf{X}})$ in point (u,t) is

$$f_{(U,g_{\mathbf{X}})}(u,t) = \begin{cases} u \text{ if } & u \in (0,1), \quad t \in R^{+}, \quad ut \leq 1, \\ 0 & \qquad\qquad \text{otherwise.} \end{cases} \tag{13.16}$$

So the density and the cdf of $g_{\mathbf{X}}$ are

$$f_{g_{\mathbf{X}}}(t) = \begin{cases} 0.5 \text{ if } 0 \leq t \leq 1 \\ \\ \frac{1}{2t^2} \text{ if } \quad t > 1 \end{cases}, \tag{13.17}$$

$$F_{g_{\mathrm{X}}}(t) = \begin{cases} \frac{t}{2} & \text{if } 0 \leq t \leq 1, \\ 1 - \frac{1}{2t} & \text{if } \quad t > 1. \end{cases} \tag{13.18}$$

It follows that the data table $\overline{\mathbf{T}}_{\mathbf{X};g_{\mathrm{X}}}$ consists of the following two columns:

$$E(U; g_{\mathrm{X}} = t) = \begin{cases} \frac{1}{3} & \text{if } 0 \leq t \leq 1 \\ \frac{1}{3t^3} & \text{if } \quad t > 1 \end{cases}, \tag{13.19}$$

$$E(V; g_{\mathrm{X}} = t) = \begin{cases} \frac{1}{3}t & \text{if } 0 \leq t \leq 1 \\ \frac{1}{3t^2} & \text{if } \quad t > 1 \end{cases}, \tag{13.20}$$

supplemented by the column with the values of density (13.17)

Inserting (13.15) to (13.19) and (13.20), we get the following two columns of $\overline{\mathbf{T}}_{(U,V)}$:

$$E(U; R_{\mathbf{X}} = t) = \begin{cases} \frac{1}{6(0.75-t)^2} & \text{if } \frac{1}{4} \leq t \leq \frac{1}{2}, \\ \frac{0.75-t}{6(t-0.25)^3} & \text{if } \frac{1}{2} \leq t \leq \frac{3}{4}, \end{cases} \tag{13.21}$$

$$E(V; R_{\mathbf{X}} = t) = \begin{cases} \frac{t-0.25}{6(0.75-t)^3} & \text{if } \frac{1}{4} \leq t \leq \frac{1}{2}, \\ \frac{1}{6(t-0.25)^2} & \text{if } \frac{1}{2} \leq t \leq \frac{3}{4}. \end{cases} \tag{13.22}$$

It is easy to check that $\overline{\mathbf{T}}_{\mathbf{X};g_{\mathrm{X}}}$ and $\overline{\mathbf{T}}_{\mathbf{X};R_{\mathrm{X}}}$ become identical when function $R_{\mathbf{X}}$ in (13.21)-(13.22) and function $g_{\mathbf{X}}$ in (13.19)-(13.20) are respectively replaced by $R^*_{\mathbf{X}}$ and $g^*_{\mathbf{X}}$ (then $\overline{\mathbf{T}}_{\mathbf{X};R^*_{\mathrm{X}}}$ will be the representation of $\overline{\mathbf{T}}_{\mathbf{X};\varphi}$ for all increasing functions of $R_{\mathbf{X}}$.) The data tables $\overline{\mathbf{T}}_{\mathbf{X};g^*_{\mathrm{X}}}$ and $\overline{\mathbf{T}}_{\mathbf{X};R^*_{\mathrm{X}}}$ both contain the same columns

$$E(U; R^*_{\mathbf{X}} = t) = \begin{cases} \frac{2}{3} & \text{if } 0 \leq t \leq 1/2, \\ \frac{4(1-t)}{3} & \text{if } 1/2 \leq t \leq 1, \end{cases} \tag{13.23}$$

$$E(V; R^*_{\mathbf{X}} = t) = \begin{cases} \frac{4t}{3} \text{ if } 0 \le t \le \frac{1}{2}, \\ \frac{2}{3} \text{ if } \frac{1}{2} \le t \le 1, \end{cases} \tag{13.24}$$

since

$$E(U; R^*_{\mathbf{X}} = t) = \frac{E(U; R_{\mathbf{X}} = r)}{f_{R_{\mathbf{X}}}(r)} \quad \text{for } t = F_{R_{\mathbf{X}}}(r),$$

$$E(V; R^*_{\mathbf{X}} = t) = \frac{E(V; R_{\mathbf{X}} = r)}{f_{R_{\mathbf{X}}}(r)} \quad \text{for } t = F_{R_{\mathbf{X}}}(r),$$

where

$$f_{R_{\mathbf{X}}}(r) = \begin{cases} \frac{1}{4(0.75-r)^2} & \text{if } 0.25 \le r \le 0.5, \\ \frac{1}{4(r-0.25)^2} & \text{if } 0.5 \le r \le 0.75. \end{cases}$$

It follows that

$$\overline{\lambda}_{\mathbf{X}}(u,v) = \begin{cases} \frac{4}{3} & \text{if } (u,v) \in (0;\frac{1}{2}] \times (0;\frac{1}{2}], \\ \frac{8}{3}(1-u) & \text{if } (u,v) \in (\frac{1}{2};1] \times (0;\frac{1}{2}], \\ \frac{8}{3}u & \text{if } (u,v) \in (0;\frac{1}{2}] \times (\frac{1}{2};1], \\ \frac{4}{3} & \text{if } (u,v) \in (\frac{1}{2};1] \times (\frac{1}{2};1], \end{cases} \tag{13.25}$$

$$\overline{\mathbf{L}}_{\mathbf{X}}(u,v) = \begin{cases} \frac{4}{3}uv & \text{if } (u,v) \in (0;\frac{1}{2}] \times (0;\frac{1}{2}], \\ -\frac{4}{3}u^2v + \frac{8}{3}uv - \frac{1}{3}v & \text{if } (u,v) \in (\frac{1}{2};1] \times (0;\frac{1}{2}], \\ \frac{4}{3}u^2v - \frac{2}{3}u^2 + \frac{2}{3}u & \text{if } (u,v) \in (0;\frac{1}{2}] \times (\frac{1}{2};1], \\ -\frac{2}{3}u^2 + \frac{4}{3}uv + \frac{2}{3}u - \frac{1}{3} & \text{if } (u,v) \in (\frac{1}{2};1] \times (\frac{1}{2};1]. \end{cases} \tag{13.26}$$

Putting $v = 1$ to (13.26), we get $\overline{\mathbf{L}}_{\mathbf{X}}(u,1) = L_{Z;R^*_{\mathbf{X}}}$, where

$$L_{Z;R^*_{\mathbf{X}}}(u) = \begin{cases} \frac{2}{3}u + \frac{2}{3}u^2 & \text{if } u \le \frac{1}{2}, \\ -\frac{2}{3}u^2 + 2u - \frac{1}{3} & \text{if } u > \frac{1}{2}. \end{cases} \tag{13.27}$$

Further, by staightforward computations,

$$\begin{aligned} ar(L_{V;g\mathbf{X}} : L_{U;g\mathbf{X}}) &= 2\int_0^1 (L_{U;g\mathbf{X}}(u) - L_{V;g\mathbf{X}}(u))\, dL_{U;g\mathbf{X}}(u) \\ &= \frac{13}{27} \end{aligned}$$

and hence

$$\mathbf{Dif}_{\mathbf{X}} = \frac{3}{4} ar(L_{V;g\mathbf{X}} : L_{U;g\mathbf{X}}) = \frac{13}{36} = 0.3611.$$

The map of the data table $\overline{\mathbf{T}}_{(U,V)}$ is shown in Figure 13.9. At the first sight, this map is almost undistinguishable from the map of the sample data table for a sample of size 1000 (Figure 13.8).

The joint density of $(R^*_{\mathbf{X}}, Z)$ is

$$f_{(R^*,Z)}(r,z) = \begin{cases} \frac{2z}{(1+2r)^2} \text{ if } 0 \le r \le \frac{1}{2},\ 0 \le z \le 1+2r, \\ \frac{2z}{(3-2r)^2} \text{ if } \frac{1}{2} \le r \le 1,\ 0 \le z \le 3-2r; \end{cases} \tag{13.28}$$

hence $E(ZR^*) = \frac{1}{2} = E(Z)E(R^*)$, which confirms that $R^*_{\mathbf{X}}$ and Z are uncorrelated.

We also have $\mathbf{Dif}_{\mathbf{X};Z} = 0$, since

$$E(U; Z = t) = E(V; Z = t).$$

Then the global differentiation of records (i.e. the maximal strength of dependence between variables and objects) is moderately large, while there is null global differentiation of records (null strength between variables and objects) when records are ordered according to Z. On the other hand, the values of Z are differentiated rather strongly if the records are ordered according to Z ($Gini(Z) = 0.233$), while there is *null* global differentiation of the values of Z when records are ordered according to $R_{\mathbf{X}}$ ($Gini(Z; R_{\mathbf{X}}) = 0$). This shows that the two trends of dependence between variables and objects: when records are associated with ordering $R_{\mathbf{X}}$ and when they are associated with ordering Z, are as different as possible.

Note that $R_{\mathbf{X}}$ orders records *so that their profiles become most strongly differentiated,* while Z orders records so that *the values of Z become most strongly differentiated.* These two orderings together provide valuable descriptive information about the records.

13.4.2 A pair of functionally dependent Lilliputian variables

According to Section 8.3, a functional relationship of two Lilliputian variables U and V is possible only if either $V = U$ or $V = 1 - U$.

In the first case,

$$Z = 2U, \quad E(Z) = 1, \quad Gini(Z) = \frac{1}{3}, \quad R_{\mathbf{X}}(u) \equiv \frac{1}{2}, \tag{13.29}$$

$$\begin{aligned} E(U; R_{\mathbf{X}} = t) = E(V; R_{\mathbf{X}} = t) = \frac{1}{2}\ ; \quad \mathbf{Dif}_{(U,V)} = 0, \\ Gini(R_{\mathbf{X}}) = Gini(Z; R_{\mathbf{X}}) = 0. \end{aligned} \tag{13.30}$$

The overrepresentation map of the data table $\overline{\mathbf{T}}_{(U,U)}$ with constant identical density $\frac{1}{2}$ in both columns is shown in Figure 13.10.

FIGURE 13.10. The map of the data table $\mathbf{T}_{U,U}$ when U is a uniform Lilliputian variable; $\mathbf{Dif}_{(U,U)} = 0$.

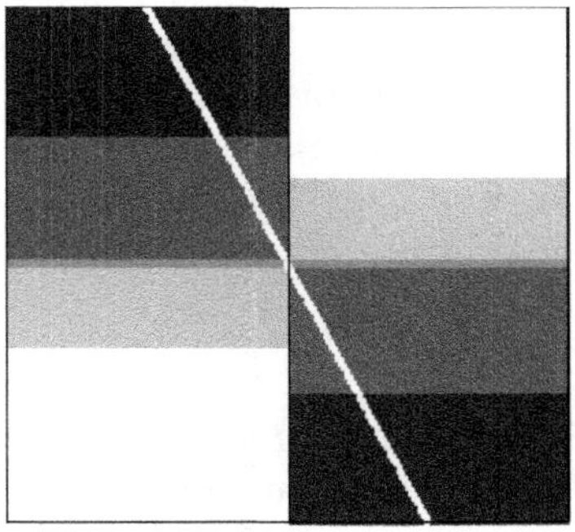

FIGURE 13.11. The map of the data table $\mathbf{T}_{U,1-U}$ when U is a uniform Lilliputian variable; $\mathbf{Dif}_{(U,1-U)} = 0.5$.

In the second case,

$$\begin{aligned} Z = U + 1 - U \equiv 1, \; Gini(Z) = 0, \; Gini(Z; R_{\mathbf{X}}) = 0, \\ R_{\mathbf{X}}(u) = \frac{3}{4} - \frac{1}{2}u, \\ E(U; R_{\mathbf{X}} = t) = 3 - 4t; \; E(V; R_{\mathbf{X}} = t) = -1 + 4t; \end{aligned}$$

$$
\begin{aligned}
\mathbf{Dif}_{(U,1-U)} &= \frac{3}{4} ar(L_{1-U;R^*_{\mathbf{X}}} : L_{U;R^*_{\mathbf{X}}}) \\
&= \frac{3}{2} \int_0^1 \left(L_{U;R^*_{\mathbf{X}}}(u) - L_{1-U;R^*_{\mathbf{X}}}(u)\right) dL_{U;R^*_{\mathbf{X}}}(u) \\
&= \int_0^1 (u - 2u^2 + u^3) du = \frac{1}{2}, \qquad (13.31)
\end{aligned}
$$

since

$$L_{U;R^*_{\mathbf{X}}}(u) = 2E(U; R^*_{\mathbf{X}} \leq u) = 2u - u^2$$

$$L_{1-U;R^*_{\mathbf{X}}}(u) = 2E(1 - U; R^*_{\mathbf{X}} \leq u) = u^2.$$

The map of the data table

$$\overline{\mathbf{T}}_{(U,1-U)} = [((3 - 4t,\ 4t - 1), f_{R_{\mathbf{X}}} \equiv 2),\ t \in (0.25, 0.75)] \qquad (13.32)$$

is shown in Figure 13.11.

Variables Z and R behave in contrast: one of them is concentrated in a single point (it is $R_{\mathbf{X}} \equiv \frac{1}{2}$ in the first case and $Z \equiv 1$ in the second) while the other is maximally differentiated. It follows that for any pair from $\mathbb{DLM}$ the global differentiation of records, i.e. $\mathbf{Dif}_{\mathbf{X}}$, changes from 0 (for $V = U$) to 0.5 (for $V = 1 - U$), while the Gini Index of Z changes from $\frac{1}{3}$ (for $V = U$) to 0 (for $V = 1 - U$).

Remark 13.3 *Figure 13.10 is the map of a uniform distribution on the unit square. Please note that this is the grade distribution of* (X, X) ***for any random variable*** X. *The equality* $\mathbf{Dif}_{(U,U)} = 0$ *means that all records have equal profiles (no differentiation of profiles).*

Figure 13.11 is the map of the grade distribution of $(X, X^{\leftarrow})$ ***for any random variable*** X *(with* $X^{\leftarrow}$ *denoting variable opposite to* X*).*

13.4.3 A family of TP_2 *distributions from* $\mathbb{BLM}$

Let (U,V) be independent and uniformly distributed Lilliputian variables, and let

$$\mathbf{X} = (X_1, X_2) = ((1-a)U + aV, aU + (1-a)V) \tag{13.33}$$

for $a \in [0, 1/2]$. It follows that $\mathbf{X} \in \mathbb{BLM}$ but not to $\mathbb{DLM}$ (unless $a = 0$ or $a = 1/2$). If $a = 0$, $\mathbf{X} = (U, V)$; the respective Lorenz Surface $\mathbf{L_X}$ was derived in Section 13.4.1. If $a = 1/2$, $X_1 = X_2$; the respective Lorenz Surface $\mathbf{L_X}$ was derived in Section 13.4.2. Generally, the density of $\mathbf{X}$ is constantly equal to $1/|2a-1|$ on the rhombus with vertices $(0,0)$, $(a, 1-a)$, $(1-a, a)$, $(1,1)$, and equal to 0 outside the rhombus. One can check that this symmetric density fulfils the condition formulated in Proposition 7.2.(iv) which determines total positivity of order two; then, for each a, vector $\mathbf{X}$ will be said to belong to $\mathbb{TP}_2$ (i.e. $\mathbf{X}$ is said to be totally positive dependent of order 2).

The correlation coefficient is $corr(\mathbf{X}) = \frac{2a(1-a)}{1-2a(1-a)}$; it is changing from 0 to 1 when a changes from 0 to 1/2.

To calculate the data table $\overline{\mathbf{T}}_{\mathbf{X}}$ one needs first to find the joint density of $(X_1, X_2, R_{\mathbf{X}})$ (where $R_{\mathbf{X}} = (0.25X_1 + 0.75X_2)/Z$) and then find $\mathrm{E}(X_j;\ R_{\mathbf{X}} = t)$ for $t \in \mathbb{R}^+$, $j = 1, 2$. These calculations are similar as in Section 13.4.1 but more complicated, and they are left to the reader as a tedious but simple probabilistic exercise.

The results of this section will be only approximate, based on samples from the distribution of $\mathbf{X}$ of size $n = 16\ 000$.

Curve (1) in Figure 13.12 presents the relationship between $\mathbf{Dif_X}$ and $corr(\mathbf{X})$ for all possible a. Note that $corr(\mathbf{X})$ evaluates how close are the components of $\mathbf{X}$, while $\mathbf{Dif_X}$ evaluates how much differ their profiles (or, according to an other interpretation of $\mathbf{Dif}$, what is the maximal strength of dependence between the components of the vector and the objects represented by the records). Therefore, $\mathbf{Dif_X}$ decreases when $corr(\mathbf{X})$ increases.

We will compare now the relationship between $\mathbf{Dif_X}$ and $corr(\mathbf{X})$ with such relationship when vector $\mathbf{X}$ is replaced by the uniform grade transforms of its components:

$$\mathbf{X}^* = (F_a((1-a)U + aV), F_a(aV + (1-a)U)) \tag{13.34}$$

where the cdf of X_1 (equal to the cdf of X_2) is given by

$$F_a(x) = \begin{cases} \frac{x}{2a(1-2)} & \text{if } \ x \in (0, a], \\ \frac{2x-a}{2(1-a)} & \text{if } \ x \in (a, 1-a], \\ \frac{-x^2+2x+2a(1-a)-1}{2a(1-a)} & \text{if } \ x \in (1-a, 1]. \end{cases} \tag{13.35}$$

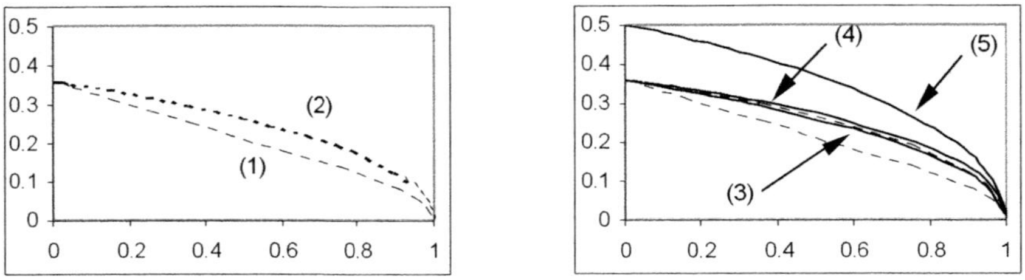

FIGURE 13.12. Plots between $\mathbf{Dif}_{\mathbf{X}}$ and $\rho^*(\mathbf{X})$ for five families of symmetric TP_2 distributions. **Horizontal** axis: the values of $\rho^*(X_1, X_2)$; **vertical** axis: values of $\mathbf{Dif}_{(X_1,X_2)}$ **Left:** (1) the family of pairs specified by formula (13.33), (2) the family of pairs specified by formula (13.34). **Right:** (3) the grade binormal family $N_2^*(\rho)$, (4) the family of grade Moran distributions, (5) the family of Moran distributions. Two dotted lines correspond to families (1) and (2). Plots for families (2), (3), and (4) are very close each other. (The curves (3), (4) and (5) will be considered in Sections 13.4.4 and 13.6)

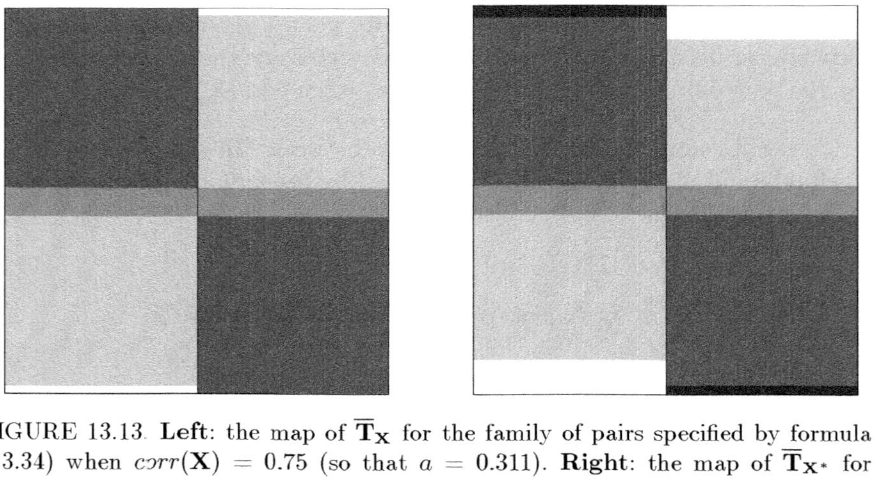

FIGURE 13.13. **Left**: the map of $\overline{\mathbf{T}}_{\mathbf{X}}$ for the family of pairs specified by formula (13.34) when $corr(\mathbf{X}) = 0.75$ (so that $a = 0.311$). **Right**: the map of $\overline{\mathbf{T}}_{\mathbf{X}^*}$ for the family of pairs specified by formula (13.33), when $corr(\mathbf{X}^*) = 0.75$ (so that $a = 0.326$).

The correlation coefficient $corr(\mathbf{X}^*)$ (i.e., $\rho^*(\mathbf{X})$ for $\mathbf{X}$ specified by (13.33)) is very close to $corr(\mathbf{X})$ and therefore it practically does not matter for curve (1) whether the horizontal axis in Figure 13.12 is related to values of $corr(\mathbf{X})$ or of $corr(\mathbf{X}^*)$. Contrarily, the difference between curves (1) and (2) implies that $\mathbf{Dif}_{\mathbf{X}^*}$ distinctly differs from $\mathbf{Dif}_{\mathbf{X}}$. It follows that the uniform grade transformations of vector components may have strong influence on $\mathbf{Dif}$.

The maps of the data tables of vector $\mathbf{X}$ and $\mathbf{X}^*$ when $corr(\mathbf{X}) = 0.75$ are presented in Figure 13.13. Generally, the maps of the data tables for $\mathbf{X}$ (and for $\mathbf{X}^*$) are gradually changing from the map for independent X_1 and X_2 (Figure 13.10 which corresponds to $a = 0$) to the map for $X_1 = X_2$ (Figure 13.11 which corresponds to $a = 1/2$).

It follows from curves (1) and (2) in Figure 13.12 that the profiles of the *grade records* ordered by GCA are more strongly differentiated than the profiles of initial records ordered by GCA.

13.4.4 Grade binormal distributions

Let $\mathbf{X} = (X_1, X_2) \sim N_2(m_1, m_2, \sigma_1, \sigma_2, \rho)$. As the normal variables assume positive and negative values, the data table $\overline{\mathbf{T}}_{\mathbf{X}}$ does not exist, but there exists the data table for $\mathbf{X}^* = (X_1^*, X_2^*)$, i.e. for the binormal copula $N_2^*(\rho)$. It will be convenient to use the notation $U = X_1^*$, $V = X_2^*$; thus in the present section (U, V) is distributed according to $N_2^*(\rho)$.

Instead of complicated exact formulas for $N_2^*(\rho)$, we will consider the sample datasets $\mathbf{D}^{(n)}_{(U,V);\rho}$ for random samples of size $n = 20$, 100 and 1000 drawn from $N_2^*(\rho)$ for $\rho = 0.75$ and $\rho = -0.75$. The aggregated data tables $\overline{\mathbf{T}}_{\mathbf{X}}$ for $\rho = 1$ and $\rho = -1$ were already constructed in Section 13.4.2 (common for all families in $\mathbb{DLM}$ with $corr(X_1, X_2) = 1$ or -1); the aggregated data table $\overline{\mathbf{T}}_{\mathbf{X}}$ for $\rho = 0$ was constructed in Section 13.4.1 (common for all families in $\mathbb{DLM}$ in which $(X_1, X_2) \in \mathbb{QD}$ and $corr(X_1, X_2) = 0$).

A few examples of the GCA maps for the sample grade binormal datasets are presented in Figure 13.14. The graph of the regression function $R^{(n)}_{(U,V);\rho}$ is inscribed to each sample dataset; these graphs are gradually more regular with increasing n.

For $n = 20$, 100, 1000, the values of $\mathbf{Dif}^{(n)}_{(U,V);\rho}$ are 0.1172, 0.1621, 0.1753 when $\rho = 0.75$, and 0.4332, 0.4871, 0.4705 when $\rho =$ -0.75. (Note that $\mathbf{Dif}^{(1000)}_{(U,V);0.75}$ is relatively close to $\mathbf{Dif}_{(U,V);1} = 0$, and $\mathbf{Dif}^{(1000)}_{(U,V);-0.75}$ is relatively close to $\mathbf{Dif}_{(U,V);-1} = 0.5$).

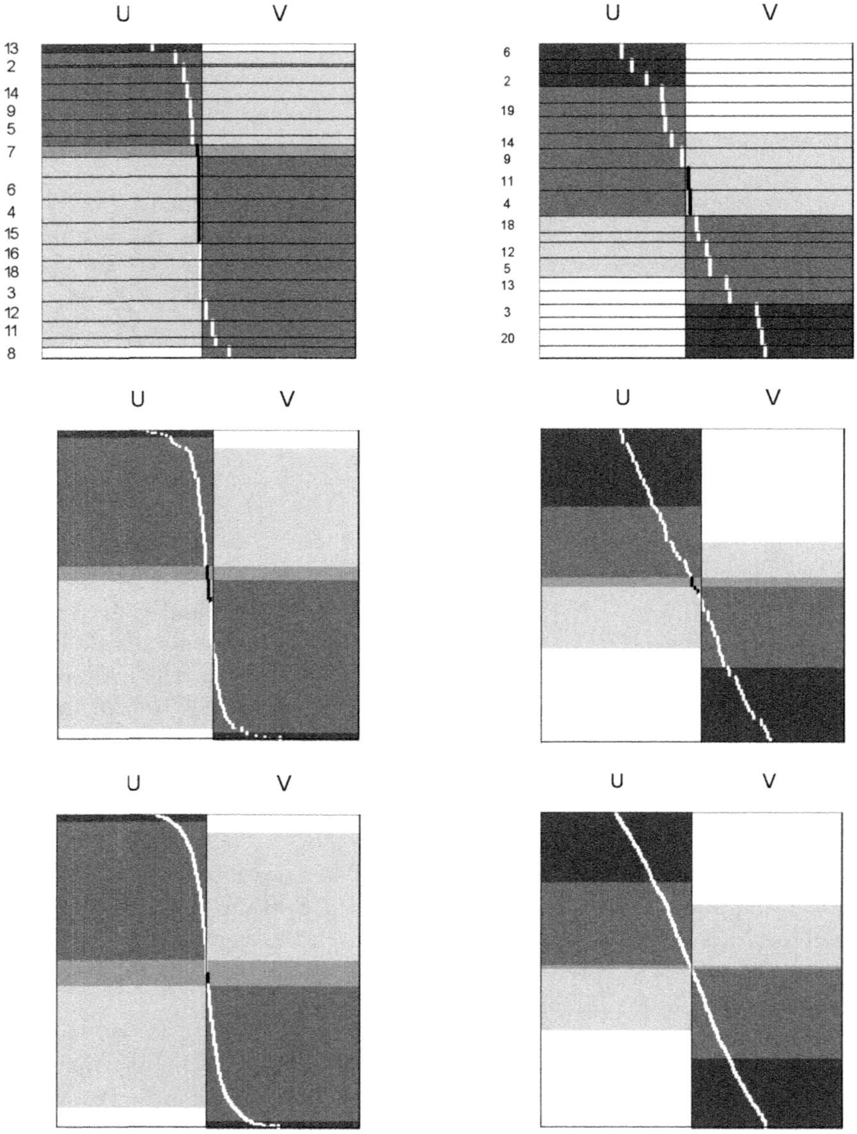

FIGURE 13.14. The maps of datasets $\mathbf{D}^{(n)}_{(U,V);\rho}$ for samples of size n drawn from the grade binormal distribution $N_2^*(\rho)$: left column: $\rho = 0.75$, right column: $\rho = -0.75$; first, second and third row: $n = 20, 100, 1000$. The grade $col|row$ regression function is shown in each map.

Let the function $f\colon [-1,1] \to [0,0.5]$ assign the differentiation $\mathbf{Dif}^{(16\,000)}_{(U,V);\rho}$ to ρ^*. The values of f are approximately equal to $\mathbf{Dif}_{(U,V);\rho}$. The graph of f is shown in Figure 13.15 and also appears (truncated to non-negative values of ρ^*) in Figure 13.12 (right) as curve (3). For non-negative ρ^* it is very close to the analogous graph derived for the grade transforms (X_1^*, X_2^*) related to the family from $\mathbb{BLM}$ considered in Section 13.4.3 (Figure 13.12 (right), curve (2) lying between curves (3) and (4)).

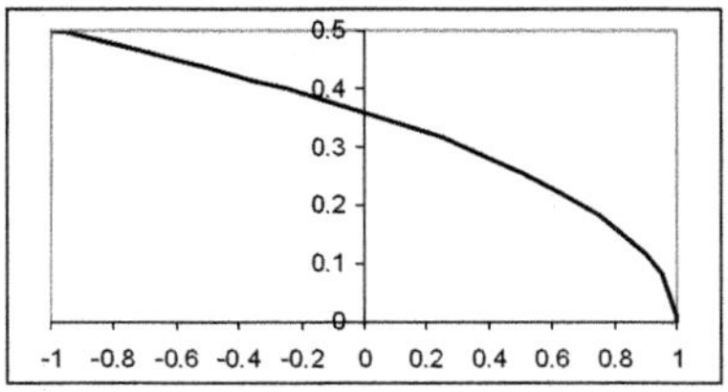

FIGURE 13.15. Averaged graph of functions $f(\rho^*) = \mathbf{Dif}_{(U,V);\rho}$ for families (2), (3), (4).

13.5 Inequality measures for grade multinormal distributions

Let

$$\mathbf{X}^* = (X_1^*, ..., X_k^*) = \left(F^*_{X_1}(X_1), ..., F^*_{X_k}(X_k)\right)$$

be the grade transform of vector $\mathbf{X}$. The following notation will be used for the correlation and grade correlation matrices of $\mathbf{X}$ and $\mathbf{X}^*$:

$$\begin{aligned} &\mathbf{CORR}(\mathbf{X}) = [corr(X_i, X_j),\ i = 1, ..., k,\ j = i, ..., k]\,, \\ &\mathbf{RHO}^*(\mathbf{X}) = [\rho^*(X_i, X_j),\ i = 1, ..., k,\ j = i, ..., k]\,, \end{aligned}$$

Note that $\mathbf{CORR}(\mathbf{X}^*) = \mathbf{RHO}^*(\mathbf{X}) = \mathbf{RHO}^*(\mathbf{X}^*)$.

Vectors $\mathbf{X}^*$ belong to the set of random vectors which will be called the *k-variate Dependence Lilliputian Model* ($\mathbb{DLM}_k$).

Remark 13.4 *The model $\mathbb{DLM}_2$ has been already considered in Chapter 8, where it was shortly denoted $\mathbb{DLM}$. As the distributions of X_i^* are uniform, the cdf of $\mathbf{X}^*$ is being called the copula of vector $\mathbf{X}$.*

In this section we deal with a family from $\mathbb{DLM}_k$ consisting of grade multinormal distributions denoted $N_k^*(\mathbf{CORR})$, where k is the number of components of vector $\mathbf{X}$, and parameter $\mathbf{CORR}$ is the grade correlation matrix of $\mathbf{X}$.

Remark 13.5 *The probability data tables* $\overline{\mathbf{T}}_{\mathbf{X}}$ *cannot be constructed for multinormal vectors distributed according to* $N_k(\mathbf{m}, \mathbf{\Sigma})$*, since the normal variables assume positive as well as negative values, but can be constructed for distributions* N_k^**. The only parameter of a distribution* N_k^* *is matrix* $\mathbf{CORR}$ *(equal to* $\mathbf{CORR}(\mathbf{X}^*)$*) which can be derived from* $\mathbf{m}$ *and* $\mathbf{\Sigma}$*).*

For any vector $\mathbf{X}^* \in \mathbb{DLM}_k$ index $\mathbf{Dif}_{\mathbf{X}^*}$ can be decomposed according to the formula

$$\begin{aligned}\mathbf{Dif}_{\mathbf{X}^*} &\stackrel{df}{=} \rho^*\left(\overline{\mathbf{T}}_{\mathbf{X}^*}\right) = \frac{8}{k^3}\Sigma_{j=1}^{k}\Sigma_{i=1}^{j-1}(j-i)\ \rho^*\left(T_{(X_i,X_j);R_{\mathbf{X}}}\right)\\ &= \frac{8}{k^3}\Sigma_{j=1}^{k}\Sigma_{i=1}^{j-1}(j-i)\mathbf{Dif}_{(X_i^*,X_j^*);R_{\mathbf{X}}}. \end{aligned} \tag{13.36}$$

This formula follows from Proposition 8.2.(i): it corresponds to formula (8.14), adjusted however to deal with columns instead of rows and with the uniform marginal horizontal distribution of the data table $\overline{\mathbf{T}}_{\mathbf{X}}$. If $\mathbf{X}^* \sim N_k^*(\mathbf{CORR})$, we know from the considerations in the binormal case (Section 13.4.4) that, for $j = 1, ..., k$ and $i = j, ..., k$,

$$\mathbf{Dif}_{(X_i^*,X_j^*)} \stackrel{df}{=} \rho^*\left(\mathbf{T}_{(X_i^*,X_j^*)}\right) = f\left(corr(X_i^*, X_j^*)\right), \tag{13.37}$$

where f is the decreasing function from [-1,1] to [0,1], introduced in Section 13.4.4 and visualized in Figure 13.15.

Note that the expressions $f(corr(X_i^*, X_j^*))$ appearing in (13.37) cannot replace $\mathbf{Dif}_{(X_i^*,X_j^*);R_{\mathbf{X}^*}}$ in formula (13.36) because function $R_{\mathbf{X}^*}$, which is related to *whole* vector $\mathbf{X}^*$, aggregates in general more records than particular functions $R_{(X_i^*,X_j^*)}$ (for any $i < j$). Therefore, *contrarily to the situation in multivariate distributions with finite sets of records,* no direct functional link between $\mathbf{Dif}_{\mathbf{X}^*}$ and $\mathbf{CORR}(\mathbf{X}^*)$ will follow from formulas (13.36) and (13.37)).

We will show however that such functional link is realized approximately when the correlation matrix is sufficiently regular: namely, it occurs when $\mathbf{CORR}$ fulfils a certain condition of regularity stronger than the property TP_2.

Let regularity of any non-negative $m \times k$ table be introduced analogously as it is introduced for a probability table.

Definition 13.3 *Table* $Q_{m\times k} = [q_{ij},\ i = 1,...,m,\ j = 1,...,k]$, $q_{ij} \geq 0$, *is said to have property* TP_2 *(total positivity of order two) if for any* $i = 1,...,m-1$ *and any* $j = 2,...,k$

$$q_{ij}q_{i+1j-1} \leq q_{ij-1}q_{i+1j};$$

and it is said to have property ETP_2 *(essential total positivity of order two) if*

$$q_{ij} \leq q_{ij-1}q_{i+1j}.$$

Remark 13.6 *The definitions of* TP_2 *and* ETP_2 *tables are motivated by the investigations of regularity of correlation matrices. This implies that generally both definitions should be extended onto tables with any real elements. However, in this section we will restrict ourselves to non-negative matrices* **CORR**.

Proposition 13.2 *Property* ETP_2 *possessed by a two-way table* $Q_{m\times k}$ *implies that* $Q_{m\times k}$ *has also property* TP_2.

The proof follows from the equalities:

$$q_{ij}q_{i+1j-1} \leq q_{ij} \leq q_{ij-1}q_{i+1j}.$$

Let us investigate simultaneous regularity of $\mathbf{CORR}(\mathbf{X}^*)$ and $\mathbf{DIFF}(\mathbf{X}^*)$, where $\mathbf{DIFF}(\mathbf{X})$ denotes the differentiation matrix related to vector $\mathbf{X}$:

$$\mathbf{DIFF}(\mathbf{X}^*) \overset{df}{=} \left[\mathbf{Dif}_{(X_i^*,X_j^*);R_{\mathbf{X}^*}},\ i = 1,...,k-1, j = i+1,...,k\right].$$

We make a conjecture that if $\mathbf{X}^* \sim N_k^*(\mathbf{CORR})$ and matrix **CORR** is ETP_2 then the aggregated table $\overline{\mathbf{T}}_{\mathbf{X}^*}$ is TP_2 (which in turn would have important implications for index $\mathbf{Dif}_{\mathbf{X}^*}$ and matrix $\mathbf{DIFF}(\mathbf{X}^*)$). This conjecture seems to be sustained by the results of a large simulation study, which additionally suggests the functional link between matrices **CORR** and **DIFF**.

The simulation study consisted in investigating a large number of various correlation matrices **CORR** with property ETP_2, i.e., satisfying

$$corr(X_i, X_j) \leq corr(X_{i+1}, X_j)corr(X_i, X_{j-1}) \tag{13.38}$$

for $k = 3, 4, 5$. For each k and each considered **CORR**, there was derived the sample data table $\overline{\mathbf{T}}^{(n)}_{\mathbf{X}^*;\overline{R}_{\mathbf{X}^*}}$ obtained from $n = 64\,000$ records drawn from

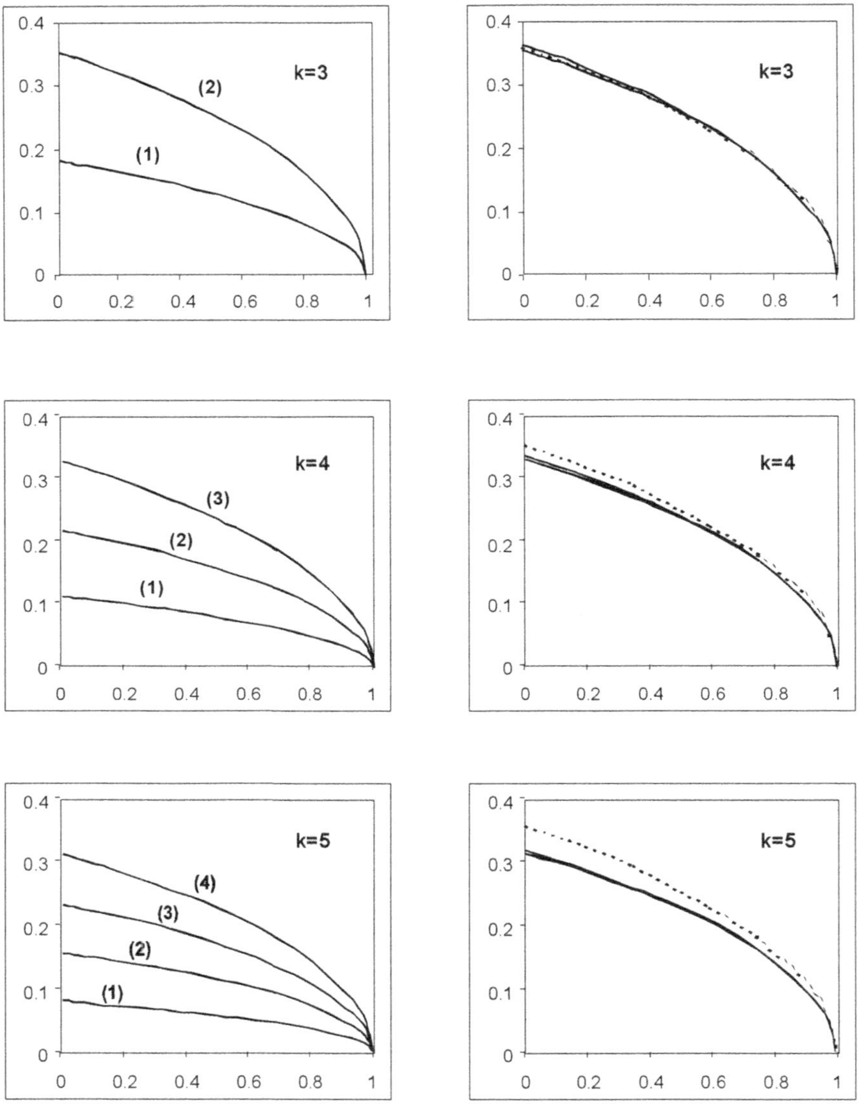

FIGURE 13.16. The results of a simulation study for grade k-normal distributions, which presents the links between the correlation matrices **CORR** having property ETP_2 and the respective differentiation matrices **DIFF** (detailed description is given in the text).

$N_k^*(\mathbf{CORR})$; the order function $\overline{R}_{\mathbf{X}^*}$ used in this table was equal to $R_{\mathbf{X}^*}^{(n)}$ *rounded to three digits* (which refers to intervals of length 0.001). Then it occured that the condition $\overline{\mathbf{T}}_{\mathbf{X}^*;\overline{R}_{\mathbf{X}^*}}^{(n)} \in \mathbb{TP}_2$ was fulfiled for almost all data tables under consideration. In the next step of the study, the matrix

$$\mathbf{DIFF}^{(n)}(\mathbf{X}^*) = \left[\mathbf{Dif}_{(X_i^*,X_j^*);\overline{R}_{\mathbf{X}^*}}^{(n)},\ i = 1,...,k, j = i+1,...,k\right]$$

was calculated for each data table $\overline{\mathbf{T}}_{\mathbf{X}^*;\overline{R}_{\mathbf{X}^*}}^{(n)}$, and its terms were plotted with respect to the values of $corr(X_i^*, X_j^*)$. The resulting plots are shown in Figure 13.5 (left part) for $k = 3, 4, 5$. The curves in these plots look quite regular (in spite of the limited number of considered matrices **CORR** and of the rounding of the values of $R_{\mathbf{X}^*}^{(n)}$). Moreover, some plots occured to be *very* close each other, namely:

- for $k = 3$, plot (1) in Figure 13.5 (left part) contains jointly the points

  $$\left(corr(X_1^*, X_2^*),\ \mathbf{Dif}_{(X_1^*,X_2^*);\overline{R}_{(X_1^*,X_2^*,X_3^*)}}\right)$$

  together with points

  $$\left(corr(X_2^*, X_3^*),\ \mathbf{Dif}_{(X_2^*,X_3^*);\overline{R}_{(X_1^*,X_2^*,X_3^*)}}\right),$$

  while plot (2) (above plot (1)) contains only points

  $$\left(corr(X_1^*, X_3^*),\ \mathbf{Dif}_{(X_1^*,X_3^*);\overline{R}_{(X_1^*,X_2^*,X_3^*)}}\right);$$

- for $k = 4$, plot (1) contains the points corresponding to (X_1^*, X_2^*), (X_2^*, X_3^*), and (X_3^*, X_4^*); plot (2) contains points corresponding to (X_1^*, X_3^*) and (X_2^*, X_4^*); plot (3) contains points corresponding just to (X_1^*, X_4^*);

- for $k = 5$, plot (1) contains points corresponding to (X_i^*, X_{i+1}^*) when $i = 1, ..., 4$; plot (2) contains points corresponding to (X_i^*, X_{i+2}^*) when $i = 1, 2, 3$; plot (3) contains points corresponding to (X_i^*, X_{i+3}^*) when $i = 1, 2$; curve (4) contains points corresponding to (X_1^*, X_5^*).

We see that all these plots are very close to thoroughly regular curves. Their particularly interesting feature is that for any k they can be transformed into plots very *close one to another*, which is shown in Figure 13.5 (right part). This transformation is given by the formula

$$\frac{k-1}{j-i}\mathbf{Dif}^{(n)}_{(X_i^*,X_j^*);\overline{R}_{\mathbf{X}^*}}, \tag{13.39}$$

applied for $k = 3, 4, 5$ when $i = 1, ..., k$ and $j = i+1, ..., k$. For $k = 3$, both transformed plots (1) and (2) were found almost equal to the dotted graph of function f, which was introduced before for the grade binormal pairs $N_2^*(\rho)$ (see curve (3) in Figure 13.15). For $k = 4$, the dotted graph corresponding to function f lies slightly *above* the transformed plots (1), (2), (3) which in turn are practically undistinguishable one from another. For $k = 5$, the distance from the dotted graph of f and the bundle of coinciding graphs of transformed plots (1)-(4) is slightly larger than for $k = 4$.

The transformed plots suggest that one can also try to experimentate in the opposite direction: that is to start from an aggregated data table $\overline{\mathbf{T}}$ with a TP_2 property, find the corresponding matrix consisting of $\mathbf{Dif}_{(X_i,X_j)}$, and reconstruct from it the correlation matrix **CORR**. Next, it would be possible to make a loop: use this correlation matrix to generate a sample of records, construct the related data table, and repeat this process until the equilibrium is attained.

This suggests the role of bivariate data tables as generators of *regular copulas* and vice versa.

* * *

We will consider now the set of distributions $N_k^*(\mathbf{CORR})$ parametrized by $\rho \in [0,1]$ so that each **CORR** satisfies

$$corr(X_i, X_j) \equiv \rho \tag{13.40}$$

for some $\rho \in [0,1]$ and $i, j = 1, ..., k$, $k = 3, 4, 5, ...$ We expect that this set will have some special properties regarding the plot of $\mathbf{Dif}_{(X_1,...,X_k)}$ with respect to ρ.

Let us start with $\rho = 0$ (for $\mathbf{X} \sim N_k^*([0])$, this implies that $X_1, ..., X_k$ are independent *en block*).

Remark 13.7 *For $k = 2$ the condition $\rho = 0$ implies independence of X_1^* and X_2^*, since binormal distributions, and hence also grade binormal distributions, are quadrant dependent. For $k > 2$ uncorrelation of (X_i^*, X_j^*) for any pair (i, j) implies independence en block of $(X_1^*, ..., X_k^*)$ since the vector of normal variables has the property called associate dependence (see e.g. Theorem 3.1 in [10]).*

Let $(\mathbf{X}^*)^{\perp}$ denote a vector $(U_1, ..., U_k)$ consisting of independent uniform Lilliputian variables (we remind that (U, V) was considered in Section 13.4.1).

Proposition 13.3 *The distribution of* $R_{(\mathbf{X}^*)^\perp}$ *is symmetric with mean 1/2 and variance tending to zero when* k *increases to infinity.*

Proof. $\frac{1}{k}\sum_{i=1}^k X_i^*$ tends to $\frac{1}{2}$ with probability 1, and

$$\frac{1}{k}\sum_{i=1}^{k} Y_i \overset{df}{=} \frac{1}{k}\sum_{i=1}^{k}\frac{2i-1}{2k}X_i^*$$

tends to $\frac{1}{4}$ with probability 1, since $\sum_{i=1}^{\infty} var(Y_i)/i^2 < \infty$; hence the ratio

$$\frac{\frac{1}{k}\sum_{i=1}^k Y_i}{\frac{1}{k}\sum_{i=1}^k X_i^*}$$

tends to 1/2 with probability 1.

Corollary 13.1 $\mathbf{Dif}_{(\mathbf{X}^*)^\perp}$ *tends to zero when* k *increases to infinity.*

The decreasing plot of $\mathbf{Dif}_{(\mathbf{X}^*)^\perp}$ versus k is presented in Fig. 13.17 (left part). Figure 13.17 (right part) shows the plots of $\mathbf{Dif}_{\mathbf{X}^*}$ versus ρ for $k = 2, 3, 4, 5$, which all decrease to zero when k infinitely increases. Figure 13.18 reveals the effect of standardizing $\mathbf{Dif}_{\mathbf{X}^*}$ by putting

$$\mathbf{stDif}_{\mathbf{X}^*} \overset{df}{=} \frac{\mathbf{Dif}_{\mathbf{X}^*}}{\mathbf{Dif}_{(\mathbf{X}^*)^\perp}}. \tag{13.41}$$

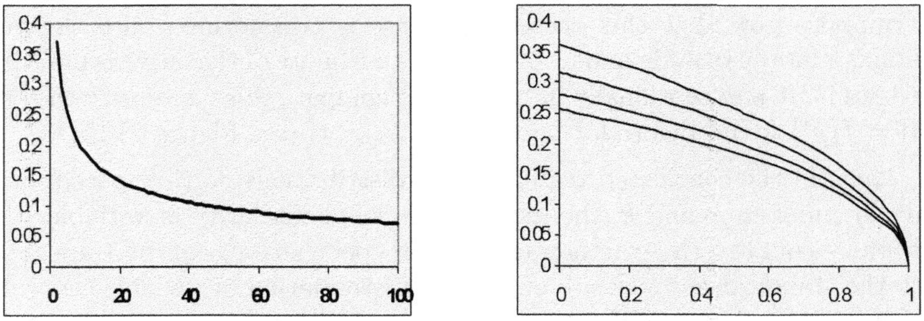

FIGURE 13.17. **Left:** the plot of $\mathbf{Dif}_{(\mathbf{X}^*)^\perp}$ for $k = 2, ..., 100$. **Right:** the plots of $\mathbf{Dif}_{\mathbf{X}^*}$ (vertical axis) versus ρ (horizontal axis) when $corr(X_i, X_j) \equiv \rho$, for $k = 2$ (at the top) through 5 (at the bottom).

For $\rho = 0$, $\mathbf{stDif}_{\mathbf{X}^*}$ is equal to 1 by definition. We conjecture that, for any $\rho \in (0, 1)$, $\mathbf{stDif}_{\mathbf{X}^*}$ is invariant on k since indeed four plots (corresponding to $\rho = 0.25$, 0.50, 0.75, 0.95) of $\mathbf{stDif}_{\mathbf{X}^*}$ versus $k = 2, 3, 4, 5$ which are shown in Figure 13.18 (left part), are straight horizontal lines. These plots were obtained

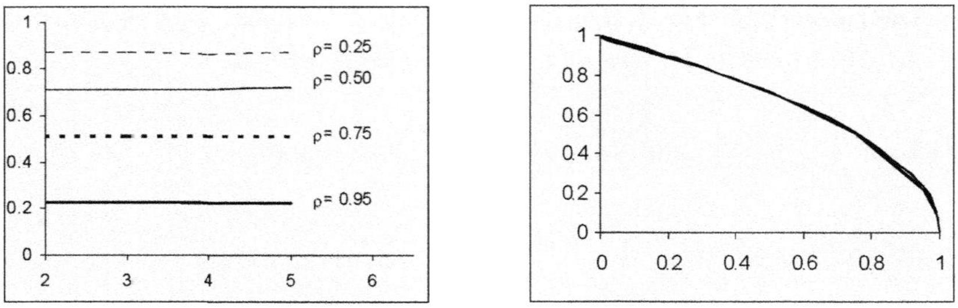

FIGURE 13.18. **Left:** the plots of $\mathbf{stDif}_{\mathbf{X}^*}$ versus k changing from 2 to 5 for grade normal vectors $(X_1^*, ..., X_k^*)$ when $\mathbf{CORR} = [\rho]$, $\rho = 0.25, 0.50, 0.75, 0.95$. **Right:** the sequence of plots for $k = 2, 3, 4, 5, 10$ of $\mathbf{stDif}_{\mathbf{X}^*}$ versus ρ when $\mathbf{CORR} = [\rho]$.

on the basis of the simulation study with sample size n=10 000 drawn for each ρ from $N_k^*([\rho])$.

Figure 13.18 (right part) presents the plots of $\mathbf{stDif}_{\mathbf{X}^*}$ versus ρ (still under the condition (13.40)). The plots form a set of points which for any k are very close to a decreasing smooth curve joining points (0,1) and (1,0).

Suppose now that this decreasing curve is transformed onto the curve joining points (0,0.3611) and (1,0) (i.e., each ordinate of the curve is multiplied by 0.3611). It is exceptionally interesting to compare this curve with the curve $\mathbf{Dif} = f(\rho^*)$ found before for families (2), (3), (4) (see Figure 13.15).

Thus, in the considered regular set of distributions $N_k^*([\rho])$ parametrized only by numbers ρ and k, the measure of global similarity of variables (here obviously equal to the common correlation coefficient ρ) can be transformed into the standardized measure of global differentiation of variables as well as of records (i.e., into **stDif**).

It is hoped that the standardized **Dif** given by formula (13.41) will have also other desirable properties. Standardization based on a reference to independence of variables promises to be a good way to neutralize the bias related to varying number of components. (Note that such bias is a drawback of various measures of global differentiation which have been introduced in the literature).

The properties of the standardized version of **Dif** for vectors $\mathbf{X} \in (\mathbb{RATIO})^k$, given by

$$\mathbf{stDif_X} = \frac{\mathbf{Dif_X}}{\mathbf{Dif_{X^\perp}}}, \tag{13.42}$$

seems to be worth further investigation as a numerical grade measure of multivariate inequality. It will be considered in Section 13.6 for the family of k-variate Moran distributions. Generally, it is hoped to be fruitfully used to compare multivariate inequality for various vectors **X**.

13.6 Inequality measures for the Moran distributions

The distribution of a random vector $\mathbf{X} = (X_1, ..., X_k)$ is usually called k-exponential if the distributions of X_j are exponential. Since the values of exponential X_j are non-negative, the data table $\overline{\mathbf{T}}_\mathbf{X}$ can be calculated without grade transformations of the components. It is only recommended to divide each variable by its expectation: according to Section 12.3, such standardization is a step towards meaningful additivity of variables.

We will consider a special family of random vectors with exponentially distributed components, called *the family of Moran distributions:*

$$\mathbf{X} \sim M_k(\mathbf{m}, \mathbf{CORR}), \tag{13.43}$$

where $\mathbf{m} = (m_1, ..., m_k)$, m_j being the expectation (and also the standard deviation) of X_j, and **CORR** being the correlation matrix. The data tables appearing in this section will be constructed for the family $M_k(\mathbf{1}, \mathbf{CORR})$, i.e. for $m_j \equiv 1$.

Vector $\mathbf{X} = (X_1, ..., X_k) \sim M_k(\mathbf{1}, \mathbf{CORR})$ can be obtained as a function of vector $\mathbf{Y} = (Y_1, ..., Y_k) \sim N_k(\mathbf{0}, \ \mathbf{CORR}_Y)$ and its copy denoted $\mathbf{Y}^c$, where N_k is a $k-$variate normal distribution, and $\mathbf{Y}^c = (Y_1^c, ..., Y_k^c)$ is a random vector independent of **Y** and distributed identically as **Y** :

$$X_j = \frac{1}{2}(Y_j^2 + (Y_j^c)^2), \, j = 1, ..., k \tag{13.44}$$

It follows that variables X_j are exponential with the density e^{-x}, $EX_j = 1$, $corr(X_i, X_j) = (corr(Y_i, Y_j))^2$.

The formula for the density of $\mathbf{X}$ is extremely complicated so that direct calculations of the data tables and its parameters are practically impossible. However, due to formula (13.44), there is no difficulty to *generate random records* from $M_k(\mathbf{1}, \mathbf{CORR})$ using any of the well known generators of random records from $N_k(\mathbf{0}, \mathbf{CORR})$. Hence we are able to investigate sample data from M_k. To simplify the notation we will write $M_2(\rho)$ instead of $M_2(\mathbf{1}, \rho)$, $M_k(\mathbf{CORR})$ instead of $M_k(\mathbf{1}, \mathbf{CORR})$, and $M_k([\rho])$ when $\mathbf{CORR}$ consists of identical correlations ρ.

TABLE 13.4. Upper part corresponding to $(X_1, ..., X_k) \sim M_k(\mathbf{CORR})$: the correlation coefficient of (X_1, X_2), the correlation matrices $\mathbf{CORR}$ of $(X_1, ..., X_k)$ and the differentiation matrices $\mathbf{DIFF}$ for non-aggregated samples of size 1000. Lower part: the same for $(X_1^*, ..., X_k^*) \sim M_k^*(\mathbf{CORR})$

		k=2	k=3		k=4		
$(X_1,\dots,X_k)$	CORR	0.25	0.25	0.0625	0.25	0.0625	0.0025
				0.25		0.25	0.0625
							0.25
	DIFF	0.4568	0.4568	0.4979	0.4568	0.4979	0.5002
				0.4473		0.4473	0.5060
							0.4496
$(X^*_1,\dots,X^*_k)$	CORR	0.1975	0.1975	0.0497	0.1975	0.0497	0.0005
				0.1975		0.1975	0.0497
							0.1975
	DIFF	0.4496	0.4496	0.4793	0.4496	0.4793	0.4808
				0.4333		0.4333	0.4905
							0.4407

First, we shall consider three Moran distributions with selected correlation matrices together with the respective distributions of their grade transforms M_k^* for $k = 2, 3$, and 4. The correlation coefficient for $k = 2$ and the correlation matrices for $k = 3$ and 4 are specified in the first row of Table 13.4. The grade correlation matrices (i.e. the correlation matrices for $\mathbf{X}^* = (X_1^*, ..., X_k^*)$ where $X_j^* = F_{X_j}(X_j) = 1 - \exp(-X_j))$ are shown in the third row of Table 13.4. The differentiation matrices $\mathbf{DIFF}$ for $\mathbf{X}$ and for $\mathbf{X}^*$ are shown in the second and fourth row.

All matrices $\mathbf{CORR}(\mathbf{X})$ in Table 13.4 are ETP_2. The maps for the sample data tables $\overline{\mathbf{T}}^{(1000)}$ are shown in Figure 13.19.

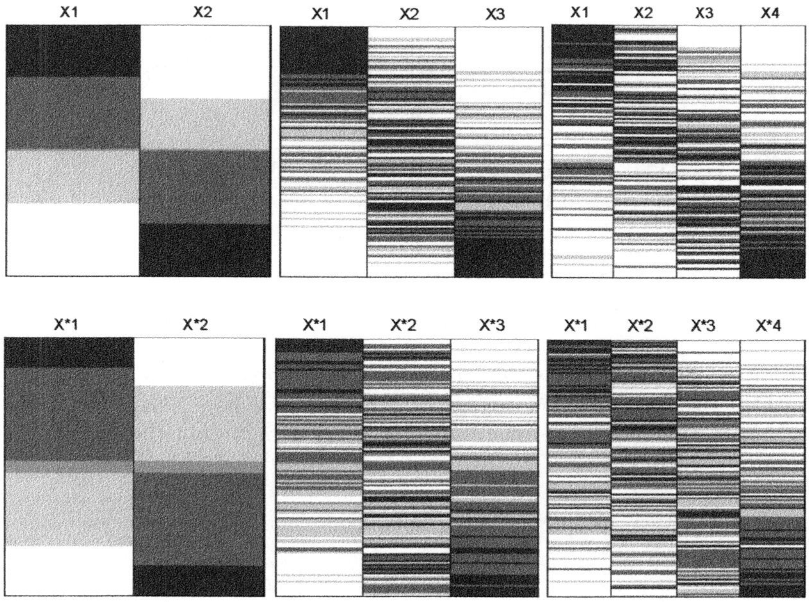

FIGURE 13.19. The maps of the sample data tables for samples of size $n = 1000$, drawn from the Moran distributions (upper part) and the grade Moran distributions (lower part) with parameters specified in Table 13.4.

The correspondence between the correlation coefficient ρ of the bivariate Moran distribution $M_2(\rho)$ and $\mathbf{Dif}_{(X_1,X_2)}$ is demonstrated by curve (5) in Fig. 13.12 (right). This curve is above other curves in Fig. 13.12, in particular above curve (4) corresponding to the grade Moran distribution $M_2^*(\rho)$. However, the difference between curves (4) and (5) almost vanishes when **Dif** is replaced by its standardized version **stDif** defined by (13.42), as presented in Fig. 13.20.

Figure 13.20 is an analog of Figure 13.18 according to which

- the standardized **Dif** for $\mathbf{X}^* \sim N_k^*([\rho])$ is invariant (or more precisely is almost invariant in sample sets $\overline{\mathbf{T}}^{(1000)}$) on the number of components k when the elements of the correlation matrix are identically equal to a number ρ from [0,1]
- there is a common decreasing function from [0,1] onto [0,1] which transforms ρ into **stDif**.

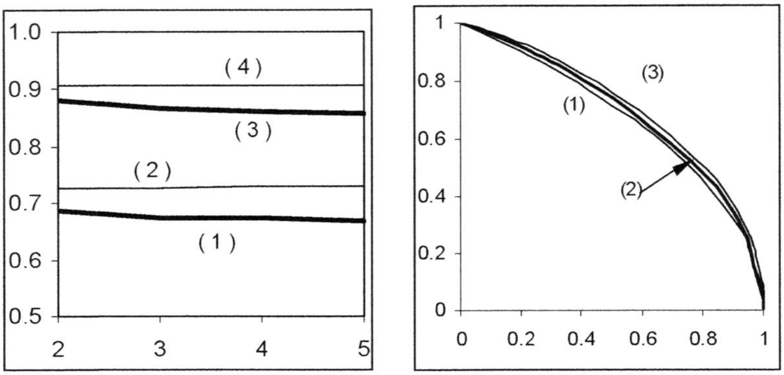

FIGURE 13.20. **Left** (almost invariance on the number of components): the plots of **stDif** vs k for: (1) $\mathbf{X} \sim M_k([\rho = 0.25])$, (2) $\mathbf{X}^* \sim M_k^*([\rho = 0.25])$, (3) $\mathbf{X} \sim M_k([\rho = 0.5625])$, (4) $\mathbf{X}^* \sim M_k^*([\rho = 0.5625])$ **Right:** the plots of **stDif** vs ρ for: (1) $(X_1, X_2) \sim N_2^*(\rho)$, (2) $(X_1, X_2) \sim M_2(\rho)$, (3) $(X_1, X_2) \sim M_2^*(\rho)$.

Figure 13.20 supplements this information:

- a similar invariance on the number of components holds (at least approximately) for Moran and grade Moran distributions when $\mathbf{CORR} = [\rho]$,
- the function transforming ρ into **stDif** is the same (or almost the same) for Moran, grade Moran and grade normal distributions of k-variate variables.

13.7 Appendix - link between grade similarity and dissimilarity of two regularly dependent random variables

The plot of function f presented in Figure 13.15 which transforms grade correlation ρ^* onto differentiation **Dif** for pairs (X, Y) distributed according to $N_2^*(\rho)$ needs some comments concerning the character of f.

First of all, we will compare function f with the following function denoted $CorDif$ which transforms the interval [-1,1] into $[1/2, 1]$ according to the formula:

$$CorDif(x) = \frac{\sqrt{2(1-x)}}{4}. \tag{13.45}$$

The graphs of f and of $CorDif$ are presented in Figure 13.21. Both plots are very close.

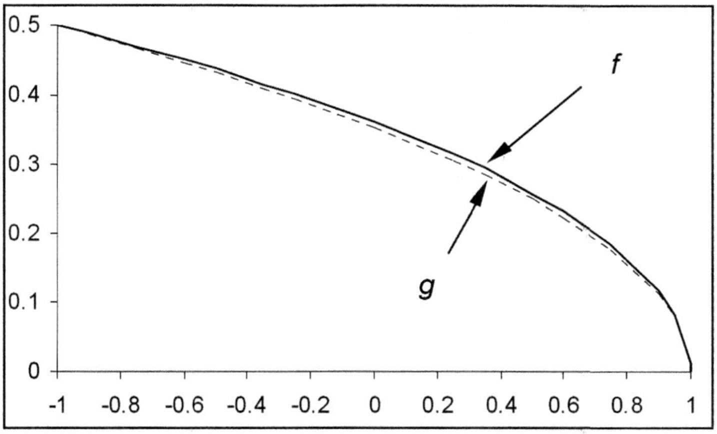

FIGURE 13.21. The graphs of f and $CorDif$ (the graph of $CorDif$ is the dotted curve denoted g).

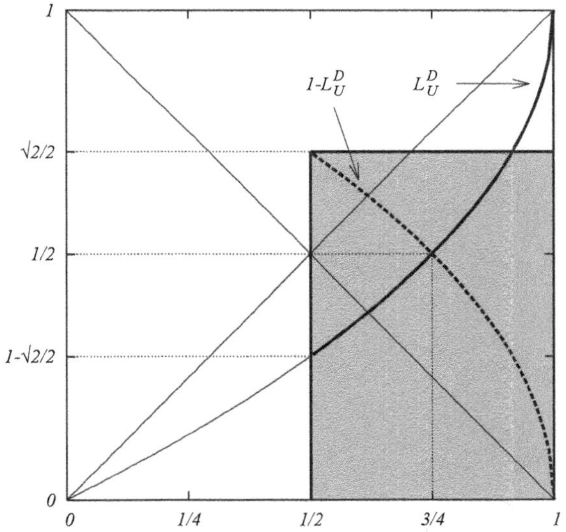

FIGURE 13.22. Construction of function $CorDif$

Now, note that the dual Lorenz Curve of a uniform Lilliputian variable U is $(L_U)^D(u) = 1 - \sqrt{1-u}$ for $u \in [0,1]$. Hence

$$CorDif(x) = \frac{\sqrt{2}}{2}\left(1 - L_U^D\left(\frac{x+3}{4}\right)\right) = \frac{\sqrt{2(1-x)}}{4}, \quad x \in [-1,1].$$

It follows from Chapter 6 that the graph of $CorDif$ forms an ideally regular curve (although not increasing but decreasing).

The construction of $CorDif$ is presented in Fig. 13.22. The graph of the dual Lorenz Curve of uniformly distributed Lilliputian variable U forms the ascending curve from (0,0) to (1,1). The "bold" part of this curve from point $(1/2, 1-\sqrt{2}/2)$ to (1,1) will be transformed into a congruent descending (dotted) curve joining points $(1/2, 1-\sqrt{2}/2)$ and (1,0). The rectangle with vertices (1/2,0), (1,0), $(1, \sqrt{2}/2)$, $(1/2, \sqrt{2}/2)$ will be now transformed into the shaded rectangle with vertices (-1,0), (1,0), (-1,1/2), (1,1/2), with the scale coefficient on the horizontal axis equal one-half of the scale coefficient on the vertical axis. The rectangule obtained after such scaling is equal to that presented in Fig. 13.21.

The contents of Sections 13.5 and 13.6 indicates that function $CorDif$ plays an important role for vectors $\mathbf{X} = (X_1, ..., X_k)$ when the distributions of $\mathbf{X}$ are sufficiently regular.

14

Case studies with multivariate data

Wiesław Szczesny, Maria Grzegorek

14.1 Introduction

These case studies demonstrate different ways of processing and visualizing data using grade tools. Two cases have been chosen which have data with interesting "demonstration" properties. The data are considered to be populations, i.e., the researcher's only concern is in these data. We will begin with an overview of the data for the two cases.

Data Overview - Case 1:
Area: Social Science.
Nature of study: Descriptive.
Scale of variables: Ordinal.
Purpose of analysis: To obtain a general understanding of the data structure, and determine the importance of each variable relative to this structure.
Background information: Variables are weakly correlated.

Data Overview - Case 2:
Area: Engineering.
Nature of study: Investigative.
Scale of variables: Ratio.
Purpose of analysis: To investigate the dependence between a response variable and a set of explanatory variables.
Background information: Variables are highly correlated.

Overview of the Analyses:

Ordered grade clustering and decomposition based on GCA is introduced in both cases. (The decomposition was described in Chapter 9 for two-way contingency tables and applied to multivariate data in a small example presented in Chapter 12.)

A main trend due to GCA is found, and then the data are decomposed into two sub-sets: one preserving the main trend, and one outlying the main trend.

In Case Study 1, the trend in the second subset is different from the main trend, while both subsets have similar marginal distributions for the variables.

In Case Study 2, most variables are still ordered according to the main trend (but a few distinctly differ from it), and the marginal distributions of some variables are rather dissimilar.

14.2 Case Study 1 - Main Trend of Questionnaire Data

(More information is available in the Appendix in Section 14.7.1)

14.2.1 The Questionnaire

These data were originally gathered in connection with a registration list for a foundation helping the disabled. The respondents were interested in rehabilitation and employment based on the learning of computer skills.

There were 601 respondents, and the questionnaire consisted of 10 items. Each item was answered on an ordinal scale using the values 1 through 6. In general, the value 1 represented the "poorer" or "worse" end of the scale.

The 10 items (and their abbreviations) are as follows:

Current Employment (employ)
Professional experience (prof-exp)
Economic situation (econ), i.e., the respondents personal economic situation
Foreign language skills (lang)
Age (age)
Education (educ)
Computer related equipment, i.e., degree of access to (equip)
Functional level of disability (func-lev)
Population size of city of residence (pop-size)
Skill level with computers (skills)

Note: For "age," scores 1 to 6 correspond to increasing age.
For "pop-size," scores 1 to 6 correspond to increasing population size.

For example - a person with all responses equal to "1" would be:

unemployed with no possibility to earn money;
lacking any kind of professional work experience;
in an extremely bad personal economic situation;
without any foreign language skills;
(Note: The respondents were living in Poland, and language skills are valuable.)
young (1 = in the teens);
minimally educated;
with no computer equipment;
with a low physical functioning level (i.e., very handicapped);
living in a village (1 = village);
very poorly skilled in the use of computers

Note: In the examples in Chapter 2, the proportions came from frequency counts. Here, the "item scores" play the role that the "frequencies" played in Chapter 2. This was discussed in Chapter 12.

14.2.2 The goal of the analysis

The goal of the analysis falls under the general category of providing descriptive information.

We will concentrate on the understanding of the analysis, and the interpretation of the output.

GCA takes the total population (601 observations and 10 variables) and determines the main trend.

14.2.3 The Overrepresentation Map for Main Trend in dataset TOTAL

Figure 14.1 presents the overrepresentation map of the main trend for the total set (i.e., N=601, and denoted TOTAL) of questionnaires.

Note: We will discuss sub-sets of this data later, therefore, we will refer to these 601 observations as data set "TOTAL."

The overrepresentation map (Figure 14.1) is composed of 601 horizontal rows or lines (very thin!), where each line represents a respondent. There are 10 columns corresponding to the 10 items (i.e., variables) in the questionnaire. The values of each variable are divided by their mean and therefore the widths of the columns are equal.

There are a total of 6,010 (i.e., 601 x 10) overrepresentation values. Their value range is represented by the shading. The lighter the shading, the closer the value is to 0; the darker the shading, the greater the ratio. (See Figure 8.2 in Chapter 8 for information on the five categories of shading.)

The observations (rows) and the items (columns) have been ordered by GCA. This ordering is such that the "trend structure" or dependency shows up as the darker shading running from the top-left to the bottom-right and concentrating at these opposite corners.

The darker areas mean "higher values than expected" on the response scale.

Note: Just as we would say: "more than expected under proportionality" in dealing with frequencies, we say "higher than expected under proportionality" in dealing with scores.

Figure 14.1 shows the trend or dependency, the strength of which is measured by the global differentiation **Dif** (where **Dif** is the maximal grade correlation ρ^* of the data table **T**). (The value is 0.224602.)

Regarding the column (variable) clusters (i.e., variable permutation groups):

There are three bold vertical lines dividing the variables into four cluster groups. These variable clusters are really permutations of the 10 variables ordered by the GCA.

Note: Sometimes you see the following notation used to represent the permutations: (employ, func-lev) (prof-exp, age) (pop-size, educ, econ) (skills, lang, equip).

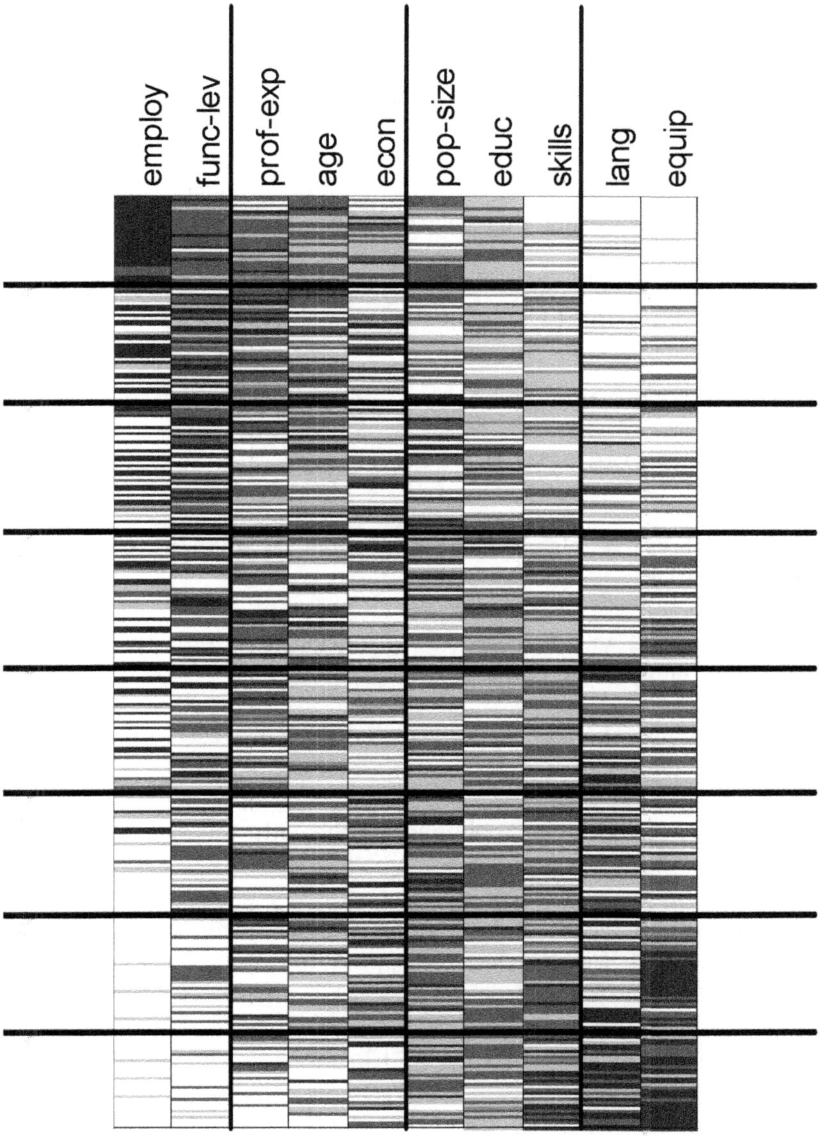

FIGURE 14.1. Case study 1: GCA overrepresentation map for dataset TOTAL (601 records, **Dif** = 0.224602). Bold vertical bars separate variable clusters; bold horizontal lines separate respondent clusters.

The ordering of the columns (i.e., the items) may be considered to indicate their "importance" relative to the structure of the data. The items near the ends (left side - right side) are more strongly associated with the trend structure than those items in the middle.

Regarding the row (observation) clusters:

GCA has ordered the rows (i.e., the respondents) relative to the structure of the data. It would be possible to display the 601 identification numbers along the left margin using a map with higher resolution.

There are seven bold horizontal lines dividing the respondents into 8 respondent cluster groups. (The number of clusters, i.e., 8, was selected by the data analyst.)

14.2.4 Interpretation of the results (with some general hints)

Note: The authors have obtained "feedback" from the researcher on the "subject area" interpretation.

We will start off with a simple interpretation using only the two extreme end variables: employ and equip.

Hint 1: Because the top cluster and the bottom cluster differ most, just look at these.

Hint 2: Because the items (columns) on the ends (left side and right side) differ most, just look at these.

Note: Relative to the size of the overrepresentation values, we will use the terms "high(er) values" and "low(er) values," as opposed to writing "higher-than-expected values," and "lower-than-expected values," which is the correct interpretation.

Looking at the top cluster - Figure 14.1:

At one end of the structural trend the respondents tend to have high values (dark shading) for employment and low values (light shading) for equipment.

Recall: High values for employment are good, i.e., employed.
Low values for equipment are bad, i.e., no computer equipment.

Looking at the bottom cluster - Figure 14.1

Here we find respondents with low values (light shading) of employment, and high values (dark shading) for equipment.

Recall: Low values for employment are bad, i.e., not employed.
High values for equipment are good, i.e., access to computer equipment.

The trend or structure of the data is such that we have employed without computer equipment at one end, and unemployed with computer equipment at the other.

Now a slightly more complicated interpretation using - rather arbitrarily - five variables from the extreme left and extreme right sides (i.e., employ, func-lev, equip, lang, and skills). The interpretation is still based primarily on the top and bottom respondent clusters.

Looking at the top cluster - Figure 14.1:

At one end of the structural trend we have those respondents who have high values (dark shading) for employ and func-lev, but at the same time have low values (light shading) for equip, lang, and skills.

Recall: High values for employment are good, i.e., employed.
High values for functional level are good, i.e., good functioning with disability.
Low values for equipment are bad, i.e., no computer equipment.
Low values for languages are bad, i.e., no foreign language abilities.
Low values for skills are bad, i.e., no computer skills.

Looking at the bottom cluster - Figure 14.1:

At the other end of this structural trend the respondents have low values (light shading) for employ and func-lev, but high values (dark shading) for computer equip, lang, and skills.

In other words, at one trend extreme we have employed disabled respondents who are working, perhaps, due to their higher functional ability, in spite of lower computer and language skills. At the other extreme are respondents with access to computer equipment, who are skilled in computer use, and who are skilled in foreign languages, but are not employed, probably due to their low functional level.

Regarding the center variables (columns) - Figure 14.1:

The center variables, are not, let us say, an important part of the structure of the data. (At least the structure of the main trend.)

For example, with "econ:" you do not see any shade "gradient" at all as you go from top of the column to bottom. It is similar for educ and pop-size.

14.3 Case Study 1 - Decomposition of the dataset into regular subpopulations

The decomposition of the dataset TOTAL into two sub-populations called: FIT-MT (those observations that FIT the Main Trend), and OUT-MT (those observations that are OUTside the Main Trend) was made according to the procedure described in Section 9.5 and adjusted in Chapter 13 to deal with data tables. Outliers from the main trend in TOTAL are classified into the subset OUT-MT. Those observations which fit the main trend are classified into subset FIT-MT. Both subsets are then transformed by GCA to find the main trends for each. The detected order of variables for FIT-MT is the same as in TOTAL. In OUT-MT the main trend changes.

The number of respondents was 320 in FIT-MT and 281 in OUT-MT.

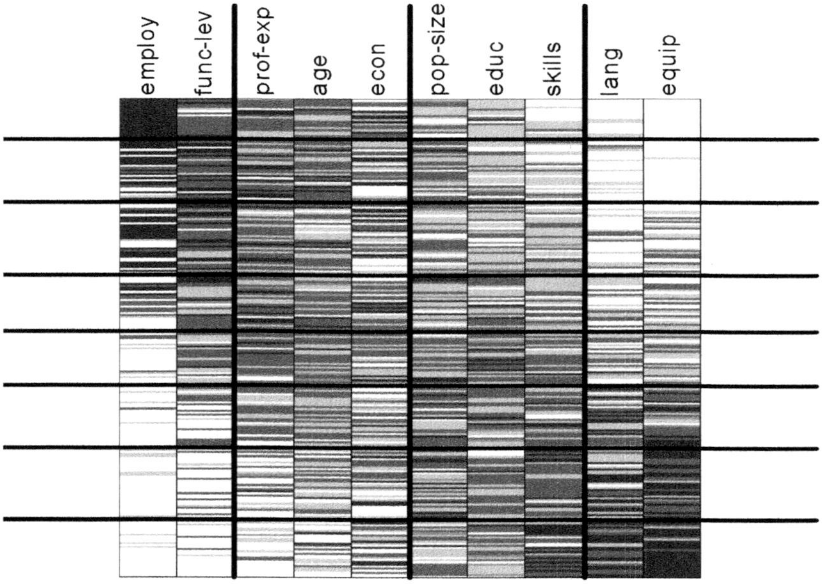

FIGURE 14.2. Case study 1: GCA overrepresentation map for dataset FIT-MT (320 records, **Dif** = 0.284768).

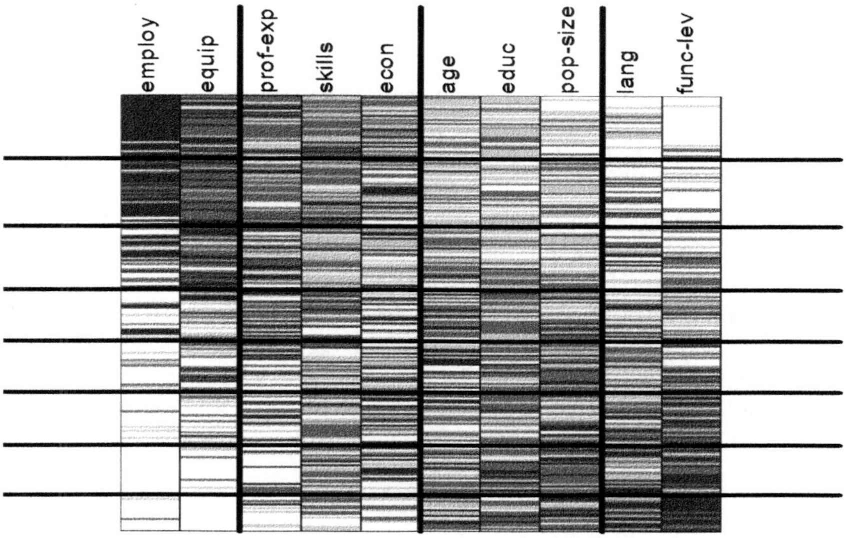

FIGURE 14.3. Case study 1: GCA overrepresentation map for dataset OUT-MT (281 records, **Dif** = 0.245965).

14.3.1 The Overrepresentation Maps for FIT-MT and OUT-MT

Figure 14.2 presents the overrepresentation map for data set FIT-MT, and Figure 14.3 presents the overrepresentation map for data set OUT-MT. The values of **Dif** in both maps have higher values than in TOTAL (Figure 14.1). In FIT-MT (i.e., Figure 14.2), the order of the variables is not changed as compared with TOTAL.

Generally, Figure 14.2 for FIT-MT (observations that fit the main trend) is very similar to that of Figure 14.1 for the TOTAL data set, but is slightly more concentrated (larger **Dif** values) around the main trend. The interpretation for Figure 14.2 is very similar to the interpretation for Figure 14.1.

Differences between Figures 14.2 and 14.3:

We have two structurally different populations represented here.

The most important difference between the map for OUT-MT (Figure 14.3) and the map for FIT-MT (Figure 14.2) is that in OUT-MT the variable func-lev is in the extreme right column, rather than the second from the left position as in FIT-MT. Those individuals in OUT-MT who are employed tend to have - at the same time - a very low functional level.

The trend in OUT-MT is different from the main trend in FIT-MT, while the marginals (i.e., the means and marginal frequencies) are quite similar. (See the Appendix in Section 14.7 for additional information.)

For Figure 14.3, looking at the first cluster of respondents: Those employed (i.e., high values or dark for first variable) are now those with equipment (high values or dark for the second variable) even though they have a very low functioning level (low values or light for rightmost variable) and they are living in small villages (low values in the eighth column). Looking at the bottom cluster (Figure 14.3): we have individuals with high function level, but no employment.

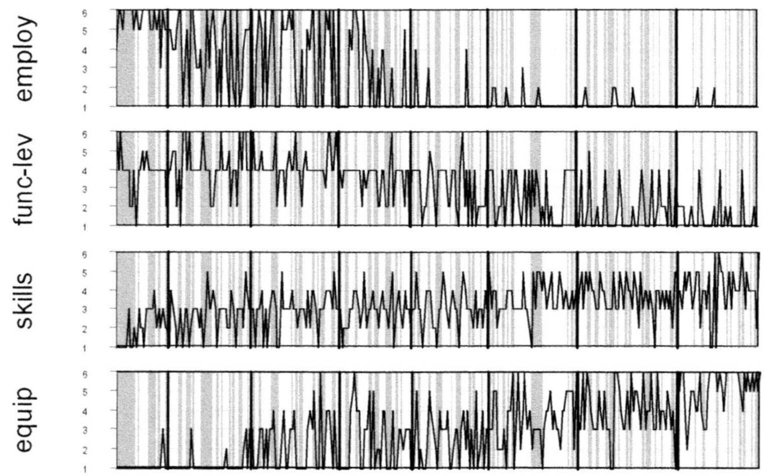

FIGURE 14.4. Case study 1: Grade strip charts for selected key variables in FIT-MT when records on the horizontal axis are ordered according to GCA. Variables are valued 1,...,6. (The gray vertical background lines are explained in the text.)

14.3.2 The grade strip charts for FIT-MT and OUT-MT

Figures 14.4 and 14.5 contain grade strip charts for the key items: employ, func-lev, skills and equip, for FIT-MT and OUT-MT. The grade strip chart in each subset plots the values of particular items against respondents ordered according to GCA. (Note: The selected items are also ordered from top to

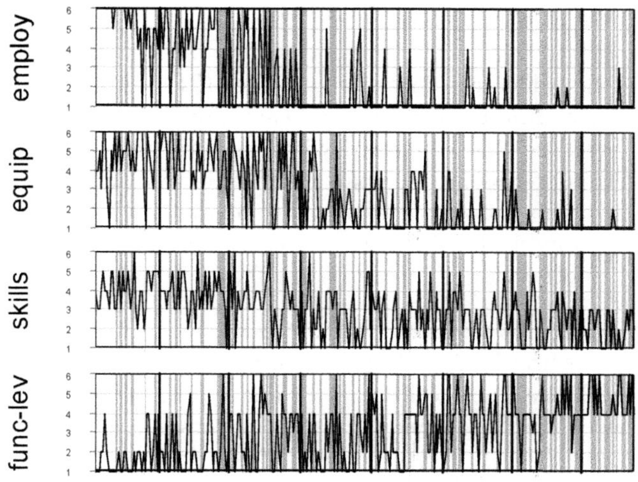

FIGURE 14.5. Case study 1: Grade strip charts for selected key variables in OUT-MT when records on the horizontal axis are ordered according to GCA. Variables are valued 1,...,6. (The gray vertical background lines are explained in the text.)

bottom according to GCA.) The horizontal strip plot for a given variable in Figure 14.4 and in Figure 14.5 correspond to the column for that variable in the GCA overrepresentation maps in Figure 14.2 and Figure 14.3.

The grade strip charts enable one to see the exact values of items, since the plots use the true initial scale (i.e., before dividing the values by their means.)

Eight clusters of respondents (the same as on the maps) are indicated by thick vertical lines. The left clusters differ visibly from the right clusters; the trend for each variable is monotone. In Figure 14.4 trends are decreasing for employ and func-lev, and increasing for skills and equip; in Figure 14.5 trends are decreasing for employ, equip, and skills, while increasing for func-lev.

Remark 14.1 *The grade strip charts in Figure 14.4 and 14.5 contain additionally thin light gray vertical lines which distinguish the respondents who arrived later (i.e., in the second half of the period of time in which the inquiry was performed). It occurred that the percent of late arrivals is similar in FIT-MT (38.2%) and OUT-MT (34.8%). However, the late arrivals are not randomly mixed with early arrivals: they tend to create distinct clusters (differently in FIT-MT and in OUT-MT).*

14.3.3 Two-way ordered clustering

To indicate the difference in the structures of FIT-MT and OUT-MT the values of each item in each cluster and each subset (i.e., FIT-MT and OUT-MT) were averaged and compared with the total mean of the item in this subset (the total means for each particular item were almost identical in each subset). Then, the cluster average was compared with the total mean, and the cluster was marked "+" if the cluster average was higher than the total mean, and "-" otherwise. The results are shown in Table 14.1.

Table 14.1 exhibits a high degree of symmetry. In each subset the left clusters 1, 2, 3 with pluses and minuses are identical, and the right clusters 6, 7, 8 are also identical and form a mirror image of those on the left. Moreover, employ behaves in FIT-MT and OUT-MT in the same way, while the "+" sign for func-lev in FIT-MT is replaced by "-" for skills and equip in OUT-MT, and the same holds for the "-" sign.

The two-way ordered clustering summarized in Table 14.1 can support decision making concerning type and extent of help needed by particular clusters of respondents.

Decomposition into more regular subsets with items and respondents ordered according to GCA, and visualisation of results provide insight into the data. A closer examination may help one to recognize trends and pecularities and interpret them.

Some details of the analysis of datasets TOTAL, FIT-MT and OUT-MT for Case Study 1 are presented in Section 14.7.1.

14.4 Case Study 2 - Analysis of Engineering Data (Strength of Concrete)

This is a study of the dependencies among the strength of various concrete mixtures and possible explanatory variables.

Comment: The data concern High Performance Concrete mixes (HPC), and have been collected from various sources by the Institute of Fundamental Technology Research of the Polish Academy of Science (see Section 14.7.3).

The data were collected from different sources during two time periods approximately ten years apart: data-1 (331 observations), and data-2 (302 observations).

14.4.1 The variables:

Each observation consists of the values of 18 variables:

Y1 to Y6 are quantities of the basic ingredients in concrete mixtures:
Y1 = cement
Y2 = water
Y3 = silica fume
Y4 = superplasticiser
Y5 = coarse aggregate
Y6 = fine aggregate

Y7 = the strength of hardened material. (This is viewed as a response variable.)

Y8 to Y15 are various ratios of Y1 to Y6 of interest to the researcher:
Y8 = Y2 / Y1
Y9 = Y2 / (Y1+Y3)
Y10 = Y4 / Y1
Y11 = Y2 / (Y1+Y3+Y4+Y5+Y6)
Y12 = (Y5+Y6) / (Y1+Y2+Y3+Y4)
Y13 = Y5 / (Y5+Y6)
Y14 = Y6 / (Y1+Y2+Y3+Y4+Y5+Y6)
Y15 = (Y1+Y2+Y3+Y4+Y5+Y6) / 2400

Y16 = Approximate volume of the components.

Y17 and Y18 were previously used prediction formulas for Y7 of interest to the researcher:
Y17 = Strength prediction formula, called here Y17
Y18 = Strength prediction formula, called here Y18

Note: All values are non-negative and on a ratio scale.

All variables (other than Y7) are considered as possible explanatory variable candidates. In that the variables were originally scaled differently (i.e., the measuring units), they have been normalized (divided by their means).

14.4.2 The goal of the analysis

The goal of the study is to investigate the degree and structure of dependence between a response variable (i.e., Y7), and a set of explanatory variables.

Comparison of two time periods is interesting from the point of view of changes in material technology over time.

We will therefore show in this section how GradeStat recognizes and compares their structures.

In the present case study Y7 is not in this case directly specified as the response variable. (That is, while the dependent variable specification is possible in the GradeStat program, it is not done here.) This provides the reader with the opportunity to see how "things fall into place," even without specifying this information.

14.4.3 The Overrepresentation Map for Main Trend in the dataset TOTAL

The GCA overrepresentation map for the TOTAL dataset is shown in Figure 14.6. The values of each variable are divided by their mean and therefore the widths of the 18 columns are equal.

The eight variable clusters (i.e., the variables separated by the bold vertical lines) are permutation sets of the 18 variables used in the GCA.

The variable clusters are:
(y4, y10, y3); (y18, y17, y7); (y1); (y13, y5); (y15, y16); (y6, y14); (y12, y2, y11); (y8, y9).

The patterns for columns y4 and y10 (i.e., in variable cluster 1) are very similar. The same occurs in cluster 2 for y7 and its two special predictors; in cluster 4 for y13 and y5; in cluster 5 for y15 and y16; in cluster 6 for y6 and y14; in cluster 7 for y2 and y11; in cluster 8 for y8 and y9. This means that variables patterns within clusters are similar in most cases. Generally, only variables y3, y1, and y12 distinctly outlie from the overall visual pattern.

There are also eight clusters of observations (record clusters or "case" clusters).

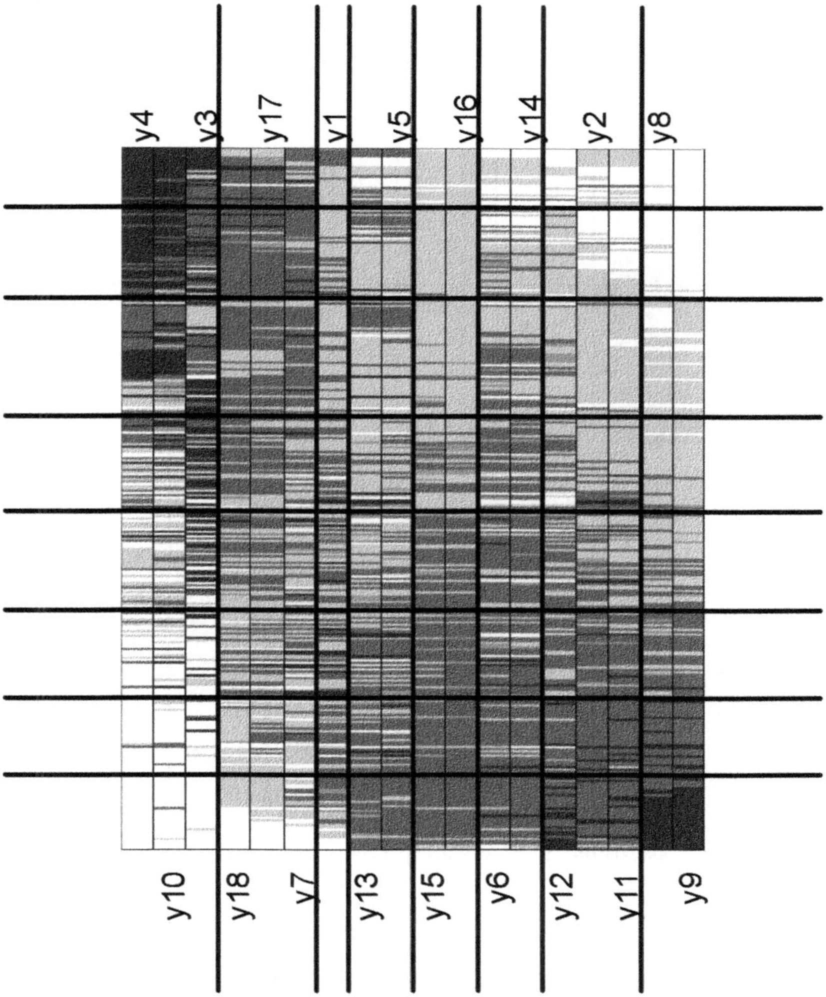

FIGURE 14.6. Case study 2: GCA overrepresentation map for dataset TOTAL (633 records, **Dif** = 0.291464). (Bold vertical lines separate the variable clusters; bold horizontal lines separate the respondent clusters.)

14.5 Case Study 2 - Decomposition of concrete mixtures into FIT-MT and OUT-MT

The decomposition of the dataset TOTAL into two sub-populations called: FIT-MT (those observations that FIT the Main Trend), and OUT-MT (those observations that are OUTside the Main Trend) was made according to the same procedure as in Case Study 1.

14.5.1 The Overrepresentation Maps for FIT-MT and OUT-MT

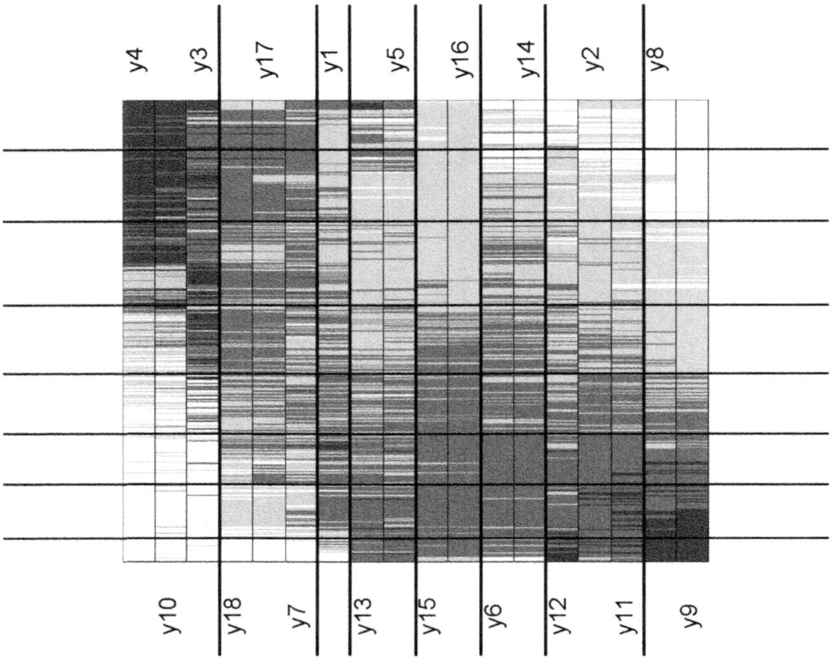

FIGURE 14.7. Case study 2: GCA overrepresentation map for dataset FIT-MT (450 records, **Dif** = 0.314467).

Figure 14.7 for FIT-MT (observations that fit the main trend) is very similar to Figure 14.6 for the TOTAL dataset, but it is slightly more concentrated (larger ρ^* value) as indicated by the main trend. The interpretation for Figure 14.7 is similar to the interpretation for Figure 14.6.

Differences between Figures 14.7 and 14.8:

These figures represent two functionally different populations.

The most important difference between the map for OUT-MT (Figure 14.8) and the map for FIT-MT (Figure 14.7) is that in OUT-MT the variable y3 is in the extreme right column, rather than in the third position in cluster 1 as in FIT-MT. Further, cluster 6 in FIT-MT moved to cluster 2 in OUT-MT; cluster 8 in FIT-MT supplemented by y5 forms cluster 6 in OUT-MT. Generally, OUT-MT is distinctly less concentrated around the main trend

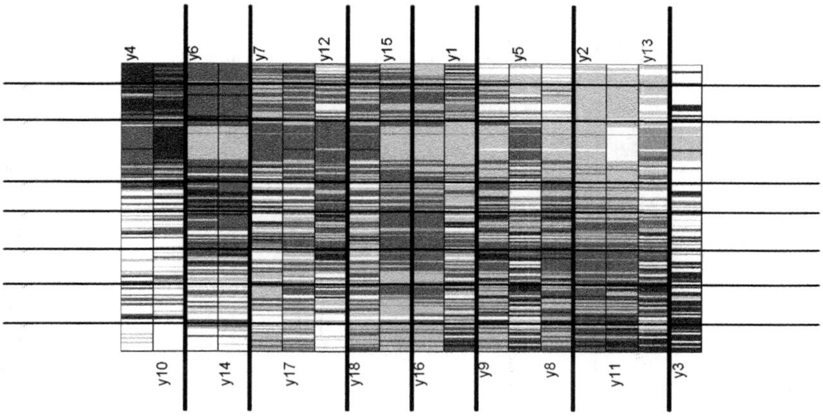

FIGURE 14.8. Case study 2: GCA overrepresentation map for dataset OUT-MT (183 records, **Dif** = 0.246205).

than FIT-MT. So in this Case Study the decomposition did not provide *two* subsets more regular than the TOTAL but one subset (FIT-MT) more regular than TOTAL, and one subset (OUT-MT) *less* regular and with a slightly different trend. The key difference concerns the position of y3 and is illustrated by the grade strip charts which accompany the GCA maps.

14.5.2 The grade strip charts for FIT-MT and OUT-MT

Eleven variables were selected to appear in the grade strip charts in order to make the structures of FIT-MT and OUT-MT clear and easy to compare with one another. Seven "twin" variables were omitted: y10 (as a twin for y4), y8 (as a twin for y9), y17 (as a twin for y18), y13 (as a twin for y5), y14 (as a twin for y6) y11 (as a twin for y2), y16 (as twin for y15).

The grade strip charts for FIT-MT and OUT-MT for selected variables are shown in Figures 14.9 and 14.10. The plots in FIT-MT are more regular than in OUT-MT (especially in the case of the variables at both ends). The plots of some variables differ in FIT-MT and OUT-MT. The main difference concerns y3, which has a decreasing trend in FIT-MT, and an increasing trend in OUT-MT. Thus, in Case Study 2, y3 plays the role played by func-lev in Case Study 1. Similarly, y4 plays the role played by employ in Case Study 1.

The thin vertical lines correspond to the data gathered in the first time period. Such data constitute one-half of FIT-MT (50.7%), and 40% of data in OUT-MT. Note that cluster 4 in FIT-MT and also cluster 4 in OUT-MT consist almost entirely of the earlier data, and that there are a lot of

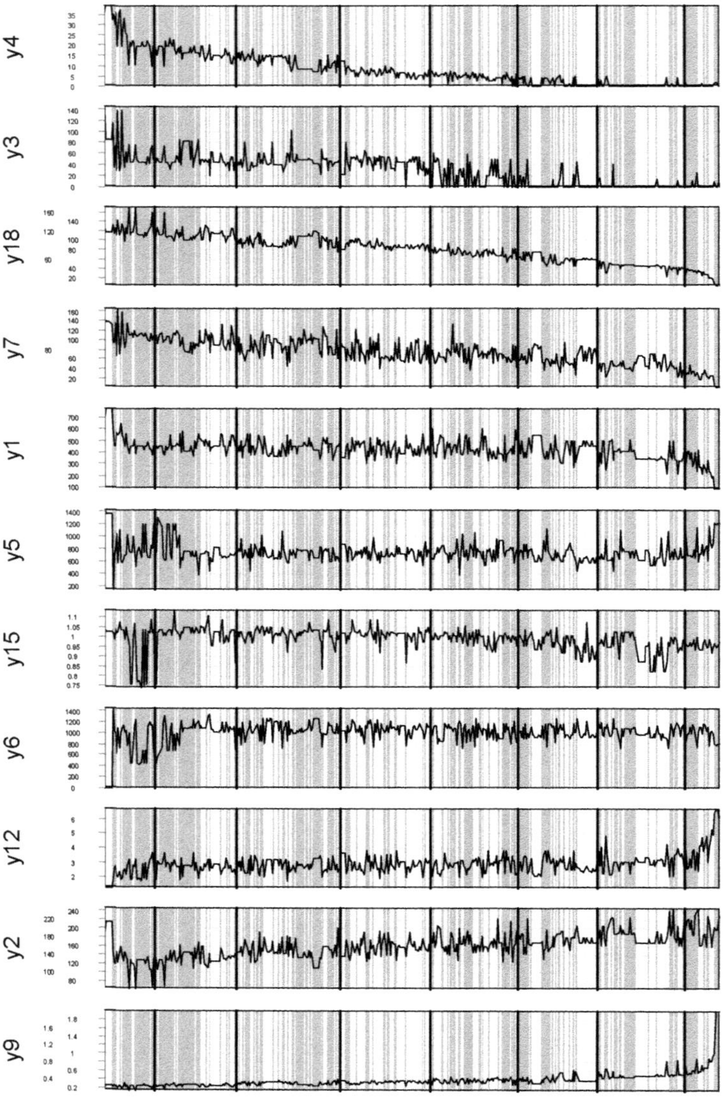

FIGURE 14.9. Case study 2: Grade strip charts for selected variables in FIT-MT when records (on horizontal axis) are ordered according to GCA. (Compare with Figure 14.7).

large groups of relatively homogenous records which are "light" or "dark." This stratification is far from the stratification plausible when both parts of the data are considered to be, for example, randomly drawn from the same population. This is something that would therefore be difficult to discover by traditional methods.

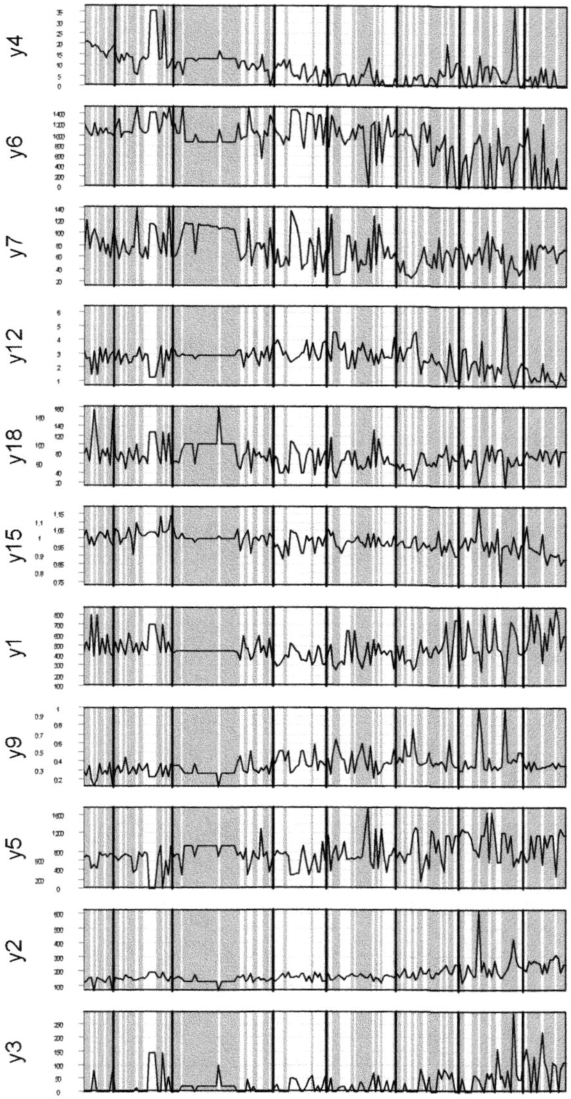

FIGURE 14.10. Case study 2: Grade strip charts for selected variables in OUT-MT when records (on horizontal axis) are ordered according to GCA. (Compare with Figure 14.8).

The large values of the response variable y7 (see first cluster in FIT-MT) are obtained mostly in the later time period (first cluster in FIT-MT). This cluster is especially diversified with respect to the values of y4, y3, y5, y6, y15.

14.6 Final remarks for the two case studies

Remember: The goal of Grade methods is to understand the data. If we have a good understanding of the data, we will be in a position to ask better questions (i.e., the questions that "should" have been asked), and to provide better answers.

The measurements for the two case studies were on different scales: one ordinal, the other ratio. The analyses have some basic operations in common, but also present the opportunity to demonstrate different Grade tool features.

We learned that the most influential variables displayed by the overrepresentation map are on the extreme left or right side, i.e., not in the center columns.

Ordered two-way grade clustering is used in both case studies. Main trends in TOTAL are found and then the data are decomposed into two subsets: one preserving the main trend and the other consisting of "outliers" from the main trend. The trends in these subsets often provide important new information for the researchers.

14.7 Appendix:

14.7.1 Case Study 1 - further details of the analysis

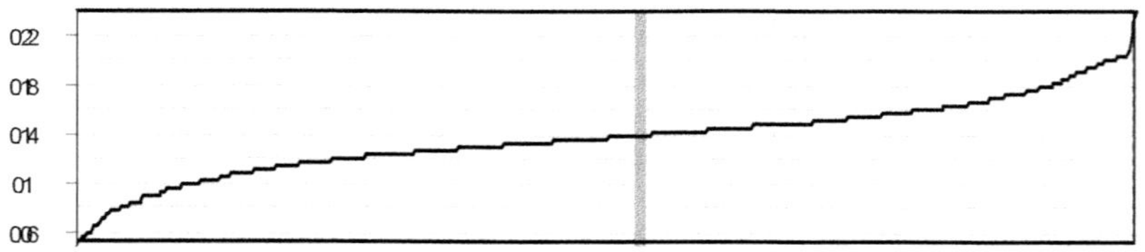

FIGURE 14.11. Case study 1: The graph of *AvgDist* (vertical axis) versus respondents (horizontal axix) ordered according to non-decreasing values of *AvgDist*. The threshold 0.14 divides the whole set of respondents into subset FIT-MT (*AvgDist*<0.14) and subset OUT-MT (*AvgDist*>0.14).

Figure 14.11 presents the graph of function $AvgDist_{row}$ and indicates the threshold used to decompose the TOTAL dataset into FIT-MT and OUT-MT. Recall that function $AvgDist_{row}$ was introduced in Chapter 9 by formula (9.13). The threshold ($AvgDist = 0.14$) is approximately the inflexion point of the graph.

* * *

The effects of the decomposition of TOTAL for the marginal distributions of key variables are shown in Figure 14.12.

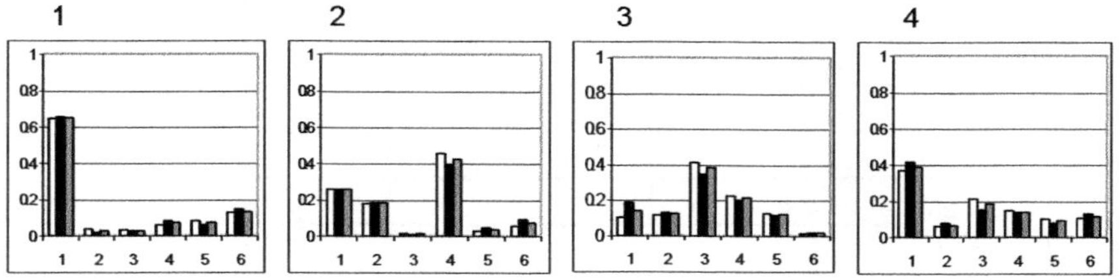

FIGURE 14.12. Case Study 1: histograms of key variables: employ (1), func-lev (2), skills (3), and equip (4), each valued 1, 2,...,6. The histograms are white for FIT-MT, black for OUT-MT, and grey for TOTAL.

White and black histograms (for FIT-MT and OUT-MT) are very similar for all variables selected for presentation (as well as for those not selected).

* * *

Contrary to the similarity of the marginal distributions, the grade correlation matrices for the selected key variables are distinctly different in particular subsets.

They are shown in Table 14.2. The key variables: employ, func-lev, skills and equip are ordered according to GCA. This order is the same in TOTAL and FIT-MT, but different in OUT-MT.

Remark 14.2 *The grade correlation matrix for TOTAL is very close to the weighted average of the grade correlation matrix for FIT-MT and of the suitably reordered grade correlation matrix for OUT-MT. For instance, for the pair (employ, func-lev) the weighted average of the values of* ρ^* *in FIT-MT and OUT-MT (with weights equal to the proportions of the subsets) is*

$$\frac{320}{601}0.367849 + \frac{281}{601}(-0.192643) = 0.1057$$

which is close to the value 0.0939 in TOTAL.

TABLE 14.1. Case study 1: structure of clusters in FIT-MT and OUT-MT due to ordered GCA clustering; "+" ("-") corresponds to cluster and item for which the cluster average is larger (smaller) than the total item average.

CLUSTERS		1	2	3	4	5	6	7	8
FIT-MT	employ	+	+	+	+	-	-	-	-
	func-lev	+	+	+	+	+	-	-	-
	equip	-	-	-	-	-	+	+	+
	skills	-	-	-	-	-	+	+	+
OUT-MT	employ	+	+	+	-	-	-	-	-
	equip	+	+	+	-	-	-	-	-
	skills	+	+	+	-	-	-	-	-
	func-lev	-	-	-	-	+	+	+	+

TABLE 14.2. Case Study 1: the grade correlation matrices for the selected key variables in TOTAL, FIT-MT and OUT-MT. (The orderings of rows in the three sets refer to GCA; therefore, the rows in the OUT-MT matrix are presented in a different order that in the case of TOTAL and FIT-MT.)

TOTAL

	employ	func-lev	skills	equip
employ	1	0.09	0.17	0.13
func-lev		1	-0.10	-0.20
skills			1	0.52
equip				1

FIT-MT

	employ	func-lev	skills	equip
employ	1	0.37	-0.07	-0.22
func-lev		1	-0.18	-0.32
skills			1	0.54
equip				1

OUT-MT

	employ	equip	skills	func-lev
employ	1	0.51	0.42	-0.19
equip		1	0.49	-0.07
skills			1	0.00
func-lev				1

This provides a good example of how misleading the level of monotone dependence assessed in an *irregular* dataset like TOTAL might be. The value 0.09 shows that in TOTAL, the employ and func-lev are almost uncorrelated; however, in an irregular dataset this fact has little to do with independence. After the decomposition into FIT-MT and OUT-MT, dependence in FIT-MT is much more regularly monotone and hence the high grade correlation 0.37 indicates (high) *positive* dependence, while -0.19 in OUT-MT indicates (weak) *negative* dependence.

It is seen that the absolute values of the grade correlations 0.37 and -0.19 are higher than 0.09 in TOTAL. A similar situation concerns the pairs (employ, skills), (employ, equip), (func-lev, skills), (func-lev, equip). For all these pairs, the absolute values of the grade correlations are higher in FIT-MT and OUT-MT than in TOTAL.

Moreover, the grade correlation matrices in FIT-MT are more regular than such matrices in TOTAL: the matrix in OUT-MT is Robinson, the matrix in FIT-MT is almost Robinson, but the matrix in TOTAL is far from being Robinson (the indices in the first row do not decrease when moving away from the main diagonal, and the same holds for the last and next to the last column).

It now only remains to calculate (in the GCA versions of TOTAL, FIT-MT and OUT-MT) the global differentiation **Dif** and its upper bound denoted *sup***Dif**.

Recall that $\mathbf{Dif}_{\mathbf{X}}$ represents the maximal global differentiation of the profiles of records and also the maximal global differentiation of the distributions of variables in vector **X**. A vector with four key components (employ, func-lev, skills, equip) is taken into account for TOTAL and FIT-MT, while in OUT-MT the considered key variables are reordered and form the vector (employ, equip, skills, func-lev). The values of $\mathbf{Dif}_{\mathbf{X}}$ are calculated from the respective matrices **DIFF** (Table 14.3) according to the formula (corresponding to the normalization under which each variable is divided by its mean):

$$\mathbf{Dif}_X = \frac{8}{k^3} \sum\nolimits_{i=1}^{k} \sum\nolimits_{j=1}^{k} (j-i)\mathbf{Dif}_{(X_i,X_j);R_{\mathbf{X}}}$$

Indices $sup\mathbf{Dif}_{\mathbf{X}}$ are calculated from the respective matrices (Table 14.3) according to the analogous formula in which matrix **DIFF** is replaced by the matrix **supDIFF** (compare similar calculations for the C/J data and A/B data in Chapter 12). Thus we have:

	$\mathbf{Dif}_X$	sup$\mathbf{Dif}_X$	$\mathbf{Dif}_X$/sup$\mathbf{Dif}_X$
TOTAL	0.327	0.430	76.0%
FIT-MT	0.395	0.452	87.4%
OUT-MT	0.350	0.425	82.3%

As mentioned in Chapter 12, the ratio $\mathbf{Dif}_X/sup\mathbf{Dif}_X$ measures how close the data table transformed by GCA is to a TP_2 table. Value 1 (100%) indicates that the table is TP_2. Subset FIT-MT is close to TP_2 (87.4%); subset OUT-MT is slightly less close (82.3%); both are closer to TP_2 than is TOTAL (76%).

TABLE 14.3. Case Study 1: the differentiation matrices **DIFF** (lower triangle) and **supDIFF** (upper triangle) for the selected key variables in TOTAL, FIT-MT and OUT-MT.

TOTAL

	employ	func-lev	skills	equip
employ	0	0.365	0.336	0.377
func-lev	0.152	0	0.290	0.377
skills	0.285	0.155	0	0.228
equip	0.371	0.258	0.108	0

FIT-MT

	employ	func-lev	skills	equip
employ	0	0.302	0.362	0.438
func-lev	0.173	0	0.287	0.385
skills	0.333	0.210	0	0.217
equip	0.432	0.338	0.137	0

OUT-MT

	employ	equip	skills	func-lev
employ	0	0.275	0.297	0.420
equip	0.147	0	0.242	0.369
skills	0.234	0.111	0	0.293
func-lev	0.411	0.317	0.207	0

14.7.2 Case Study 2 - further details of the analysis

In Case Study 2 the decomposition of TOTAL into FIT-MT and OUT-MT is based not only on the function $AvgDist_{row}$ (as in Case Study 1), but also on the function $MaxDist_{row}$, defined in Chapter 9 by formula (9.15) (see also Figure 12.17 for the A/B data). The graphs of both functions are presented in Figure 14.13: in the upper graph the records are ordered according to the increasing values of $AvgDist$, in the lower graph according to the increasing values of $MaxDist$.

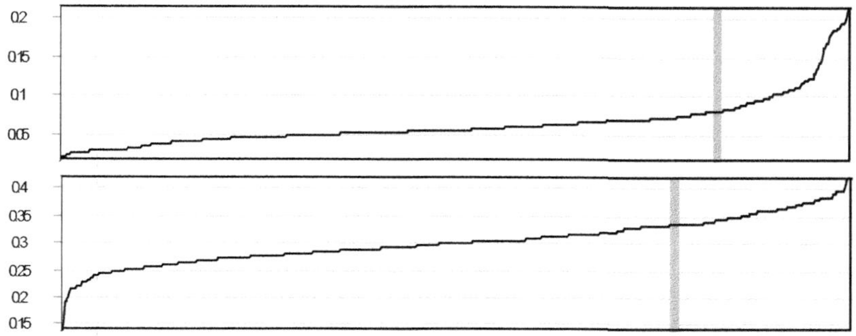

FIGURE 14.13. Case study 2: graphs of *AvgDist* (upper curve) and *MaxDist* (lower curve), with thresholds 0.080 and 0.3364.

Several decompositions were tried. The first was based on *AvgDist* only, but the trends in the resulting subsets were too weak and not distinctly diversified. Finally, subset OUT-MT was specified by the condition

$$either \quad AvgDist > 0.08 \quad or \quad MaxDist > 0.3364.$$

* * *

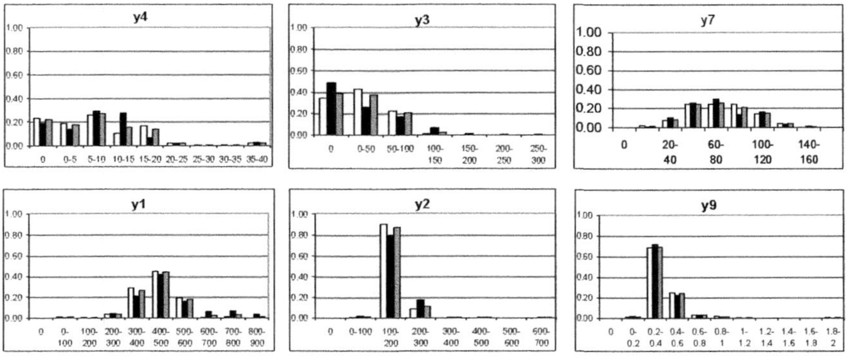

FIGURE 14.14. Case study 2: histograms for y4, y3, y7, y1, y2, y9 in FIT-MT (white), OUT-MT (black), and TOTAL (grey).

The marginal distributions in FIT-MT are not much different from the respective marginal distributions in OUT-MT. This is illustrated in Figure 14.14 for six key variables. The similarity of the distributions in FIT-MT and OUT-MT is especially strong for variables y7 and y9; extreme values are more frequent in OUT-MT than in FIT-MT for variables y1 and y2; the distributions in FIT-MT and OUT-MT are particularly different for y3 and also for y4.

* * *

TABLE 14.4. Case study 2: the grade correlation matrices for selected key variables in TOTAL, FIT-MT and OUT-MT. (Rows and columns within each matrix arranged according to GCA.)

TOTAL

	y4	y3	y7	y1	y2	y9
y4	1	0.48	0.58	0.27	-0.54	-0.66
y3		1	0.47	0.20	-0.22	-0.49
y7			1	0.46	-0.57	-0.81
y1				1	0.05	-0.63
y2					1	0.65
y9						1

FIT-MT

	y4	y3	y7	y1	y2	y9
y4	1	0.72	0.68	0.35	-0.65	-0.78
y3		1	0.58	0.28	-0.42	-0.65
y7			1	0.45	-0.64	-0.80
y1				1	-0.07	-0.66
y2					1	0.72
y9						1

OUT-MT

	y4	y7	y1	y9	y2	y3
y4	1	0.35	0.13	-0.38	-0.33	-0.08
y7		1	0.49	-0.84	-0.39	0.24
y1			1	-0.59	0.30	0.05
y9				1	0.48	-0.13
y2					1	0.17
y3						1

The grade correlation matrices for y4, y3, y7, y1, y2, y9 in TOTAL, FIT-MT and OUT-MT are shown in Table 14.4. The grade correlations of the key variables y4 and y3 vary from 0.48 in TOTAL through 0.72 in FIT-MT to -0.08 in OUT-MT; it is therefore evident that the division of TOTAL into FIT-MT and OUT-MT is based mainly on the differences in the distributions of the pair (y3, y4). Thus the grade correlation matrices confirm the observations implied previously by the GCA overrepresentation maps and grade strip charts.

The grade correlation matrix for FIT-MT has the following desirable property: if variables y2 and y4 were to be reversed, all grade correlation coefficients would be positive (so that all variables would be positively dependent).

Remark 14.3 *There is a tendency that regular data tables can be transformed into data matrices of positively interdependent variables when some of the variables are reversed. (A description of the corresponding selection procedure and reversing technique are in progress by the authors.)*

Table 14.5 contains the differentiation matrices **DIFF** (upper triangles) and **supDIFF** (lower triangles) for TOTAL, FIT-MT and OUT-MT. The global differentiations **Dif**, their upper bounds sup**Dif** and the ratios resulting from Table 14.5 are shown below:

	$\mathbf{Dif}_X$	$\text{sup}\mathbf{Dif}_X$	$\mathbf{Dif}_X/\text{sup}\mathbf{Dif}_X$
TOTAL	0.386	0.466	83%
FIT-MT	0.411	0.453	91%
OUT-MT	0.357	0.508	70%

TABLE 14.5. Case study 2: the differentiation matrices **DIFF** (lower triangle) and **supDIFF** (upper triangle) for six selected key variables in TOTAL, FIT-MT and OUT-MT. (Rows and colums within each matrix arranged by GCA.) Please note that the matrix for FIT-MT is almost anti-Robinson.

TOTAL

	y4	y3	y7	y1	y2	y9
y4	0	0.39	0.32	0.36	0.42	0.47
y3	0.04	0	0.38	0.40	0.43	0.49
y7	0.25	0.21	0	0.14	0.22	0.28
y1	0.32	0.29	0.09	0	0.12	0.22
y2	0.39	0.37	0.16	0.08	0	0.12
y9	0.44	0.43	0.23	0.16	0.08	0

FIT-MT

	y4	y3	y7	y1	y2	y9
y4	0	0.28	0.31	0.36	0.43	0.49
y3	0.11	0	0.31	0.34	0.40	0.46
y7	0.28	0.20	0	0.13	0.21	0.28
y1	0.35	0.28	0.08	0	0.12	0.22
y2	0.42	0.37	0.17	0.09	0	0.11
y9	0.48	0.43	0.25	0.17	0.09	0

OUT-MT

	y4	y7	y1	y9	y2	y3
y4	0	0.32	0.33	0.41	0.38	0.58
y7	0.16	0	0.15	0.27	0.23	0.51
y1	0.18	0.03	0	0.21	0.13	0.50
y9	0.24	0.09	0.06	0	0.13	0.54
y2	0.25	0.10	0.07	0.01	0	0.49
y3	0.54	0.43	0.39	0.36	0.34	0

These results confirm that subset FIT-MT (91% of closeness to $\mathbf{TP}_2$) is more regular than OUT-MT (70% of closeness). However, the poor regularity of OUT-MT might be much improved by a *repeated* search for outliers and their exclusion. (The repetition consists in treating OUT-MT in the same way as TOTAL when it was searched for outliers which formed OUT-MT). The operation is illustrated by Figure 14.15, where the plot of $AvgDist_{row}$ supplements the grade strip charts of the four key variables y3, y1, y7, y4. Nineteen records with the highest values of $AvgDist_{row}$ (distinguished in Figure 14.15 by the thin vertical lines) were selected. All possess extreme values of the four key variables. Note that ten of these nineteen records form a cluster of almost adjacent records and probably have some common feature (e.g., specific source) which might be recognized by the data analyst. Regularity of the subset consisting of the remaining 164 records becomes as high as 81%.

A successful decomposition of TOTAL into FIT-MT (318 records) and refined OUT-MT (164 records) is helpful in performing successful predictions of the values of the response variable y7, based on the explanatory variables. The prediction procedure initially reorders the records in FIT-MT and OUT-MT according to the GCA orderings obtained when y7 is excluded. A new record with unknown value of y7 is then added to each subset according to its value of the respective col|row regression function, and the value of y7 is obtained as though it was a "missing value." Many different versions of prediction procedures are proposed with interesting results for various missing value algorithms based on GCA (cf Chapter 9).

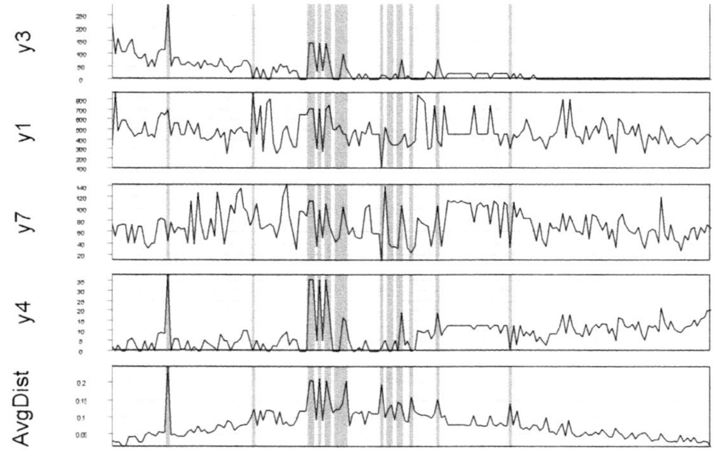

FIGURE 14.15. Case Study 2, subset OUT-MT with records ordered according to GCA: the grade strip charts for y3, y1, y7, y4, and the related $AvgDist_{row}$. The nineteen records which are most outlying from the main trend according to $AvgDist$ are indicated by the thin gray vertical lines.

14.7.3 Bibliographical remarks

The questionnaire data used in Case Study 1 were gathered by the Foundation Supporting Physically Disabled Mathematicians and Computer Specialists in Poland from 1997 to 2001. The first 400 questionnaires were analysed in [98], all questionnaires were analysed in [85]; the ordered grade clustering was developed in [87].

The data HPC (High Performance Concrete mixes) used in Case Study 2 were gathered for many years (see [48]). The authors of the present book analysed the HPC data in cooperation with Professor Kasperkiewicz (one of the authors of [48]) and used them to exemplify various grade methods (see e.g. [103]).

15

The GradeStat program

Olaf Matyja

15.1 Introduction

The GradeStat application[1] has been created using Microsoft Visual C++ with MFC. This data analysis package implements many grade methods.

The program is very complex, its source code occupies 932 KB (133 files), and consists of about 35000 lines of code. Complete description of such a program is beyond the scope of this book.

The program may be downloaded on the page http://gradestat.ipipan.waw.pl.

The GradeStat table is a two-dimensional matrix. Each cell of the table may contain a real value or be empty when the data are missing.

Rows and columns with additional parameters are attached to the table on the screen — some concern normalization, others present common statistics derived from the data.

15.2 Main implemented features

15.2.1 Data overview

- Choosing variables and objects to process
- History of the operations
- Tables of dependence between variables and objects (Spearman's *rho*, Spearman's rank *rho*, Kendall's *tau*, Pearson's *rho*, SSP/deviance table)
- Dependence indices between row and column variables (Spearman's *rho*, Kendall's *tau*, Kendall's $tau - b$, Kendall's $tau - c$, Pearson's *rho*, τ_{sgn}, τ_{abs}, regularity index, number of concordant and discordant pairs, Kruskal's *gamma*, Sommer's *d*, Gini index for rows, columns and with respect to binormal table).
- Over-representation table
- Bivariate and univariate distribution

---

[1] The first version has been called GRAD.STAT

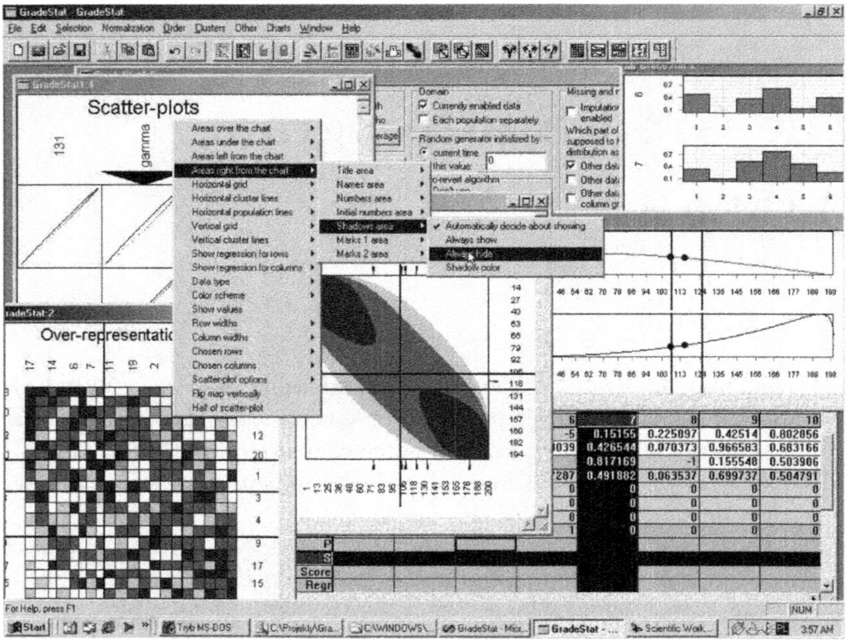

FIGURE 15.1. Exemplary screenshot of the GradeStat application.

- Looking for identical rows (with 12 options)
- Looking for outliers (also ar, and ar_{max} tables)
- Grade regression and correlation curve

15.2.2 Charts

- Data maps (overrepresentation map, ranks, missing and negative values)
- Line plot
- Bar plot
- Histogram
- Scatter-plot and scatter-map
- Three different markers

15.2.3 Preprocessing

- Creating contingency table from a sample of two variates

- Sampling of the data from contingency or probability table
- Decomposing nominal variables
- Missing and negative data imputation
- Normalization (groups and weights, simple replacing, ranks, linear transformation)
- Data smoothing
- Table transposition
- Linear regression

15.2.4 Ordering

- Sorting
- CA
- GCA (for ρ^*, τ, τ_{sgn}, in populations, random permutation sampling, checking all permutations)
- SGCA
- Significance testing for Spearman's *rho*, Kendall's *tau*, τ_{sgn}, GCA.

15.2.5 Clustering

- Grade clustering
- Quantiles
- Aggregation of rows and columns (sum, average, minimum, maximum, standard deviation, correlation, ρ^*, τ, moving average)
- Clustering comparison

References

[1] Agresti A (1984) Analysis of Ordinal Categorical Data. J. Wiley, New York

[2] Ali SM, Silvey SD (1966) A General Class of Coefficient of Divergence of One Distribution from Another. J. Roy. Statist. Soc. Ser. B, 28, 131–142

[3] Arabie P, Hubert LJ, De Soete G (Eds) (1996) Clustering and Classification, World Scientific Publ., River Edge, NJ

[4] Arens Y, Chee CY, Hsu CN, Knoblock CA (1993) Retrieving and integrating data from multiple information sources. Inter. J. Intell. Cooper. Inf. Systems 2(2), 127–158

[5] Arnold B (1987) Majorization and the Lorenz Order: a Brief Introduction. Lecture Notes in Statist. 43, Dekker

[6] Averous J, Meste M (1997) Skewness for multivariate distributions: two approaches. The Annals of Statistics, Vol. 25, No. 5, 1984–1997

[7] Bartoszyński R, Pleszczyńska E (1980) Reducibility of Statistical Structures and Decision Problems. Banach Center Publ. 6, 29-38

[8] Bickel PJ, Lehmann EL (1975) Descriptive statistics for nonparametric models. The Annals of Statistics, Vol. 3, No. 5, 1038–1069

[9] Blasius J, Greenacre M (Eds) (1998) Visualization of categorical data. Academic Press

[10] Block HW, Fang Z (1990) Setwise Independence for Some Dependence Structures. J. Mult. Analysis 32, 103–119

[11] Bock H-H (2002) Clustering methods: from classical models to new approaches. Statistics in Transition, Vol. 5 No. 5, 725–758

[12] Bock H-H (1996) Probability Models and Hypotheses Testing in Partitioning Cluster Analysis. In: Arabie P, Hubert LJ, De Soete G (Eds) Clustering and Classification, World Scientific Publ., River Edge, NJ

[13] Bondarczuk Z, Kowalczyk T, Pleszczyńska E, Szczesny W (1994) Evaluating departures from fair representation, Applied Stochastic Models and Data Analysis 10, 169–185

[14] Bradburd RM, Ross DR (1988) A General Measure of Multidimensional Inequality. Oxford Bulletin of Economics and Statistics 50, 429–433

[15] Cifarelli DM, Regazzini E (1987) On a general definition of concentration function, Sankhya Ser. B 49, 307–319

[16] Ciok A (2002) Grade Analysis of Repeated Multivariate Measurements. In: Grzegorzewski P, Hryniewicz O, Gil MA (Eds), Soft Methods in Probability, Statistics and Data Analysis, Physica-Verlag, 266–273

[17] Ciok A (2002) Grade approach to skewness. ICS PAS Report No. 944, Warsaw, May 2002

[18] Ciok A (2002) Grade Correspondence-Cluster Analysis Applied to Separate Components of Reversely Regular Mixtures. In: Jajuga K, Sokołowski A, Bock H-H (Eds), Classification, Clustering and Data Analysis, Recent Advances and Applications, Springer, 211–218

[19] Ciok A (2000) Double versus optimal grade clusterings. In: Kiers HAL, Rasson J-P, Groenen PJF, Schader M (Eds), Data Analysis, Classification, and Related Methods, Springer, 41–46

[20] Ciok A (1999) Trend analysis in qualitative business survey in Polish retail trade. Statistics in Transition, Vol. 4, No. 3, 383–390

[21] Ciok A (1998) A comparative study of exploratory methods applied to car switching data. Bulletin of the Polish Academy of Sciences, Technical Sciences 46(1), 133–148

[22] Ciok A (1998) Discretization as a tool in cluster analysis. In: Rizzi A., Vichi M., Bock H.-H (Eds), Advances in Data Science and Classification, Proceedings of the 6th Conference of the International Federation of Classification Societies (IFCS-98), Rome, July 21-24 1998, Springer, 349–354

[23] Ciok A (1996) An estimator of Gini Separation Index which is nearly median-unbiased. Comput. Statist. Data Anal. Vol. 21, 533–546

[24] Ciok A, Kowalczyk T, Matyja O, Pleszczyńska E, Szczesny W (2000) Visualization of grade statistical parameters and methods implemented in GRAD.STAT, In: Jansen W, Bethlehem JG (Eds), Proceedings in Computational Statistics COMPSTAT'2000, 14 Symposium held in Utrecht, Short Communication and Posters, Statistics Netherlands, 21–22

[25] Ciok A, Kowalczyk T, Matyja O, Pleszczyńska E, Szczesny W (2000) Applying grade data analysis and system GradeStat v 1.0 to dataset on candidates to Air Force College in Dęblin (in Polish). A manuscript.

[26] Ciok A, Kowalczyk T, Pleszczyńska E (1998) How a New Statistical Infrastructure Induced a New Computing Trend in Data Analysis. In: Polkowski L, Skowron A (Eds) Rough Sets and Current Trends in Computing, Lecture Notes in Artificial Intelligence 1424, Proceedings of the First International Conference, RSCTC'98, Warsaw, Poland, June 22-26, 1998, Springer, 75–82

[27] Ciok A, Kowalczyk T, Pleszczyńska E, Szczesny W (1995) Algorithms of grade correspondence-cluster analysis. The Collected Papers of Theoretical and Applied Computer Science, Vol. 7, No. 1–4, 5–22

[28] Ciok A, Kowalczyk T, Pleszczyńska E, Szczesny W (1994) Inequality measures in data analysis. The Collected Papers of Theoretical and Applied Computer Science, Vol. 6, No. 1–4, 3–20

[29] Ciok A, Kowalczyk T, Pleszczyńska E, Szczesny W (1994) Visualisation of a two-way table as a Data Mining Tool, Proceedings of the Workshop on Intelligent Information Systems, Wigry, Poland, Jun 6–10, 1994

[30] Cuesta-Albertos J (2002) Shape of a distribution through the L_2-Wasserstein distance. In: Cuadras C, Fortiana J, Rodriquez-Lallena J (Eds) Distributions With Given Marginals and Statistical Modelling. Kluwer Academic Publishers, Dordrecht/Boston/London, 51–61

[31] Ćwik J, Pleszczyńska E (1991) Nonidentifiability of components in a mixture of distributions corresponding to repeated observations. ICS PAS Report No. 710, Warsaw, November 1991

[32] Doksum KA (1975) Measures of location and asymmetry. Scand. J. Statist. 2, 11–22

[33] Doksum K, Blyth S, Bradlow E, Meng X, Zhao H (1994) Correlation curves as local measures of variance explained by regression. J. Amer. Statist. Assoc. 89, 571–582

[34] Duncan OD, Duncan B (1955) A methodological analysis of segregation indexes. Amer. Sociological Rev., 210–217

[35] Everitt B.S. (1993) Cluster analysis. Edward Arnold, London.

[36] Famili A, Shen W-M, Weber R, Simoudis E (1997) Data Preprocessing and Intelligent Data Analysis. IDA, 1(1)

[37] Fogelson S (1933) Concentration measures and their applications (in Polish). Kwartalnik Statystyczny X(1), 149–197

[38] Foster J (1985) Inequality measurement. Proceedings of Symposia in Applied Mathematics 33, 31–68

[39] Gafrikova W, Kowalczyk T (1993) Links between measuring divergence and inequality. Metron, Vol. LI, No. 3–4, 119–138

[40] Goodman LA (1996) A Single General Method for the Analysis of Cross-Classified Data: Reconciliation and Synthesis of Some Methods of Pearson, Yule, and Fisher, and Also Some Methods of Correspondence Analysis and Association Analysis. Amer. Statist. Assoc. Vol. 91, No. 433, 408–428

[41] Hampel F (2002) Some Thoughts about Classification. In: Jajuga K, Sokolowski A, Bock H-H (Eds) Classification, Clustering and Data Aanalysis. Recent Advances and Applications, Springer, 5–26

[42] Hand DJ (1997) Intelligent Data Analysis: Issues and Opportunities. IDA 1997, 1–14

[43] Harshman RA, Green PE, Wind Y, Lundy ME (1982) A model for the analysis of assymetric data in marketing research. Marketing Science, Vol. 1, No. 2, 1982

[44] Heijden PGM, van der (1987) Correspondence analysis on longitudinal categorical data, DSWO Press, Leiden

[45] Joe H (1993) Multivariate dependence measures and data analysis. Comput. Statistics & Data Analysis 16, 279–297

[46] Joe H (1990) Multivariate Concordance (1990) J. Mult. Analysis 35, 12–30

[47] Jolliffe IT (1986) Principal Component Analysis. Springer-Verlag, 1–222

[48] Kasperkiewicz J, Racz J, Dubrawski A (1995) HPC strength prediction using artificial neural network. Journal of Computing in Civil Engineering, Vol. 9, No. 4, October 1995, 279–284

[49] Kimeldorf G, Sampson AR (1989) A Framework for Positive Dependence. Ann. Inst. Statist. Math. 41, 31–45

[50] Kimeldorf G, Sampson AR (1975) Uniform representation of bivariate distributions. Comm. Statist. Vol. 4, 617–627

[51] Korhonen P, Siljamaki A (1998) Ordinal principal component analysis, theory and an application. Comput. Statistics & Data Analysis 26, 411–424

[52] Koshevoy GA (1997) Multivariate Gini Indices. J. Mult. Anal. 60, 252–276

[53] Koshevoy GA (1995) Multivariate Lorenz majorization. Soc. Choice Welfare 12, 93–102

[54] Kowalczyk T (2000) Link between grade measures of dependence and of separability in pairs of conditional distributions. Statistics and Probability Letters 46, No. 4, 371–379

[55] Kowalczyk T (2000) On decomposition of $m \times m$ statistical tables based on grade dependence measures. Control and Cybernetics, Vol. 29, No. 4, 1043–1058

[56] Kowalczyk T (1999) Decomposition of $m \times m$ tables for father/son occupational status. In: Kłopotek M, Michalewicz M (Eds) Intelligent Information Systems VIII, Proceedings of the Workshop held in Ustroń, June 14–18, 1999, 60–64

[57] Kowalczyk T (1998) Decomposition of the Concentration Index and its implications for the Gini inequality and dependence measures. Statistics in Transition, Vol. 3, No. 3, 503–521

[58] Kowalczyk T (1997) Properties of the Standard Concepts of Monotone Dependence Based on Separation Curves. Discussiones Mathematicae, Algebra and Stochastic Methods 17(1997) 143–156

[59] Kowalczyk T (1994) A unified Lorenz-type approach to divergence and dependence. Dissertationes Mathematicae, No. 135, 1–54

[60] Kowalczyk T (1982) On the shape of the monotone dependence functions of bivariate distributions. Math. Operationsforsch. Statist. Ser. Statist. Vol. 13, 183–192

[61] Kowalczyk T (1977) General definition and sample counterparts of monotone dependence functions of bivariate distributions. Math. Operationsforsh. Statist. Vol. 8, 351–365

[62] Kowalczyk T, Kowalski A, Matuszewski A, Pleszczyńska E (1979) Screening and monotonic dependence functions in the multivariate case. Ann. Statist. 7, 607–614

[63] Kowalczyk T, Niewiadomska-Bugaj M (2002) A new grade measure of monotone multivariate separability. In: Cuadras C, Fortiana J, Rodriquez-Lallena J (Eds) Distributions with Given Marginals and Statistical Modelling. Boston/Dordrecht/London: Kluwer Academic Publishers, 143–151

[64] Kowalczyk T, Niewiadomska-Bugaj M (1998) Grade Correspondence Analysis Based on Kendall's tau. Data Science Classification and Related Methods. VI Conference of the International Federation of Classification Societes, Rome, July 21–24, Institutio Nazionale di Statistica, Roma, 182–185

[65] Kowalczyk T, Pleszczyńska E (1998) Absolute Concentration Index: graphical decomposition. In: Payne RW, Lane PW (Eds), COMPSTAT 1998, Proceedings in Computational Statistics, Short Communications and Posters, Harpenden, IACR-Rothamsted

[66] Kowalczyk T, Pleszczyńska E (1977), Monotonic dependence functions of bivariate distributions. Ann. Statist. 5, 1221–1227

[67] Kowalczyk T, Pleszczyńska E, Szczesny W (1991) Evaluation of stochastic dependence. In: Bromek T, Pleszczyńska E (Eds) Statistical Inference, Theory and Practice. PWN — Polish Scientific Publishers Warszawa, Kluwer Academic Publishers Dordrecht/Boston/London, 106–136

[68] Krzanowski WJ, Marriott FHC Multivariate analysis (1994) Part 1: Distributions, ordination and inference. London Edward Arnold 8 s. IX, 280 Kendall's Library of Statistics, 1 (III.62–01,62Hxx)

[69] Krzanowski WJ, Marriott FHC Multivariate analysis (1996) Part 2: Classification, covariance structures and repeated measurements. London Edward Arnold 8 s. VIII,280 Kendall's Library of Statistics, 2 (III.62–01,62Hxx)

[70] Lehmann EL (1966) Some concepts of dependence. Ann. Math. Statist., No. 37, 1137–1153.

[71] Lehmann EL, Rojo J (1992) Invariant directional orderings. The Annals of Statistics, Vol. 20, No. 4, 2100–2110

[72] Marriott FHC (Ed.) (1990) Dictionary of Statistical Terms. 5th edition, Longman Scientific & Technical, Wiley, New York

[73] Matyja O (2003) Smooth Grade Correspondence Analysis and Related Computer System. Ph.D. dissertation, Institute of Computer Science, Polish Academy of Science, Warszawa 2003

[74] Matyja O (2002) Decomposition of a Probability Table into a Mixture of Permuted Binormal Discretized Tables. Proceedings in Computational Statistics, COMPSTAT 2002, 16 Symposium, Berlin, Short Communication (published on CD-ROM)

[75] Matyja O (2002) Grade Decomposition of a Contingency Table into Binormal Discretized Tables. Proceedings of the IIS'2002 Symposium, Sopot, June 3–6, 2002, Springer-Verlag, 251–258

[76] Matyja O, Szczesny W (2000) Visualisation in prediction based on grade correspondence analysis, In Kłopotek MA, Michalewicz M, Wierzchoń ST (Eds), Intelligent Information Systems, Proceedings of the IIS'2000 Symposium, Bystra, Poland, June 12–16, 2000, Physica-Verlag, 289–301

[77] McLachlan G, Peel D (2000) Finite Mixture Models, Willey Series in Probability and Statistics, John Wiley and Sons, Inc.

[78] Mooijaart A, van der Heijden PGM, van der Ark LA (1999) A least squares algorithm for a mixture model for compositional data. Comput. Statist. Data Anal. No. 30, 359–379

[79] Mustonen S (1997) A measure for total variability in multivariate normal distribution. Comput. Statist. Data Anal. 23, 321–334

[80] Nelsen RB (1999) An Introduction to Copulas. Lecture Notes in Statistics No. 139, Springer-Verlag, New York

[81] Nelsen RB (1992) On measures of association as measures of positive dependence. Statistics and Probability Letters No. 14, 269–274

[82] Nelsen RB, Quesada-Molina JJ, Rodriquez-Lallena JA, Ubeda-Flores M (2001) Distribution functions of copulas: a class of bivariate probability integral transforms. Statist. Prob. Letters 54, 277–282

[83] Niewiadomska-Bugaj M (1991) Discriminant analysis In: Bromek T, Pleszczyńska E (Eds) Statistical Inference. Theory and Practice, Kluwer Academic Publishers, Warszawa, 53–85

[84] Pleszczyńska E, Ciok A, Kowalczyk T (1997) On parameter decomposition in statistical data analysis. Intelligent Information Systems VI, Proceedings of the Workshop held in Zakopane, Poland, June 9-13, 1997, 74–85

[85] Pleszczyńska E, Grzegorek M (2002) Data analysis concerning physically disabled people (I) (in Polish). ICS PAS Report No. 945, Warsaw, July 2002

[86] Pleszczyńska E, Szczesny W (2002) Grade explanatory methods applied to some medical data sets. Biocybernetics and Biomedical Engineering, Vol. 22, 1, 17–30

[87] Pleszczyńska E, Szczesny W, Grzegorek M (2002) Clustering Respondents in Clinical Databases Using Ordered Grade Clustering. In: Bobrowski L, Doroszewski J, Victor N (Eds), Lecture Notes of the ICB Seminars "Statistics and Clinical Practice", International Center of Biocybernetics, Warsaw, June 2002, 64–69

[88] Pyatt G (1976) On the interpretation and disagregation of Gini coefficient. Econ. J. 86, 243–255

[89] Radzikowski S, Pleszczyńska E, Gołembiewska-Skoczylas M, Sapetto-Rebow B (1995) Pattern of total protein content regulation in Trithigmostoma steini (Chilodonella steini). European Journal of Protistology, No. 31, 174–181

[90] Rao CR (1982) Analysis of Diversity: A Unified Approach, Statistical Decision Theory and Related Topics III, Vol. 2, 233–250

[91] Regazzini E (1990) Concentration comparisons between probability measures. Instituto per le Applicazioni della Matematica e dell'Informatica, preprint 90.15, Milan

[92] Rietveld P (1990) Multidimensional Inequality Comparisons. Economics Letters 32, 187–192

[93] Schriever BF (1987) An ordering for positive dependence. Annals of Statistics 15, 1208–1214

[94] Schriever BF (1985) Order Dependence. Ph.D. dissertation, Centrum voor wiskunde en informatika, Vrije Universiteit te Amsterdam

[95] Shaked M (1977) A family of concepts of dependence for bivariate distributions. J. Amer. Statist. Assoc. 72, 642–650

[96] Silber J (1989) Factor components, population subgroups and the computation of the Gini index of inequality. The review of Economics and Statistics, 71, 107–115

[97] Sutherland H, Taylor R, Gomulka J (2001) Combining household income and expenditure data in policy simulations. Microsimulation Unit Discussion Paper MU0101, Department of Applied Economics, University of Cambridge, January 2001, http://www.econ.cam.ac.uk/dae/mu/publications/mu0101.pdf

[98] Szczesny W (2002) Grade correspondence analysis applied to contingency tables and questionnaire data. Intelligent Data Analysis, 6, 17–51

[99] Szczesny W (2000) Detecting rows and columns of contingency table, which outlie from a total positivity pattern. Control and Cybernetics, Vol. 29, No. 4, 1059–1073

[100] Szczesny W (1991) On the performance of a discriminant function, J. Classification 8, 201–215

[101] Szczesny W, Ciok A, Kowalczyk T, Pleszczyńska E, Wysocki W (1998) Grade correspondence analysis of contingency tables applied to analysis of Polish parliament election results in 1993 and 1997 (in Polish), Studia Socjologiczne, No. 3 (150), 49–68

[102] Szczesny W, Kowalczyk T (2002) On regularity of multivariate data sets. In: Kłopotek M, Wierzchoń TS, Michalewicz M (Eds), Intelligent Information Systems'2002, Proceedings of the IIS'2002, Sopot, Poland, June 3–6, 2002, Advances in Soft Computing, Physica-Verlag, 237–246

[103] Szczesny W, Matyja O (2001) Using grade correspondence analysis to merge populations and detect latent orders. In: Kłopotek M, Michalewicz M, Wierzchoń TS (Eds), Intelligent Information Systems'2001, Proceedings of the IIS'2001, June 18–22, 2001, Advances in Soft Computing, Physica-Verlag, 111–120

[104] Szczesny W, Pleszczyńska E (1999) Another look at the regression graphs in an experiment on glucose control. Biocybernetics and Biomedical Engineering, Vol. 19, No 4, 31–40

[105] Taguchi T (1987) On the structure of multivariate concentration — some relationships among the concentration surface and two variate mean difference and regressions. Comput. Statist. Data Anal. 6, 307–334

[106] Taguchi T (1972) On the two-dimensional concentration surface and extensions of concentration coefficient and Pareto distribution to the two dimensional case. Part I. Ann. Instit. Statist. Math. 25, 355–381, part II (1972) Vol. 24, 599–619, part III (1972) Vol. 25, 215–237

[107] Taylor R (2000) Guidelines for Identifying Clusters Using Grade Correspondence Analysis: Practical and Technical Issues. Microsimulation Unit Research Note MU/RN/39, Department of Applied Economics, University of Cambridge

[108] Taylor R, Gomulka J, Sutherland H (2000) Creating order out of chaos? Identifying homogeneous groups of households across multiple datasets. Paper presented at the 26th General Conference of the International Association for Research in Income and Wealth, Cracow, Poland, August 27, 2000

[109] White N (1986) Segregation and diversity measures in population distribution. Population Index 52, 198–221

[110] Wysocki W (1999) Classes of generators of Archimedean copulas. ICS PAS Report No. 886, Warsaw, July 1999

[111] Wysocki W (1998) On hierarchical structure of positive dependence orderings. Discussiones Mathematicae, Algebra and Stochastic Methods, no. 18, 227–239

[112] Xiaohui Liu, Cohen PR, Berthold MR (1997) Advances in Intelligent Data Analysis, Reasoning about Data. Proceedings of the Second International Symposium IDA'97, London, UK, August 4–6, 1997, Springer

[113] Xiao-Li M, Rubin DB (1992) Performing Likelihood-ratio Tests with Multiply Imputed Data Sets. Biometrika 79, 103–111

[114] Yanagimoto T (1972) Families of positively dependent random variables. Ann. Inst. Statist. Math. 24, 559–573

Index

Printing: Strauss GmbH, Mörlenbach
Binding: Schäffer, Grünstadt